Mesozoic Vertebrate Life

LIFE OF THE PAST
James O. Farlow, Editor

Darren H. Tanke *&* Kenneth Carpenter, editors
Michael W. Skrepnick, art editor

Mesozoic Vertebrate Life

*New Research Inspired
by the Paleontology of
Philip J. Currie*

NRC·CNRC
NRC Research Press

Indiana University Press
Bloomington *&* Indianapolis

This book is a publication of

Indiana University Press

601 North Morton Street

Bloomington, IN 47404-3797 USA

http://iupress.indiana.edu

Telephone orders 800-842-6796

Fax orders 812-855-7931

Orders by e-mail iuporder@indiana.edu

The paper used in this publication meets
the minimum requirements of American
National Standard for Information
Sciences—Permanence of Paper for
Printed Library Materials, ANSI
Z39.48-1984.
Manufactured in the United States of
America

**Library of Congress Cataloging-in-
Publication Data**

Mesozoic vertebrate life / Darren Tanke
    and Kenneth Carpenter, editors.
        p.   cm. — (Life of the past)
    Includes bibliographical references
and index.
    ISBN 0-253-33907-3 (alk. paper)
        1. Vertebrates, Fossil.
    2. Paleontology—Mesozoic. I. Tanke,
Darren.   II. Carpenter, Kenneth,
date III. Series.
QE841.M574   2001
566—dc21   00-053534

1 2 3 4 5 06 05 04 03 02 01

Dedicated to
Philip J. Currie

*In celebration of his
25 years in vertebrate
paleontology*

CONTENTS

Contributors

William L. Abler, Department of Geology, Field Museum of Natural History, Chicago, IL 60605, USA.

M. K. Brett-Surman, National Museum of Natural History, Mail Stop NHB-121, Smithsonian Institution, Washington, DC 20560-0121, USA.

Kathleen Brill, Denver Museum of Natural History, 2001 Colorado Blvd., Denver, CO 80205, USA.

Kenneth Carpenter, Denver Museum of Natural History, 2001 Colorado Blvd., Denver, CO 80205, USA.

Robert L. Carroll, Redpath Museum, McGill University, 859 Sherbrooke St. W, Montreal, PQ H3A 2K6, Canada.

Dan Chure, Dinosaur National Monument, P.O. Box 128, Jensen, UT 84035, USA.

Rodolfo A. Coria, Direccion Provincial de Cultura de Neuquén—Museo Carmen Funes, Av. Cordoba 55 (8318) Plaza Huincul, Neuquén, Argentina.

Clive Coy, 120 - 7th Ave. S.W., Drumheller, AB T0J 0Y6, Canada.

Tony DiCroce, Department of Earth Sciences, Denver Museum of Natural History, 2001 Colorado Boulevard, Denver, CO 80205, USA.

Dong Zhiming, Institute of Vertebrate Paleontology and Paleoanthropology, Academia Sinica, P.O. Box 643, Beijing 100044, China.

James O. Farlow, Department of Geosciences, Indiana University–Purdue University, Fort Wayne, IN 46805, USA.

Tracy L. Ford, 13503 Powers Rd., Poway, CA 92064, USA.

Karl F. Hirsch (deceased), Geology Section, Museum, University of Colorado, Boulder, CO 80309, USA.

Thomas R. Holtz Jr., Department of Geology, University of Maryland, College Park, MD 20742, USA.

Jørn H. Hurum, Paleontologisk Museum, Universitetet i Oslo, Sars'gate 1, N-0562 Oslo, Norway.

A. R. Jacobsen, Steno Museum, University Campus, C.F. Moellers Alle, Build 100, 8000 Aarhus C, Denmark.

Paul Janke, Pan Terra, Inc., P.O. Box 556, 103 Park Avenue, Hill City, SD 57745, USA.

Ji Qiang, National Geological Museum of China, Yangrou Hutong 15, Xisi, 10034 Beijing, China.

Ji Shu-an, National Geological Museum of China, Yangrou Hutong 15, Xisi, 10034 Beijing, China.

Hans C. E. Larsson, Department of Organismal Biology and Anatomy, University of Chicago, 1027 E. 57th St., Chicago, IL 60637, USA.

Thomas M. Lehman, Department of Geosciences, Texas Tech University, Lubbock, TX 79409-1053, USA.

Martin G. Lockley, Department of Geology, University of Colorado at Denver, Campus Box 172, P.O. Box 173364, Denver CO 80217, USA.

Peter J. Makovicky, Division of Paleontology, American Museum of Natural History, Central Park West at 79th St., New York, NY 10024-5192, USA.

Kim Manley, Los Alamos National Laboratory, 4691 Ridgeway, Los Alamos, NM 87544, USA.

Kevin C. May, Department of Geology and Geophysics, University of Alaska Fairbanks, 308 NSF, Box 755780, Fairbanks, AK 99775, USA.

Richard T. McCrea, Department of Earth and Atmospheric Sciences, University of Alberta, Edmonton, Alberta T6G 2E3, Canada.

Lorrie McWhinney, Denver Museum of Natural History, 2001 Colorado Blvd., Denver, CO 80205, USA.

Susanne Meyer, Department of Earth Sciences, Denver Museum of Natural History, 2001 Colorado Blvd., Denver, CO 80206, USA.

R. E. Molnar, Queensland Museum, Queensland Cultural Centre, P.O. Box 3300, S. Brisbane, Queensland 4101, Australia.

G. C. Nadon, Department of Geological Sciences, Ohio University, Athens, OH 45701, USA.

Kevin Padian, Museum of Paleontology, 1101 Valley Life Sciences, University of California, Berkeley, CA 94720, USA.

Anne D. Pasch, Department of Geology, University of Alaska, Anchorage, 3211 Providence Dr., Anchorage, AK 99508, USA.

Bruce Rothschild, Arthritis Center of Northeast Ohio, 5500 Market St., Suite 119, Youngstown, OH 44512, USA.

Anthony P. Russell, Department of Biological Sciences, University of Calgary, Calgary, AB T2N 1N4, Canada.

Michael J. Ryan, Department of Biological Sciences, University of Calgary, Calgary, Alberta T2N 1N4, Canada.

Scott D. Sampson, Utah Museum of Natural History and Department of Geology and Geophysics, University of Utah, 1390 East Presidents Circle, Salt Lake City, UT 84112, USA.

Frank Sanders, Department of Earth and Space Sciences, Denver Museum of Natural History, 2001 Colorado Blvd., Denver, CO 80205, USA.

William A. S. Sarjeant, Department of Geological Sciences, University of Saskatchewan, 114 Science Place, Saskatoon, SK, S7N 5E2, Canada.

Matt Smith, Smith Studios, 122 East Park St., Livingston, MT 59047, USA.

David A. E. Spalding, 1105 Ogden Road, R.R. 1, Pender Island, BC, V0N 2M1, Canada.

Darren H. Tanke, Royal Tyrrell Museum of Palaeontology, P.O. Box 7500, Drumheller, AB T0J 0Y0, Canada.

Leon Theisen, Custom Paleo, P.O. Box 348, Hill City, SD 57745, USA.

Virginia Tidwell, Department of Earth Sciences, Denver Museum of Natural History, 2001 Colorado Blvd., Denver, CO 80206, USA.

David Trexler, Timescale Adventures, P.O. Box 356, Chouteau, MT 59422, USA.

David J. Varricchio, Museum of the Rockies, Montana State University, Bozeman, MT 59717, USA.

Joanna L. Wright, Department of Geology, University of Colorado at Denver, Campus Box 172, P.O. Box 173364, Denver, CO 80217, USA.

Dinosaurs are probably the most frequent objects to kindle children's interest in science. So it was with Phil Currie, who first expressed his intention to study dinosaurs when he was 11 years old. However, Phil is unusual in never ceasing his pursuit of dinosaurs, and for the scope and excellence of his contributions to their discovery and description. Phil is also unique in his contributions to public education, with numerous television documentaries and lectures illustrating his expeditions, ranging from the deserts of Mongolia and the badlands of Argentina to the margins of the Arctic Ocean, and on scientific subjects as distinct as herding behavior in dinosaurs and their implication in the ancestry of birds.

I first come to know Phil while he was a graduate student at McGill, beginning in 1972. Neither his M.Sc. or Ph.D. concerned dinosaurs, but rather involved primitive mammal-like reptiles and early aquatic diapsids. From the beginning, he demonstrated the ability to do independent research, essentially without a graduate supervisor, for I was on sabbatical in South Africa when he began his master's degree. Later, he left McGill to accept a curatorial position at the Alberta Provincial Museum in Edmonton at the very beginning of his Ph.D. program. The excellence of his M.Sc. thesis was such that Phil was granted a special exemption from the usual two years of residence for the Ph.D. Indicative of the nature of his subsequent research, Phil's Ph.D. thesis went well beyond the required illustration and description of specimens to detailed analysis of relationships, patterns of growth, and the function of the stones that filled their abdominal cavities. It was nominated by the Department of Biology for the Canadian Society of Zoologists' prize for the best thesis of the year.

Phil and I collaborated on two other projects: the possibility of relationship between caecilians (elongate, limbless, tropical salamanders) and Paleozoic microsaurs, and the interrelationships of the major lineages of Late Permian and Early Mesozoic diapsids. In all his work at McGill, Phil demonstrated an extraordinary capacity for conceptualizing important scientific questions, and the dedication and energy to carry their solution to rapid fruition.

The nature of Phil's research changed dramatically in the context of the opportunities for fieldwork in Alberta. The two scientists he succeeded as Curator of Earth Sciences at the Provincial Museum in Edmonton had had very limited success in collecting dinosaurs in Alberta, but within two or three field seasons, Phil had collected so many specimens that the province began to plan for an entire new museum to house the riches. In 1985 this resulted in the completion of the Royal Tyrrell Museum in Drumheller, which rapidly achieved world status for its paleontological research and as a mecca for tourists and serious amateurs who were willing to pay for the privilege of spending long hours in the hot summer sun, crawling on their hands and knees, looking for bones. Phil was finding not only a wealth of individual dinosaur skeletons, but bone beds preserving hundreds of bones in close proximity. Because the bone beds included primarily the remains of single species, he concluded that they represented migrating herds.

From 1986 much of Phil's efforts were involved in the joint Canada-China Dinosaur Project, which he directed, together with Dale Russell, then at the Canadian Museum of Nature, Ottawa, and Dong Zhiming, Institute of Vertebrate Paleontology and Paleoanthropology, Beijing. This work included expeditions, with crews from both countries, in Xinjian, Inner Mongolia, Alberta, and the Northwest Territories. Early results were published in a special volume of the *Canadian Journal of Earth Sciences:* "Results from the Sino-Canadian Dinosaur Project," of which Phil was the editor (Currie 1993, 1996). His own papers included descriptions of five theropod dinosaurs and an overview of the correlation, stratigraphy, sedimentary geology, and paleontology of the Djadokhta Formation in Inner Mongolia.

More recently, Phil has been involved in the detailed study and analysis of a number of small, feathered animals from the Early Cretaceous of China (Ji et al. 1998). This fauna includes both early birds and more primitive genera that are not birds, but clearly nonflying dinosaurs, in which the feathers were apparently used for behavior other than flight, such as display, as well as for insulation. This material unequivocally demonstrates that feathers evolved among small, bipedal dinosaurs, which certainly included the ancestors of birds. This work was featured in *Time* magazine, with Phil's picture on the cover (Lemonick 1998; Purvis 1998).

Phil has been extremely successful in keeping a balance between extraordinarily ambitious and successful field programs, and a history of solid professional research and publications. At the same time, he has been a leading force behind the scientific activities of the Tyrrell Museum since its inception. Phil's own research, the prodigious success of his field programs and organization of three major scientific conferences, have raised the Royal Tyrrell Museum of Palaeontology to the rank of one of the leading natural history museums of the world.

Phil's field work and descriptive studies of dinosaurs have exposed him to extensive media coverage. As a result, he has become a powerful spokesman for geology, paleontology, and scientific research in general. He has handled this in a modest, but knowledgeable manner, providing the best possible role model for anyone who might be attracted to a career in science. He has avoided the trap that so often befalls scientists with high public profiles of letting the science slip as they become increasingly involved in public education. Rather, the number and importance of his publications increase, year after year.

In 1999, Phil was honored by election to the Royal Society of Canada, with the following citation:

> Philip J. Currie is a paleontologist whose work focuses on the detailed anatomy, mode of life, and evolutionary relationship of dinosaurs in North America, South America and Eurasia (particularly China and Mongolia). His scientific studies have changed the direction of research in his field and pioneered new and fresh insights, ideas and theories about how dinosaurs became established and how they flourished in Mesozoic times. Currie's recent find, with Chinese colleagues, of bipedal dinosaurs with feathers in northeastern China virtually establishes that theropod dinosaurs are most likely to be the ancestors of birds. His discoveries on the evolution and life habits of Jurassic and Cretaceous dinosaurs have appealed to young and old.

Maclean's magazine featured Phil among 12 outstanding Canadians in 1998 (Bergman 1998). He is quoted as saying: "It's the nature of the game. It [paleontology] doesn't let you lose interest or your excitement. It's like having a career of going out and finding buried treasure."

Phil's impact on the field of vertebrate paleontology is appropriately honored by the great diversity of papers contributed to this volume, covering all major groups of dinosaurs from all parts of the world.

References

Bergman, B. 1998. *Maclean's* honour roll: Philip Currie. *Maclean's* December 21, p. 65.

Currie, P. J. (ed.). 1993. Results from the Sino-Canadian Dinosaur project. *Canadian Journal of Earth Sciences* 30: 1997–2272.

Currie, P. J. (ed.). 1996. Results from the Sino-Canadian Dinosaur project, Part 2. *Canadian Journal of Earth Sciences* 33: 511–648

Ji Q., P. J. Currie, M. A. Norell, and Ji S.-A. 1998. Two feathered dinosaurs from northeastern China. *Nature* 393: 753–762.

Lemonick, M. D. 1998. Dinosaurs of a feather. *Time* 151: 48–50.

Purvis, A. 1998. Call him Mr. Lucky. *Time* 151: 52–55.

Acknowledgments

Several individuals and companies made financial contributions for the dinosaur art reproduction and for the publication of this book. John Lanzendorf (Chicago) was an enthusiastic supporter throughout the book's conception, planning, and development. He not only provided seed money to get this book off the ground, but also was a major contributor to the art budget. Gerry Neville (National Research Council, Ottawa) arranged financial contribution to the art budget as well. Other major contributors were Canada Fossil Sales (Calgary), Korite Minerals, Ltd. (Calgary), Black Hills Institute of Geological Research (Hill City, S.D.), Eric Felber (Troodon Oil, Calgary), Pam and Tony Ashton, and Nathan P. Myrhvold. Additional financial support was provided by Dinosaur Provincial Park (Patricia, Alta.), Dinosaur Natural History Association (Brooks, Alta.), Joe and Leonne Stuart (Patricia Hotel, Patricia, Alta.), Hans Larsson (University of Chicago), Joe Vipond, Patricia E. Ralrick, and the late Sam Girouard (Bellingham, Wash.).

Thanks also to Kevin Seymour (Royal Ontario Museum), Bob Sloan (Indiana University Press), and Jim Farlow (Indiana University–Purdue University, Fort Wayne) for their support of the project from its earliest inception. And, of course, to all the contributors for their papers.

Finally, we would like to thank Phil's wife, Eva Koppelhus, our "partner in crime," who helped us keep this book a surprise by making sure Phil knew nothing of its planning, development, and execution. In all fairness we should also thank Phil himself—on several occasions during the projects development he willingly (and unknowingly) acted as a courier for us, hand-delivering letters of invitation and manuscripts to and from some of the contributing authors.

Darren H. Tanke
Ken Carpenter
Drumheller, Alberta, and
Denver, Colorado
2001

Philip Currie is the personification of everything I wanted to grow up to be and dreamed of as a child. As an adult, having the privilege of being Philip's friend has made a great impact in my life. My love of dinosaur art has been elevated because of all the inspiration I have received from him. I thank Philip for writing the foreword for my book *Dinosaur Imagery—The Lanzendorf Collection* and for the prize treasure in my collection, his drawing of *Monolophosaurus jiangi.*

My vacations with Philip and Eva in Dinosaur Provincial Park and everywhere have indeed been memorable experiences.

Philip is a gentleman and a scholar and I am proud to be his friend.

John J. Lanzendorf
Project Exploration
Chicago, Illinois
2000

Section I.
Theropods

1. New Theropod from the Late Cretaceous of Patagonia

RODOLFO A. CORIA

Abstract

The first theropod is recorded for the Upper Cretaceous Allen Formation of Argentina. The specimen is composed of a distal right femur and a complete right tibia, and represents a new taxon. It is characterized by a femur with a strong and well-developed mediodistal crest, a tibia with a hook-shaped cnemial crest, and a lateral maleolus twice the size of the medial one. A low facet for the ascending process of the astragalus and a fossa on the distal articular surface of the tibia link this new taxon with other tetanuran South American theropods such as *Giganotosaurus*. The new taxon is part of a dinosaur assemblage composed of titanosaurs, hadrosaurs, and ankylosaurs.

Introduction

In the last few years, our knowledge of South American theropods has increased greatly. One Triassic and several Cretaceous taxa have been described (Arcucci and Coria 1997, 1998; Coria and Salgado 2000; Novas et al. in press; Calvo et al. in press). Most of this record is composed of carcharodontosaurids and abelisaurids (Coria and Salgado 1995; Bonaparte and Novas 1985; Bonaparte et al. 1990; Coria and Salgado 2000). These theropods seem to be the dominant carnivores in the faunas during the Cretaceous in South American, although there are a few maniraptoran-related forms as well (Novas 1998; Novas et al. in press).

A new, fragmentary, but very peculiar theropod has been found in the Allen Formation (Campanian-Maastrichtian) of Rio Negro Province, Argentina; it represents the first theropod from this strata. The specimen was collected in the fluvial sandstones of the lower part of the Allen Formation. Previous dinosaur remains found include titanosaurs (Salgado and Coria 1993), hadrosaurs (Powell 1987) and ankylosaurs (Salgado and Coria 1996; Coria and Salgado, in press). The new theropod was collected in late 1980s by a field crew headed by Dr. Jaime Powell for the Universidad Nacional de Tucumán.

Institutional Abbreviation: MPCA-PV, Museo Provincial Carlos Ameghino, Vertebrate Paleontology Collection, Cipolletti, Rio Negro Province, Argentina.

Systematic Paleontology
Saurischia
Theropoda
Quilmesaurus gen. nov.

Holotype: MPCA-PV-100, distal right femur, complete right tibia.
Diagnosis: same as for the species.
Horizon: Allen Formation, Campanian-Maastrichtian.
Locality: Salitral Ojo de Agua, 40 km south of Roca City, Rio Negro Province, Argentina.

Quilmesaurus curriei sp. nov.

Diagnosis: Medium-size theropod; femur with strong, well-developed mediodistal crest; tibia with hook-shaped cnemial crest, lateral maleolus twice the size of medial, asymmetrical distal end.

Description

Femur

Approximately 50% of the distal half of a right femur is preserved (fig. 1.1). The shaft is slender and apparently slightly sigmoidal (fig. 1.1C). The mediodistal crest is remarkable in that it is unusually well developed. It extends as a broad, anteromedially projecting lamina (fig. 1.1C,D). This thickened structure connects distally with the anterior side of the medial condyle (fig. 1.1C). In medial view, the width of the crest is about the same as the femur shaft itself. In distal view (fig. 1.1E), the axis of both the medial and lateral condyles slightly diverge posteriorly. The internal condyle is larger than the external one. The external condyle does not show any well-developed condylid because the area is weathered. Nevertheless, there is a distinct neck dividing the base of the condylid from the rest of the articular condyle. The condylid was medially placed with respect to the medial condyle. The anterior intercondylar groove is not well developed. Both anterior and distal articular surfaces are well defined by a transversal ridge.

On the posterior side of the femur, above the posterior intercondylar groove, there is a deep and roughened fossa, probably for a muscle origin.

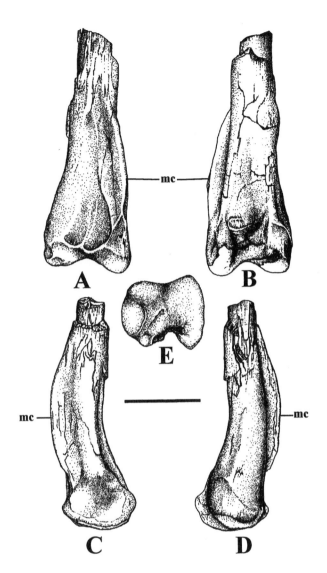

Figure 1.1. Quilmesaurus curriei, *holotype. Right femur in (A) anterior, (B) posterior, (C) medial, (D) lateral, and (E) distal views. Scale: 10 cm.*

TABLE 1.1
Measurements of the Femur and Tibia of *Quilmesaurus curriei* (mm)

	Shaft length
Femur	350 (preserved part)
Tibia	520
anteroposterior proximal width	192
anteroposterior distal width	91
(lateral condyle)	29
Transverse proximal width	74
(excluding cnemial crest)	
Transverse distal width	105
	129
Mid-shaft width	59
	64
Shaft circumference	124

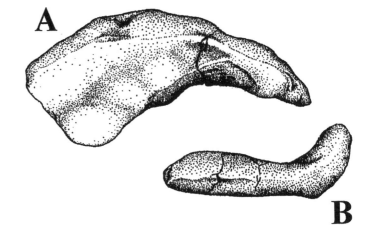

Figure 1.2. Quilmesaurus curriei, holotype. Right tibia in (A) proximal and (B) distal views. Scale: 10 cm.

Figure 1.3. Quilmesaurus curriei, Holotype. Right tibia in (A) lateral, (B) posterior, (C) medial, and (D) anterior views. Scale: 10 cm.

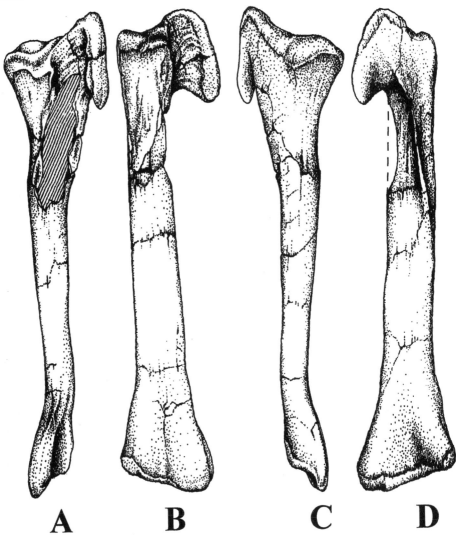

A   B   C   D

Tibia

The tibia is complete, and is long and slender (see table 1.1 for measurements). Most of the proximal end, which contacted the fibula, is weathered, including the crista fibularis). The proximal end is anteroposteriorly expanded and transversely compressed (fig. 1.2A). The proximal lateral condyle is placed posteriorly and separated from the bigger medial condyle by a notch. This notch is deeper and more closed than in other theropods (e.g., *Giganotosaurus* [Coria and Salgado 1995], *Sinraptor* [Currie and Zhao 1993]). The articular surface for the femur is quite weathered but it appears to have been more narrow than in other theropods like *Allosaurus, Sinraptor, Giganotosaurus,* and *Carnotaurus.* The distal end (fig. 1.2B) is slightly expanded. The distal articular surface is narrow anteroposteriorly, and shows a notch on the articular surface of the medial condyle as in *Sinraptor* and *Giganotosaurus.*

The most conspicuous feature of the proximal end is the morphology of the cnemial crest. In lateral view (fig. 1.3A), the anterior end of the crest projects upward as in some abelisaurs. The shaft is flattened on both extensor and flexor sides. In posterior view (fig. 1.3B), the lateral projection of the cnemial crest may have overlapped the fibula entirely. In medial view (fig. 1.3C), the distal end of the cnemial crest is noticeably expanded, with a distinctive hooklike shape.

In anterior view (fig. 1.3D), the lateral maleolus projects distally more than the internal maleolus, resulting in an asymmetrical profile. There is no fusion of the proximal tarsals as in some abelisaurs and ceratosaurs. The facet for the ascending process of the astragalus indicates that it was low as in *Giganotosaurus, Ceratosaurus,* and *Allosaurus.* The facet for the ascending process is 16% of the tibia length, similar to *Sinraptor* (Currie and Zhao 1993), and a new Triassic theropod recently reported (Arcucci and Coria 1997). This low percentage indicates a more primitive condition as compared to 20% in allosaurids and 33% in tyrannosaurids (Molnar et al. 1990). This area for the ascending process is deep due to the anteroposterior thickness of the internal maleolus.

Discussion

Quilmesaurus represents a new record for the Cretaceous of South America, as well as a very unusual theropod with unique features in the knee. The specimen bears several features that place this taxon in the Theropoda, such as highly pneumatized bone shafts, an expanded mediodistal crest on the femur, a tibia with a well-developed cnemial crest, and the distal end of the tibia expanded for a triangular ascending process of the astragalus.

Quilmesaurus is not a ceratosaur (sensu Rowe and Gauthier 1990) because the tibia shows no evidence of fusion with proximal tarsals. On the other hand, it shares having a very well-developed cnemial crest with *Ceratosaurus* and *Xenotarsosaurus* (Martínez et al. 1986; Coria and Rodríguez 1993), although this feature could be a primitive condition among the Theropoda. *Quilmesaurus* also shares with *Gigano-*

tosaurus (Coria and Salgado 1995) and *Sinraptor* (Currie and Zhao 1993) the presence of a notch on the distal articular surface of the tibia.

Quilmesaurus is the first record for a theropod in the Allen Formation, Malarge Group, Neuquén Basin, Argentina. This unit has been explored for several years and has yielded a diverse dinosaur fauna, including the titanosaur *Aeolosaurus* (Salgado and Coria 1993), a lambeosaurine hadrosaur (Powell 1987), a possible nodosaurid (Salgado and Coria 1996; Coria and Salgado in press b), and dinosaur eggs, possibly of titanosaurs. This association involves the coexistence of both North American and South American forms. Interestingly, the new form does not show any unquestionable feature related with Laurasian forms, which would be expected since it was found at the same levels where North American related fauna is common (e.g., hadrosaurs and ankylosaurs). In contrast, it bears many plesiomorphic characters more similar to typical South American dinosaurs (e.g., absence of a well-developed anterior intercondylar groove on femur, facet for the ascending process of astragalus less than 20% of tibia shaft length). These features are present in the South American *Xenotarsosaurus, Giganotosaurus, Piatnitzkysaurus,* and several undescribed forms (Coria and Currie 1997; Calvo et al. in press).

Acknowledgments: I thank Mr. Carlos Muñoz for allowing me to study the specimen under his care. I am indebted to A. Arcucci and L. Salgado for their comments on early drafts of this chapter, and to Mr. Darren Tanke for his invitation to participate in this book. The illustrations were skillfully made by Aldo Beroisa. Lastly, I thank Dr. Philip Currie for his steadfast scientific guidance and kind friendship.

References

Arcucci, A. B., and R. A. Coria. 1997. Primer registro de theropoda (Dinosauria—Saurischia) de la Formacion Los Colorados (Triasico Superior, La Rioja, Argentina). *Ameghiniana* 34: 531.

Arcucci, A. B., and R. A. Coria. 1998. Skull features of a new primitive theropod from Argentina. *Journal of Vertebrate Paleontology, Abstracts* (suppl. to no. 3) 18: 24A.

Bonaparte, J. F., and F. E. Novas. 1985. *Abelisaurus comahuensis*, n. gen., n. sp., Carnosauria del Cretacico Tardo de la Patagonia. *Ameghiniana,* 21: 259–265.

Bonaparte, J. F., F. E. Novas, and R. A. Coria. 1990. *Carnotaurus sastrei,* the horned lightly built carnosaur from the Middle Cretaceous of Patagonia. *Natural History Museum of Los Angeles County, Contributions in Science* 416: 1–42.

Calvo, J. O., D. Rubilar, and K. Moreno. In press. Report of a new theropod dinosaur from Northwest Patagonia. *Abstracts 15 Jornadas Argentinas de Paleontología de Vertebrados, Ameghiniana.*

Coria, R. A., and P. J. Currie. 1997. A new theropod from the Ro Limay Formation. *Journal of Vertebrate Paleontology, Abstracts* (suppl. to no. 3) 17: 40A.

Coria, R. A., and J. Rodríguez. 1993. Sobre *Xenotarsosaurus bonapartei* Martínez et al. 1986; un problem tico Neoceratosauria (Novas 1989) del Cretacico del Chubut. *Ameghiniana* 30: 326–327.

Coria, R. A., and L. Salgado. 1995. A new giant carnivorous dinosaur from the Cretaceous of Patagonia. *Nature* 377: 224–226.

Coria, R. A., and L. Salgado. 2000. A basal Neoceratosauria (Theropoda-Ceratosauria) from the Cretaceous of Patagonia, Argentina. *Gaia* 15: 89–102.

Coria, R. A., and L. Salgado. In press. South American Ankylosaurs. In K. Carpenter (ed.), *The Armored Dinosaurs*. Bloomington: Indiana University Press.

Currie, P. J., and X. Zhao. 1993. A new carnosaur (Dinosauria, Theropoda) from the Jurassic of Xinjiang, People's Republic of China. *Canadian Journal of Earth Sciences* 30: 2037–2081.

Martínez, R., O. Gimènez, J. Rodriguez, and G. Bochatey. 1986. *Xenotarsosaurus bonapartei:* nov. gen. et sp. (Carnosauria, Abelisauridae), un neuvo Theropoda de la Formacion Bajo Barreal Chubut, Argentina. *Cuarto Congresso Argentino de Paleontología y Biostratigraphía, Mendoza, Argentina* 2: 23–31.

Molnar, R. E., S. M. Kurzanov, and Dong Z. 1990. Carnosauria. In D. B. Weishampel, P. Dodson, and H. Osmólska (eds.), *The Dinosauria,* pp. 169–209. Berkeley: University of California Press.

Novas, F. E. 1998. *Megaraptor namunhuaiquii,* gen. et sp. nov., a large-clawed, Late Cretaceous theropod from Patagonia. *Journal of Vertebrate Paleontology* 18: 4–9.

Novas, F. E., S. Apesteguia, D. Pol, and A. Cambiaso. In press. Un probable troodontido (Theropoda-Coelurosauria) del Cretacico Tardio de Patagonia. xv Jornadas Argentinas de Paleontologia de Vertebratos, Ameghiniana.

Powell, J. E. 1987. Hallazgo de un dinosaurio hadrosaurido (Ornithischia, Ornithopoda) en la Formación Allen (Cretacico Superior) de Salitral Moreno, Provincia de Río Negro, Argentina. *Décimo Congreso Geologico Argentino, Actas* 3: 149–152.

Rowe, T., and J. A. Gauthier. 1990. Ceratosauria. In D. B. Weishampel, P. Dodson, and H. Osmólska (eds.), *The Dinosauria,* pp. 151–168. Berkeley: University of California Press.

Salgado, L., and R. A. Coria. 1993. Un nuevo titanosaurino (Sauropoda-Titanosauridae) de la Fm. Allen (Campaniano-Maastrichtiano) de la Provincia de Rio Negro, Argentina. *Ameghiniana* 30 (2): 119–128.

Salgado, L., and R. A. Coria. 1996. First evidence of an ankylosaur (Dinosauria, Ornithischia) in South America. *Ameghiniana* 33: 367–371.

2. On the Type and Referred Material of *Laelaps trihedrodon* Cope 1877 (Dinosauria: Theropoda)

DAN CHURE

Abstract

The material of the theropod *Laelaps trihedrodon* named in the 19th century has long been thought lost. The rediscovery of some referred material, along with the recent discovery of both O. W. Lucas's shipping records of the material and Cope's notebook on the Garden Park Quarries clarifies the type and referred specimens of *L. trihedrodon*, and provides insight into what happened to much of that material. AMNH 5780 has recently been suspected of being a fragment of the type, but morphological and historical data shows that it cannot be. The morphology of AMNH 5780, the only existing material of *L. trihedrodon*, strongly suggests that this specimen is referable to *Allosaurus*.

Introduction

During his long career, Professor E. D. Cope made enormous contributions to our understanding of the fossil record of vertebrates and amassed an immense personal collection that was ultimately purchased by Henry Fairfield Osborn for the Department of Vertebrate Paleontology of the American Museum of Natural History in New York City. Cope's collection of fossil reptiles was purchased in 1902, and William D. Matthew went to Philadelphia to oversee the packing and shipment of the collection to New York, where it arrived in 1903 (Osborn 1931).

For a variety of reasons, many of Cope's original descriptions were brief, failed to cite any specimen numbers, and were usually not illustrated. This problem ultimately led to difficulties in recognizing some of the types and other specimens in his collection. Even worse, it appears that some specimens described by Cope could not even be located in Philadelphia.

Laelaps trihedrodon is one of these unfortunate taxa. This theropod dinosaur was collected by O. W. Lucas in 1877 and 1878 from the Morrison Formation of Garden Park, Colorado. The type specimen, a dentary with teeth, appears to be lost, and the history of much of the referred material (also mostly lost) has been difficult to reconstruct. However, the recent discovery of both Lucas's shipping records and Cope's notebook from his 1879 visit to Garden Park have greatly clarified the record of the sauropod dinosaurs from that area, and led to the suggestion that AMNH 5780 may be part of the lost holotype of *L. trihedrodon* (McIntosh 1998).

McIntosh (1998) provides a redrawing of Cope's sketch map of the quarries around his Saurian Hill (a.k.a. Cope's Nipple) in Garden Park. Monaco (1998) presents a modern Brunton-and-tape-measure map of these quarries.

Institutional Abbreviations: AMNH, American Museum of Natural History, New York City; UUVP, Museum of Natural History, University of Utah, Salt Lake City.

Description of AMNH 5780

AMNH 5780 consists of five tooth crowns in seven pieces: two anterior, two anterolateral, and one lateral. None are complete and some still have significant amounts of matrix adhering to them. The teeth were collected by O. W. Lucas from Quarry I, Saurian Hill, Garden Park, Colorado.

The most anterior tooth is missing its tip and is not recurved (fig. 2.1A–E). In basal cross-section, both the lingual and labial surfaces are convex, the labial more strongly so. Mesial and distal serrations extend to the base of the preserved crown, with the individual serrations becoming smaller toward the base. Serrations are well defined and have marked grooves between them. The crown is 36 mm tall as preserved. The base is 21 mm wide and 21 mm long.

The second anterior tooth is more posterior in position (fig. 2.1F–J). The crown is recurved and missing its tip. Both the labial and lingual surfaces are convex, with the former more so. The mesial and distal serrations extend to the crown base and are similar to those of the more anterior tooth. The crown is 47 mm tall as preserved. The base is 21 mm long and 19 mm wide.

There are two anterolateral tooth crowns and both are recurved (fig. 2.2). The more complete of these two crowns (fig. 2.2E, F) has mesial and distal serrations extending to the crown base. The distal row has serrations similar to those in the above-mentioned teeth. However, the mesial serrations are poorly defined, lacking well-defined grooves between them, and are smaller than the serrations on the distal tooth

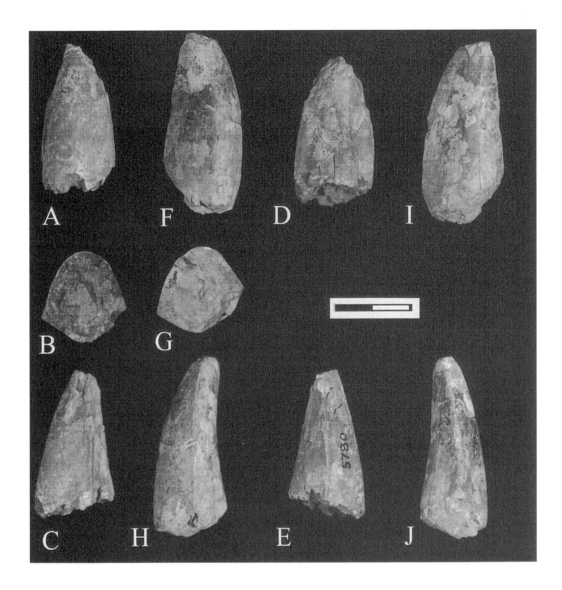

Figure 2.1. AMNH 5780, referred teeth of Laelaps trihedrodon. *Anterior tooth in (A) labial, (B) basal, (C) one side, (D) lingual, and (E) view of other side. The other anterior tooth in (F) labial, (G) basal, (H) one side, (I) lingual, and (J) view of other side. Scale: 1 cm.*

margin. The small mesial serrations are fused for the proximal half of the crown height and form a bumpy carina that curves strongly lingually at the crown base. This crown measures 41 mm tall as preserved. Its base measures 19 mm long and 14 mm wide.

The second anterolateral tooth is in two pieces and is very incomplete. Little can be said other than it is recurved, has mesial and distal serrations, and that these serrations are separated by grooves (fig. 2.2 C, D).

The lateral tooth is in two pieces, but is nearly complete (fig. 2.2A,B). The tooth is laterally compressed and recurved, although the distal margin is straighter. Mesial and distal serrations are present and extend to the base. The distal serrations are more pronounced and the mesial row forms a scalloped carina. The crown is 62 mm tall. It is 28 mm long and 17 mm wide at its base.

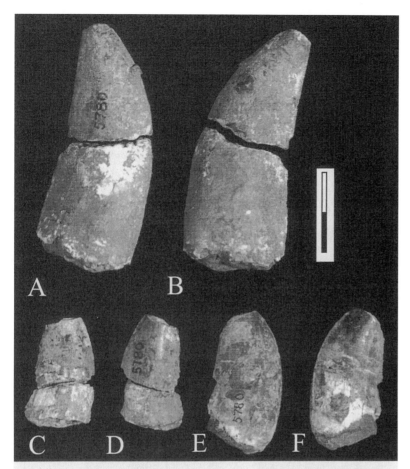

Figure 2.2. AMNH 5780, referred teeth of Laelaps tri-hedrodon. *Large lateral tooth in side views (A, B). Smaller anterolateral tooth in side views (C, D) and another anterolateral tooth in side views (E, F). (G) Museum label associated with AMNH 5780. (H) O. W. Lucas's handwritten label associated with AMNH 5780.*

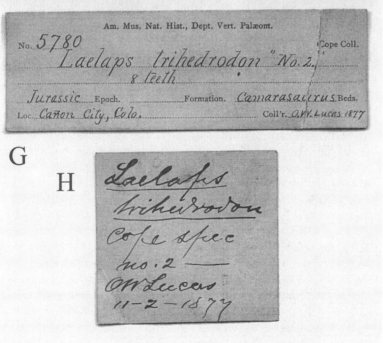

Systematic Paleontology

Laelaps trihedrodon Cope, 1877

Type Specimen: A right dentary which "supports eight teeth, and contains a cavity at the anterior extremity, from which one tooth was probably shed." (Cope 1877, 805), from Cope's Quarry I, Saurian Hill, Garden Park (McIntosh 1998, 492).

Referred Specimens:

1. Five teeth, AMNH 5780 (Osborn and Mook 1921, 258). It is not known from which of Cope's Saurian Hill, Garden Park, quarries this specimen was collected.

2. Femur, AMNH, catalog number unknown (Cope 1878a; Osborn and Mook 1921, 261). Specimen cannot be identified in the AMNH collections. Cope's Quarry II, Saurian Hill, Garden Park (McIntosh 1998, 492).

3. O. W. Lucas's bones B, D, E, part of bone G, and fragments of skull. No repository or catalog number given. These specimens cannot be identified in the AMNH collections and are presumably lost. It is not known from which of Cope's Saurian Hill, Garden Park, quarries these specimens came from.

Locality: Cañon City, Colorado

Horizon: Morrison Formation

Age: Late Jurassic, Kimmeridgian–early Tithonian (Peterson and Turner 1998).

Discussion

The Type Specimen of L. trihedrodon

Cope (1866) erected the genus and species *Laelaps aquilunguis* for theropod material from the Late Cretaceous of New Jersey. Subsequently, Cope created several additional species of *Laelaps,* including *L. trihedrodon.* Marsh (1877) noted that the generic name was preoccupied and replaced it with *Dryptosaurus,* although Cope (1877, 806) disputed the need for this. The subsequent history of that genus is extraneous to the present study.

Laelaps trihedrodon was established for a right dentary with teeth (Cope 1877). Unfortunately, Cope did not illustrate any of the type or any of the subsequently referred material. Although the holotype has been considered lost, McIntosh (1998, 502) recently suggested that AMNH 5780 may be a part of the missing type. However, morphological comparison of AMNH 5780 with Cope's description of the teeth of the type and a handwritten note by O. W. Lucas concerning AMNH 5780 strongly suggest that it is not.

In the type description Cope writes, "At present, I only describe a portion of the right dentary bone, which supports eight teeth . . ." (1877, 805). Cope states that these were "[f]ive successional and two functional teeth [which] exhibit crowns complete, or nearly so" (805). All five of the tooth crowns of AMNH 5780 are functional teeth and even the smallest shows wear on the serrations. Furthermore the most anterior tooth crown has a concave base, suggesting that it might be a shed crown. Cope also notes, "The enamel is smooth and with a fine silky luster." However, many of the teeth in AMNH 5780 have matrix adhering to them and only a little of the enamel is visible. AMNH 5780 does not compare well with the descriptions of the type teeth.

Further confusion comes from Osborn and Mook (1921, 258), who refer AMNH 5780 to *Laelaps trihedrodon* (but clearly do not believe it to be the type), and state that it contains *eight* teeth! This discrepancy concerning the number of teeth in AMNH 5780 cannot be reconciled.

There are two labels associated with AMNH 5780. The first is a museum label of standard AMNH issue (fig. 2.2G), which states that the specimen consists of eight teeth! The second has important bearing on the question of the specimen being part of the type.

It is handwritten on the back of a business card which is torn in half. It reads "*Laelaps trihedrodon* Cope spec no. 2 O W Lucas 11-2-1877" (fig. 2.2H). Comparison of this signature with known examples of Lucas's signature shows that the second AMNH label is in Lucas's handwriting. It is uncertain if the date "11-2-1877" is the date of collection, but it indicates that AMNH 5780 was in Colorado at least as late as that date. This is critical, because the type description of *L. trihedrodon* was published on August 15, 1877 (Osborn 1931, 645), nearly *three months before* the date associated with AMNH 5780, and Lucas's shipping records document that the type was sent in the summer of 1877 (McIntosh 1998, 485). This conclusively shows that AMNH 5780 cannot be part of the type and the holotype must be considered lost.

The conclusion that the type is lost is further supported by Osborn and Mook (1921, 258), who report that when W. D. Matthew went to Philadelphia to organize the shipment of the Cope Collection to the AMNH, he identified and catalogued Cope's types, but could not find the type of *Laelaps trihedrodon*.

Referred Specimens of L. trihedrodon

Lucas gave a separate number for each individual in a quarry. His "Fossil 2" was assigned to the type and referred specimens of *L. trihedrodon*, and was sent in three shipments. The records for Fossil 2 are as follows (from McIntosh 1998, 485–486 and McIntosh, pers. comm. 1998):

Shipment 1 (sent summer of 1877): Type dentary with teeth.
Shipment 5 (sent October 22, 1877): Part of Fossil 2 containing "fragments of head found not far from the jaw sent in the first shipment."
Shipment 8 (sent January 6, 1878):
  Box 27: 2 bundles of Fossil 2.
  Box 28: 5 bundles of bone G of Fossil 2, 1 bundle of bone B of Fossil 2, 1 bundle of bone D of Fossil 2.
  Box 31: Bone G, leg bone of Fossil 2.
  Box 32: Bone H, 8 bundles of femur from Fossil 2.
  Box 43: 1 bundle of bone E of Fossil 2.

Teeth are only mentioned in the first shipment. Because AMNH 5780 is not part of shipment 1 and could not have been shipped before

November 2, 1877, it must have arrived in shipment 8. The "fragments of head" (i.e., skull) in shipment 5 were not more specifically identified and cannot be located in the collections of the AMNH. There is no evidence that they ever arrived at the AMNH.

The femur in shipment 8 has a complicated history. Cope (1878a) erected the genus and species *Hypsirophus discurus,* but suggested that it might turn out to be the same taxon as *Laelaps trihedrodon.* The type of *H. discurus* (AMNH 5731) consists of two dorsal vertebrae and a caudal neural arch fragment, all of which is stegosaurian (Galton 1990, 450). Cope (1878a) interpreted *Hypsirophus* as a theropod because he had recently examined a theropod femur from the Morrison that he felt was similar to that of *Megalosaurus bucklandii* and different from that of *Laelaps.* Cope even listed the taxon *Hypsirophus trihedrodon* (a combination never published) as coming from Quarry 1 in his notes made during his visit to Garden Park in 1879 (see McIntosh 1998, 492 for transcription). The provenance of the femur was not given in Cope 1878a, but Cope's notebook lists a femur of *Hypsirophus* as coming from Quarry 2, a quarry some 10 feet from Quarry 1, the source of the type of *L. trihedrodon.* (McIntosh 1998, 492). This is the femur mentioned by Cope (1878a) and it is undoubtedly the one sent in shipment 8. However, it is clear from Cope's remarks that the femur was *not* part of the type of *Hypsirophus discurus.* Cope (1878b) reiterated his belief that *Hypsirophus discurus* was carnivorous, apparently because of the referred femur. Glut (1997, 111) is in error when he states that *Hypsirophus discurus* is based in part on a neural spine of *Allosaurus.* The supposed *Allosaurus* spine in Glut (1997, 113) is clearly stegosaurian.

To further complicate matters, Osborn and Mook mention that in the Cope collection "there is a theropodous femur [which] may be provisionally referred to *Epanterias*" (1921, 261). Inexplicably, they do not even mention this femur in their description of *Epanterias amplexus* later in the same work (Osborn and Mook 1921, 282–284). Cope (1878b) did not mention a femur as part of the type (only known specimen) of *Epanterias amplexus* (AMNH 5767), and no such bone is now with the type of *E. amplexus.* This femur is almost certainly the femur of *Hypsirophus* mentioned by Cope (see above) as coming from Quarry 2, the type locality for *Epanterias,* and the femur identified in shipment 8. Thus, it is clear from Osborn and Mook (1921) that the femur did arrive at the AMNH after the Cope collection was shipped in 1902. Unfortunately they did not cite an AMNH catalog number for the femur and it cannot now be identified in the collections.

When Matthew prepared the material in Philadelphia for shipping to the AMNH, he made two lists for the material: one by box shipped (to New York) and the other by Cope's catalog number. The list by box shipped gives the contents of box 6 as *Creosaurus trigonodon,* part of femur, tibia, and two vertebrae (the name *C. trigonodon,* which occurs in Osborn 1931 [452–453], is a *lapsus calami* for *L. trihedrodon*). In his list by Cope's catalog number, specimen 1013 is given as *C. trigonodon,* femur and tibia (McIntosh, pers. comm. 1998). The discrepancy concerning two vertebrae between the two lists cannot be reconciled, but they may well be part of the type of *Epanterias amplexus*. The tibia could be the leg bone (bone G) in the box 31 of the January 6, 1878, shipment to Cope. The tibia is not in the list of Cope's types in Osborn and Mook 1921 (258), nor is it mentioned anywhere in that work, and it cannot be identified in the AMNH collections.

There is no indication as to the identity of Lucas's bones B, D, E, G (of box 28, not the bone G of box 31), nor the two bundles of bones in box 27, all of which were in the January 6, 1878, shipment to Cope. If they still exist, they cannot be identified in the AMNH collections. However, W. D. Matthew's shipping data suggests that they were never sent to the AMNH.

Affinities of AMNH 5780

AMNH 5780 is the only material of *Laelaps trihedrodon* that currently can be located. It resembles *Allosaurus* in lacking tall, blade-like lateral teeth, and lacking striations on the lingual face of the anterior teeth. Lingual striations are an autapomorphy for *Ceratosaurus* (UUVP 674; Madsen 1976, 17; Madsen and Welles 2000). *Ceratosaurus* and *Torvosaurus* have large and compressed, blade-like lateral teeth, unlike AMNH 5780 (Britt 1991; Gilmore 1920; Madsen and Welles 2000). However, the two similarities between AMNH 5780 and *Allosaurus* are primitive features for theropods and may be present in other poorly known Morrison theropods, such as *Marshosaurus, Stokesosaurus,* and *Coelurus.* However, on the basis of these features and the overwhelming abundance of *Allosaurus* among theropods in the Morrison Formation (Foster and Chure 1998), it is likely that AMNH 5780 is a specimen of *Allosaurus.*

Acknowledgments: This study is part of a Ph.D. dissertation reviewing the systematics of the theropod family Allosauridae. I thank Mark Norell (American Museum of Natural History) for allowing me to study and borrow specimens under his care and Ms. Charolette Holton (AMNH) for processing the loan of AMNH 5780. Dr. J. S. McIntosh (Weslyan University, Conn.) provided much unpublished information concerning the shipping records and packing lists and gladly discussed them with me. Dr. Kenneth Carpenter (Denver Museum of Natural History) provided copies of letters and other historical documents with O. W. Lucas's signature. Both J. S. McIntosh and K. Carpenter read a first draft of the manuscript. Travel was supported by the National Park Service.

References

Britt, B. B. 1991. Theropods of the Dry Mesa Quarry (Morrison Formation, Late Jurassic), Colorado, with emphasis on the osteology of *Torvosaurus tanneri. Brigham Young University Geology Studies* 37: 1–72.

Cope, E. D. 1866. On the discovery of the remains of a gigantic dinosaur in the Cretaceous of New Jersey. *Proceedings of the Academy of Natural Sciences, Philadelphia* 18: 275–279.

Cope, E. D. 1877. On a carnivorous dinosaurian from the Dakota Beds of

Colorado. *Bulletin of the United States Geological and Geographical Survey of the Territories,* series 3, no. 4: 805–806.

Cope, E. D. 1878a. A new genus of Dinosauria from Colorado. *American Midland Naturalist* 12: 188–189.

Cope, E. D. 1878b. A new opisthocoelous dinosaur. *American Midland Naturalist* 12: 406.

Foster, J. R., and Chure, D. J. 1998. Patterns of theropod diversity and distribution in the Late Jurassic Morrison Formation, western USA. *Abstracts and Program for the Fifth International Symposium on the Jurassic System, International Union of Geological Sciences, Subcommission on Jurassic Stratigraphy,* pp. 30–31. Vancouver.

Galton, P. M. 1990. Stegosauria. In D. B. Weishampel, P. Dodson, and H. Osmólska (eds.), *The Dinosauria,* pp. 435–455. Berkeley: University of California Press.

Gilmore, C. W. 1920. Osteology of the carnivorous Dinosauria in the United States National Museum, with special reference to the genera *Antrodemus* (*Allosaurus*) and *Ceratosaurus*. *United States National Museum Bulletin* 110: 1–159.

Glut, D. F. 1997. *Dinosaurs: The Encyclopedia.* Jefferson, N.C.: McFarland.

Madsen, J. H. 1976. *Allosaurus fragilis:* A revised osteology. *Utah Geological and Mineralogical Survey Bulletin* 109: 1–163.

Madsen, J. H., and S. P. Welles. 2000. *Ceratosaurus* (Dinosauria. Theropoda), a Revised Osteology. *Utah Geological Survey, Miscellaneous Publication* 00-2: 1–80.

Marsh, O. C. 1877. Notice of a new and gigantic dinosaur. *American Journal of Science,* 3d ser., 14: 87–88.

McIntosh, J. S. 1998. New information about the Cope collection of sauropods from Garden Park, Colorado. In K. Carpenter, D. J. Chure, and J. I. Kirkland (eds.), The Upper Jurassic Morrison Formation: An interdisciplinary study. *Modern Geology* 23: 481–506.

Monaco, P. 1998. A short history of dinosaur collecting in the Garden Park Fossil Area, Cañon City, Colorado. In K. Carpenter, D. J. Chure, and J. I. Kirkland (eds.), The Upper Jurassic Morrison Formation: An interdisciplinary study. *Modern Geology* 23: 465–480.

Osborn, H. F. 1931. *Cope: Master Naturalist.* Princeton: Princeton University Press. New York: Arno Press.

Osborn, H. F., and C. C. Mook 1921. *Camarasaurus, Amphicoelias,* and other sauropods of Cope. *Memoirs of the American Museum of Natural History,* n.s., 3 (part 3): 245–387.

Peterson, F., and C. E. Turner. 1998. Stratigraphy of the Ralston Creek and Morrison Formations (Upper Jurassic) near Denver, Colorado. In K. Carpenter, D. J. Chure, and J. I. Kirkland (eds.), The Upper Jurassic Morrison Formation: An interdisciplinary study. *Modern Geology* 22: 3–38.

3. Endocranial Anatomy of *Carcharodontosaurus saharicus*
(Theropoda: Allosauroidea) and Its Implications for Theropod Brain Evolution

Hans C. E. Larsson

Abstract

Complete theropod endocrania are scarce but hold promise to elucidate the early evolution of the avian brain from the ancestral reptilian brain. A complete endocast of *Carcharodontosaurus saharicus* was obtained using computed tomography (CT) scan data and edited with three-dimensional (3-D) reconstruction software. Endocranial and inner ear anatomies are described. Allometric regressions for the proportion of the brain composed of the cerebrum are calculated for extant birds and nonavian reptiles. These regressions are used to compare the fossil data to the allometric scaling of birds and reptiles. The degree to which selected theropods approach the bird regression from the reptile regression is calculated and placed into a phylogenetic context. Results suggest that noncoelurosaur theropods have a similar cerebrum-to-total-brain ratio as modern nonavian reptiles. The increase in cerebral proportions that characterize modern birds are hypothesized to have occurred at Coelurosauria and continued throughout the evolution of maniraptorans and early birds.

Introduction

The endocranium describes the space within the braincase occupied by the brain and its supporting tissues, vasculature, and cerebrospinal fluid. Numerous fossil endocasts have been described and have been the focus of a broad comparative research program (Romer and Edinger 1942; Jerison 1955, 1961, 1963, 1968, 1969, 1973; Edinger 1964, 1966; Hopson 1977, 1979). Few specific studies of nonavian theropods and early fossil avian endocasts have been presented due to the paucity of available specimens (Osborn 1912; Edinger 1926, 1951; Russell 1969, 1972; Dechaseaux 1970; Hopson 1979; Currie and Zhao 1993; Currie 1995; Rogers 1998). Many small theropods have delicate braincases that are often incomplete, whereas the largest, such as *Tyrannosaurus* and *Tarbosaurus,* have massive braincases that were previously examined from specimens hemisected with a diamond saw (Osborn 1912, Maleev 1965).

An expedition lead by Paul Sereno, of the University of Chicago, in 1995 to the Kem Kem region in southeastern Morocco recovered a partial skull of an adult *Carcharodontosaurus saharicus* (Sereno et al. 1996). The specimen included a complete and undistorted braincase. The braincase was CT scanned and the digital data was used to create a 3-D copy of the endocast with no harm to the fossil. A recent comparison of the total endocranial volume of this taxon with *Tyrannosaurus rex,* a theropod of similar body size but phylogenetically closer to birds (Holtz 1994; Sereno 1999), concluded that the endocranial space is approximately 150% larger in the latter taxon (Larsson et al. in press). This volume difference suggests an increase in total brain size at the level of Coelurosauria because the similar body size of the two taxa avoids problems of the allometry of this index. The present study will describe the endocranial and inner-ear anatomy of *C. saharicus* and follow with a discussion of a refined comparative technique of endocranial volumes with preliminary results applied to the early evolution of the avian brain.

Institutional Abbreviations: AMNH, American Museum of Natural History, New York; SGM, Ministère de l'Energie et des Mines, Rabat, Morocco; UUVP, University of Utah, Salt Lake City.

Materials and Methods: A complete description of the three-dimensional scanning and reconstruction of the *Carcharodontosaurus* and *Tyrannosaurus* endocasts is described in Larsson et al. in press.

Description

The braincase of *Carcharodontosaurus* completely encloses the endocranial region. This high degree of ossification allows for a thorough description of the endocranial anatomy. Whereas endocasts of birds and mammals tend to match closely the brain's surface anatomy, most reptiles have only subtle surface associations. Hopson (1979) points out that the forebrain in reptiles closely matches the endocranium, but the remaining volume is generally filled with large vascular sinuses and cerebrospinal fluid. In this regard, the endocast of *Carcharodontosaurus* appears quite reptilian with the forebrain fairly well demarcated from an otherwise uniform endocast.

The endocast of *Carcharodontosaurus* is quite similar to its close relative, *Allosaurus fragilis* (Hopson 1979; Rogers 1998). The olfactory bulbs and peduncles lie on approximately the same horizontal plane as the forebrain, the midbrain is angled posteroventrally from the forebrain, and the hindbrain parallels the forebrain at a more ventral level (fig. 3.1A,B). Using the terminology of Hopson (1979), the cephalic flexure (between the fore- and midbrain) is approximately 45°, and the pontine flexure (between the mid- and hindbrain) is approximately 40°.

Figure 3.1. Carcharodontosaurus saharicus *(SGM-Din 1). Endocast in dorsal (A) and left lateral views (B); and a cutaway view of the left inner ear in lateral view (C). Abbreviations: acc, cerebral carotid artery; ao, ophthalmic artery; ca, anterior semicircular canal; cc, crus commune; cer, cerebrum; ch, horizontal semicircular canal; ci, ventral margin of the crista interfenestralis; cl, cavum labyrinthicum; cp, posterior semicircular canal; end, endolymphatic duct; faf, fossa acustico-facialis; floc, flocculus; fm, plane of the foramen magnum; fp, foramen perilymphaticum; fpr, fenestra pseudorotunda; l, lagena; ls, longitudinal sinus; pit, pituitary; rst, recessus scalae tympani; s, sacculus; ts, transverse sinus; u, utriculus; vc?, possible vascular canal; vcd, vena capitis dorsalis; vf, vagus foramen. Roman numerals represent cranial nerves. A and B are to same scale, 5 cm; scale in C is 2 cm.*

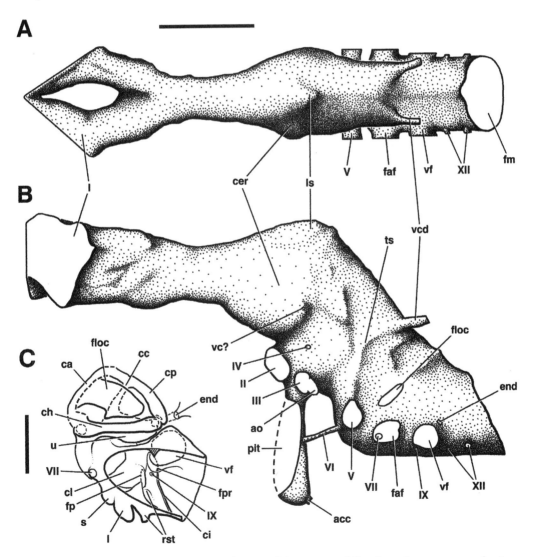

The dural covering of the forebrain is relatively thin in extant crocodilians and birds and allows for a fairly accurate interpretation of the enclosed soft tissue anatomy (Hopson 1979). The olfactory bulbs and tracts (I) occupy the anteriormost extent of the endocast. The bulbs exit the anterior margin of the ossified braincase and are cupped posteriorly by a mesethmoid bone (sensu Witmer 1996). The mesethmoid separates each bulb with a median septum. The olfactory tracts extend posteriorly from the bulbs and are not separated by a bony septum. The tracts continue posteriorly to the cerebral hemispheres. The junction between the tracts and the cerebral hemispheres is indicated by a sharp dorsoventral expansion of the endocast and approximates the anterior extent of the cerebral hemispheres. A large ridge lies dorsal to the cerebral hemispheres and probably marks the size and extent of the large median longitudinal venous sinus and smaller paired arteries found in crocodiles (Hochstetter 1906). The ridge extends posteriorly through the length of the endocast and probably terminated in the sinus longitudinalis medullae spinalis, which lies over the medulla oblongata.

The lateral expansion of the cerebral hemispheres reaches its maximum at a level just dorsolateral to the exit of the optic nerve. This region also marks the exit of a small vascular element. Janensch (1936) interpreted a similar foramen in sauropods to be equivalent to the embryonic foramen epioptica and suggested that it transmitted a vestigial anterior cerebral vein.

The optic (II) nerve exits the endocast near the ventral margin of the cerebrum. This nerve is large and separated from its counterpart by an ossified orbitosphenoid. The oculomotor (III) nerve departs the endocast just behind the optic nerve and is also separated by the orbitosphenoid. The posterior margin of the right exit of the oculomotor nerve is partially separated from the foramen by a thin bony projection from the laterosphenoid. The ophthalmic artery probably occupied this region.

A small trochlear (IV) nerve exits dorsal to the oculomotor nerve and posteroventral to the small anterior cerebral vein. The pituitary exits ventrally between the oculomotor nerve foramina. The anterior and lateral surfaces of the pituitary were probably enclosed with cartilage during life and the posterior and ventral surfaces were cradled in the sella turcica of the basisphenoid. The abducens (VI) nerve enters the hypophyseal fossa through the posterior wall of the sella turcica. The proximal exit of this nerve from the endocast is located on the ventral margin of the endocast near the exit of the trigeminal (V) nerve. As in most vertebrates, the paired internal carotid arteries enter the endocranial region through the basin of the hypophyseal fossa. The entry of the arteries is partially separated by a thin bony septum that probably completely separated the pair during life.

The trigeminal nerve exits the endocast dorsal to the abducens entry into the basisphenoid. The ophthalmic, maxillary, and mandibular rami of this nerve exit through a common foramen and must have diverged from each other outside the braincase. A prominent ridge extends posterodorsally over the endocast from the dorsal edge of the trigeminal foramen to the foramen of the vena capitis dorsalis. This ridge represents the transverse sinus into which the vena capitis dorsalis drains (O'Donoghue 1920). This vein drains the anterior neck musculature into the endocranium through a pair of long canals on the posterior surface of the endocast. In many reptiles, the transverse sinus drains via a middle cerebral vein that exits the braincase through its own foramen. This anatomy is found in theropods such as *Allosaurus* (Hopson 1979) and *Dromaeosaurus albertensis* (Currie 1995). However, in *Carcharodontosaurus,* the transverse sinus probably drained into a middle cerebral vein that exited the braincase in the ridge present on the dorsal edge of the trigeminal foramen. A similar anatomy has been described for *Troodon formosus* (Currie and Zhao 1993).

The facial (VII) and acoustic (VIII) nerves exit the endocranium in a fossa acustico-facialis located just behind the trigeminal foramen. The facial nerve exits the anteroventral corner of the fossa and continues out the braincase through a single foramen. The exits for the cochlear and vestibular rami of the acoustic nerve from the endocranium are not preserved. The bony edge of the fossa appears complete and suggests that the acoustic nerves exited through a cartilaginous septum forming the hiatus acusticus. The hiatus acusticus medially borders the cavum labyrinthicum, a recess that would have housed the utriculus, saccule, and perilymphatic cistern of the inner ear.

The roof of the cavum labyrinthicum opens into a canal that would have housed the crus commune, visible in the endocast (fig. 3.1C). The canal projects dorsally and slightly posteriorly and terminates at the junction of the anterior and posterior semicircular canals. The chamber for the posterior semicircular canal is completely preserved and slopes posterolaterally with little arching from the crus commune to its contact with the chamber for the horizontal semicircular canal. This junction connects with the utriculus anteromedially. Although the canal joining the junction with the utriculus is undivided, the posterior and horizontal semicircular canals would have remained separate in life. The chamber for the horizontal semicircular canal arches laterally to join the chambers for the anterior semicircular canal and the utriculus. The ampullae for the horizontal and anterior semicircular canals would have been located at this junction. Only the anterior border of the chamber housing the anterior semicircular canal is preserved. The braincase was fractured along this region. However, the available anatomy indicates the anterior semicircular canal passed in a relatively linear posterodorsomedial direction.

The subtriangular outline of the three semicircular canals in lateral view is present in *Allosaurus* (Hopson 1979). Rogers (1998) noted that this anatomy is also present in nonavian reptiles, such as turtles and lizards, but not in birds. Rogers (1999) also pointed out that the sharp apex at the junction of the anterior and posterior semicircular canals of *Allosaurus* is most similar to that in modern crocodilians. The acute apex is the result of the nearly linear anterior and posterior semicircular canals. A similar anatomy is also present in *Carcharodontosaurus* and may represent a basal archosauromorph condition.

A large floccular recess projects into the region surrounded by the semicircular canals. The recess has a ventral fingerlike projection. The recess enters medial to the anterior semicircular canal, passes lateral to the crus commune, and extends near the junction of the posterior and horizontal semicircular canals. This recess, which in life housed the floccular lobe of the brain, is present in extant birds and has been described in numerous theropods (Hopson 1979; Currie and Zhao 1993; Currie 1995) and pterosaurs (Newton 1888; Edinger 1927).

A small foramen is visible on both sides of the endocast just posterodorsal to the vagal foramen. The foramen opens into a canal that extends anterolaterally for a short distance. The remainder of the canal may be lost from the scanning resolution due to its small size. However, the position of the endocranial opening and direction of the canal suggest that it housed the endolymphatic duct. If this identification is correct, the canal would continue under the utriculus and into the saccule, as in extant amniotes.

The anteroventral surface of the cavum labyrinthicum forms a shallow depression that probably housed the anteroventral expansion of the saccule, common in tetrapods. Behind this depression is a deeper columnar fossa that housed the lagena. The fossa projects posteroventrally. This orientation of the lagena is most similar to those found in extant crocodiles and the proximal region of the lagena of some birds (Wever 1978; Lowenstein 1974).

The perilymphatic duct would have exited the perilymphatic cistern, at the round window, and extended along the basilar papilla where it formed the lagena, as in extant amniotes. In *Carcharodontosaurus*, the distal end of this duct probably extended along the lateral surface of the distal ramus of the opisthotic bone, as in *Varanus*, crocodilians, and birds (de Beer and Barrington 1934; pers. obs.).

The perilymphatic duct would have exited through the foramen perilymphaticum that appears to have been bordered medially, ventrally, and ventrolaterally by the ventral ramus of the opisthotic, and dorsally and dorsolaterally, probably, by a cartilaginous supraperilymphatic strut of the opisthotic (de Beer 1937) (Fig. 3.1C). After passing through the foramen perilymphaticum, the perilymphatic duct entered the recessus scalae tympani, found in all tetrapods (de Beer 1937). This recess is a sharp fossa posterior to the fossa housing the lagena. The posterior margin of the recessus scalae tympani is bounded by a posteroventrolaterally oriented edge of bone that is probably part of the basisphenoid.

The lateral edge of the recessus scalae tympani appears to taper along a shallow groove situated anterior to the posteroventrolaterally oriented edge of bone described above. The recessus scalae tympani was probably roofed in life by a cartilaginous or thin bony septum to separate it from the anteriorly adjacent middle ear cavity.

The medial extent of the recessus scalae tympani appears to terminate at a subcircular crest that is confluent with the anterolateral border of the vagal foramen. The crest extends ventromedially from the bony ridge bounding the posterior wall of the recessus scalae tympani. The crest gently curves dorsomedially along the ventral ramus of opisthotic and curls ventrolaterally at its dorsalmost extent. This semicircular rim appears to form a fenestra that would have separated the recessus scalae tympani from a cavity occupied by the structures exiting the vagal foramen. The fenestra does not appear to face the endocranium, but, rather faces posteromedially. The identity of the fenestra is probably the fenestra pseudorotunda. The lateral edge of the fenestra pseudorotunda was presumably bounded by a cartilaginous or thin bony septum stretched vertically from the sharp edge on the basisphenoid, described above. This septum, the crista interfenestralis, would have supported the lateral edge of the secondary tympanic membrane stretched over the fenestra pseudorotunda. The septum may have been partially or completely ossified as a delicate crista interfenestralis or remained cartilaginous. Similar anatomies exist in *Dromaeosaurus* (Currie 1995), where a bony crista is not present (or preserved), and *Troodon* (Currie and Zhao 1993), where the dorsal and ventral regions of the crista are ossified but the remainder is either cartilaginous or not preserved.

If the identification of the fenestra pseudorotunda is correct, the bone at the medial edge of the fenestra pseudorotunda must be the metotic strut. This strut is laterally hypertrophied in nonavian maniraptorans, such as *Dromaeosaurus* and *Troodon* (Currie and Zhao 1993; Currie 1995), and birds, such as *Archaeopteryx* and *Hesperornis* (Witmer 1990) and extant birds. Further comparisons may indicate this reduced and medial position of the metotic strut to be present in more basal theropods as well.

The glossopharyngeal (IX) nerve exited the endocranium just posterior to the fossa acustico-facialis. The nerve exits near the ventral margin of the endocranium and is separated from the posterodorsal region of the fissura metotica by a thin horizontal strut of bone. The nerve appears to travel partway through the recessus scalae tympani because it exits the braincase just ventral to the fenestra pseudorotunda to enter the recess.

The vagus (X) nerve most likely exited the endocranium with the posterior cerebral vein through a large vagus foramen located posterior and dorsal to the exit of the glossopharyngeal nerve. Two foramina are located posterior to the vagus foramen. The anterior of the two foramina is smaller and probably housed an anterior branch of the hypoglossal (XII) nerve while the larger posterior foramen housed the posterior branch(es) of the hypoglossal nerve.

The foramen magnum is represented as a large subcircular opening. The ventral surface of the endocranium near the foramen magnum appears relatively smooth with no trace of ventral branches of the basilar artery. The remainder of the endocast bears little discrete anatomical detail.

Allometric Comparison

Past comparisons of brain volumes among different animals have typically employed total endocranial volumes and total body masses to make bivariate comparisons (Jerison 1969, 1973). Many problematic issues arise with this technique. Dendy (1911) reported that the brain of

Sphenodon filled only about half of the endocranial space and numerous workers have used this 50% estimate for fossil endocasts (Osborn 1912; Jerison 1973). Hopson (1979) cautioned that some fossil reptile brains appear to have filled their endocrania more completely than others based on the details preserved on the endocranial surface. He also reported that in a small range of *Caiman* specimens, the disparity between brain and endocranial volume increases with increasing body size. Clearly, the transition from a reptilian-type brain to an avian-type brain occurred and suggests that a simple 50% or nearly 100% ratio cannot be used for all fossil endocasts.

Body mass estimates are also problematic. Peczkis (1994) examined possible body mass estimates for a broad range of dinosaurian taxa. The range of body mass estimates for each specimen typically spanned a fourfold range, such as *Spinosaurus* with a range of 1000 to 4000 kg. The broad ranges of body mass estimates, combined with the ambiguous ratio of endocranial volume occupied by the brain, present a high degree of uncertainty for this index of brain size (see Larsson et al. in press, fig. 3).

In an attempt to reduce the effects of uncertainties while comparing fossil endocrania, Larsson et al. (in press) compared the volumes of the forebrain and the total endocranial space. The only assumption this technique requires is that the relative thickness of the surrounding dura over different parts of the brain is consistent between different endocasts. That is, the ratio of the dural thickness over the forebrain to the dural thickness over the medulla oblongata is assumed to be equal among different taxa. Unfortunately no empirical data exist for these assumptions. Hopson (1979) pointed out that in *Caiman,* the dural thickness over the fore- and hindbrain increase with age. The dura over the medullary region appeared to increase in thickness to a greater extent than that over the forebrain, but it is unknown if the relative ratio between the two regions is maintained. However, in light of the poorly resolved brain and body mass estimates, I believe this method to be a better technique until data from extant taxa are obtained.

With the above assumption, a ratio of the volume of a specific region of the endocast to the volume of the entire endocast should approximate the same ratio of the formerly enclosed brain. The cerebrum within the forebrain is one of the most clearly demarcated regions within fossil endocrania (Hopson 1979). Although the actual total brain and cerebral volumes cannot be known for fossil taxa, the technique used here is an attempt to minimize the possible errors involved. The endocranial volume may overestimate the actual brain volume by as much as a factor of two. The cerebral region impressed on the endocast will overestimate the cerebral volume by a less than a factor of two because this region most closely fits the endocranium in living reptiles than other parts of the brain (Hopson 1977). These two indices, in a logarithmically transformed comparison, should have a greater accuracy than a comparison of brain and body masses for fossil taxa.

The cerebrum is involved primarily with sensory integration and nervous control. The ancestral condition of the avian brain lies within the nonavian reptiles. The convergent evolution of enlarged brains in birds and mammals has been widely associated with the relative intelligence of each group. The remainder of this discussion will attempt to trace the early evolution of the avian cerebral proportions from its reptilian ancestry.

First, the cerebral and total brain masses for extant nonavian reptiles and birds were obtained. Data for extant nonavian reptiles were collected from Platel (1976) and Gans (1980). Extant bird data were compiled from Ebinger and Löhmer (1984, 1987) and Rehkämper et al. (1988, 1991). Data from domesticated birds were excluded because domesticated animals generally have reduced brain volumes compared to their wild counterparts (for a brief review see Ebinger and Löhmer 1987). All data expressed as volumes were transformed to masses using a specific gravity of 1.0 g/ml. A similar volume to mass transformation was used with the fossil data, following Jerison (1973). These data were graphed in a log-transformed bivariate plot (fig. 3.2). Each set of data falls near well supported power regressions. The nonavian reptile and bird regressions are $y = 0.332x^{0.95}$ ($R^2 = 0.993$) and $y = 0.484x^{1.12}$ ($R^2 = 0.982$), respectively.

These regressions indicate that as brain mass increases, cerebral mass increases with slight negative allometry in nonavian reptiles and with positive allometry in birds. That is to say, as bird brains increase in size, their cerebra increase at a faster rate. In contrast, as nonavian reptile brains increase in size, their cerebra increase at a slower rate. These regressions cross each other at a brain mass of approximately

Figure 3.2. Log-transformed bivariate plot of the cerebral mass as a function of total brain mass for extant nonavian reptiles (triangles) and birds (squares). The regressions (solid lines) are bounded by their 95% confidence limits (dashed lines). Data from table 3.1 (Xs) are plotted for the endocrania of Allosaurus *(Al),* Archaeopteryx *(Ar),* Carcharodontosaurus *(C),* Numenius gypsorum *(Ng), N.* tahitiensis *(Nt),* Sebecus *(S),* Troodon *(Tr) (upper and lower limits), and* Tyrannosaurus *(Ty).*

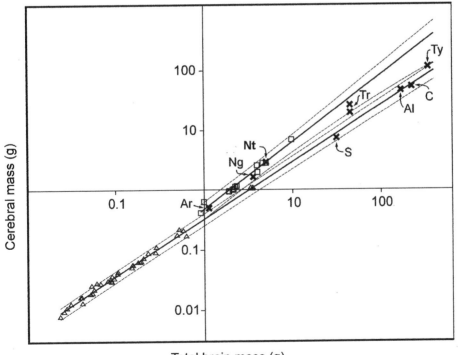

0.11 grams, a value typical for a mid-sized adult skink (*Eumeces*) and water snake (*Nerodia*) but half the size of a hummingbird (*Amazilia*) (Criule and Quiring 1940; Platel 1976).

Unfortunately, few complete endocrania are available for the majority of nonavian theropods. The limited fossil data used here are listed in table 3.1. Following Jerison (1969, 1973), the total endocranial volumes were measured between the narrowest transverse point of the olfactory tracts and the posterior exit of the hypoglossal nerve. The volume for *Allosaurus* was obtained by water displacement of a cast of UUVP 294. Figures of the endocasts of *Sebecus* (Colbert 1946), *Archaeopteryx* (Jerison 1979), *Numenius gypsorum* (an Eocene bird), and *Numenius tahitiensis* (an extant sandpiper) (Dechaseaux 1970) were used to calculate the total endocranial volume. These calculations were performed using a modified version of Jerison's "graphical double integration" method, discussed in Larsson et al. (in press). The volume for *Troodon* is taken from Russell (1969) and Currie and Zhao (1993). Volumes for *Carcharodontosaurus* and *Tyrannosaurus* were calculated directly from digital data using Surfacer (version 8.1) software. The *Carcharodontosaurus* data is from the 3-D scan data discussed above, while the *Tyrannosaurus* data is from a digitally reconstructed scan of AMNH 5029 (Larsson et al. in press).

The cerebral volumes for each fossil were approximated by fitting the largest pair of ellipsoids into the region identified as the cerebral hemispheres. Total cerebral length, width, and height for each taxon were measured from the above references and are listed in table 3.1.

The ventral margin of the endocast for *Troodon* is unknown. The endocasts of AMNH 6174 (figured in Russell 1969 and Hopson 1979) and RTMP 86.36.457 (figured in Currie and Zhao 1993) are similar in size. Currie and Zhao (1993) reconstructs the ventral margin of the endocast to give a cerebral height of approximately 32 mm. A range of values will be used here to span the upper and lower limits of this unknown. The total cerebral width in *Allosaurus* and *Tyrannosaurus* is 122% the cerebral height. The same ratio in *Archaeopteryx* is 166%. Using these two ratios, leaving allometry aside, to approximate upper and lower bounds for the cerebral height in *Troodon* results in values of 33.9 mm and 24.9 mm, respectively. The former value closely matches the estimation figured in Currie and Zhao (1993).

The resulting volumes were transformed to masses with the same specific gravity used above. These values are plotted with the extant nonavian reptile and avian data. The regressions for the extant data are extrapolated to the limits of the fossil data to graphically estimate their relationships. The R^2 values for the regressions are extremely high and give some support for this extrapolation. A percentage distance based on fixed x-values was used to estimate the relationship of each fossil datum to the regressions. The x-values were constrained at the total endocast mass to fix the ratio relationships between the total and cerebral masses. This technique is similar to calculating the least squares residuals since they are uncorrelated with the independent variable. The percentage distance was simply calculated as the percent of vertical space, from one regression to the other, occupied by the datum. The

TABLE 3.1
Table 3.1. Data Calculated for Fossil Endocrania.

| Taxon | Endocranium mass (g) | | Least squares residuals % toward the avian regression from the nonavian reptile regression |
	Total	Cerebrum	
Sebecus	31.1	7.41	-9.45
Allosaurus	169.0	46.73	2.61
Carcharodontosaurus	224.4	53.67	-2.46
Tyrannosaurus	338.6	111.84	10.71
Troodon (lower limit)	45.0	19.49	31.54
Troodon (upper limit)	45.0	26.53	63.06
Archaeopteryx	1.12	0.51	78.17
Numenius gypsorum	3.54	1.67	63.08
Numenius tahitiensis	5.01	2.97	101.07

percentage toward the bird regression from the reptile regression is listed for each fossil in table 3.1.

The results indicate that the fossil crocodilian *Sebecus* and the two allosauroid theropods, *Allosaurus* and *Carcharodontosaurus,* lie within 10% of the nonavian reptile regression. These fossil taxa also lie within the 95% confidence limits of the nonavian reptile regression. *Numenius tahitiensis,* an extant bird with data calculated using figures of its endocast, lies approximately 1% above the bird regression and within the 95% confidence limits of this regression. It is interesting to note that some of these taxa are up to two orders of magnitude larger than any of the extant taxa. Such relationships may support the robustness of using endocast ratio data.

Tyrannosaurus lies approximately 11% toward the bird regression from the nonavian reptile regression and lies just outside the 95% confidence limit of the nonavian reptile regression. Many recent systematic analyses of theropod phylogeny have nested tyrannosaurids within the Coelurosauria, a clade more proximal to birds than the least inclusive clade including the Allosauroidea, the Neotetanurae (Holtz 1994; Sereno et al. 1994). Such a position supports the hypothesis that a trend in nonavian theropod brain enlargement was initiated near the level of Coelurosauria (Larsson et al. in press).

Troodon and other nonavian maniraptorans have long been recognized as close relatives of birds (Ostrom 1969; Currie 1985, 1987; Gauthier 1986; Sereno 1999). Their unusually large endocasts have been used to argue both for the close relationship of these theropods with birds and their probable endothermy (Bakker 1974; Dodson 1974; Hopson 1977). The range of cerebral hemisphere sizes (based on the range of cerebral height discussed above) places *Troodon* 31.5% to 63% toward the bird regression from the nonavian reptile regression. The latter percentage is calculated from the larger range estimate that, as noted above, most closely matches the reconstruction by Currie and Zhao (1993).

The early bird *Archaeopteryx* lies approximately 78% toward the bird regression from the nonavian reptile regression. Its close relationship to the extant avian regression no doubt reflects its phylogenetic position. However, the Eocene bird *Numenius gypsorum* lies only 63% toward the bird regression. The low value for this taxon may be a reflection of the slightly distorted nature of the endocast. Dechaseaux (1970) figures the endocast as slightly anteroposteriorly sheared and laterally compressed.

Conclusions

The above data reflect the early evolution of the avian brain in spite of the paucity of complete endocranial data. The endocranial and inner ear anatomy of *Carcharodontosaurus* is comparable to that of extant crocodilians. These similarities suggest that more basal archosaurs also share these anatomies. These results, combined with the similarities of some nonavian maniraptoran endocrania with birds, also suggest that more avian endocranial and inner ear features evolved within the nonavian coelurosaurs.

Using a cerebrum to total brain volume comparison, it was found that the extant nonavian reptile regression extrapolates near the fossil crocodile, *Sebecus,* and both allosauroid theropods, *Allosaurus* and *Carcharodontosaurus.* These taxa do not plot significantly from the nonavian reptile regression. These relationships support a hypothesis that the cerebrum to total brain volume ratios of noncoelurosaur theropods do not differ significantly from the extant nonavian reptile ratio. Coelurosaur theropods have long been considered the closest relatives of birds (Ostrom 1976; Gauthier 1986). One of the most basal coelurosaurs, *Tyrannosaurus,* exhibits a cerebrum to total endocranium ratio that approaches approximately 11% toward a bird-type ratio from a reptile-type ratio. The same data for *Tyrannosaurus* is also significantly different from the nonavian reptile regression. This relationship suggests that a trend toward a bird-type ratio had begun near the origin of the Coelurosauria. *Troodon,* a coelurosaur even more closely related to birds, has a ratio that lies somewhere within a range of approximately 32% to 63% toward the extant bird condition. And the basal bird, *Archaeopteryx,* lies approximately 78% toward the bird regression. Clearly these data are not phylogenetically independent. But, until more data are available, a phylogenetically independent method, such as Felsenstein's (1985) independent contrasts method, cannot be used with such a low sample size. With the promise of more material, future work can certainly examine trends in the evolution of the theropod endocranium.

Acknowledgments: Most of all I thank Phil Currie for his sincere encouragement and interest in my work and early career. His enthusiasm and love for paleontology inspires everyone and certainly has and continues to influence many of my decisions. I also thank Paul Sereno for allowing me to participate in all his expeditions to Africa and South America. The manuscript was improved during its many stages as a result of conversations with Paul Sereno and Jeff Wilson and critiques from Kenneth Carpenter, Felis Lando, Darren Tanke, and an anonymous reviewer.

References

Bakker, R. T. 1974. Dinosaur bioenergetics: A reply to Bennett and Dalzell, and Feduccia. *Evolution* 28: 497–503.

Colbert, E. H. 1946. *Sebecus*, representative of a peculiar suborder of fossil Crocodylia from Patagonia. *Bulletin of the American Museum of Natural History* 87: 217–270.

Criule, G., and D. P. Quiring. 1940. A record of the body weight and certain organ and gland weights of 3690 animals. *Ohio Journal of Science* 40: 219–259.

Currie, P. J. 1985. Cranial anatomy of *Stenonychosaurus inequalis* (Saurischia, Theropoda) and its bearing on the origin of birds. *Canadian Journal of Earth Science* 22: 1643–1658.

Currie, P. J. 1987. Bird-like characteristics of the jaws and teeth of troodontid theropods (Dinosauria, Saurischia). *Journal of Vertebrate Paleontology* 7: 72–81.

Currie, P. J. 1995. New information on the anatomy and relationships of *Dromaeosaurus albertensis* (Dinosauria: Theropoda). *Journal of Vertebrate Paleontology* 15: 576–591.

Currie, P. J., and X. Zhao. 1993. A new troodontid (Dinosauria, Theropoda) braincase from the Dinosaur Park Formation (Campanian) of Alberta. *Canadian Journal of Earth Science* 30: 2231–2247.

de Beer, G. 1937. *The Development of the Vertebrate Skull.* Oxford: Oxford University Press.

de Beer, G., and E. J. E. Barrington. 1934. The segmentation and chondrification of the skull of the duck. *Philosophical Transactions of the Royal Society of London,* series B 223: 411–467.

Dechaseaux, C. 1970. Cérébralisation croissante chez le courlis (*Numenius*) au cours de la période quiba de l'Eocene supérieur à l'époque actuelle. *Comptes rendus des séances de l'Académie des sciences,* series D 270: 771–773.

Dendy, A. 1911. On the structure, development and morphological interpretation of the pineal organs and adjacent parts of the brain in the tuatara (*Sphenodon punctatus*). *Philosophical Transactions of the Royal Society of London,* series B 201: 227–331.

Dodson, P. 1974. Dinosaurs as dinosaurs. *Evolution* 28: 494–496.

Ebinger, P., and R. Löhmer. 1984. Comparative quantitative investigations on brains of rock doves, domestic and urban pigeons (*Columba l. livia*). *Zeitschrift für Zoologische Systematik und Evolutionsforschung* 22: 136–145.

Ebinger, P., and R. Löhmer. 1987. A volumetric comparison of brains between greylag geese (*Anser anser L.*) and domestic geese. *Journal für Hirnforschung* 28: 291–299.

Edinger, T. 1926. The brain of *Archaeopteryx*. *Annual Magazine of Natural History* 18: 151–156.

Edinger, T. 1927. Das Gehirn der Pterosaurier. *Zeitschrift für Anatomie und Entwicklungsgeschichte* 83: 106–112.

Edinger, T. 1951. The brain of the Odontognathae. *Evolution* 5: 6–24.

Edinger, T. 1964. Recent advances in paleoneurology. *Progress in Brain Research* 6: 147–160.

Edinger, T. 1966. Brains from forty million years of Camelid history. In R. Hassler and H. Stephan (eds.), *Evolution of the Forebrain*, pp. 153–161. Stuttgart: Thieme.

Felsenstein, J. 1985. Phylogenies and the comparative method. *American Naturalist* 125: 1–15.

Gans, C. 1980. Allometric changes in the skull and brain of *Caiman crocodilus*. *Journal of Herpetology* 14: 295–297.

Gauthier, J. 1986. Saurischian monophyly and the origin of birds. In K. Padian (ed.), *The Origin of Birds and the Evolution of Flight*, pp. 1–55. San Francisco: California Academy of Sciences.

Hochstetter, F. 1906. Beiträge zur Anatomie und Entwicklungsgeschichte des Blutgefässsystems der Krokodile. In A. Voeltzkow (ed.), *Reise in Ostafrika, 1903–1905*, 4:1–139. Stuttgart.

Holtz, T. R., Jr. 1994. The phylogenetic position of the Tyrannosauridae: Implications for theropod systematics. *Journal of Paleontology* 68: 1100–1117.

Hopson, J. A. 1977. Relative brain size and behavior in archosaurian reptiles. *Annual Review of Ecology and Systematics* 8: 429–448.

Hopson, J. A. 1979. Paleoneurology. In C. Gans (ed.), *Biology of the Reptilia*, 9:39–146. New York: Academic Press.

Janensch, W. 1936. Über Bahnen von Hirnvenen bei Saurischiern und Ornithischiern, sowie einigen anderen fossilen und rezenten Reptilien. *Paläontologische Zeitschrift* 18: 181–198.

Jerison, H. J. 1955. Brain to body ratios and evolution of intelligence. *Science* 121: 447–449.

Jerison, H. J. 1961. Quantitative analysis of evolution of the brain in mammals. *Science* 133: 1012–1014.

Jerison, H. J. 1963. Interpreting the evolution of the brain. *Human Biology* 35: 263–291.

Jerison, H. J. 1968. Brain evolution and *Archaeopteryx*. *Nature* 219: 1381–1382.

Jerison, H. J. 1969. Brain evolution and dinosaur brains. *American Naturalist* 103: 575–588.

Jerison, H. J. 1973. *Evolution of the Brain and Intelligence*. New York: Academic Press.

Larsson, H. C. E., P. C. Sereno, and J. A. Wilson. In press. Forebrain enlargement among non-avian theropod dinosaurs. *Journal of Vertebrate Paleontology*.

Lowenstein, O. E. 1974. Comparative morphology and physiology. In H. H. Dornhuber (ed.), *The Vestibular System 1: Basic Mechanisms*, pp. 1–120. New York: Springer Verlag.

Maleev, E. A. 1965. [On the brain of carnivorous dinosaurs.] *Paleontologeskii zhurnal* 2: 141–143. (In Russian.)

Newton, E. T. 1888. On the skull, brain, and auditory organ of a new species of pterosaurian (*Scaphognathus purdoni*), from the Upper Lias near Whitby, Yorkshire. *Philosophical Transactions of the Royal Society of London*, series B 179: 503–537.

O'Donoghue, C. H. 1920. The blood vascular system of the tuatara, *Sphenodon punctatus*. *Philosophical Transactions of the Royal Society of London*, series B 210: 175–252.

Osborn, H. F. 1912. Crania of *Tyrannosaurus* and *Allosaurus*. *Memoirs of the American Museum of Natural History* 1: 1–30.

Ostrom, J. H. 1969. Osteology of *Deinonychus antirrhopus,* an unusual theropod from the Lower Cretaceous of Montana. *Bulletin of Yale Peabody Museum of Natural History* 30: 1–165.

Ostrom, J. H. 1976. *Archaeopteryx* and the origin of birds. *Biological Journal of the Linnaean Society of London* 8: 91–182.

Peczkis, J. 1994. Implications of body-mass estimates for dinosaurs. *Journal of Vertebrate Paleontology* 14: 520–533.

Platel, R. 1976. Analyse volumétrique comparée des principales subdivisions encéphaliques chez les reptiles sauriens. *Journal für Hirnforschung* 17: 513–537.

Rehkämper, G., J. D. Frahm, and K. Zilles. 1991. Quantitative development of brain and brain structures in birds (Galliformes and Passeriformes) compared to that in mammals (insectivores and primates). *Brain, Behavior and Evolution* 37: 125–143.

Rehkämper, G., E. Haase, and H. D. Frahm. 1988. Allometric comparison of brain weight and brain structure volumes in different breeds of the domestic pigeon, *Columba livia f.d.* (fantails, homing pigeons, strassers). *Brain, Behavior and Evolution* 31: 141–149.

Rogers, S. W. 1998. Exploring dinosaur neuropaleobiology: Computed tomography scanning and analysis of an *Allosaurus fragilis* endocast. *Neuron* 21: 673–679.

Rogers, S. W. 1999. *Allosaurus,* crocodiles, and birds: Evolutionary clues from spiral computed tomography of an endocast. *The Anatomical Record* 257: 162–173.

Romer, A. S., and T. Edinger. 1942. Endocranial casts and brains of living and fossil Amphibia. *Journal of Comparative Neurology* 77: 355–389.

Russell, D. A. 1969. A new specimen on *Stenonychosaurus* from the Oldman Formation (Cretaceous) of Alberta. *Canadian Journal of Earth Sciences* 6: 595–612.

Russell, D. A. 1972. Ostrich dinosaurs from the Late Cretaceous of western Canada. *Canadian Journal of Earth Sciences* 9: 375–402.

Sereno, P. C. 1999. The evolution of dinosaurs. *Science* 284: 2137–2147.

Sereno, P. C., D. B. Duthiel, M. Iarochene, H. C. E. Larsson, G. H. Lyon, P. M. Magwene, C. A. Sidor, D. J. Varricchio, and J. A. Wilson. 1996. Predatory dinosaurs from the Sahara and Late Cretaceous faunal differentiation. *Science* 272: 986–991.

Sereno, P. C., J. A. Wilson, H. C. E. Larsson, D. B. Duthiel, and H. D. Sues. 1994. Early Cretaceous dinosaurs from the Sahara. *Science* 266: 267–271.

Wever, E. G. 1978. *The Reptile Ear.* Princeton: Princeton University Press.

Witmer, L. 1990. The craniofacial air sac system of Mesozoic birds (Aves). *Zoological Journal of the Linnean Society* 100: 327–378.

Witmer, L. 1996. The skull of *Deinonychus* (Dinosauria: Theropoda): New insights and implications. *Journal of Vertebrate Paleontology* 16: 73A.

4. Lower Jaw of *Gallimimus bullatus*

JØRN H. HURUM

Abstract

An almost complete lower jaw of *Gallimimus bullatus* reveals new features not previously recognized. Medially, it differs from the previous reconstruction in numerous details: the splenial does not extend to the symphysis and it has a large ventral mylohyoid foramen, the intramandibular joint indicates that separate movement of the anterior part of the lower jaw was not possible, the prearticular is large and covers the articular in medial view, the prearticular is not covered by the splenial anterodorsally, but the two bones have a close fit, and finally the coronoid and the supradentary are absent.

Introduction

Lower jaws of Late Cretaceous North American ornithomimosaurs have been known for a long time; *Struthiomimus* (Osborn 1917) and *Ornithomimus* (Parks 1933), but were only described from the lateral side. Osmólska et al. (1972) described both lateral and medial views of the Late Cretaceous *Gallimimus bullatus* from the Gobi Desert, Mongolia. Barsbold later (1983) described the lower jaw of the Late Cretaceous Mongolian *Garudimimus brevipes* in both lateral and medial view, and Barsbold and Perle (1984) described the lower jaw of the Middle Cretaceous *Harpymimus okladnikovi* in medial view. Russell (1972) revised the ornithomimosaurs from the Late Cretaceous of western Canada and erected the genus *Dromiceiomimus* for *Struthiomimus brevetertius* and *S. samueli,* and figured lower jaws of *D. brevetertius* and *D. samueli* in lateral view. Pérez-Moreno et al. (1994) described *Pelecanimimus polyodon,* an Early Cretaceous toothed ornithomimosaur from Spain with the lower jaw preserved in lateral view.

The lower jaw of ornithomimosaurs was described by Barsbold and Osmólska as: "slender, tapers slightly forward, and is very shallow for most of its length. The surangular portion of the mandible is gently convex dorsally, but the adductor prominence is usually not pronounced. The external mandibular fenestra is small and elongate. The mandibular symphysis is relatively long and inclined caudoventrally" (1990, 229).

A further preparation of the juvenile skull of *Gallimimus bullatus* described by Osmólska et al. (1972), provides new data requiring an emending of the observations of Osmólska et al. (1972), Barsbold (1983) and Barsbold and Osmólska (1990).

Materials and Methods

The paper-thin, right lower jaw of *Gallimimus bullatus* (ZPAL MgD-I/1) was freed from the skull by careful preparation, and partly embedded in Carbowax peg 2000. This was later dissolved after preparation of the delicate surfaces.

Institutional Abbreviations: RTMP, Royal Tyrrell Museum of Palaeontology, Drumheller, Alberta; ZPAL, Institute of Paleobiology, Polish Academy of Sciences, Warsaw.

Description

The toothless lower jaw is slender and was covered by a horny beak along its anterodorsal part. The constituent elements fit together perfectly and there does not appear to have been movement between them.

Dentary (figs. 4.1, 4.2)

The anterior one-third of the dentary was described as shovel-like by Osmólska et al. (1972), because of the laterally curved dorsal border of the dentary. The curvature begins above the anteriormost end of the splenial. In MgD-I/1, the symphysed region to the left dentary is missing. The posterolateral part of the dentary is also missing, but this can be reconstructed from the impression in the sandstone that covers the angular and prearticular in lateral view.

In lateral view, the dentary differs from the reconstruction in Osmólska et al. (1972). There is a small foramen situated anteriorly. The posterior border is interpreted to be more like in other ornithomimosaurs by forming the anterior border of the external mandibular fenestra.

In medial view, the posterior two-thirds of the dentary is covered by the splenial, the Meckelian groove is deep and extends to the anteroventral border of the preserved part. The splenial covers most of the Meckelian groove and fits perfectly with the dorsal and ventral parts of the dentary. The posterodorsal contact with the surangular is divided into two short processes, one lateral and one medial. They form a groove for the insertion of the intramandibular process of the surangular. The dorsomedial process of the dentary fits into a shallow groove in the surangular.

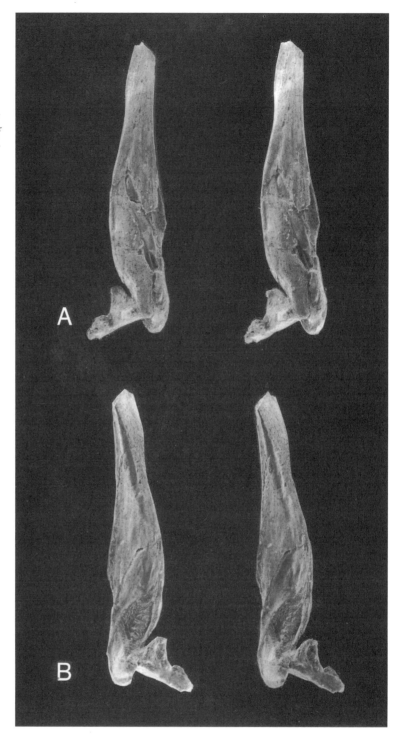

Figure 4.1. Stereo photographs of the lower jaw and the quadrate of Gallimimus bullatus *(ZPAL MgD-I/4): (A) lateral view, (B) medial view.*

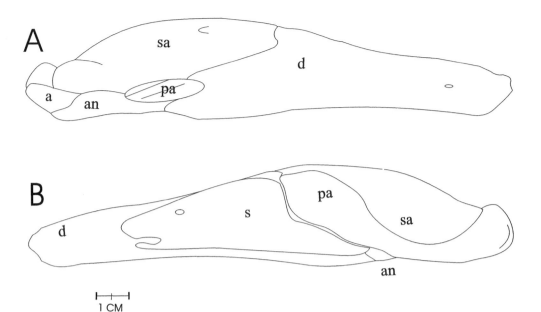

A

B

1 CM

Figure 4.2. Schematic drawings of the lower jaw of Gallimimus bullatus *(ZPAL MgD-I/4): (A) lateral view, (B) medial view. Anatomical abbreviations:*
a-articular, an-angular, d-dentary, pa-prearticular, s-splenial, sa-surangular.

Splenial (figs. 4.1, 4.2)

The splenial, only visible in medial view, is roughly triangular in shape. Contrary to Osmólska et al. (1972), it does not reach anteriorly to the symphysis, but only covers the posterior two-thirds of the dentary. The bone contains two foramina, the large anteroventral mylohyoid foramen and a smaller dorsal foramen. The mylohyoid foramen is surrounded by the splenial, except for a small anterior slit. The posteroventral part fits into a shallow groove in the angular.

Surangular (figs. 4.1, 4.2)

The surangular is the second largest bone of the mandible, which is the usual condition in theropods. It covers the anterior part of the articular posterolaterally and has a long, partly preserved, ventral border to the angular. The dorsal part of the intramandibular joint is covered by the splenial and the prearticular, but as far as can be determined, it is a normal theropod joint with a process of the surangular that fits between the lateral and medial processes of the dentary (see Hurum and Currie in press). The surangular has a small anterior surangular foramen. A shallow groove extends anteriorly from the foramen. The posterior border of the external mandibular fenestra is reconstructed.

Angular (figs. 4.1, 4.2)

The angular is somewhat broken, but is possible to reconstruct in lateral view. The bone covers the anteroventral part of the articular and forms the ventral border of the external mandibular foramen. The most anterior part is covered laterally by the dentary. In medial view, only a small part is visible between the splenial and the prearticular, where it has a shallow groove for the splenial.

Prearticular (figs. 4.1, 4.2)

The prearticular is visible in lateral view due to the preparation of the external mandibular foramen and some damage to the dentary. In the external mandibular foramen, it is possible to see a sharp ridge on the prearticular. In medial view, the bone has an extensive anterior part covering the posterior of the dorsal intramandibular joint, and has a relatively close fit to the posterior border of the splenial. Medially, the posterior part of the prearticular totally covers the articular and extends to the posterior end of the retroarticular process. The prearticular forms the ventral and lateral margins of the adductor fossa.

Articular (figs. 4.1, 4.2)

Only the posterior part of the articular is visible because the quadrate still covers the glenoid fossa. In lateral view, the retroarticular process is relatively short compared to other ornithomimids. The articular is wide in dorsal view, but is unfortunately covered by the quadrate. On the medial side the bone is covered completely by the prearticular.

Coronoid and Supradentary

The coronoid and supradentary were never present in the lower jaw of *Gallimimus*. There is no sign of an attachment between a coronoid and the surangular or prearticular, and no groove for a supradentary on the dorsolateral side of the dentary, as in, for example, tyrannosaurids (see Hurum and Currie in press).

Discussion

The shape of the bill in ornithomimosaurs has been under constant debate. Osborn (1917) suggested that the premaxillaries and dentaries of *Struthiomimus altus* were sheathed in narrow horny beaks somewhat similar to those of the extant ostrich, *Struthio*. Russell stated, "The development of a bony vault over the anterior part of the oral cavity (secondary palate) and a transverse axis of flexure in the skull roof anterior to the orbits, together with the general shape of the muzzle, recall the morphology of the bill in modern insectivorous birds" (1972, 399–400). Later, Nicholls and Russell (1985) suggested a flat "herbivorous" beak. Barsbold and Osmólska stated, "the lower and upper jaws of ornithomimosaurians did not form a flattened beak comparable to that in hadrosaurids, for example. Instead, the beak, although broad, was relatively deep rostrally, at least in these species in which it was well preserved" (1990, 244).

The mandibles in Ornithomimosauria are best known in *Gallimimus bullatus* and *Garudimimus brevipes*. They both have a "shovel-like" anterior end of the dentary (Osmólska et al. 1972). The shape of the dentary in *Gallimimus* is comparable to that of the front of the dentary in the common seagull (*Larus*) and indicates a similar shaped bill. The seagull-like lower jaw suggests that *Gallimimus*, like seagulls, had an opportunistic, possibly omnivorous diet as suggested by Gregory in Osborn 1917.

The angular is the most flexible bone of the lower jaw in ornithomimosaurs (fig. 4.3). *Struthiomimus altus* has a small angular and *Ornithomimus edmontonensis* has an angular covering the articular nearly to the posterior end of the mandible. The extent of the angular in *Gallimimus* is similar to *Dromiceiomimus brevetertius*.

The anterior expansion of the prearticular in *Gallimimus* is not seen in any other theropod, except to some degree in *Ceratosaurus* (Bakker et al. 1988). The usual manner, as seen in tyrannosaurids (Hurum and Currie in press), for example, is for the prearticular to taper into a thin anterodorsal end covering a small portion of the intramandibular joint. The anterodorsal area covered by the prearticular in *Gallimimus* is the same as covered by the prearticular and coronoid in other theropods. Because of the lack of the coronoid this widening of the anterior portion might be a secondary specialization to cover the same area medial to the adductor fossa. The coronoid and supradentary are missing in all theropods that evolved toothless beaks (ornithomimosaurs and oviraptors), *Segnosaurus* (Perle 1979), in *Erlikosaurus* (Clark et al. 1994), and in birds. *Garudimimus* is described with a peculiar prearticular that borders the ventral side of the adductor fossa (Barsbold 1983; Barsbold and Osmólska 1990). This is very different from the more normal theropod prearticular in *Gallimimus*. If this is not an artifact of preservation, this makes the prearticular of *Garudimimus* more like the structure observed in oviraptorids (see, e.g., Barsbold 1983, fig. 13) than in *Gallimimus*.

In *Gallimimus*, the fit between the prearticular and splenial is tight and allows no movement between the anterior (dentary + splenial) and posterior part (surangular + angular + articular + prearticular) of the mandible. Theropods like tyrannosaurids (Hurum and Currie in press), dromaeosaurids (Currie 1995) and large theropods (e.g., *Monolophosaurus* Zhao and Currie 1993), have a relatively wide opening between the prearticular and the splenial. The close fit is possibly a plesiomorphy also observed in *Archaeopteryx* (Elzanowski and Wellnhofer 1996), *Dilophosaurus wetherilli*, *Syntarsus rhodesiensis*, *Liliensternus* (Huene 1934), *Segnosaurus*, and *Erlikosaurus*.

The articular in *Gallimimus* has a small retroarticular process compared to other ornithomimosaurs (fig. 4.3). On the medial side the bone is covered completely by the prearticular, an unusual condition seen only in *Carnotaurus* (Bonaparte et al. 1990). The prearticular covers the medioventral part of the articular and extends to the posterior end of the lower jaw in *Dilophosaurus wetherilli* (Welles 1984), *Syntarsus rhodesiensis* (Raath 1977), *Sinraptor* (Currie and Zhao 1993), *Allosaurus* (Madsen 1976), and oviraptors (Barsbold 1983).

Conclusions

Even though it is peculiar in lacking teeth and the prearticular is widened anteriorly, the lower jaw of *Gallimimus* shows the common theropod structure.

The following theropod plesiomorphic characters are recognized in the lower jaw:

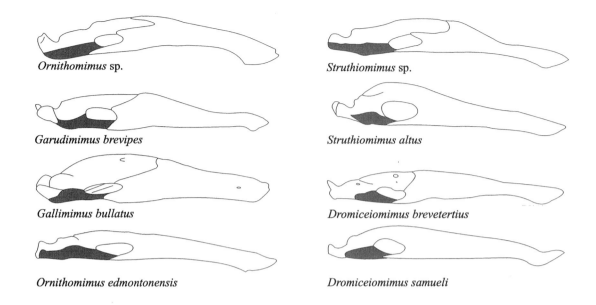

Ornithomimus sp.

Struthiomimus sp.

Garudimimus brevipes

Struthiomimus altus

Gallimimus bullatus

Dromiceiomimus brevetertius

Ornithomimus edmontonensis

Dromiceiomimus samueli

Figure 4.3. Lateral view of the lower jaw in different ornithomimosaurs, the angular highlighted in black. Struthiomimus altus, Dromiceiomimus brevetertius, and D. samueli redrawn from Russell (1972); Garudimimus brevipes redrawn from Barsbold (1983); Ornithomimus edmontonensis redrawn from Paul (1988). Ornithomimus sp. reconstructed from RTMP 95.110.1, Struthiomimus sp. from RTMP 96.05.09. Juvenile Gallimimus bullatus (this study). Not to scale.

- prearticular covers the articular in medial view,
- simple intra-mandibular joint,
- close fit between the splenial and prearticular.

Suggested apomorphy of the lower jaw of *Gallimimus bullatus* is:
- large widening of the anterior end of the prearticular.

Following Sereno's (1999) phylogeny of dinosaurs, the lack of coronoid and supradentary might be an apomorphy for the therizinosaur-ornithomimosaur clade.

Acknowledgments: I especially thank H. Osmólska for making the specimen *Gallimimus* available for my study and reading an early version of the manuscript. Photographs were taken by Per Aas. The review by Thomas R. Holtz Jr. also contributed to this chapter. This chapter has benefited much from the hospitality and friendly help of Phil Currie during my postdoctorate 1998–2000. From my first meeting with Phil in 1988 as a volunteer in Dinosaur Provincial Park, to the field seasons in Argentina and Canada in 1999, he has always been a support and a dear friend. This work was supported by the Norwegian Research Council (grant no. 122898/410).

References

Bakker, R. T., M. Williams, and P. J. Currie. 1988. *Nanotyrannus,* a new genus of pygmy tyrannosaur, from the latest Cretaceous of Montana. *Hunteria* 5: 1–29.

Barsbold, R. 1983. [Carnivorous dinosaurs from the Cretaceous of Mongolia.] *Transactions of the Joint Soviet-Mongolian Paleontological Expeditions* 19: 5–119. (In Russian.)

Barsbold, R., and H. Osmólska. 1990. Ornithomimosauria. In D. B. Weishampel, P. Dodson, and H. Osmólska (eds.), *The Dinosauria,* pp. 225–244. Berkeley: University of California Press.

Barsbold, R., and A. Perle. 1984. The first record of a primitive ornitho-mimosaur from the Cretaceous of Mongolia. *Paleontological Journal* 18: 118–120.

Bonaparte, J. F., F. E. Novas, and R. A. Coria. 1990. *Carnotaurus sastrei* Bonaparte, the horned, lightly built carnosaur from the Middle Cretaceous of Patagonia. *Contributions in Science* 416: 2–41.

Clark, J. M., A. Perle, and M. A. Norell. 1994. The skull of *Erlicosaurus andrewsi*, a Late Cretaceous "Segnosaur" (Theropoda: Therizinosauridae) from Mongolia. *American Museum Novitates* 3115: 1–39.

Currie, P. J. 1995. New information on the anatomy and relationships of *Dromaeosaurus albertensis* (Dinosauria: Theropoda). *Journal of Vertebrate Paleontology* 15 (3): 576–591.

Currie, P. J., and X. Zhao. 1993. A new carnosaur (Dinosauria, Theropoda) from the Jurassic of Xinjiang, People's Republic of China. *Canadian Journal of Earth Sciences* 30: 2037–2081.

Elzanowski, A., and P. Wellnhofer. 1996. Cranial morphology of *Archaeopteryx:* Evidence from the seventh skeleton. *Journal of Vertebrate Paleontology* 16 (1): 81–94.

Huene, F. 1934. Ein neuer Coelosaurier in der thüringischen Trias. *Paläontologische Zeitschrift* 16: 145–170.

Hurum, J. H., and P. J. Currie. In press. The crushing bite of tyrannosaurids. *Journal of Vertebrate Paleontology.*

Madsen, J. H. 1976. *Allosaurus fragilis:* A revised osteology. *Bulletin of Utah Geological and Mineral Survey* 109: 1–163.

Nicholls, E. L., and A. P. Russell. 1985. Structure and function of the pectoral girdle and forelimb of *Struthiomimus altus* (Theropoda: Ornithomimidae). *Palaeontology* 28: 643–677.

Osborn, H. F. 1917. Skeletal adaptions of *Ornitholestes, Struthiomimus, Tyrannosaurus. Bulletin of the American Museum of Natural History* 35: 733–771.

Osmólska, H., E. Roniewicz, and R. Barsbold. 1972. A new dinosaur; *Gallimimus bullatus* n.gen., n.sp. (Ornithomimidae) from the Upper Cretaceous of Mongolia. *Palaeontologia polonica* 27: 103–143.

Parks, W. A. 1933. New species of dinosaurs and turtles from the Upper Cretaceous formations of Alberta. *University of Toronto Studies, Geological Series* 34: 3–33.

Pérez-Moreno, B. P., J. L. Sanz, A. D. Buscalioni, J. J. Moratalla, F. Ortega, and D. Rasskin-Gutman. 1994. A unique multitoothed ornithimimosaur dinosaur from the Lower Cretaceous of Spain. *Nature* 370: 363–367.

Perle, A. 1979. [Segnosauridae: A new family of theropods from the Upper Cretaceous of Mongolia.] *Transactions of the Joint Soviet-Mongolian Paleontological Expeditions* 8: 45–55. (In Russian.)

Raath, M. A. 1977. The anatomy of the Triassic theropod *Syntarsus rhodesiensis* (Saurischia: Podokesauridae) and a consideration of its biology. Ph.D. dissertation, Rhodes University, Grahamstown, South Africa.

Russell, D. A. 1972. Ostrich dinosaurs from the Latest Cretaceous of Western Canada. *Canadian Journal of Earth Sciences* 9: 375–402.

Sereno, P. C. 1999. The evolution of dinosaurs. *Science* 284: 2137–2147.

Welles, S. P. 1984. *Dilophosaurus wetherilli* (Dinosauria, Theropoda) osteology and comparisons. *Palaeontographica* A 185: 85–180.

Zhao, X., and P. J. Currie. 1993. A large crested theropod from the Jurassic of Xinjiang, People's Republic of China. *Canadian Journal of Earth Sciences* 30: 2027–2036.

5. Late Cretaceous Oviraptorosaur (Theropoda) Dinosaurs from Montana

DAVID J. VARRICCHIO

Abstract

Two new specimens represent the first oviraptorosaur material from Montana. An articular region from the lower jaw of *Caenagnathus sternbergi* comes from the Campanian Two Medicine Formation of western Montana. The occurrence of *C. sternbergi*, previously known only from Alberta, Canada, further emphasizes that despite differences in the herbivore faunas, the late Campanian of Alberta and Montana possessed nearly identical theropod assemblages. A partial foot from the Maastrichtian Hell Creek Formation of eastern Montana represents *Elmisaurus elegans*. Elmisaurid oviraptorosaurs possessed unusual feet with an arctometatarsalian construction, a long phalanx II-1, elongated penultimate phalanges on digits III and IV so that III-3 exceeds III-2 and IV-4 exceeds IV-3 and IV-2 in length, and tall ginglymoid articular surfaces allowing a wide range of extension and flexion but only in a dorsoventral plane. These specializations suggest a strong grasping foot perhaps adapted for climbing or prey capture.

Introduction

The clade Oviraptorosauria (sensu Barsbold et al. 1990) represents a group of relatively rare, small theropod dinosaurs, most notable for their stout and toothless bills. Currently, the most complete specimens come from the Cretaceous of Asia. These include *Oviraptor philocera-tops, Ingenia yanshini,* and *Caudipteryx zoui* (Osborn 1924; Barsbold 1981; Barsbold et al. 1990; Qiang et al. 1998; Smith et al. 1998; Sereno 1999). Recent discoveries, although fragmentary, show that the oviraptorosaur distribution, once considered limited to the Northern Hemisphere, extended to the Gondwana areas of South America and Australia (Frey and Martill 1995; Currie et al. 1996; Frankfurt and Chiappe 1999). Unfortunately, the North American record for the group remains relatively sparse.

North American oviraptorosaurs were first described in the 1920s and 1930s: *Chirostenotes pergracilis* (Gilmore 1924), "*Macrophalangia canadensis*" (Sternberg 1932), and "*Ornithomimus*" *elegans* (Parks 1933). These early specimens consist only of articulated hands, an ankle and foot, and a metatarsus, respectively. Although recent and more complete discoveries (Currie and Russell 1988; Sues 1997) have greatly clarified the morphology of these forms, the lack of overlapping elements among the specimens has prevented full resolution of the taxonomy. For example, the type specimen of "*Ornithomimus*" *elegans* (Parks 1933) has borne the names of *Chirostenotes pergracilis* (Currie and Russell 1988), *Elmisaurus elegans* (Currie 1989, 1990), and *Chirostenotes elegans* (Sues 1997). Table 5.1 lists the described oviraptorosaur specimens of North America and their taxonomy.

Currently, there are either three or five recognized and named species from North America. The earliest, the Lower Cretaceous *Microvenator celer* (Ostrom 1970), represents a primitive form and the possible sister taxon to all other oviraptorosaurs (Makovicky and Sues 1998). Most recently Sues (1997) synonymized a number of specimens (table 5.1). He recognizes only one family, the Caenagnathidae, with two species from the Late Cretaceous. *Chirostenotes pergracilis* includes *Macrophalangia canadensis* (Sternberg 1932) and *Caenagnathus collinsi* (Sternberg 1940), and *Chirostenotes elegans* represents "*Ornithomimus*" *elegans* (Parks 1933) and *Caenagnathus sternbergi* (Cracraft 1971). Currie (1989, 1990, 1997) takes a more cautious approach and retains four Late Cretaceous species in two families: *Chirostenotes pergracilis* and *Elmisaurus elegans* in the Elmisauridae and *Caenagnathus collinsi* and *Caenagnathus sternbergi* of the Caenagnathidae. These classifications differ for two reasons. First, Currie (1989, 1997) maintains generic distinction for *Chirostenotes pergracilis* and *Elmisaurus elegans* because among other characters, the latter shows fusion of the tarsometatarsus, a feature also found in the Asian *Elmisaurus rarus* (Osmólska 1981). Second, because the types of *Caenagnathus collinsi* and *Caenagnathus sternbergi* consist only of isolated jaws, elements not found with other specimens, Currie (1990) retains them as distinct taxa. Currie (1997) nevertheless considers *Caenagnathus* probably synonymous with *Chirostenotes* and actually Currie and Russell (1988) first suggested that the two *Caenagnathus* species might correspond to *Chirostenotes pergracilis* and *Chirostenotes* (= *Elmisaurus*) *elegans*. They further proposed that these two species might represent sexual morphs of one species.

The North American Late Cretaceous oviraptorosaurs thus in-

TABLE 5.1
North American Described Oviraptorosaur Specimens and Their Assigned Taxonomy

SPECIMEN	ELEMENTS	FORMATION	AGE	ORIGINAL NAME	Osmólska 1981
NMC 2367	Two articulated hands	Judith River Group, Alberta	Campanian	*Chirostenotes pergracilis* Gilmore 1924	*Chirostenotes pergracilis* Elmisauridae
NMC 8538	Partial foot	Judith River Group, Alberta	Campanian	*Macrophalangia canadensis* Sternberg 1932	*Macrophalangia canadensis* Elmisauridae
ROM 781	Articulated metatarsus	Judith River Group, Alberta	Campanian	*Ornithomimus elegans* Parks 1933	
NMC 8776	Complete lower jaws	Judith River Group, Alberta	Campanian	*Caenagnathus collinsi* Sternberg 1940	
NMC 2690	Articular region of lower jaw	Judith River Group, Alberta	Campanian	*Caenagnathus sternbergi* Cracraft 1971	
AMNH 3041	Partial skeleton, missing feet and most of skull	Cloverly, Montana	Aptian-Albian	*Microvenator celer* Ostrom 1970	
RTMP 79.14.499	Manual phalanx II-3	Judith River Group, Alberta	Campanian		
RTMP 79.20.1	Partial skeleton lacking skull	Judith River Group, Alberta	Campanian		
NMC 9570	Metatarsal II	Horseshoe Canyon, Alberta	Maastrichtian		
RTMP 82.39.4	Partial tarsometatarsus	Judith River Group, Alberta	Campanian		
ROM 37163	Partial metatarsal II	Judith River Group, Alberta	Campanian		
RTMP 79.8.622, 90.56.6, 91.144.1, 92.36.360	Partial dentaries	Judith River Group, Alberta	Campanian		
RTMP 92.36.53	Caudal vertebra	Judith River Group, Alberta	Campanian		
BHM 2033	Articular region of lower jaw	Hell Creek, South Dakota	Maastrichtian		
ROM 43250	Partial skeleton missing limbs and most of skull	Horseshoe Canyon, Alberta	Maastrichtian		
MOR 1107	Articular region of lower jaw	Two Medicine, Montana	Campanian		
MOR 752	Partial foot	Hell Creek, Montana	Maastrichtian		

TABLE 5.1 (cont.)

Currie & Russell 1988	Paul 1988	Currie 1989, 1990, 1997	Currie et al. 1993	Sues 1997	Makovicky & Sues 1998	This Chapter
Chirostenotes	*Chirostenotes*	*Chirostenotes*		*Chirostenotes*		*Chirostenotes*
pergracilis	*pergracilis*	*pergracilis*		*pergracilis*		*pergracilis*
Caenagnathidae	Caenagnathidae	Elmisauridae		Caenagnathidae		Elmisauridae
Chirostenotes	*Chirostenotes*	*Chirostenotes*		*Chirostenotes*		*Chirostenotes*
pergracilis	*pergracilis*	*pergracilis*		*pergracilis*		*pergracilis*
Caenagnathidae	Caenagnathidae	Elmisauridae		Caenagnathidae		Elmisauridae
Chirostenotes		*Elmisaurus*		*Chirostenotes*		*Elmisaurus*
pergracilis		*elegans*		*elegans*		*elegans*
Caenagnathidae		Elmisauridae		Caenagnathidae		Elmisauridae
possibly	possibly	*Caenagnathus*	*Caenagnathus*	*Chirostenotes*		*Caenagnathus*
Chirostenotes	*Chirostenotes*	*collinsi*	*collinsi*	*pergracilis*		*collinsi*
	pergracilis	Caenagnathidae	Caenagnathidae	Caenagnathidae		Caenagnathidae
possibly	possibly	*Caenagnathus*	*Caenagnathus*	*Chirostenotes*		*Caenagnathus*
Chirostenotes	*Chirostenotes*	*sternbergi*	*sternbergi*	*elegans*		*sternbergi*
	pergracilis	Caenagnathidae	Caenagnathidae	Caenagnathidae		Caenagnathidae
possibly primitive	possibly			not a	primitive	
caenagnathid	caenagnathid			caenagnathid	oviraptorosaur	
		Chirostenotes				
		pergracilis				
		Elmisauridae				
Chirostenotes	*Chirostenotes*	*Chirostenotes*		*Chirostenotes*		*Chirostenotes*
pergracilis	*pergracilis*	*pergracilis*		*pergracilis*		*pergracilis*
Caenagnathidae	Caenagnathidae	Elmisauridae		Caenagnathidae		Elmisauridae
Chirostenotes		*Chirostenotes*		*Chirostenotes*		*Chirostenotes*
pergracilis		cf. *pergracilis*		*pergracilis*		*pergracilis*
Caenagnathidae		Elmisauridae		Caenagnathidae		Elmisauridae
		Elmisaurus				*Elmisaurus*
		elegans				*elegans*
		Elmisauridae				Elmisauridae
		Elmisaurus				*Elmisaurus*
		elegans				*elegans*
		Elmisauridae				Elmisauridae
			Caenagnathus			
			cf. *sternbergi*			
			Caenagnathidae			
			Caenagnathus			
			sp.			
			Caenagnathidae			
			Caenagnathus			*Caenagnathus*
			sp.			sp.
			Caenagnathidae			Caenagnathidae
				Chirostenotes		*Chirostenotes*
				pergracilis		*pergracilis*
				Caenagnathidae		Elmisauridae
						Caenagnathus
						sternbergi
						Caenagnathidae
						Elmisaurus
						elegans
						Elmisauridae

clude: *Chirostenotes pergracilis* in the Campanian to early Maastrichtian and either *Chirostenotes elegans* sensu Sues (1997) or *Elmisaurus elegans, Caenagnathus collinsi,* and *Caenagnathus sternbergi* in the Campanian. All these specimens come from Alberta or Saskatchewan (Weishampel 1990). Furthermore, a recently described but unnamed specimen (Currie et al. 1993) from the Hell Creek of South Dakota represents a new, larger species of *Caenagnathus* (or *Chirostenotes* sensu Sues 1997).

Two new oviraptorosaur specimens have been found, an articular region of the lower jaw and a partial left foot, both from the Late Cretaceous of Montana. Because of their fragmentary nature, these specimens cannot help resolve the current taxonomic issue. They do, however, extend geographic and temporal ranges. The foot also provides some new and novel morphologic information on these still poorly known taxa.

Although the synonymies of Sues (1997) are likely correct, in order to maintain clarity in discussion of fragmentary specimens, the classification of Currie (1989, 1990, 1997) is used (table 5.1).

Institutional Abbreviations: AMNH, American Museum of Natural History, New York; BHM, Black Hills Museum of Natural History, Hill City, South Dakota; MOR, Museum of the Rockies, Bozeman, Montana; NMC, National Museum of Canada (now Canadian Museum of Nature), Ottawa; ROM, Royal Ontario Museum, Toronto; RTMP, Royal Tyrrell Museum of Paleontology, Drumheller, Alberta; ZPAL, Institute of Paleontology, Polish Academy of Sciences, Warsaw.

Systematic Paleontology

Dinosauria
Saurischia
Theropoda
Oviraptorosauria Barsbold 1976
Caenagnathidae Sternberg 1940
Caenagnathus Sternberg 1940
Caenagnathus sternbergi Cracraft 1971

Holotype: NMC 2690, posterior end of right mandibular ramus including the articular region, Judith River Group, Steveville, Red Deer River, Alberta.

Newly Referred Specimen: MOR 1107, posterior end of right mandibular ramus including the articular region, Two Medicine Formation, MOR locality TM-013, sec. 27, T37N, R8W, Landslide Butte, Glacier County, Montana.

Stratigraphy: Judith River Group and Two Medicine Formation (Campanian, Upper Cretaceous).

Diagnosis: *Caenagnathus sternbergi* differs from other *Caenagnathus* species in having (1) a smaller adult size and an articular with (2) a higher and more arched anteroposterior ridge on the articulating surface, (3) a medial glenoid that is relatively short anteroposteriorly (Cracraft 1971), and (4) no well-developed chorda tympani foramen or slot.

Description

MOR 1107 comes from the Upper Two Medicine Formation of Landslide Butte, Glacier County, Montana. Its stratigraphic position suggests a late Campanian age (Rogers et al. 1993). This specimen, with a total length of 18 mm, represents the posterior, articular region of the right lower jaw (fig. 5.1). Currie et al. (1993) described this region in oviraptorosaurs as representing the fusion of three bones, the articular, surangular, and coronoid, and dubbed it the ASC complex. MOR 1107 exhibits no visible signs of fusion and presumably represents an adult animal. The anterior ramus, with the coronoid process (the surangular contribution) and the retroarticular process of the ASC complex have both been lost. MOR 1107 preserves only the articular surface of the jaw and a portion of bone immediately ventral to it. Where broken, the anterior ramus had a diamond-shaped cross-section. The cross-section of the retroarticular process is narrow (1.0 to 3.5 mm) and laterally compressed. Both breaks reveal relatively thin cortical and irregularly chambered trabecular bone.

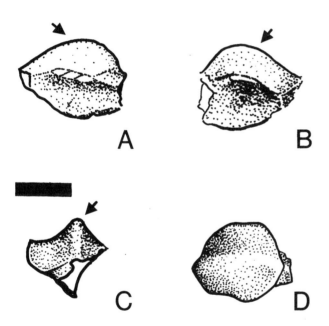

Figure 5.1. MOR 1107, Caenagnathus sternbergi articular region of the right lower jaw in (A) lateral, (B) medial, (C) posterior, and (D) dorsal views. Anterior is to the right in D. Arrow indicates the anteroposterior ridge of the articulating surface. Scale: 1 cm.

In dorsal view the articular surface has a roughly trapezoidal outline, but small portions of the lateral, medial, and posteriormost edges have been lost. As in other caenagnathid specimens, a large anteroposterior ridge divides the articular surface in two, a narrow lateral and a wider medial glenoid (the external and internal processes of the mandible in Cracraft 1971). The ridge measures 16.6 mm but was slightly longer when complete. In lateral view the ridge is strongly arched and its apex occurs about one-third along its length from its anterior limit.

The medial glenoid measures 12.9 mm anteroposteriorly, roughly

70% the length of the ridge. It is convex in medial view and concave in cross-section. The lateral glenoid appears significantly shorter and its preserved portion is a planar dorsolateral-facing surface.

Currie et al. (1993) describe an emargination in the posterior edge of the articular surface of the jaw in the Hell Creek specimen, BHM 2033, and also figure a similar slot in the type of *Caenagnathus collinsi,* NMC 8776. They interpret this as the chorda tympani foramen. Although the posteriormost portion of the articular ridge is incomplete in MOR 1107, the slot is clearly absent. This also seems true in *Caenagnathus sternbergi* (Cracraft 1971, fig. 2b). Beneath the articular surface, MOR 1107 is slightly concave medially and quite concave laterally. The lateral depression may represent part of the insertion point for the pterygomandibularis muscle (Currie et al. 1993).

MOR 1107 possesses an articular surface with a strong, rounded anteroposterior ridge separating two subcircular and horizontal processes, features unique to oviraptorosaurs (Barsbold et al. 1990). Currently, there are only three other North American specimens preserving the ASC complex: NMC 8776, the type of *Caenagnathus collinsi* (Sternberg 1940); NMC 2690, the type of *Caenagnathus sternbergi* (Cracraft 1971); and BHM 2033, the unnamed Hell Creek *Caenagnathus* (Currie et al. 1993). MOR 1107 matches *Caenagnathus sternbergi* in: (1) smaller adult size, (2) a higher and more arched anteroposterior ridge, (3) a medial glenoid that is relatively short anteroposteriorly, and (4) no well-developed chorda tympani foramen. Cracraft (1971) used features 1–3 to distinguish *C. sternbergi* from *C. collinsi.* These and feature 4 also separate *C. sternbergi* from the Hell Creek specimen. Consequently, MOR 1107 is best assigned to *Caenagnathus sternbergi* or *Chirostenotes elegans* if the synonymy of Sues (1997) proves correct.

Elmisauridae Osmólska 1981

Elmisaurus Osmólska 1981
Elmisaurus elegans (Park 1933)

Holotype: ROM 781, complete left metatarsals II and IV and the incomplete remains of distal tarsal III, distal tarsal IV and metatarsal III, Judith River Group, Little Sandhill Creek, Dinosaur Provincial Park, Alberta. Originally described as *Ornithomimus elegans.*
Newly Referred Specimen: MOR 752, a partial left foot including a fragment of the astragalus, an unidentified metatarsal fragment, a partial metatarsal II, the distal end of phalanx II-1, phalanx II-2, and complete digits III and IV, Hell Creek Formation, MOR locality HC-147, sec. 32, T16N, R56E, Dawson County Montana.
Other Referred Material: TMP 82.39.4, proximal end of right metatarsus, Judith River Group, Dinosaur Provincial Park, Alberta; ROM 37163, partial left metatarsal II, Dinosaur Provincial Park, Alberta.
Stratigraphy: Judith River Group and Hell Creek Formation (Campanian and Maastrichtian, Upper Cretaceous).
Diagnosis: *Elmisaurus elegans* is more gracile than *Elmisaurus rarus* and *Chirostenotes pergracilis. Elmisaurus elegans* possesses: (1) in dorsal view, a tarsometatarsus with a more deeply emarginated

posteromedial corner; (2) a longitudinal, ridgelike posterolateral margin on metatarsal IV which is more weakly developed than on *Elmisaurus rarus;* and (3) metatarsals II and IV that near their distal articular surfaces have small processes that overlap metatarsal III (Currie 1989).

Description

MOR 752, a partial left foot from the Maastrichtian Hell Creek Formation of Dawson County, Montana includes a fragment of the astragalus, an unidentified metatarsal fragment, a nearly complete metatarsal II, and most of the pedal phalanges. Neither the astragalus nor the unidentified metatarsal fragments currently provide any morphological information.

The preserved portion of metatarsal II lacks the proximal articulation and the medial half of the first third of the shaft (fig. 5.2). Thus, it is impossible to determine if this metatarsal was fused as part of a tarsometatarsus in life. As preserved, the specimen measures 130 mm; when complete it was not probably longer than 135 mm. The shaft is

Figure 5.2. MOR 752, Elmisaurus elegans left metatarsal II in (A) dorsal, (B) medial, (C) ventral, and (D) lateral views. Scale: 1 cm.

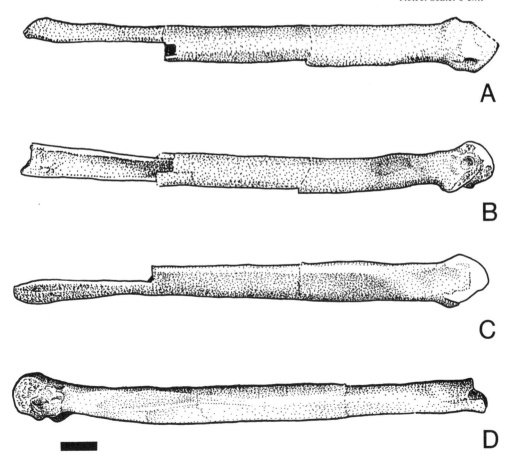

straight and hollow. Its cross-section remains a rounded triangle proximally; over the distal quarter it becomes rectangular. A roughened longitudinal ridge marks the medioposterior edge of the distal region of the shaft. This ridge projects posteriorly and suggests the metatarsus was strongly concave behind. A rough and irregular attachment scar marks the medial portion of the shaft 20 mm above the distal articulation. The flat contact for metatarsal III extends the length of the shaft and shifts from a broad and lateral surface distally to a lateroposterior one proximally. Here the metatarsal shaft expands laterally. Apparently metatarsal II closely adhered to metatarsal III throughout its length and may have excluded metatarsal III from the anterior margin of the metatarsus proximally. The distal articulation is broadly convex, angles slightly medially, and bears deep ligament fossae. A small, roughened anterolateral projection marks the shaft adjacent to the distal articulation.

The dimensions of the distal end of phalanx II-1 suggest its length matched or more likely exceeded that of III-1. Overall, digits II and IV appear to have been subequal in length and roughly 80% that of digit III (76 mm for digit IV and 94 mm for digit III). Proximal articular surfaces of first phalanges, III-1, IV-1 and presumably II-1 based on the articular surface of the metatarsal, are shallow undivided concavities (fig. 5.3). That of III-1 narrows ventrally whereas that of IV-1 narrows dorsally. All other phalanges have a divided proximal articulation with a flat ventral border and a tapering and proximally directed dorsal border. These correspond with the ginglymoid distal articulations found on all phalanges (fig. 5.3). Each phalanx also exhibits proximal rugosities on their ventromedial, ventrolateral, and sometimes ventral borders. These represent the insertions of the collateral ligaments and the flexor digitalis brevis (Currie and Zhao 1993).

The more proximal phalanges (II-1, III-1, III-2, IV-1, IV-2, IV-3) possess a well-defined pit on their dorsal (extensor) surface just proximal to the distal articulation. Unlike the irregular bone lining the ligament fossae, the surface bone of these dorsal pits is smooth. The pits of IV-2 and IV-3 exist as extensions off the central groove of the distal trochlea. Collateral ligament fossae and distal condyles are subequal on phalanges III-1 and III-2. On the more proximal phalanges of digits II and IV, the fossae and condyles are larger on the side nearest digit III (i.e., the lateral side of II-1 and the medial side of IV-1, IV-2 and IV-3). The lateral fossae of IV-2 and IV-3 consist only of shallow, poorly defined depressions. On all phalanges, fossae typically exhibit additional pitting or rugosities or both along their proximal margin proportionate to the size of the fossa (fig. 5.3).

The penultimate or ungual-bearing phalanges (II-2, III-3, IV-4) share several features making them distinct from those more proximal (fig. 5.3). The penultimate phalanges are long and slender. Phalanx III-3 exceeds III-2, and IV-2 exceeds both IV-3 and IV-2 in length (table 5.2). Each also has a tall (dorsoventrally) proximal end and a low distal end. None possesses any sign of the dorsal (extensor) pits found just proximal to the distal articulation of other phalanges. All have well-

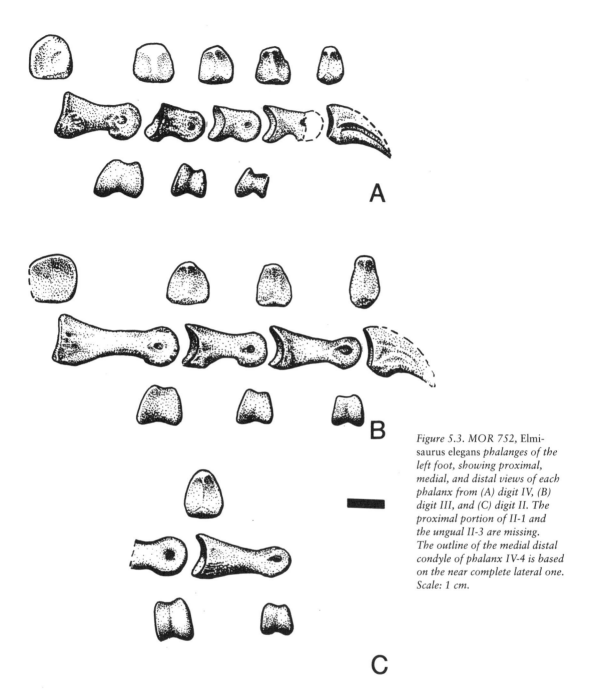

Figure 5.3. MOR 752, Elmisaurus elegans *phalanges of the left foot, showing proximal, medial, and distal views of each phalanx from (A) digit IV, (B) digit III, and (C) digit II. The proximal portion of II-1 and the ungual II-3 are missing. The outline of the medial distal condyle of phalanx IV-4 is based on the near complete lateral one. Scale: 1 cm.*

developed and symmetrical collateral ligament fossae. These occur more dorsally and are consequently more visible in dorsal view. The two condyles of the distal articulation are equally developed but the trochlea does not extend as far dorsally (fig. 5.3). Thus, the articular surface forms an arc of roughly 180° rather than the 220° to 240° found on the trochlea of more proximal phalanges.

Comparison of elmisaurid pedal phalanges to those of some contemporary theropods shows them to be most similar to those of dromaeosaurids. Height-width ratios of articular surfaces are similar; for example, the ratio for the distal surface of III-1 is 0.88 in MOR 752 as compared to 0.96 in *Saurornitholestes* (MOR 660); 0.91 in golden eagle, *Aquila* (MOR osteology 116); 0.79 in *Troodon formosus* (MOR 553S-7.7.91.20); 0.72 in an unidentified ornithomimid (MOR 450); 0.72 in *Daspletosaurus* (MOR 590); 0.67 in emu, *Dromiceius;* and 0.63 in *Allosaurus* (Madsen 1976). Those taxa with the narrowest articulations (high ratios) also show the most ginglymoid surfaces with well-grooved trochlea. These features may correspond to a greater grasping rather than cursorial function of the foot; note the ratios for the eagle and emu. Although all the above theropods bear a dorsal pit just proximal to the distal articulation, those of MOR 752, and to a lesser extent *Saurornitholestes,* are more consistently developed on all more proximal phalanges.

MOR 752 preserves two near complete unguals. The smaller of the two belongs to digit IV. The larger presumably comes from digit III but alternatively could come from digit II. The unguals are short, tall dorsoventrally, and taper rapidly to a sharp point. Both have divided, tall articular surfaces, distinct flexor tubercles, and well-defined lateral grooves. The ventral surface is only slightly expanded. Broken surfaces reveal large medullary cavities within each.

Several North American oviraptorosaur specimens preserve significant portions of metatarsal II (table 5.1). They represent both *Chirostenotes pergracilis* (NMC 8538, the "*Macrophalangia canadensis*" type; RTMP 79.20.1; NMC 9570) and *Elmisaurus elegans* (ROM 781, ROM 37163). However only a few pedal phalanges exist, all from *Chirostenotes pergracilis*. The few previously described *Elmisaurus* phalanges represent one Mongolian *Elmisaurus rarus* specimen (Osmólska 1981).

MOR 752 is clearly distinct from most other contemporary theropods. Metatarsal II lacks such features as the ginglymoid distal articulation of dromaeosaurids, the laterally compressed shaft of troodontids, the elongate proportions of ornithomimosaurs, and the antero-laterally angled contact for metatarsal III as in tyrannosaurids. Instead, metatarsal II possesses several distinctive elmisaurid features, a straight shaft with a tight contact with the remaining metatarsals, and a strong longitudinal ridge along its posteromedial edge contributing to the deeply emarginated posterior of the metatarsal.

The overall proportions of the phalanges also match known elmisaurid specimens (table 5.2; also compare fig. 5.3 and Osmólska 1981, fig. 3). Phalanx II-1 of MOR 752, although incomplete, appears to have been longer than III-1 (fig. 5.3), a feature also occurring in both *Chirostenotes pergracilis* (Currie 1990) and *Elmisaurus rarus* (Osmólska 1981, table 1). The only other theropods to display a long II-1 are some ornithomimosaurs. Only MOR 752 and NMC 8538 preserve a complete digit III, and both have an elongated III-3 that exceeds III-2 in length. MOR 752 shows further elongation in the penultimate phalanx of digit IV with IV-4 being longer than both IV-3 and IV-2 (table 5.2). How widespread this feature is among elmisaurids remains unknown, for only MOR 752 preserves a complete digit IV. Elongation of the ungual-bearing phalanges in the pes of nonavian theropods is very rare. In addition to these elmisaurids, *Compsognathus longipes* (Ostrom 1978) and two troodontids, *Sinornithoides youngi* (Russell and Dong 1993) and *Troodon formosus* (Russell 1969), show some elongation in digit IV, with IV-4 slightly longer than IV-3. Within birds, long penultimate phalanges reflect a grasping foot used for climbing, perching, or prey capture (Clark et al. 1998; Hopson and Chiappe 1998). Similar modifications of manus phalanges are pleisiomorphic for almost all theropods (Sereno et al. 1993). Possible synapomorphies of elmisaurid pedal digits include a long phalanx II-1 (longer than III-1 and the longest of the foot), phalanx III-3 longer than III-2, and phalanx IV-4 longer than both IV-3 and IV-2.

TABLE 5.2.

Greatest Lengths of Metatarsals and Phalanges in Some Elmisaurids (mm)

	MOR 752	ROM 781	ZPAL MgD-I/172&98	RTMP 79.20.1	NMC 8538
	E. elegans	*E. elegans*	*E. rarus*	*C. pergracilis*	*C. pergracilis*
metatarsal II	131.	155	147	181	205
metatarsal III			157	207	230
phalanx II-1			44		78
phalanx II-2	26.1		33		63
ungual II-3					60
phalanx III-1	34		43	58	75
phalanx III-2	23.3		32		52
phalanx III-3	25.1				58
ungual III-4	24.3				60
phalanx IV-1	23.1				59
phalanx IV-2	16.2				
phalanx IV-3	14.4				
phalanx IV-4	16.5				
ungual IV-5	21				

MOR 752 lacks the superelongate phalanges (the "macro-pha-langia") of *Chirostenotes*, where III-1 is well over 30% the length of metatarsal III or II (table 5.2). Its small size and gracile proportions compare most closely to *Elmisaurus elegans*. The maximum shaft diameter to metatarsal length is 0.08 in MOR 752; 0.09 in ROM 781, the type of *Elmisaurus elegans;* but 0.12 in *Elmisaurus rarus* (Currie 1989). However, all these proportions are likely to have changed with growth; phalanx to metatarsal length for the available specimens shows a linear increase with metatarsal size (fig. 5.4). Given the unknown ontogenetic state of MOR 752, the taxonomic usefulness of these features remains somewhat dubious. Currie and Russell (1988) proposed that the gracile *Elmisaurus elegans* and more robust *Chirostenotes pergracilis* might represent sexual morphs of one species because the two species appear very similar and co-occur within Dinosaur Provincial Park. Thus, the more robust proportions and relatively longer phalanges of larger specimens may simply reflect allometric scaling and not taxonomic differences. *Troodon formosus* shows similar changes with ontogeny (pers. obs.).

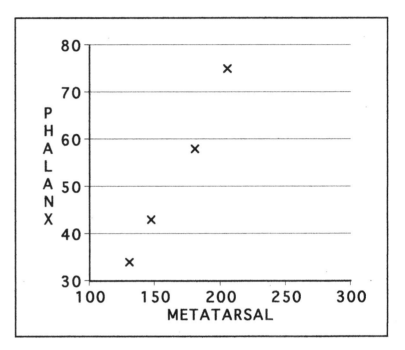

Figure 5.4. Graph comparing the length of phalanx III-1 to the length of metatarsal II (mm). Metatarsal II was used rather than metatarsal III in order to include MOR 752 data. Species and specimens represented from smallest to largest are Elmisaurus elegans *(MOR 752),* Elmisaurus rarus *(MgD-I/172 and 98), and two* Chirostenotes pergracilis *(RTMP 79.20.1 and NMC 8536). Table 5.2 lists measurements.*

The metatarsal II of MOR 752 does possess one discrete and taxonomically significant feature, a small but distinct anterolateral process just proximal to the distal articulation (fig. 5.2A). This process occurs only on *Elmisaurus elegans* (Currie 1989) and suggests MOR 752 can be assigned to this taxon.

Summary

Although ornithischian dinosaurs, such as *Orodromeus, Maiasaura, Achelousaurus,* and *Einiosaurus,* distinguish the Two Medicine Formation herbivore fauna from that of the contemporary Judith River Wedge (Eberth 1997), their theropod faunas are virtually identical. MOR 1107 indicates the presence of oviraptorosaurs in the Campanian of Montana and adds *Caenagnathus sternbergi* to the list of theropods shared by both the Two Medicine and Judith River of Montana and Alberta.

MOR 752 is the first record of *Elmisaurus elegans* in the Maastrichtian. Currie and Russell (1988) suggested that *Elmisaurus elegans* and *Chirostenotes pergracilis* might represent sexual morphs. Although this isolated foot cannot verify this hypothesis, it is consistent with it in that both forms are now known from the Maastrichtian.

Elmisaurids exhibit a number of pedal features that together make their feet unique among nonavian theropods. These features include: a tight fitting arctometatarsalian construction with the potential for fusion and formation of a tarsometatarsus; elongated phalanges with II-1 the longest; elongation of penultimate phalanges so that III-3 is longer than III-2 and IV-4 is longer than IV-3 and IV-2; and phalanges with tall, ginglymoid articulations that restrict motion to a strictly dorsoventral plane but allow a wide range of extension and flexion. Combined, these features suggest a strong foot with good grasping capabilities. Elmisaurids possess a number of other unusual features, including large flexor tubercles on manual unguals, a short postacetabular blade on the ilium, and a short ischium (Currie and Russell 1988). Currie and Russell (1988) suggested that some of these features might be adaptations for wading. Given the apparent grasping capabilities of the feet, these unique elmisaurid features may instead reflect climbing or some specialized feeding behavior.

Acknowledgments: Ken Olson of Lewistown, Montana, discovered MOR 752, one of his many excellent paleontological contributions. Frankie Jackson provided the fine illustrations. Many thanks to Montana State University, including Jack Horner, the Museum of the Rockies, Jim Schmitt, and the Department of Earth Sciences. Greg Erickson and Anthony Mongelli were absentee contributors. Finally, special thanks to Phil Currie for always making all his fine specimens available for study.

References

Barsbold, R. 1981. Predatory toothless dinosaurs from Mongolia. *Transactions of the Joint Soviet-Mongolian Paleontological Expedition* 15: 28–39.

Barsbold, R., T. Maryanska, and H. Osmólska. 1990. Oviraptorosauria. In D. B. Weishampel, P. Dodson, and H. Osmólska (eds.), *The Dinosauria*, pp. 249–258. Berkeley: University of California Press.

Clark, J. M., J. A. Hopson, R. Hernandez, D. E. Fastovsky, and M. Montellano. 1998. Foot posture in a primitive pterosaur. *Nature* 391: 886–889.

Cracraft, J. 1971. Caenagnathiformes: Cretaceous birds convergent in jaw mechanisms to dicynodont reptiles. *Journal of Paleontology* 45: 805–809.

Currie, P. J. 1989. The first records of *Elmisaurus* (Saurischia, Theropoda) from North America. *Canadian Journal of Earth Sciences* 26: 1319–1324.

Currie, P. J. 1990. Elmisauridae. In D. B. Weishampel, P. Dodson, and H. Osmólska (eds.), *The Dinosauria*, pp. 244–248. Berkeley: University of California Press.

Currie, P. J. 1997. Elmisauridae. In P. J. Currie and K. Padian (eds.), *Encyclopedia of Dinosaurs*, pp. 209–210. San Diego: Academic Press.

Currie, P. J., and D. A. Russell. 1988. Osteology and relationships of *Chirostenotes pergracilis* (Saurischia, Theropoda) from the Judith

River (Oldman) Formation of Alberta, Canada. *Canadian Journal of Earth Sciences* 25: 972–986.

Currie, P. J., and X. Zhao. 1993. A new carnosaur (Dinosauria, Theropoda) from the Jurassic of Xinjiang, People's Republic of China. *Canadian Journal of Earth Sciences* 30: 2037–2081.

Currie, P. J., S. J. Godfrey, and L. Nessov. 1993. New caenagnathid (Dinosauria: Theropoda) specimens from the Upper Cretaceous of North America and Asia. *Canadian Journal of Earth Sciences* 30: 2255–2272.

Currie, P. J., P. Vickers-Rich, and T. H. Rich. 1996. Possible oviraptorosaur (Theropoda, Dinosauria) specimens from the Early Cretaceous Otway Group of Dinosaur Cove, Australia. *Alcheringa* 20: 73–79.

Eberth, D. A. 1997. Judith River wedge. In P. J. Currie and K. Padian (eds.), *Encyclopedia of Dinosaurs,* pp. 379–388. San Diego: Academic Press.

Frankfurt, N. G., and L. M. Chiappe. 1999. A possible oviraptorosaur from the Late Cretaceous of northwestern Argentina. *Journal of Vertebrate Paleontology* 19: 101–105.

Frey, E., and D. M. Martill. 1995. A possible oviraptorosaurid theropod from the Santana Formation (Lower Cretaceous, ?Albian) of Brazil. *Neues Jahrbuch für Geologie und Paläontologie Monatshefte* 7: 397–412.

Gilmore, C. W. 1924. A new coelurid dinosaur from the Belly River Cretaceous of Alberta. *Canadian Geological Survey Bulletin* 38: 1–12.

Hopson, J. A., and L. M. Chiappe. 1998. Pedal proportions of living and fossil birds indicate arboreal or terrestrial specializations. *Journal of Vertebrate Paleontology* 18 (suppl. to no. 3): 52A.

Madsen, J. H. 1976. *Allosaurus fragilis:* A revised osteology. *Utah Geological Survey Bulletin* 109: 1–163.

Makovicky, P. J., and H.-D. Sues. 1998. Anatomy and phylogenetic relationships of the theropod dinosaur *Microvenator celer* from the Lower Cretaceous of Montana. *American Museum Novitates* 3240: 1–27.

Osborn, H. F. 1924. Three new Theropoda, *Protoceratops* zone, central Mongolia. *American Museum Novitates* 144: 1–12.

Osmólska, H. 1981. Coossified tarsometatarsi in theropod dinosaurs and their bearing on the problem of bird origins. *Palaeontologica polonica* 42: 79–95.

Ostrom, J. H. 1970. Stratigraphy and paleontology of the Cloverly Formation (Lower Cretaceous) of the Bighorn Basin Area, Wyoming and Montana. *Peabody Museum of Natural History* 35: 1–234.

Ostrom, J. H. 1978. The osteology of *Compsognathus longipes* Wagner. *Zitteliana* 4: 73–118.

Parks, W. A. 1933. New species of dinosaurs and turtles from the Upper Cretaceous Formations of Alberta. *University of Toronto Studies,* Geological Series, 34: 1–33.

Paul, G. S. 1988. *Predatory Dinosaurs of the World.* New York: Simon and Schuster.

Qiang, J., P. J. Currie, M. A. Norell, and J. Shu-an. 1998. Two feathered dinosaurs from northeastern China. *Nature* 393: 753–761.

Rogers, R. R., C. C. Swisher, and J. R. Horner. 1993. 40Ar/39Ar age and correlation of the non-marine Two Medicine Formation (Upper Cretaceous), northwestern Montana. *Canadian Journal of Earth Science* 30: 1066–1075.

Russell, D. A. 1969. A new specimen of *Stenonychosaurus* from the

Oldman Formation (Cretaceous) of Alberta. *Canadian Journal of Earth Science* 6: 595–612.

Russell, D. A., and Dong Z. 1993 A nearly complete skeleton of a new troodontid dinosaur from the Early Cretaceous of the Ordos Basin, Inner Mongolia, People's Republic of China. *Canadian Journal of Earth Science* 30: 2163–2173.

Sereno, P. C. 1999. The evolution of dinosaurs. *Science* 284: 2137–2147.

Sereno, P. C., C. A. Forster, R. R. Rogers, and A. M. Monetta. 1993. Primitive dinosaur skeleton from Argentina and the early evolution of Dinosauria. *Nature* 361: 64–66.

Smith, J. B., H. You, and P. Dodson. 1998. The age of the Sihetun Quarry in Liaoning Province, China, and its implications for early bird evolution. *Geological Society of America, Abstracts with Programs* 30 (7): 38.

Sternberg, C. M. 1932. Two new theropod dinosaurs from the Belly River Formation of Alberta. *Canadian Field-Naturalist* 46: 99–105.

Sternberg, R. M. 1940. A toothless bird from the Cretaceous of Alberta. *Journal of Paleontology* 14: 81–85.

Sues, H.-D. 1997. On *Chirostenotes*, a Late Cretaceous oviraptorosaur (Dinosauria: Theropoda) from western North America. *Journal of Vertebrate Paleontology* 17: 698–716.

Weishampel, D. B. 1990. Dinosaurian distribution. In D. B. Weishampel, P. Dodson, and H. Osmólska (eds.), *The Dinosauria*, pp. 63–139. Berkeley: University of California Press.

6. Tooth-Marked Small Theropod Bone: An Extremely Rare Trace

A. R. JACOBSEN

Abstract

Tooth-marked dinosaur bones provide insight into feeding behaviors and biting strategies of theropod dinosaurs. The majority of theropod tooth marks reported to date have been found on herbivorous dinosaur bones, although some tyrannosaurid bones with tooth marks have also been reported. In 1988 a partial skeleton of the dromaeosaurid *Saurornitholestes* was collected from southern Alberta, Canada, that bore tooth marks on one dentary. The location and morphology of the tooth marks suggests that a theropod (possibly a juvenile tyrannosaurid) included a *Saurornitholestes* in its diet.

Introduction

Ecological and behavioral aspects of dinosaur research have received increased interest in recent years (Farlow and Brett-Surman 1997; Currie and Padian 1997), with studies of theropod teeth and theropod tooth-marked dinosaur bones being used to determine clues as to the potential feeding behavior and predator-prey or intraspecific interactions of theropods (Abler 1992, 1999; Chin 1997; Chure et al. 1998; Currie and Jacobsen 1995; Tanke and Currie 1995, in press; Erickson 1999; Erickson et al. 1996; Erickson and Olson 1995; Fiorillo 1991; Jacobsen 1995, 1997, 1998; Larson 1999; Mongelli et al. 1999).

The morphology of tooth marks on a bone can be correlated with specific theropod taxa by comparing the serration marks to the denticle size and shape of known taxa (Jacobsen 1995). Based on such comparisons, tyrannosaurid tooth marks on bone are the most common and have been identified on a variety of bones of prey, including hadrosaurids, ceratopsids, and other tyrannosaurids (Erickson and Olson 1995; Jacobsen 1995, 1997; Tanke and Currie 1998). These tyrannosaurid bones comprise only 2% of the tooth-marked bones known (Jacobsen 1998). It is rarely possible to correlate tooth marks of small theropods to known taxa, probably due to the small size and similarity of denticles on some small theropod teeth (Currie and Jacobsen 1995). Two exceptions include a single ornithomimid caudal vertebra (TMP85.6.158) that exhibits *Saurornitholestes* tooth drag marks (Jacobsen 1995), and a partial skeleton of a *Troodon* (MOR 748) with puncture marks (D. J. Varricchio, pers. comm., 2000).

In the Dinosaur Park Formation of southern Alberta, small theropods are rare (Currie 1997) and comprise only a small percentage of the dinosaur fauna (Brinkman 1990; Brinkman et al. 1999). Their thin-walled bones are found broken or poorly preserved. The discovery of a partial skeleton of *Saurornitholestes* is therefore significant, especially because it also bears tooth marks.

Institutional Abbreviations: TMP, Royal Tyrrell Museum of Palaeontology, Drumheller, Alberta; MOR, Museum of the Rockies, Bozeman, Montana.

Description

TMP88.121.39 is a partial *Saurornitholestes* skeleton collected from the Campanian Dinosaur Park Formation, along the Milk River in southern Alberta. The ontogeny and anatomy are currently being studied by Drs. P. J. Currie and D. J. Varricchio (in prep.). The specimen has been used previously in anatomical studies and relationships of theropods (Currie 1995; Makovicky 1995; Britt 1993; Xu et al. 1999). The skeleton consists of several cranial elements (including a left dentary), right scapula, right coracoid, right humerus, several ribs, gastralia, right femur, right tibia, fibula, right metatarsus, pedal phalanges, several unguals, and an articulated distal tail section. The skeleton was examined for tooth marks, but marks were only found on the dentary.

The dentary is about 12 cm long and very well preserved (fig. 6.1). There are 15 tooth positions, with 10 teeth visible; five are fully erupted, and three are partially erupted. Two other teeth (nos. 4 and 7) are broken and show wear facets, indicating that they were functional even after they were broken. Three tooth marks were found on the lingual surface of the dentary. Two of them bear serration marks as parallel grooves or striae and have similar morphology. One of these tooth marks is located on the bone and the other is located on the crown of the seventh tooth.

The first tooth mark consists of 6 or 7 parallel striae covering an area 4 mm x 1.3 mm. The striae are positioned below the alveolus for the third tooth, and just above the Meckelian groove. The striae are orientated 45° from the longitudinal axis of the bone. Two of the striae

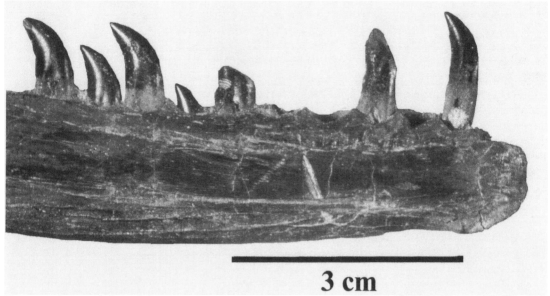

Figure 6.1. Saurornitholestes *left dentary (TMP88.121.39); lingual view, with three tooth marks.*

are slightly turned at one end. The sizes of the striae are between 0.37 mm and 0.40 mm, and they are cuboidal in cross-section.

The second tooth mark is below the fifth and sixth alveoli. It consists of two smaller marks separated 1.8 and 1.6 mm respectively from a larger, central mark; all three are arranged in a straight line and lack serration marks. The dorsalmost bite mark is 1.3 mm long and resembles an inverted teardrop. Below this is a prominent V-shaped groove with its axis approximately 60° from the longitudinal axis of the bone. This mark is 6 mm long. It gradually expands from 0.2 mm to 1.3 mm, where it cuts across the dorsal margin of the Meckelian groove.). Within the tooth mark the majority of bone fibers are broken; a few fibers point ventrally, indicating the direction of tooth movement. The ventralmost mark is a circular impression 1.3 mm in diameter. These three marks probably formed as the tooth skipped across the bone.

The third tooth mark is located on the lingual side of the seventh tooth. This mark covers an area 2 mm × 2 mm, and contains four prominent parallel striae oriented at right angles to the longitudinal axis of the tooth.

Discussion

The size and shape of the serration in the tooth marks are not like those of *Saurornitholestes*, therefore excluding intraspecific face-biting behavior (sensu Tanke and Currie 1995, in press) as a possible etiology for the marks. The small theropod *Dromaeosaurus* has denticles that are cuboidal in cross-section, as do tyrannosaurids, but these denticles would have produced finer serrations (Currie et al. 1990). The size and cuboid shape of the tooth marks on TMP88.121.39 are most consistent with those that could have been produced by a tyrannosaurid. Their small size indicates the biter was small, possibly a juvenile. But whether these marks were produced by *Gorgosaurus, Daspletosaurus,* or *Aublysodon* cannot be determined at this time. The placement and perpendicular orientation of the striae in relation to the upper dentition indicates that the marks were not produced by occluding teeth.

The similar serration morphology between two of the tooth marks indicates that they were made by a similar tooth. The lack of serration impressions in one of the tooth marks makes it difficult to assign it to a specific theropod taxa. Nevertheless, the absence of other type of serration pattern on the specimen, makes it most probable that it was also made by the same animal.

Preservational biases and methods of carcass consumption may explain why tooth-marked small theropod bones are extremely rare. The bones are small, and their thin, hollow construction is easily destroyed. Furthermore, the small bones might have simply been swallowed whole. Such factors makes the discovery of the *Saurornitholestes* skeleton all the more remarkable, especially one with tooth marks.

Conclusions

Based on the morphology of the serrated tooth marks found on a *Saurornitholestes* dentary, the trace maker may have been a juvenile tyrannosaurid. This feeding trace is significant because it shows that tyrannosaurids did not feed exclusively on herbivorous dinosaurs (such as ceratopsids and hadrosaurs), but also included carnivorous dinosaurs in their diet.

Acknowledgments: I am indebted to Dr. P. J. Currie for his many years of encouragement and inspiration. Also, I thank D. J. Varricchio, M. P. Ryan, D. H. Tanke, and P. Ralrick for reviewing and editing the manuscript. Support by staff of the Royal Tyrrell Museum (Canada) and the Steno Museum (Denmark) is gratefully acknowledged.

References

Abler, W. L. 1992. The serrated teeth of tyrannosaurid dinosaurs, and biting structures in other animals. *Paleobiology* 18 (2): 161–183.

Abler, W. L. 1999. The teeth of the tyrannosaurs. *Scientific American* 281 (3): 40–41.

Brinkman, D. B. 1990. Paleoecology of the Judith River Formation (Campanian) of Dinosaur Provincial Park, Alberta, Canada: Evidence from microfossil localities. *Palaeogeography, Palaeoclimatology, Palaeoecology* 78: 37–54.

Brinkman, D. B., M. J. Ryan, and D. A. Eberth. 1999. The paleogeographic and stratigraphic distribution of Ceratopsia (Ornithischia) in the Upper Judith River Group of western Canada. *Palaios* 13: 160–169.

Britt, B. B. 1993. Pneumatic postcranial bones in dinosaurs and other archosaurs. Ph.D. thesis, University of Calgary, Canada.

Chin, K. 1997. What did dinosaurs eat? Coprolites and other direct evidence of dinosaur diets. In J. O. Farlow and M. K. Brett-Surman (eds.),

The Complete Dinosaur, pp. 371–382. Bloomington: Indiana University Press.

Chure, D. J., A. R. Fiorillo, and A. R. Jacobsen. 2000. Prey bone utilization by predatory dinosaurs in the Late Jurassic of North America, with comments on prey bone use by dinosaurs throughout the Mesozoic. *Gaia* 15: 227–232.

Currie, P. J. 1995. New information on the anatomy and relationships of *Dromaeosaurus albertensis* (Dinosauria: Theropoda). *Journal of Vertebrate Paleontology* 15 (3): 576–591.

Currie, P. J. 1997. Theropoda. In P. J. Currie and K. Padian (eds.), *Encyclopedia of Dinosaurs,* pp. 731–737. San Diego: Academic Press.

Currie, P. J., and A. R. Jacobsen. 1995. An azhdarchid pterosaur eaten by a velociraptorine theropod. *Canadian Journal of Earth Sciences* 32: 922–925.

Currie, P. J., and K. Padian (eds.). 1997. *Encyclopedia of Dinosaurs.* San Diego: Academic Press.

Currie, P. J., K. J. Rigby Jr., and R. E. Sloan. 1990. Theropod teeth from the Judith River Formation of southern Alberta, Canada. In K. Carpenter and P. J. Currie (eds.), *Dinosaur Systematics, Approaches, and Perspectives,* pp. 107–127. Cambridge: Cambridge University Press.

Erickson, G. M. 1999. Breathing life into *Tyrannosaurus rex. Scientific American* 281 (3): 32–39.

Erickson, G. M., and K. Olson. 1995. Bite marks attributable to *Tyrannosaurus rex:* Preliminary description and implications. *Journal of Vertebrate Paleontology* 16 (1): 175–178.

Erickson, G. M., S. D. Van Kirk, J. Su, M. E. Levenston, and W. E. Caler. 1996. Bite-force estimated for *Tyrannosaurus rex* from tooth-marked bones. *Nature* 382: 706–708.

Farlow, J. O., and M. K. Brett-Surman. 1997. *The Complete Dinosaur.* Bloomington: Indiana University Press.

Fiorillo, A. R. 1991. Prey bone utilization by predatory dinosaurs. *Palaeogeography, Palaeoclimatology, Palaeoecology* 88: 157–166.

Jacobsen, A. R. 1995. Predatory behavior of carnivorous dinosaurs: Ecological interpretations based on tooth marked dinosaur bones and wear patterns of theropod teeth. M. Sci. thesis, University of Copenhagen.

Jacobsen, A. R. 1997. Toothmarks. In P. J. Currie and K. Padian (eds.), *Encyclopedia of Dinosaurs,* pp. 738–739. San Diego: Academic Press.

Jacobsen, A. R. 1998. Feeding behavior of carnivorous dinosaurs as determined by tooth marks on dinosaur bones. *Historical Biology* 13: 17–26.

Larson, P. L. 1999. Guess who's coming to dinner; *Tyrannosaurus* vs. *Nanotyrannus:* Variance in feeding habits. *Journal of Vertebrate Paleontology* 19 (3): 58A.

Makovicky, P. J. 1995. Phylogenetic aspects of the vertebral morphology of Coelurosauria (Dinosauria: Theropoda). M.Sc. thesis, University of Copenhagen, Denmark.

Mongelli, A., Jr., D. J. Varricchio, and J. J. Borkowski. 1999. Wear surfaces and breakage of tyrannosaurid (Theropoda: Coelurosauria) teeth. *Journal of Vertebrate Paleontology* 19 (3): 65A.

Tanke, D. H., and P. J. Currie. 1995. Intraspecific fighting behavior inferred from toothmark trauma on skulls and teeth of large carnosaurs (Dinosauria). *Journal of Vertebrate Paleontology* 15 (3): 55A.

Tanke, D. H., and P. J. Currie. 2000. Head-biting behavior in theropod dinosaurs: paleopathological evidence. *Gaia* 15: 167–184.

Xu X., Wang X.-L., and Wu, X.-C. 1999. A dromaeosaurid dinosaur with a filamentous integument from the Yixian Formation of China. *Nature* 401: 262–266.

7. The Phylogeny and Taxonomy of the Tyrannosauridae

THOMAS R. HOLTZ JR.

Abstract

A phylogenetic analysis of tyrant dinosaurs reveals a basal division between a weakly supported Aublysodontinae (characterized by unserrated premaxillary teeth, but otherwise plesiomorphic relative to other tyrannosaurids) and Tyrannosaurinae. Monophyly of *Albertosaurus* and *Gorgosaurus* outside of other tyrannosaurines was supported in only some of the most parsimonious trees, so that retention of a distinct generic name for the latter is advised at present. A newly discovered tyrannosaurine from the upper Two Medicine Formation of Montana demonstrates unresolved phylogenetic affinities with both *Daspletosaurus* and *Tyrannosaurus*. *Siamotyrannus*, recently described from material of the Barremian of Thailand, shows some tyrannosaurid synapomorphies, but lies outside Tyrannosauridae proper (i.e., the clade of aublysodontines and tyrannosaurines). *Shanshanosaurus*, a Late Cretaceous Chinese form, seems to be a tyrannosaurid, and is supported in the present study as a tyrannosaurine.

Introduction

The tyrant dinosaurs (Tyrannosauridae) represent a major radiation of Late Cretaceous theropods in western North America and eastern and central Asia. Previous studies on the intrafamily relationships of Tyrannosauridae (Matthew and Brown 1922; Russell 1970; Paul 1988; Bakker et al., 1988; Carpenter 1992; Olshevsky et al. 1995a,b) have not employed explicit numerical cladistic analyses.

Institutional Abbreviations: CMNH, Cleveland Museum of Natural History, Cleveland; FMNH, Field Museum of Natural History, Chicago.

Methods and Materials

Recent phylogenetic studies (Novas 1992; Pérez-Moreno et al. 1993, 1994; Holtz 1994, 2000; Sereno 1997, 1999; Makovicky and Sues 1998; Forster et al. 1998) have established that tyrannosaurids lie within Coelurosauria, as previously proposed by Matthew and Brown (1922), Huene (1923, 1926), and Currie (1989). However, the closest relatives to tyrannosaurids among the coelurosaurs are unclear. Three different general hypotheses have been offered (fig. 7.1): (1) Tyrannosaurids lie outside the clade Maniraptoriformes (Holtz 1996), the latter comprised of ornithomimosaurs and maniraptorans (oviraptorosaurs, deinonychosaurs, and birds) (Pérez-Moreno et al. 1994; Makovicky and Sues 1998; Forster et al. 1998); (2) Tyrannosaurids are closer to ornithomimosaurs than they are to maniraptorans (Pérez-Moreno et al. 1993; Holtz 1994, 2000); (3) Tyrannosaurids are closer to oviraptorosaurs, deinonychosaurs, and birds than to ornithomimosaurs (Sereno 1997, 1999). Because of this uncertainty, a hypothetical outgroup was chosen, coded after the morphological condition of relatively unspecialized coelurosaurs such as *Scipionyx, Coelurus, Ornitholestes,* and Compsognathidae. These taxa have been found in recent phylogenetic analyses (e.g., Sereno 1999; Holtz 2000) to lie close to the basal divergences within Coelurosauria and furthermore lack the numerous trophic and locomotory specializations found in other coelurosaurs such as ornithomimosaurs, oviraptorosaurs, and dromaeosaurids, so it

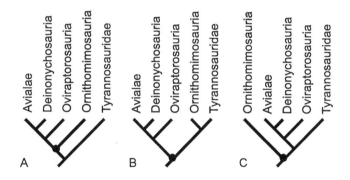

Figure 7.1. Alternative outgroup relationships for Tyrannosauridae. (A) tyrannosaurids lie outside the ornithomimosaur-maniraptoran clade Maniraptoriformes (indicated by the solid circle) (Pérez-Moreno et al. 1994; Makovicky and Sues 1998; Forster 1998); (B) tyrannosaurids shared a more recent common ancestor with ornithomimosaurs than with maniraptorans (Pérez-Moreno et al. 1993; Holtz 1994, 2000); (C) tyrannosaurids shared a more recent common ancestor with advanced maniraptorans than with ornithomimosaurs (Sereno 1997, 1999).

is assumed here that these generalized forms more closely approximate the ancestral coelurosaurian condition. The conditions of the character states for the possible proximate outgroups Ornithomimosauria and advanced Maniraptora (comprised of Oviraptorosauria, Dromaeosauridae, and Avialae) are provided.

Eight named tyrannosaurid species were included as ingroups, as were two additional tyrannosaurid taxa, currently unnamed. Furthermore, the positions of two taxa considered by some to be primitive tyrant dinosaurs, *Siamosaurus* of the Lower Cretaceous of Thailand (Buffetaut et al., 1996) and *Shanshanosaurus* of the Upper Cretaceous

(?Maastrichtian) Subashi Formation of China (Dong 1977), are examined.

The character descriptions for this study are provided as appendix 7.1, the data matrix as appendix 7.2. The total number of characters evaluated was 111; however, 24 of these are autapomorphies (limited to a single operational taxon in the study). Most characters are binary, but some are multistate. Some multistate characters were coded as "ordered" where the states represent gradational series (e.g., character 5, with state 5.0 "prefrontal present," state 5.1 "prefrontal reduced," and state 5.2 "prefrontal absent"), as noted in the appendix: however, when the analyses were run with these same characters considered "unordered" the same sets of most parsimonious trees were obtained. Where multistate characters do not necessarily represent a gradational series (e.g., character 53, with state 53.0 "lacrimal horn absent," 53.1 "lacrimal horn directly dorsal to descending ramus of lacrimal," and 53.2 "lacrimal horn rostral to descending ramus of lacrimal") they were considered "unordered."

Results

The most parsimonious trees were determined using PAUP* 4.0 (Swofford 1999), using the Exhaustive Search option. The tree metrics tree length (TL), consistency index (CI), retention index (RI), reduced consistency index (RC), and homoplasy index (HI) were calculated by PAUP*. MacClade 3.07 (Maddison and Maddison 1997) was used afterward to examine character distribution within the trees. The strict consensus of the 15 equally most parsimonious trees is presented in solid lines in figure 7.2. Characters supporting the nodes on the most parsimonious trees are shown in appendix 7.3. It is interesting to note that although the basal synapomorphies for Tyrannosauridae as a whole are divided between the skull and the postcranial skeleton, almost all the potential synapomorphies within the clade are cranial. Similar situations exist for other dinosaur taxa (Ceratopsidae, Hadrosauridae, etc.), where the skulls may be quite distinctive, but the postcrania very constant within the clade. However, there has been little direct study of the variation in tyrannosaurid postcrania, and future analysis may indeed reveal more diagnostic characters outside of the skull.

As in some previous studies (Matthew and Brown 1922; Olshevsky et al. 1995a,b; Currie in press), Tyrannosauridae was found to comprise two clades: Aublysodontinae and Tyrannosaurinae. Potential phylogenetic taxonomic definitions for the clades in question might be: Tyrannosauridae, all descendants of the most recent common ancestor of *Tyrannosaurus* and *Aublysodon;* Aublysodontinae, *Aublysodon* and all taxa sharing a more recent common ancestor with it than with *Tyrannosaurus;* and Tyrannosaurinae, *Tyrannosaurus* and all taxa sharing a more recent common ancestor with it than with *Aublysodon.* However, these definitions must be provisional, as the type species of *Aublysodon, A. mirandus,* is known only from isolated premaxillary teeth, while the somewhat more complete *A. molnari* is known only from skull elements and may eventually prove to be a different genus. Similar problems would result from using *Alectrosaurus* rather than *Aublysodon* as the anchor taxon for Aublysodontinae.

(Note that the phylogenetic taxonomy proposed by Sereno [1998] is problematic as well: his "Tyrannosauridae" is defined as all taxa closer to *Tyrannosaurus* than to *Alectrosaurus, Aublysodon,* and *Nanotyrannus.* The latter specimen is very likely a juvenile *Tyrannosaurus rex* [Carr 1999; see also below], rendering his "Tyrannosauridae" as a subgroup within the species *T. rex* [all specimens sharing a more recent common ancestor with the type specimen of *T. rex* than with the "type" of *Nanotyrannus*]. Furthermore, Sereno's [1998] "Tyrannosaurinae," all taxa closer to *Tyrannosaurus* than to *Albertosaurus, Daspletosaurus,* or *Gorgosaurus,* would be limited to the genus *Tyrannosaurus* itself following the phylogeny presented here. It is recommended that until such time as the more complete Mongolian specimens currently referred to *Alectrosaurus olseni* [see below], which seem to represent the best material of primitive tyrannosaur, are more adequately described, that the provisional phylogenetic taxonomy proposed here be used for tyrannosaurid systematics.)

Aublysodontinae is comprised at present of only incompletely known taxa. These forms are united by two dental synapomorphies: 76.1 and 77.1 (unserrated premaxillary teeth with prominent vertical ridges on the caudal surface). These derived features are not present on other tyrannosaurids, including the oldest confirmed tyrannosaurid premaxillary tooth (from the Early Cretaceous Jobu Formation of Japan: Manabe 1999). Although Molnar and Carpenter (1989) separated *Aublysodon* from Tyrannosauridae as Family Aublysodontidae, the presence of *Aublysodon*-like teeth in specimens of *Alectrosaurus olseni,* a taxon with tyrannosaurid synapomorphies (Mader and Bradley 1989) led Currie et al. (1990) to include this group as a subfamily within Tyrannosauridae, a nomenclatural practice followed here. Other than the specialized premaxillary teeth, aublysodontines generally retain more primitive coelurosaurian features lost in other tyrannosaurids: for example, the skulls of aublysodontines are longer and lower, the tooth count higher, and the lateral teeth less labiolingually expanded than in derived tyrannosaurines (Currie in press). *Aublysodon* and *Alectrosaurus* (but not the Kirtland Shale taxon) are smaller than typical tyrannosaurines: however the ontogenetic status of known aublysodontine specimens is hindered by the incompleteness of their fossils, and these might represent juvenile individuals of taxa which reached larger sizes. Although *Aublysodon* is provisionally recognized as a valid name here, Currie et al. (1990) caution that this may not be a taxon, but instead an ontogenetic stage or a sexual dimorph. It might be further added that tooth wear in life, postmortem abrasion, and digestion might also conceivably render a typical tyrannosaurid premaxilla tooth into a nonserrated one. In such a case, the synapomorphies uniting "Aublysodontinae" would be lost, and the remaining resemblances between these taxa would be symplesiomorphies.

Aublysodon molnari Paul 1988: The type species of *Aublysodon mirandus* is a set of isolated premaxillary teeth from the Judith River

Formation of north-central Montana. Currie (1987) referred primitive triangular tyrannosaurid frontals from Dinosaur Park Formation to cf. *Aublysodon.* A second species of *Aublysodon, A. molnari,* was proposed for material from the Hell Creek Formation of eastern Montana, formally known under the label "the Jordan theropod" (Molnar 1978). Currie (1987), Paul (1988) and Molnar and Carpenter (1989) referred this material to *Aublysodon.* Olshevsky et al. (1995b) went further to use this specimen as the type of a new genus, "*Stygivenator,*" distinguishing it from *A. mirandus* in having premaxillary teeth which are narrower in lateral view: whether this is taxonomically significant or due to allometry or even different tooth positions in the premaxilla has yet to be determined.

Alectrosaurus olseni Gilmore 1933: A tyrannosaurid from the Iren Dabasu Formation of Inner Mongolia (People's Republic of China) and the Bayn Shire Formation of Mongolia, units of uncertain Late Cretaceous age (most probably younger than Cenomanian, and possibly as young as Campanian: Currie and Eberth 1993). Forelimb elements in Gilmore's type material have since proven to come from a therizinosauroid (Mader and Bradley 1989). The type specimen from China is difficult to diagnose relative to other tyrannosaurids: however, material from the Mongolian Republic referred to this taxon by Perle (1977) is more complete. This material indicates that the skull was long and low and that the premaxilla contained *Aublysodon*-like teeth. The skull is generally plesiomorphic relative to other tyrannosaurids, but Currie (in press) documents the following synapomorphy: 78.2 Cranialmost 2 to 3 maxillary teeth incisiform. This material is currently under review and Currie (in press) states that the postcranium demonstrates additional (undescribed) diagnostic features. Additional study may indicate that the Mongolian material may not be referable to *Alectrosaurus olseni* (Mader and Bradley 1989).

Kirtland Shale aublysodontine: Lehman and Carpenter (1990) described a partial tyrannosaurid skeleton from the Upper Campanian Kirtland Shale of the San Juan Basin, northwestern New Mexico. Because of the possession of (plesiomorphically) triangular frontals and unserrated premaxillary teeth with a prominent caudomedian ridge, they referred it to *Aublysodon* cf. *A. mirandus.* In this analysis the Kirtland Shale form is found to be an aublysodontine, of uncertain relationship with either *Aublysodon molnari* or *Alectrosaurus olseni.* However, the specimen also possesses a convex tablike process on the dorsolateral surface of the postorbital (59.1), as in *Daspletosaurus torosus.* If an aublysodontine it is by far the largest specimen discovered at present, with an estimated femur length of 1080 mm, thus comparable in size to large individuals of *Gorgosaurus* and *Daspletosaurus.* The tibia demonstrates an autapomorphic medial embayment along the facet for the ascending process of the astragalus (88.1) not seen in *Alectrosaurus* nor in tyrannosaurines.

Alioramus remotus Kurzanov 1976: In some ways, this taxon is relatively primitive, retaining a higher tooth count, lower snout, and long, more slender dentary than other tyrannosaurines. However, the ontogenetic stage of this specimen is uncertain, and these features are found in juvenile specimens of other tyrannosaurs to be lost in adulthood (Carr 1999). In the present study *Alioramus* was most parsimoniously supported in a position outside the better-known and larger tyrannosaurines; however, Currie (in press) suggests it may have shared a more recent common ancestor with *Tyrannosaurus* and *Daspletosaurus* than with *Gorgosaurus* and *Albertosaurus*. It possesses an interesting suite of primitive tyrannosaurid, derived tyrannosaurine, and autapomorphic character states (appendix 7.3). It is known only from a partial skull and associated metatarsals from the Nogon Tsav beds of the Ingeni Khoboor valley, Mongolia. The reconstruction of Kurzanov (1976) fails to correct for dorsolateral crushing of the braincase; restorations in Paul 1988 and Olshevsky 1995a correct for this and use a short rather than a pointed premaxilla.

The remaining tyrannosaurines are all represented by more complete and larger material than the previous taxa. They are also restricted to the Campanian and Maastrichtian of western North America and eastern and central Asia. In some of the most parsimonious trees, *Gorgosaurus libratus* and *Albertosaurus sarcophagus* were united outside other taxa. Russell (1970), Paul (1988), and Carpenter (1992) had previously united these two (as the genus *Albertosaurus*), but the features used to do so were symplesiomorphies relative to the derived conditions in *Tyrannosaurus* and *Daspletosaurus*. Thus, the expanded *"Albertosaurus"* was simply composed of advanced tyrannosaurines that were neither *Daspletosaurus* nor *Tyrannosaurus* (in present usage). In this study two cranial features potentially unite *Gorgosaurus* and *Albertosaurus*: 53.1 Lacrimal horn rostral to descending ramus of lacrimal; 68.1 Ventral pocket to ectopterygoid chamber greatly reduced. However, in other trees *Gorgosaurus* is closer to the *Daspletosaurus-Tyrannosaurus* clade than to *Albertosaurus*. Given the weak support for this clade, and given that additional features in their morphology support potential alliances with other tyrannosaurid taxa, this study uses the original generic name for each species.

Albertosaurus sarcophagus Osborn 1905: A taxon at present only known from the Early Maastrichtian Horseshoe Canyon Formation of Alberta. The juvenile tyrannosaurid specimen from the Horseshoe Canyon Formation referred by Russell (1970) to *Daspletosaurus* (a taxon otherwise unknown from the Maastrichtian) may in fact be a juvenile *Albertosaurus*. More recently discovered material, as yet unpublished, helps to better document the rest of the anatomy of this dinosaur. *Albertosaurus* is comparable to *Gorgosaurus* in size, and like that taxon lacks many derived features shared by *Daspletosaurus* and *Tyrannosaurus*.

Gorgosaurus libratus Lambe 1914: This taxon is known from more numerous and more complete specimens than any other North American species of tyrannosaurid. The ontogeny of this species includes many parts of the growth series (Carr 1999). It is presently only confirmed from the Late Campanian Dinosaur Park Formation of Alberta: as the isolated postcranial material from other formations referred to this taxon do not show features unique to *Gorgosaurus libratus*, these assignments are tentative at best. A large but incomplete tyrannosaurid skull from the Judith River, FMNH PR308, has formed the basis of many restorations of *Gorgosaurus libratus* (Russell 1970, fig. 1; Paul 1988, 335; Carpenter 1992, figs. 1, 2E), but lacks *Gorgosaurus* synapomorphies and in fact almost certainly represents a specimen of *Daspletosaurus torosus* (Carr 1999).

Daspletosaurus torosus, the Two Medicine tyrannosaurine, and *Tyrannosaurus* share several derived features lacking in other tyrannosaurids. Many of the synapomorphies of this clade suggest a more forcefully built and muscular skull and neck than in other tyrannosaurids. In some of the most parsimonious trees *Daspletosaurus torosus* and a presently unnamed tyrannosaurine from the upper Two Medicine Formation were united by a potential synapomorphy: 53.1 Lacrimal horn directly dorsal to descending ramus. However, in other trees *Daspletosaurus* is closer to *Tyrannosaurus* than to the Two Medicine form, and in still others the Two Medicine form is closer to *Tyrannosaurus.*

Daspletosaurus torosus Russell 1970: From the Dinosaur Park Formation of Alberta, this form is more robustly and powerfully constructed than the sympatric *Gorgosaurus libratus* (Russell 1970).

Two Medicine tyrannosaurine: Horner et al. (1992) briefly describe a new taxon of tyrannosaurid from the upper part of the Two Medicine Formation of Montana, stratigraphically higher in the Late Campanian than the typical Judith River and Dinosaur Park Formation dinosaur-bearing horizons. They interpreted this new specimen as intermediate between *Daspletosaurus* and *Tyrannosaurus rex* (the Asian *Tyrannosaurus bataar* was not included in that preliminary study). This material remains unpublished, and study of the taxon is ongoing. The present analysis supports an intermediate position in some of the most parsimonious trees, but in others it is the sister taxon to *Daspletosaurus torosus,* and in still others it is outside a *Daspletosaurus-Tyrannosaurus* clade.

Tyrannosaurus bataar Maleev 1955a: As used here, this species includes several specimens previously referred to other taxa: *Tarbosaurus efremovi* Maleev 1995b, *Gorgosaurus lancinator* Maleev 1955b, and *Maleevosaurus novojilovi* (Maleev 1955b). As with Currie (in press), and Carr (1999), the present study considers these taxa a growth series of a single species, rather than two (Carpenter 1992) or three (Olshevsky et al., 1995a,b) different genera. *Tyrannosaurus bataar* is from the Nemegt Formation (Early Maastrichtian) of Mongolia: numerous isolated elements and teeth from comparable aged units in China might be referable to *T. bataar.*

The hypothesis of Olshevsky et al. (1995a,b) that *T. bataar* is less closely related to *Tyrannosaurus rex* than the latter is to other North American tyrannosaurines is not supported: instead numerous synapomorphies strongly unite the Asian taxon with *Tyrannosaurus rex.* Given the number of these similarities, the original name *Tyrannosaurus bataar* is retained. However the use of the name *Tarbosaurus bataar* (as in Russell 1970; Molnar et al. 1990; Currie in press) would be no less appropriate phylogenetically. The juvenile and subadult material demonstrates some of the autapomorphies found in the adults: the type skull of "*Gorgosaurus lancinator*" shows 12.0 and 73.1; the type material of "*Maleevosaurus*" shows 84.2 and 102.1. *T. bataar* is characterized by the most reduced forelimbs known within Tyrannosauridae: the general theropod reduction in digital and metacarpal elements from digit V toward digit I (Wagner and Gauthier 1999) is seen developed further in this species than in other tyrannosaurids.

Tyrannosaurus rex Osborn 1905: The last and largest known tyrannosaurid, *T. rex* is represented by numerous skulls and postcrania from the late Maastrichtian Hell Creek Formation of Montana, Wyoming, and South Dakota, the Lance Formation of Wyoming, and equivalent beds in Saskatchewan, Alberta, and other localities in the North American West. This species is characterized by numerous autapomorphies. Gilmore (1946) described CMNH 7541, a 572 mm long skull from the Hell Creek Formation of Montana, as a new species of *Gorgosaurus, G. lancensis*. This taxon was later (Bakker et al. 1988) referred to its own genus, *Nanotyrannus*. Because of some similarities with adult *Tyrannosaurus rex*, these authors and others (Russell 1970; Carpenter 1992) have voiced suspicion that this skull might represent a juvenile of that larger sympatric species. Carr (1999) documents the presence of juvenile striated cortical bone over most of the skull's surface, and cannot verify the presence of cranial fusions previously used to indicate the adult nature of this skull. Additionally, the changes in lateral tooth shape and maxillary tooth number used to distinguish "*Nanotyrannus*" from *Tyrannosaurus* also occur in the growth series of *Gorgosaurus*. Furthermore, the skull of "*Nanotyrannus*" demonstrates several *T. rex* autapomorphies: 103.1, 104.1, 105.1, 106.1, 108.1, and 109.1. In light of this, and pending the discovery of a skull of different morphology which can be more clearly demonstrated to be a juvenile *T. rex*, "*Nanotyrannus*" is here considered to be a young individual of *Tyrannosaurus* and not a distinct taxon.

Addition of *Siamotyrannus* and *Shanshanosaurus* results in a set of nine equally parsimonious trees: the consensus of these trees resembles those of the previous analysis, in which *Gorgosaurus* and *Albertosaurus* are sister taxa. The positions of these taxa are indicated in figure 7.2 with dotted lines.

Siamotyrannus isanensis Buffetaut, Suteethorn, and Tong 1996: Buffetaut et al. (1996) considered this fragmentary form from the Barremian Sao Khua Formation of northeastern Thailand to be a primitive tyrannosaurid. It was found here to lie outside Tyrannosauridae proper (aublysodontines and tyrannosaurines), but shares with it the following synapomorphies: 24.1, 25.1, 27.1, 28.1, and 31.1. Instead of the single midline crest on the ilium in tyrannosaurids, however, the Thai taxon has a pair of crests. *Siamotyrannus* may indeed be an ancestral member of the tyrannosaur lineage, but lacking additional material (in particular, the skull) such a position remains uncertain.

Shanshanosaurus huoyanshanensis Dong 1977: This small dinosaur is known only from a partial skull and associated postcrania from the Subashi Formation of the Turpan Basin, Xinjiang, People's Republic of China, a unit thought to be from the Maastrichtian Age by Lucas

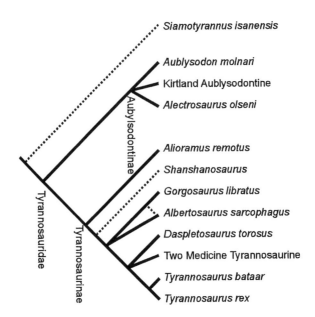

Figure 7.2. Phylogeny of Tyrannosauridae. Solid lines represent the strict consensus of the fifteen most parsimonious trees excluding Siamotyrannus *and* Shanshanosaurus *(tree metrics including all autapomorphic characters: TL 150, CI 0.860, RI 0.761, RC 0.655, HI 0.160; if autapomorphic characters are excluded, TL 126, CI 0.833, RI 0.761, RC 0.634, HI 0.273). Dashed lines indicate position of* Siamotyrannus *and* Shanshanosaurus *when included (tree metrics including autapomorphic characters: TL 152, CI 0.855, RI 0.768, RC 0.657, HI 0.164).*

and Estep (1998). Paul (1988) and Olshevsky et al. (1995a,b) considered it to be aublysodontine (in the present taxonomy). The material does document several tyrannosaurid synapomorphies, but because the premaxillary teeth are unknown all remaining similarities with Aublysodontinae are symplesiomorphies, and thus are not helpful in establishing positive phylogenetic relationships. When included in the analysis, it is found to lie as a tyrannosaurine more advanced than *Alioramus* in the possession of a reduced maxillary (45.1) and dentary (75.1) tooth count: at 8 and 12, respectively, these values are as low or lower than *Tyrannosaurus rex.* If a tyrannosaurid, however, it is the only form with a retroarticular process (-13.0) and with somewhat procoelous cervical vertebrae (89.1).

Conclusions

As with all such studies, the results of the phylogenetic analysis presented here are tentative and subject to change with addition of new characters and new taxa. Nevertheless, it is hoped that this preliminary analysis will serve to advance the understanding of the tyrant dinosaurs.

Acknowledgments: I would like to thank the editors for their invitation to participate in this volume honoring the contributions of Phil Currie to the field of Mesozoic vertebrate paleontology. Like the tyrannosaurids before him, Phil has been a most successful hunter of dinosaurs in both western North America and Asia. His incomparable work has greatly expanded our knowledge of the anatomy, systematics, and behavior of fossil reptiles in general, and of theropods in particular. His advice, support, and criticism have helped many (myself included) who are following in his footsteps in the study of dinosaurs and other fossil vertebrates.

References

Bakker, R. T., M. Williams, and P. J. Currie. 1988. *Nanotyrannus*, a new genus of pygmy tyrannosaur, from the latest Cretaceous of Montana. *Hunteria* 1 (5): 1–30.

Buffetaut, E., V. Suteethorn, and H. Tong. 1996. The earliest known tyrannosaur from the Lower Cretaceous of Thailand. *Nature* 381: 689–691.

Carpenter, K. 1992. Tyrannosaurids (Dinosauria) of Asia and North America. In N. Mateer and P.-J. Chen (eds.), *Aspects of Nonmarine Cretaceous Geology*, pp. 250–268. Beijing: China Ocean Press.

Carr, T. D. 1999. Craniofacial ontogeny in Tyrannosauridae (Dinosauria, Coelurosauria). *Journal of Vertebrate Paleontology* 19: 497–520.

Currie, P. J. 1987. Theropods of the Judith River Formation of Dinosaur Provincial Park, Alberta. In P. J. Currie and E.H. Koster (eds.), *Fourth Symposium on Mesozoic Terrestrial Ecosystems: Short Papers,* pp. 52–60. Drumheller: Tyrrell Museum of Palaeontology.

Currie, P. J. 1989. Theropod dinosaurs of the Cretaceous. In K. Padian and D. J. Chure (eds.), *The Age of Dinosaurs,* pp. 113–120. *Short Courses in Paleontology Number 2,* The Paleontological Society.

Currie, P. J. in press. Theropods from the Cretaceous of Mongolia. In M. Benton, E. Kurochkin, M. Shiskin, and D. Unwin (eds.), *The Age of Dinosaurs in Russia and Mongolia.* Cambridge: Cambridge University Press.

Currie, P. J., and D. A. Eberth. 1993. Palaeontology, sedimentology, and palaeoecology of the Iren Dabasu Formation (Upper Cretaceous), Inner Mongolia, People's Republic of China. *Cretaceous Research* 14: 127–144.

Currie, P. J., J. K. Rigby Jr., and R. E. Sloan. 1990. Theropod teeth from the Judith River Formation of southern Alberta, Canada. In K. Carpenter and P. J. Currie (eds.), *Dinosaur Systematics: Approaches and Perspectives,* pp. 107–135. Cambridge: Cambridge University Press.

Dong, Z.-M. 1977. [On the dinosaurian remains from Turpan, Xinjiang]. *Vertebrata PalAsiatica* 15: 59–66. (In Chinese, with English summary.)

Forster, C. A., S. D. Sampson, L. M. Chiappe, and D. W. Krause. 1998. The theropod ancestry of birds: New evidence from the Late Cretaceous of Madagascar. *Science* 279: 1915–1919.

Gilmore, C. W. 1933. On the dinosaurian fauna of the Iren Dabasu Formation. *Bulletin of the American Museum of Natural History* 67: 23–78.

Gilmore, C. W. 1946. A new carnivorous dinosaur from the Lance Formation of Montana. *Smithsonian Miscellaneous Collections* 106: 1–19.

Holtz, T. R., Jr. 1994. The phylogenetic position of the Tyrannosauridae: Implication for theropod systematics. *Journal of Paleontology* 68: 1100–1117.

Holtz, T. R., Jr. 1996. Phylogenetic taxonomy of the Coelurosauria (Dinosauria: Theropoda). *Journal of Paleontology* 70: 536–538.

Holtz, T. R., Jr. 2000. A new phylogeny of the carnivorous dinosaurs. *Gaia* 15: 5–62.

Horner, J. R., D. J. Varricchio, and M. B. Goodwin. 1992. Marine transgressions and the evolution of Cretaceous dinosaurs. *Nature* 358: 59–61.

Huene, F. 1923. Carnivorous Saurischia in Europe since the Triassic. *Bulletin of the Geological Society of America* 34: 449–458.

Huene, F. 1926. The carnivorous Saurischia in the Jura and Cretaceous formations, principally in Europe. *Revista Museo de La Plata* 29: 35–167.

Kurzanov, S. M. 1976. [A new Late Cretaceous carnosaur from Nogon-Tsav, Mongolia.] *Sovmestnaia sovetsko-mongol'skaia paleontologi-cheskaia ekspeditsiia, trudy* 3: 93–104. (In Russian, with English summary.)

Lambe, L. B. 1914. On a new genus and species of carnivorous dinosaur from the Belly River Formation of Alberta with a description of the skull of *Stephanosaurus marginatus* from the same horizon. *Ottawa Naturalist* 28: 13–20.

Lehman, T. M., and K. Carpenter. 1990. A partial skeleton of the tyrannosaurid dinosaur *Aublysodon* from the Upper Cretaceous of New Mexico. *Journal of Paleontology* 64: 1026–1032.

Lucas, S. G., and J. W. Estep. 1998. Vertebrate biostratigraphy and biochronology of the Cretaceous of China. In S. G. Lucas, J. I. Kirkland, and J. W. Estep (eds.), *Lower and Middle Cretaceous Terrestrial Ecosystems. New Mexico Museum of Natural History and Science Bulletin* 14: 1–20.

Maddison, W. P., and D. R. Maddison. 1997. MacClade: Analysis of phylogeny and character evolution. Version 3.07. Sinauer Associates, Sunderland, Massachusetts.

Mader, B. J., and R. L. Bradley. 1989. A redescription and revised diagnosis of the syntypes of the Mongolian tyrannosaur *Alectrosaurus olseni*. *Journal of Vertebrate Paleontology* 9: 41–55.

Makovicky, P. J., and H.-D. Sues. 1998. Anatomy and phylogenetic relationships of the theropod dinosaur *Microvenator celer* from the Lower Cretaceous of Montana. *American Museum Novitates* 3240: 1–27.

Maleev, E. A. 1955a. [Gigantic carnivorous dinosaurs of Mongolia.] *Dokladi akademii Nauk S.S.S.R.* 104: 634–637. (In Russian.)

Maleev, E.A. 1955b. [New carnivorous dinosaurs from the Upper Cretaceous of Mongolia.] *Dokladi akademii Nauk S.S.S.R.* 104: 779–782. (In Russian.)

Manabe, M. 1999. The early evolution of the Tyrannosauridae in Asia. *Journal of Paleontology* 73: 1176–1178.

Matthew, W. D., and B. Brown. 1922. The family Deinodontidae, with notice of a new genus from the Cretaceous of Alberta. *Bulletin of the American Museum of Natural History* 46: 367–385.

Molnar, R. E. 1978. A new theropod dinosaur from the Upper Cretaceous of central Montana. *Journal of Paleontology* 52: 73–82.

Molnar, R. E., and K. Carpenter. 1989. The Jordan theropod (Maastrichtian, Montana, U.S.A.) referred to the genus *Aublysodon. Geobios* 22: 445–454.

Molnar, R. E., S. M. Kurzanov, and Z.-M. Dong. 1990. Carnosauria. In D. B. Weishampel, P. Dodson, and H. Osmólska (eds.), *The Dinosauria*, pp. 169–209. Berkeley: University of California Press.

Novas, F. E. 1992. La evolución de los dinosaurios carnivoros. In J. L. Sanz and A. D. Buscalioni (eds.), *Los dinosaurios y su entorno biotico: Actas del segundo curso de paleontologia en Cuenca*, pp. 125–163. Cuenca: Instituto "Juan de Valdes."

Olshevsky, G., T. L. Ford, and S. Yamamoto. 1995a. [The origin and evolution of the tyrannosaurids, part 1]. *Kyoryugaku saizensen* 9: 92–119. (In Japanese.)

Olshevsky, G., T. L. Ford, and S. Yamamaoto. 1995b. [The origin and evolution of the tyrannosaurids, part 2]. *Kyoryugaku saizensen* 10: 75–99. (In Japanese.)

Osborn, H. F. 1905. *Tyrannosaurus* and other Cretaceous carnivorous dinosaurs. *Bulletin of the American Museum of Natural History* 21: 259–265.

Paul, G. S. 1988. *Predatory Dinosaurs of the World*. New York: Simon and Schuster.

Pérez-Moreno, B. P., J. L. Sanz, J. Sudre, and B. Sigé. 1993. A theropod dinosaur from the Lower Cretaceous of southern France. *Revue de paléobiologie* 7 (special vol.): 173–188.

Pérez-Moreno, B. P., J. L. Sanz, A. D. Buscalioni, J. J. Moratalla, F. Ortega, and D. Rasskin-Gutman. 1994. A unique multitoothed ornithomimosaur dinosaur from the Lower Cretaceous of Spain. *Nature* 370: 363–367.

Perle, A. 1977. [On the first finding of *Alectrosaurus* (Tyrannosauridae, Theropoda) in the Late Cretaceous of Mongolia]. *Problemi geologii mongolii* 3: 104–113. (In Russian.)

Russell, D. A. 1970. Tyrannosaurs from the Late Cretaceous of western Canada. *National Museum Natural Sciences Publications in Palaeontology* 1: 1–34.

Sereno, P. C. 1997. The origin and evolution of dinosaurs. *Annual Review of Earth and Planetary Sciences* 25: 435–489.

Sereno, P. C. 1998. A rationale for phylogenetic definitions with application to the higher-level taxonomy of Dinosauria. *Neues Jahrbuch für Geologie und Paläontologie Abhandlungen* 210: 41–83.

Sereno, P. C. 1999. The evolution of dinosaurs. *Science*: 2137–2147.

Swofford, D. L. 1999. PAUP*: Phylogenetic analysis using parsimony (and other methods), version 4.0 b2. Sinaeur Associates, Sunderland, Massachusetts.

Wagner, G. P., and J. A. Gauthier. 1999. 1,2,3 = 2,3,4: A solution to the problem of the homology of the digits in the avian hand. *Proceedings of the National Academy of Sciences USA* 96: 5111–5116.

List of Characters

Characters used for phylogenetic analysis of Tyrannosauridae. Multistate characters are followed by the letter O if ordered, and UO if unordered. Scoring: 0 = primitive state; 1, 2, or 3 = derived character states.

Synapomorphies of Tyrannosauridae

(1) Ventral ramus of the premaxilla: 0, as long or longer rostrocaudally than tall dorsoventrally; 1, taller dorsoventrally than long rostrocaudally.

(2) Premaxillary tooth row arcade: 0, more rostrocaudally than mediolaterally oriented; 1, more mediolaterally than rostrocaudally oriented.

(3) Nasals: 0, unfused; 1, fused.

(4) Squamosal-quadratojugal flange constricting infratemporal fenestra: 0, absent; 1, present.

(5) Prefrontals: 0, large; 1, reduced; 2, absent. O

(6) Well-formed sagittal crest on dorsal surface of parietals: 0, absent; 1, present.

(7) Lateral nuchal crest formed by parietals: 0, absent or small; 1, present (at least twice as tall as foramen magnum vertical height).

(8) Pair of tablike processes on supraoccipital wedge: 0, absent; 1, present.

(9) Basisphenoid sphenoidal sinus: 0, shallow, foramina small or absent; 1, deep, foramina large.

(10) Prominent muscular fossae on dorsal surface of palatines: 0, present; 1, absent.

(11) Rostral portion of the fused vomers: 0, small expansion (less than twice shaft width); 1, expanded to greater than twice shaft width to form a rhomboid (diamond) shape.

(12) Caudal surangular foramen: 0, small or absent; 1, very large.

(13) Retroarticular process of articular: 0, present; 1, absent.

(14) Premaxillary teeth cross section: 0, asymmetrical ovals with a rostral and a caudal carina; 1, D-shaped or U-shaped, with both carinae placed along the same plane perpendicular to the skull axis.

(15) Premaxillary tooth size: 0, subequal to lateral teeth; 1, much smaller than lateral teeth.

(16) Distal caudal neural spines: 0, axially short or absent; 1, axially elongate.

(17) Acromial expansion: 0, small, less than twice scapula midshaft width; 1, well developed, more than twice scapula midshaft width.

(18) Scapula contribution to glenoid: 0, half; 1, greater than half.

(19) Femur-humerus ratio: 0, less than 2.5; 1, between 2.8 and 3.5; 2, greater than 3.5. O

(20) Scapula-humerus ratio: 0, less than 2.1; 1, between 2.2 and 2.5; 2, greater than 2.5. O

(21) Distal carpals of adults; 0, well formed with transverse trochlea; 1, poorly formed and lack trochlear surfaces.

(22) Metacarpal III: 0, bears a digit; 1, very reduced and bears no digit.

(23) Ilium length: 0, clearly shorter than femur; 1, slightly shorter than femur; 2, longer than femur. O

(24) Horizontal medial shelf from preacetabular blade to sacral ribs: 0, absent; 1, present.

(25) Broad, ventral hooklike projection from preacetabular blade of ilium: 0, absent; 1, present.

(26) Notch on cranial surface of preacetabular blade of ilium: 0, absent; 1, present.

(27) Dorsal surfaces of iliac blades: 0, well separated in dorsal view; 1, converge closely along midline.

(28) Pronounced midline crest on ilium: 0, absent; 1, present.

(29) Supracetabular crest on ilium: 0, prominent; 1, reduced.

(30) Pubic boot length: 0, one-third or less pubis length (or femur length); 1, approximately one-half pubis length (or femur length); 2, approximately two-thirds or more pubis length (or femur length). O

(31) Pronounced semicircular scar on caudolateral surface of ischium, just ventral to the iliac process: 0, absent; 1, present.

(32) Shaft of ischium: 0, almost as long and thick as pubis; 1, long but more slender and shorter than pubis; 2, very short (66% or less pubis length). UO

(33) Ischium termination: 0, small expansion; 1, pointed tip.

(34) Lesser trochanter height: 0, less than dorsalmost point of femoral head; 1, as tall or taller than dorsalmost point of femoral head.

(35) Tibia proportions: 0, moderate (falls along main allometric trend of nonavian theropods); 1, elongate relative to comparable-sized theropods (other than ornithomimosaurs).

(36) Fibular cranial tubercle distal to cranial expansion: 0, absent or composed of single bulge; 1, composed of two longitudinal ridges.

(37) Arctometatarsus: 0, absent; 1, present.

(38) Metatarsal proportions: 0, moderate (falls along main allometric trend of nonavian theropods); 1, elongate and slender relative to comparable-sized theropods (other than ornithomimosaurs).

(39) Dorsal surface of metatarsal III: 0, oval or hourglass-shaped; 1, crescentic and limited to the caudal portion of the metatarsus dorsal surface.

Tyrannosaurid ingroup characters

(40) Lateral flange of maxilla obscuring cranialmost portion of maxillary antorbital fossa in lateral view: 0, absent; 1, present as small crest; 2, large shelf. O

(41) Size of maxillary fenestra: 0, small, less than one-half area of

eyeball-bearing portion of orbit; 1, expanded, approximately two-thirds area of eyeball-bearing portion of orbit.

(42) Maxillary fenestra cranial margin: 0, terminates caudal to cranial margin of antorbital fossa, promaxillary fenestra present; 1, terminates along cranial margin of antorbital fossa, promaxillary fenestra lost (or absorbed).

(43) Internal antorbital fenestra proportions: 0, longer than tall; 1, as tall or taller than long.

(44) Ventral curvature of maxilla: 0, absent; 1, present; 2, pronounced. O

(45) Maxillary tooth count in adults: 0, 18; 1, 13 or fewer.

(46) Nasal rugosity: 0, absent; 1, present.

(47) Nasal caudal width: 0, nearly as wide between lacrimal as rostral to lacrimals; 1, pinched between lacrimals, thinnest point approximately one-half mediolateral width of thickest point; 2, extremely pinched, thinnest point approximately one-sixth mediolateral width of thickest point. O

(48) Nasal caudal suture shape: 0, medial projection extends as far or further caudally than lateral projections; 1, lateral projections extend further caudally than medial projections.

(49) Dorsal surface of antorbital fossa: 0, restricted to maxilla and lacrimal; 1, contacts the nasal margin.

(50) Dorsal ramus of lacrimal: 0, slender; 1, inflated appearance.

(51) Triangular hornlet on lacrimal: 0, absent; 1, present.

(52) Lacrimal horn orientation: 0, absent; 1, dorsal; 2, rostrodorsal. UO

(53) Lacrimal horn position: 0, absent; 1, directly dorsal to descending ramus of lacrimal; 2, rostral to descending ramus of lacrimal. UO

(54) Angle of the descending and dorsal rami contact of lacrimal: 0, right angle; 1, strongly acute angle.

(55) Margin of external antorbital fenestra on craniolateral surface of descending ramus of lacrimal: 0, continued as clearly demarcated margin on lateral surface of jugal; 1, flattens out and is not continued on surface of jugal.

(56) Postorbital dorsal surface in adults: 0, smooth; 1, enlarged bump; 2, large rugose boss. UO

(57) Suborbital prong of postorbital: 0, absent or very small; 1, prominent.

(58) Contact between dorsal surface of lacrimal and postorbital in lateral view: 0, absent; 1, present, no intergrowth; 2, intergrowth of bone to form supraorbital torus. O

(59) Convex tablike prominence on dorsolateral surface of postorbital: 0, absent; 1, present.

(60) Ventral termination of postorbital descending ramus: 0, nearly as ventral as ventralmost margin of orbit, and clearly ventral to the squamosal-quadratojugal contact; 1, well dorsal to ventralmost margin of orbit, and approximately the same level as the squamosal-quadratojugal contact.

(61) Shape of frontals in adults: 0, triangular; 1, caudal end expanded laterally; 2, main body rectangular, only small triangular cranial prong remains. O

(62) Supratemporal fossa on frontal: 0, absent; 1, occupies laterocaudal third; 2, occupies laterocaudal half; 3, occupies most of caudal frontal, meet along midline to form frontal sagittal crest. O

(63) Nuchal crest mediolateral width: 0, less than twice height; 1, more than twice height.

(64) Nuchal crest rostrocaudal thickness: 0, relatively thin, dorsal margin smooth; 1, much thicker, dorsal margin rugose.

(65) Orientation of occipital region: 0, caudal; 1, caudoventral.

(66) Supraoccipital contribution to foramen magnum: 0, present, 1, absent (exoccipitals form complete dorsal margin).

(67) Basitubera size: 0, large (comparable to ventral ends of basipterygoid processes); 1, reduced.

(68) Ventral pocket to ectopterygoid chambers: 0, present; 1, greatly reduced.

(69) Ectopterygoid sinus: 0, moderate; 1, inflated.

(70) Palatine shape: 0, triradiate; 1, inflated trapezoid.

(71) Foramina on ventral surface of palatine: 0, one; 1, two or more.

(72) Vertical depth of caudal portion of dentary: 0, only half again as deep or less of vertical depth at symphysis; 1, twice or more as deep as depth at symphysis.

(73) Caudal termination of angular: 0, caudal or ventral to caudal surangular foramen; 1, rostral to caudal surangular foramen.

(74) Surangular shelf: 0, horizontal; 1, slightly pendant, overhangs dorsal margin of caudal surangular foramen.

(75) Dentary tooth count in adults: 0, 16 or more; 1, 15 or fewer.

(76) Premaxillary teeth serrations: 0, present; 1, absent.

(77) Vertical ridge on caudal surface of premaxillary teeth: 0, weakly developed; 1, strongly developed.

(78) Number of incisiform teeth in rostral end of maxilla: 0, none; 1, first maxillary tooth incisiform; 2, first two to three maxillary teeth incisiform.

(79) Lateral teeth: 0, ziphodont; 1, incrassate (cross-section greater than 60% wide mediolaterally as long craniodistally).

(80) Cervical centra: 0, longer than height of cranial face (neck length only slightly shorter than length of dorsal series); 1, less than half as long as height of vertical face (neck much shorter than dorsal series).

(81) Tallest cervical neural spines: 0, less than vertical diameter of centrum; 1, more than vertical diameter of centrum.

(82) Distal end of scapula: 0, not expanded; 1, greatly expanded cranially and caudally to more than twice midshaft width.

(83) Deltapectoral crest: 0, relatively large; 1, reduced.

(84) Metacarpal II–metacarpal I ratio: 0, 200%; 1, 170%; 2, 160% or shorter. O

(85) Phalanx 1 of manual digit I: 0, longer than metacarpal II; 1, subequal to metacarpal II; 2, shorter than metacarpal II. O

(86) Distal end of manual unguals: 0, tapers to point; 1, blunt.

Aublysodon molnari

(87) Rostral end of dentary tooth row: 0, along same curve as rest of tooth row; 1, placed on a dorsally raised step.

Kirtland Shale aublysodontine

(88) Medial margin of tibial facet for ascending processes of astragalus: 0, smooth curve; 1, small embayment.

Shanshanosaurus huoyanshanensis

(89) Cervical vertebrae centra: 0, amphicoelous or amphiplatyan; 1, slightly procoelous.

Alioramus remotus

(90) Nasal surface: 0, no individual hornlets; 1, double row of five vertical blades.

(91) Prootic rostral expansion: 0, absent; 1, present.

(92) Trigeminal foramen: 0, in laterosphenoid; 1, in prootic.

Gorgosaurus libratus

(93) Promaxillary fenestra: 0, rostral to maxillary fenestra (or absent); 1, rostrodorsal to maxillary fenestra in adults.

(94) Postorbital-lacrimal ventral contact: 0, do not contact below orbit; 1, contact below orbit in adults, orbit more circular (dorsoventral axis not twice or more rostrocaudal axis) than in other large theropods.

Albertosaurus sarcophagus

(95) Rostral margin of postorbital suborbital prong: 0, smooth (or prong absent); 1, jagged.

(96) Basisphenoid foramina in sphenoidal sinus: 0, lies within same surface; 1, each foramen lies within a distinct fossa.

Daspletosaurus torosus

(97) Premaxillary symphysis: 0, no intergrowth; 1, intergrowth.

(98) Premaxilla-nasal contact: 0, premaxilla contacts nasal ventral to external nares (although not always visible in lateral view); 1, premaxilla does not contact nasal ventral to external nares.

(99) Number of lacrimal apertures: 0, one; 1, two or more.

Tyrannosaurus bataar

(100) Ulnar-humeral ratio: 0, 60%; 1, 45%.

(101) Metacarpal III: 0, longer than metacarpal I; 1, shorter than metacarpal I.

(102) Postacetabular blade of ilium: 0, squared end; 1, tapered end.

Tyrannosaurus rex

(103) Mediolateral width of rostrum at caudal end of maxillary tooth row: 0, twice or less width of nasals; 1, approximately three times width of nasals.

(104) Maximum postorbital skull width: 0, less than one half premaxilla-occipital condyle length; 1, more than two-thirds premaxilla-occipital condyle length.

(105) Orbits: 0, directed mostly laterally; 1, directed more rostrally.

(106) Nasal processes of premaxilla: 0, slightly divergent at dorsal end; 1, tightly appressed throughout entirely length, terminate as single tip.

(107) Premaxillary dental arcade: 0, teeth closely appressed; 1, teeth less closely appressed (although teeth retain U-shaped cross-section).

(108) Maxillary palatal shelves: 0, contact vomer for length one-half or less length of tooth row; 1, contact vomer for length greater than three-quarters the length of tooth row.

(109) Jugal contribution to margin of internal antorbital fenestra: 0, extensive; 1, restricted between maxilla and lacrimal to very small surface.

(110) Shape of vomer cranial end: 0, lanceolate (lateral margins parallel sided); 1, diamond.

(111) Iliac antitrochanter just dorsal to iliac-ischial articulation: 0, absent; 1, present.

Data Matrix

Character states 0–3 as above. ? = uncertainty, numbers in paranetheses indicate more than one character state observed in that taxon.

Outgroup

00000 00000 00000 00000 00000 00000 00000 00000 00000 00000 00000 00000 00000
00000 00000 00000 00000 00000 00000 00000 00000 00000 0

Aublysodon molnari

??1?? ????? ??11 ????? ????? ????? ????? ????0 ???1? 0???? ????? ????? 01???
????? ????? 1100? ????? ?1??0 ??0?? ????? ??0?? ????? ?

Alectrosaurus olseni

1?11? ????? ?1111 ???1? ????? ????? ????1 11110 00010 0??0? ????? 00?0? ????0
????? ?0??0 1120? ????? 000?0 ??000 ????? ????? ???0? ?

Kirtland Shale aublysodontine

????? 1???? ??11 ????? ????? ????1 ???1? ?11?? ????? ????? ????? 1??1? 01???
????? ????? 11??? ????? ??1?? ???0 ????? ????? ????? ?

Alioramus remotus

??11? 1111? ?11?? ????? ????? ????? ????? ?11?? ??010 1??0 000?? 00100 12101
01??? ?0000 ???0? ????? ????1 11?00 0???? ??00 ????? ?

Gorgosaurus libratus

11111 11111 11111 11111 11111 11111 11111 11111 00011 10100 12200 10000 12000
10100 01001 00110 00101 00000 00110 00000 00000 00000 0

Albertosaurus sarcophagus

11111 11111 ?1111 11111 11111 11111 11111 11111 00011 10000 11200 11000 12001
11100 11001 00010 01000 00000 00001 10000 00000 00000 0

Daspletosaurus torosus

11111 11111 11111 11111 11211 11112 11111 11111 10121 11000 11101 10011 12000
10010 01011 00011 10100 00000 00000 01110 00000 0000? 0

Two Medicine tyrannosaurine

1111? 11??? ?1111 ????? ????? ????? ????? ????1 10111 1??00 11101 20001 ????0
????? ?1111 ??01? ????? ?0??0 ??000 ???0? ????? ?0?0? ?

Tyrannosaurus bataar

11111 11111 ?0111 11122 11211 11112 11111 11112 11021 11011 00011 21001 23111
11011 11111 00011 10120 10000 00000 00001 11000 0000? 0

Tyrannosaurus rex

11111 11111 11111 11111 11211 11112 11111 11112 11121 12011 00011 21(02)00
23111 11011 (01)1011 00011 11012 10000 00000 00000 ?0111 11111 1

Shanshanosaurus huoyanshanensis

1???? ????? ?1011 ?1?11 ????? ????1 ???11 ????0 ?0?11 ???0? ????? ????? ?????
????? ?0??1 ?200? ?01?? ?001? ????? ??0?? ??0?? ????? ?

"*Maleevosaurus novojilovi*"

1?1?? ????? ???11 11??? ??211 11112 11111 11110 00021 1??0 0001? 0010? ?????
????? ????1 ??010 1??1? ?000? ??000 ???0? ?1000 ????? 0

"*Nanotyrannus lancensis*"

11111 1111? ?1111 ????? ????? ????? ????? ????0 00020 1?110 00011 00000 13101
?1010 ?101? 0010? ????? ????0 ??000 0000? ??111 1011? ?

Siamotyrannus isanensis

```
????? ????? ????? ????? ???11 ?1100 10??? ????? ????? ????? ????? ????? ?????
????? ????? ????? ????? ????? ????? ????? ?0??? ????? 0
```

Ornithomimosauria

```
00000 00000 00010 00000 10000 11000 10011 01100 00000 00000 00000 00000 00001
00000 00000 00000 00000 00000 00000 00000 00000 00000 0
```

Maniraptora

```
00001 00000 00000 00000 00000 00000 02110 00000 00000 00000 00000 00000 00000
00000 00000 00000 00000 00000 00000 00000 00000 01000 0
```

APPENDIX 7.3.
Character States Supporting Nodes

Tyrannosauridae: 1.1, 3.1, 4.1, 6.1, 12.1, 13.1, 14.1 (also in Ornithomimosauria), 15.1, 19.1, 34.1 (also in Ornithomimosauria and Maniraptora, but not basal coelurosaurs), 35.1 (also in Ornithomimosauria), 36.1, 37.11 (also in Ornithomimosauria), 38.11 (also in Ornithomimosauria), 39.1, 44.1, 62.1. Because of the incompleteness of Aublysodontinae and *Alioramus,* the following might be synapomorphies for all Tyrannosauridae, or might be more restricted: 2.1, 5.1 (also in Maniraptora), 7.1, 8.1, 9.1, 10.1, 11.1, 16.1, 17.1, 18.1, 20.1, 21.1, 22.1, 23.1, 24.1, 25.1, 26.1 (also in Ornithomimosauria), 27.1 (also in Ornithomimosauria), 28.1, 29.1, 30.1, 31.1 (also in Ornithomimosauria), 32.1, 33.1 (also in Maniraptora).

Aublysodontinae: 76.1, 77.1.

Aublysodon molnari: 87.1.

Alectrosaurus olseni: 78.2.

Kirtland shale aublysodontine: 59.2 (also in *Daspletosaurus*), 88.1.

Tyrannosaurinae: 46.1, 61.1, 62.2.

Alioramus remotus: 58.1 (also in juvenile *T. bataar*), 64.1 (also in *Tyrannosaurus*), 90.1, 91.1, 92.1.

Albertosaurus, Gorgosaurus, Daspletosaurus, Two Medicine tyrannosaurine, and *Tyrannosaurus:* 45.1, 56.1, 66.1, 72.1, 75.1, 79.1.

Albertosaurus sarcophagus: 52.1 (also in *Daspletosaurus* and the Two Medicine tyrannosaurine), 53.1 (also in *Gorgosaurus*), 57.1 (also in *Tyrannosaurus*), 65.1 (also in *Tyrannosaurus*), 67.1 (also in *Tyrannosaurus*), 68.1 (also in *Gorgosaurus*), 71.1 (also in *Tyrannosaurus*), 82.1 (also in *T. rex*), -83.0 (also in *T. rex*), 95.1, 96.1.

Gorgosaurus libratus: 48.1 (also in juvenile *T. rex*), 53.1 (also in *Albertosaurus*), 68.1 (also in *Albertosaurus*), 78.1 (also in juvenile *T. rex*), 85.1, 93.1, 94.1.

Daspletosaurus, Two Medicine tyrannosaurine, and *Tyrannosaurus:* 23.2, 30.2, 41.1, 47.1, 55.1, 69.1, 74.1, 80.1, 81.1.

Daspletosaurus torosus: 44.2 (also in *Tyrannosaurus*), 53.1 (also in Two Medicine tyrannosaurine), 59.1 (also in Kirtland Shale aublysodontine), 97.1, 98.1, 99.1.

Two Medicine tyrannosaurine: 53.1 (also in *Daspletosaurus*), 56.2 (also in *Tyrannosaurus*), 73.1 (also in *T. bataar*).

Tyrannosaurus: 40.2, 42.1, 44.2, 49.1, 50.1, 54.1, 56.2 (also in Two Medicine tyrannosaurine), 57.1 (also in *Albertosaurus*), 62.3, 63.1, 64.1 (also in *Alioramus*), 65.1 (also in *Albertosaurus*), 67.1 (also in *Albertosaurus*), 70.1, 71.1 (also in *Albertosaurus*), 84.1, 86.1.

Tyrannosaurus bataar: -12.0, 19.2, 20.2, 73.1 (also in Two Medicine tyrannosaurine), 84.2, 100.1, 101.1, 102.1.

Tyrannosaurus rex: 47.2, 82.1 (also in *Albertosaurus*), -83.0 (also in *Albertosaurus*), 85.2, 103.1, 104.1, 105.1, 106.1, 107.1, 108.1, 109.1, 110.1, 111.1.

Shanshanosaurus huoyanshanensis: -13.0, 89.1.

8. A Kerf-and-Drill Model of Tyrannosaur Tooth Serrations

WILLIAM L. ABLER

Abstract

Neighboring serrations on the teeth of the theropod dinosaur *Albertosaurus* meet at an angle so small that their junction amounts to a crack in the tooth. A round void (ampulla) at the base of the junction distributes force over an increased area, preventing the crack from propagating through the tooth and breaking it. Serrations on teeth of other ancient reptiles (phytosaur, *Dimetrodon*) exhibit the junction "crack," but lack the protective ampulla.

Introduction

The biological function of serrations in the tyrannosaur tooth is well understood (Abler 1999). The posterior edge of the tooth is lined, from base to tip, with serrations that contact their neighbors but are not attached to them. Because the exposed edge of the serrations forms a curved ridge, neighboring serrations meet at an angle that approaches zero degrees (fig. 8.1a). Each pair of neighboring serrations, then, essentially forms a crack in the edge of the tooth. As currently understood, the teeth of tyrannosaurs functioned as pegs for gripping food, rather than as knives for cutting it (Abler 1992, 1997, 1999). Under the peg model, a tyrannosaur would have reached forward with its head, gripped a section of a carcass in its jaws, using the teeth as holdfasts, and detached a chunk of flesh by pulling (and shaking) its head. Evidence for shaking is taken from multidirectional surface scratches on the teeth, suggesting that head movements during feeding may have been complex (Abler 1992, 176, 179).

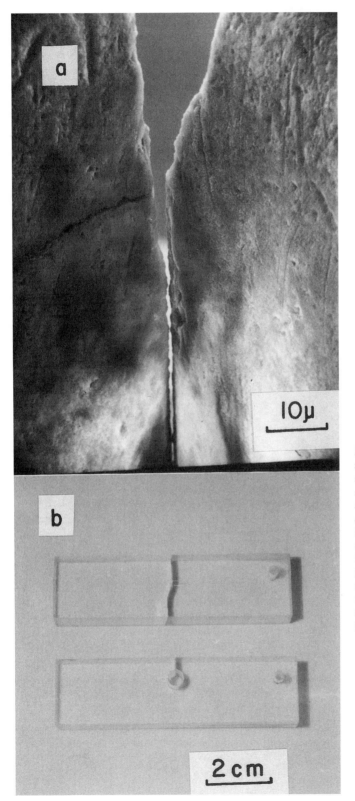

Figure 8.1. The kerf-and-drill mechanism. (a) Junction between neighboring serrations in tooth of Albertosaurus. The two serrations meet at an angle that approaches zero degrees. (b) Demonstration of the kerf-and-drill mechanism. Above: Plexiglas bar with simple kerf, broken by a force applied at the tip. Below: identical plastic bar with kerf-and-drill. This bar was not broken by a significantly larger force. See text for explanation.

*Figure 8.2. (facing page)
The kerf-and-drill mechanism in
nature and technology. (a) At left,
kerfed wooden stick of the kind
used by guitar makers for
attaching top and back of a
guitar to the sides. Saw cuts
(kerfs) allow the stick to bend
without breaking. At right,
similar wooden stick showing
drill holes at the end of each kerf.
Force applied in the direction of
the arrow would tend to open up
the kerfed slots. (b) Close-up
view of wooden sticks showing
simple kerf (at left) and kerf-and-
drill (at right). (c) Drill hole used
for stopping the progress of a
crack through an airplane surface
(Federal Aviation Agency 1971,
p. 213). (d) Unique photograph
of drill-hole, made with a copper
tube, stopping the progress of a
crack in the 40-inch telescope
lens, Yerkes Observatory,
Williams Bay, Wisconsin. (e)
Neighboring serrations on
Albertosaurus tooth. Horizontal
line at center of photograph is
the narrow gap ("kerf") where
two neighboring serrations rest
against each other. The dark
mass at the left end of the kerf is
the finely disrupted void ("drill
hole") inside the tooth. The
serration row presents a
succession of such structures
similar to the kerfed stick seen
at right in (a). See text for
explanation.*

The pulling action would have caused the teeth to act as beams under stress (Farlow et al. 1991), with the potential of breaking. The inevitable pulling or tugging action, which would have been part of the tyrannosaur's feeding process, would have acted to make the tooth tip rotate toward the front of the mouth. Because the tooth is anchored at its base, such an action would have introduced compression in the anterior edge of the tooth, and tension in the posterior edge. Cracks would thus have appeared in the posterior edge. Teeth of the theropod dinosaur *Albertosaurus* apparently were protected from cracking by a subtle adaptation in the structure of their serrations. At the base of each crack is a space where the material of the tooth is so finely disrupted as to form a void in the tooth structure (fig. 8.2e).

Kerf-and-Drill Model

In a hard brittle material, such as a tyrannosaur tooth, breaking begins with the formation of a crack. The force applied to the tooth would then have been concentrated at the leading edge of the crack. Because (in principle) the area of the leading edge approaches zero, the local force there approaches infinity (Federal Aviation Agency 1971, 213). Thus, once started, a crack in a tooth would easily propagate to the other side causing the tooth to break in two. For tyrannosaur teeth, then, the best protection against breaking is to prevent a crack from starting in the first place.

The best protection from breaking is by a mechanism similar to, but more sophisticated than, one used by guitar makers to impart alternating regions of flexibility and rigidity to a stick of wood. The division of the posterior edge of the tooth into a stack of approximately cubic regions is mechanically identical to the kerfing (sawing) of a wooden stick by a guitar maker to create a stick that has alternating regions of rigidity ("webs") and flexibility. The kerfs (see "carve," Oxford English Dictionary) are saw cuts into the stick (fig. 8.2a,b). The flexible regions allow the stick to be curved to fit the curves of the guitar, while the rigid regions form a stable base for attaching the top and back to the sides of the guitar (Doubtfire 1981; Hromek 1984; Cumpiano and Natelson 1987).

Once in place, the kerfed stick is no longer subjected to force. But the tyrannosaur tooth was necessarily subjected to force that would tend, as a result of rotation of the tooth tip toward the front of the mouth, to open up the narrow gaps between neighboring serrations. By distributing the applied force over a relatively large area, the voids at the base of the gaps would protect the tooth against cracking. While guitar makers do not need to use drill holes to protect a stationary kerfed stick, a drill hole is the standard method for stopping the propagation of cracks in airplane surfaces (fig. 8.2c; Federal Aviation Agency 1971, 213) and telescope lenses (fig. 8.2d).

To demonstrate the kerf-and-drill principle, I cut a slot at the middle of two Plexiglas bars approximately 82 mm x 22 mm x 12 mm (Fig. 8.1b). Both slots were approximately 9.5 mm deep, but one terminated in a drill hole 5 mm in diameter, while the other was a simple

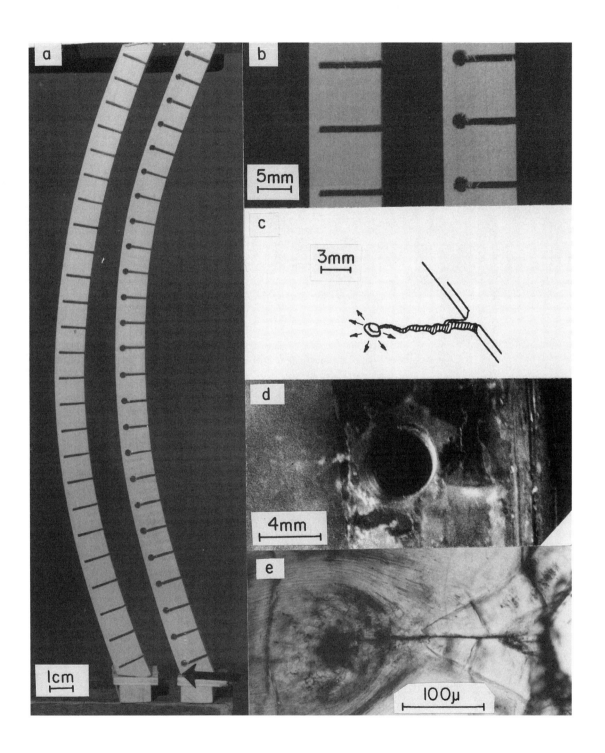

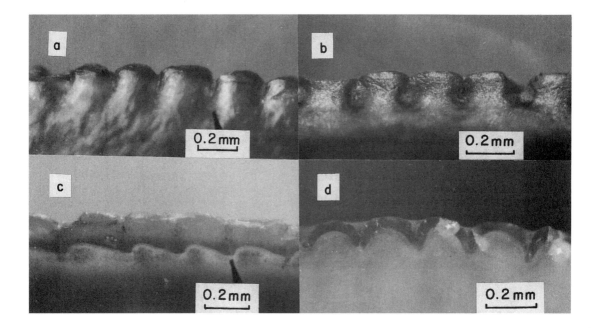

Figure 8.3. Serrations on teeth of ancient reptiles. Exterior (a,b); interior (c,d). Phytosaur (a,c), and Dimetrodon *(b,d). The exterior surface of the serrations shows the same kind of high-pressure slot described in* Albertosaurus *(Abler 1992), but the interior lacks the ampulla. Pointer in (a) and (c) indicates junctions between neighboring serrations.*

slot approximately 0.2 mm wide. I then supported each plastic bar cantilever-fashion, and applied force at a hole drilled in the free end, opening up the slot at the middle. The bar with the simple slot broke when a force of 227 newtons was applied, while the bar with the drilled slot did not break when a force 25% greater (283 newtons) was applied.

From this experiment, I conclude that a kerf-and-drill structure would protect the posterior edge of the *Albertosaurus* tooth against the propagation of cracks formed by the junction between neighboring serrations. However, other ancient reptile teeth, which have serrations of approximately the same surface shape as *Albertosaurus* serrations, may lack the interior drill feature. For example, phytosaur teeth and *Dimetrodon* teeth (sp. indet.) lacked the drill feature (fig. 8.3). The kerf-and-drill feature therefore cannot be assumed for all teeth with serrations, but must be considered on a species-by-species basis.

Acknowledgments: I thank Dr. Philip J. Currie for granting me permission to publish on data collected while working for the Tyrrell Museum of Palaeontology (1982), without which none of my dinosaur studies would have been possible. Figure 8.2d, photograph by Richard Dreiser, University of Chicago Yerkes Observatory. Figure 8.1a, SEM photomicrograph by William F. Simpson, Department of Geology, Field Museum of Natural History (Chicago); reprinted with permission of the Paleontological Society.

References

Abler, W. L. 1992. The serrated teeth of tyrannosaurid dinosaurs, and biting structures in other animals. *Paleobiology* 18 (2): 161–183.

Abler, W. L. 1997. Tooth serrations in carnivorous dinosaurs. In P. J. Currie and K. Padian (eds.), *Encyclopedia of Dinosaurs,* pp. 740–743. San Diego: Academic Press.

Abler, W. L. 1999. The teeth of the tyrannosaurs. *Scientific American* 281 (3): 50–51.

Cumpiano, W. R., and J. D. Natelson. 1987. *Guitar Making: Tradition and Technology.* Amherst: Rosewood Press.

Doubtfire, S. 1981. *Make Your Own Classical Guitar.* New York: Schocken Books.

Federal Aviation Agency. 1971. *Acceptable Methods, Techniques, and Practices—Aircraft Inspection and Repair.* FAA AC no. 43.13–1. Washington, D.C.: Government Printing Office.

Farlow, J. O., D. L. Brinkman, W. L. Abler, and P. J. Currie. 1991. Size, shape, and serration density of theropod dinosaur lateral teeth. *Modern Geology* 16:161–198.

Hromek, P. 1984. *Build Your Own Guitar.* Cologne: Amsco.

9. Forelimb Osteology and Biomechanics of *Tyrannosaurus rex*

Kenneth Carpenter and Matt Smith

Abstract

Although proportionately the forelimb is very small, the mechanical advantage reveals an efficiently designed force-based system (vs. a velocity-based system) used for securing its prey during predation. In addition, the M. biceps is shown to be 3.5 times more powerful than the same muscle in the human, the straight, columnar humerus provides maximum strength to mass ratio to counter the exertion of the M. biceps, and the thick cortical bone indicates bone selected for ultimate strength. Such mechanical adaptations can only indicate that the arms were not useless appendages, but were used to hold struggling prey while the teeth dispatched the animal. *Tyrannosaurus rex* was therefore an active predator and not a mere scavenger, as has been suggested.

Introduction

"[W]hat, if anything, did *Tyrannosaurus* do with its puny front legs anyway?" is a question that has been asked by many individuals (Gould 1980, 12). The hypotheses range from an aid during mating (Osborn 1906), assisting the animal to get up from a prone position (Newman 1970), a vestigial organ (Paul 1988), or an organ whose function decreased with maturity of the animal (Mattison and Griffin 1989).

The discovery in recent years of several specimens of *Tyrannosaurus rex* preserves the first arms of this animal, allowing a testing of the various hypotheses by biomechanical analysis. Because the entire fore-limb of *Tyrannosaurus* has not been described before, a description is presented before the biomechanical analysis.

TABLE 9.1
Measurements for the Forearm of *Tyrannosaurus rex* (cm)

	MOR 555	FMNH PR 2081
humerus length	37.7	37.3
humerus distal width	8.4	8.9
ulna length	>19.5	21.9
ulna prox. anteropost. length	6.7	7.1
ulna width prox.	6.4	4.4
radius length	>15.1	17.3
radius prox. anteropost. length	4.2	5.2
radius prox. width	2.9	3.7
radiale height	1.6	
radiale anteropost. length	2.1	
radiale width	~1.8	
ulnare height	~1.5	
ulnare anteropost. length	2.1	
ulnare width	1.5	
metacarpal I length	-	6.4
metacarpal I prox. anteropost. length	3.0	
metacarpal I prox. width		1.7
metacarpal I distal width		3.0
metacarpal II length	9.4	10.9
metacarpal II prox. anteropost. length	-	4.1
metacarpal II prox. width	-	4.9
metacarpal II distal width	-	3.9
phalanx I-1 length	>8.4	
phalanx I-2 (ungual) length		>7.5
phalanx I-2 (ungual) prox. length		3.9
phalanx I-2 (ungual) prox. width		2.5
phalanx II-1 length	5.7	5.5
phalanx II-1 prox. anteropost length	3.5	4.3
phalanx II-1 prox. width	4.0	3.9
phalanx II-1 distal width	3.7	3.5
phalanx II-2 length	7.8	7.9
phalanx II-2 prox. anteropost length	~3.4	3.4
phalanx II-2 prox. width	3.2	3.4
phalanx II-2 distal width	2.5	2.9
phalanx II-3		-

Materials: Parts of specimens used in this study include: CMNH 9380 (holotype, formerly AMNH 973) scapula and humerus; DMNH 30665 (casts of FMNH PR 2081, formerly BHI 2033) right humerus, ulna, radius, metacarpals I and II, phalanges II-1 and 2, unguals I-2 and II-3; MOR 555 parts or most of the left scapula, coracoid, humerus, ulna, radius, metacarpals I and II, carpals, and phalanges I-1, II-1, and 2. Measurements are given in table 9.1. on page 91.

Institutional Abbreviations:
AMNH, American Museum of Natural History;
BHI, Black Hills Institute of Geological Research;
CMNH, Carnegie Museum of Natural History;
DMNH, Denver Museum of Natural History;
MOR, Museum of the Rockies.

Anatomical Abbreviations:

Lc	Lig. collaterale
Liru	Lig. interosseum radioulnare
Mb	M. brachialis
Mbi	M. biceps
Mbit	M. "biceps" tubercle
Mcd	M. coracobrachialis dorsalis (two heads)
Mcv	M. Coracobrachialis ventralis
Mdm	M. deltoideus minor
Mdmj	M. deltoideus major
Mdms	M. deltoideus major superficialis
Mdmv	M. deltoideus major ventralis
Mecr	M. extensor carpi radialis
Mecu	M. extensor carpi ulnaris
Meld-I	M. extensor longus digiti I
Meld-II	M. extensor longus digiti II pars distalis
Memu	M. extensor metacarpi ulnaris
Menu	M. entepicondylo-ulnaris
Meu	M. ectepicondylo-ulnaris
Mfcr	M. flexor carpi radialis
Mfcu	M. flexor carpi ulnari
Mfd-I	M. flexor digiti I
Mflu	M. flexor carpi ulnaris
Mfld-II	M. flexor longes digiti II
Mi	M. interosseus
Mld	M. latissimus dorsi
Mpp	M. pectoralis profundus
Mps	M. pectoralis superficialis
Mpsp	M. pronator superficialis et profundus
Ms	M. spinatus
Msh	M. scapulohumeralis
Mss	M. subscapularis
Msu	M. supinator
Msvc	M. serratus ventralis caudalis
Msvcr	M. serratus ventralis cranialis
Mt	M. triceps

Mth	M. triceps humeralis
Mts	M. triceps scapularis
Mumd	M. ulnometacarpi dorsalis
Mumv	M. ulnometacarpi ventralis
nas	nonarticular surface
Ref	Retinaculum m. extensor et flexor metacarpi ulnaris

Osteology and Myology

The osteology and myology of the forelimb are intimately connected and are treated together. Most myological studies of dinosaurs have been done with the hind limbs (e.g., Romer 1923; Galton 1969). Only a few studies have included the forelimbs. These include reconstructions of *Anatosaurus* (= *Edmontosaurus*) (Lull and Wright 1942), *Apatosaurus* (Filla and Redman 1994), *Chasmosaurus* (Russell 1935), *Euoplocephalus* (Coombs 1978), *Iguanodon* (Norman 1986), *Plateosaurus* (Huene 1907–1908), *Syntarsus* (Raath 1977), *Deinonychus* (Ostrom 1974), and numerous dinosaurs by Paul (1986). Muscle terminology used below is that of Berge (1979) for birds because of the close phyletic relationship between theropods and birds (e.g., Gauthier 1986; Holtz 1994). Use of avian muscle terminology is a departure from previous forelimb studies, which rely on crocodilian muscle patterns.

Furcula. A fused furcula is widespread among theropods (Chure and Madsen 1996; Makovicky and Currie 1998), including *Tyrannosaurus bataar* (Sabath pers. comm.), so is expected for *T. rex*. The furcula is similar in most theropods, being a widely open U- or V-shaped structure. Where found articulated, it connects the anterodorsal edge of

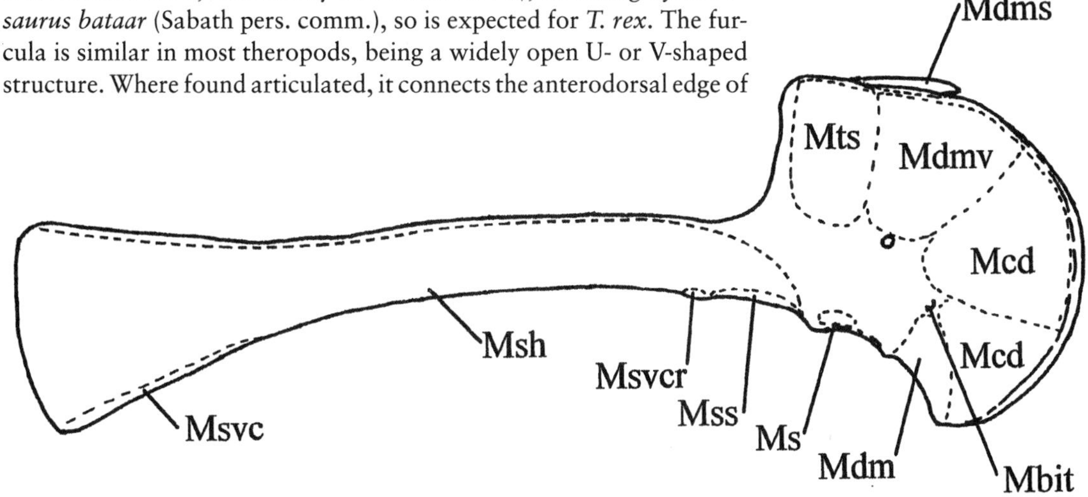

Figure 9.1. Scapula-coracoid and furcula for Tyrannosaurus rex showing muscle origin sites. The furcula is not yet known for T. rex, but is known for T. bataar. See list of anatomical abbreviations for muscle names.

the acromion plate scapulae. As in birds, one branch of the M. deltoideus major probably originated there (fig. 9.1).

Scapula (fig. 9.2). The scapula closely resembles that of the tyrannosaurids *Gorgosaurus* and *Albertosaurus*, although there are important differences. The acromial process of the scapula is proportionally taller and the posterior edge more round than in *Gorgosaurus* and *Albertosaurus*. The process is concave above the glenoid for the M. triceps scapularis, although it is not certain how much of the area the muscle occupied. The bone is thin and easily damaged, and is best preserved in CMNH 9380 and FMNH PR 2081. The bone thickens along the margins, especially along the posterior margin and along the contact with the coracoid. This thickening along the coracoid margin is mostly developed on the medial side, immediately above the glenoid. The anterodorsalmost part of the acromial process overhangs the coracoid (best seen in CMNH 9380 Osborn 1906, fig. 6). The sutural surface of the acromion, where it articulates with the coracoid, is very rugose in individuals where the two bones are not coossified. The glenoid is discussed below.

In lateral view, the scapular blade is long and straplike, a condition typical of that of many large theropods (e.g., *Allosaurus* and *Albertosaurus*). The blade is bowed dorsally, although the degree of this bow is variably developed; it is more prominent in CMNH 9380, than in MOR 555. Functionally, the bow countered the stresses of the M. scapulohumeralis that occupies most of the lower half of the blade. Distally, the scapular blade flares, especially along its dorsal margin, for the caudal portion of the M. trapezius. This flair is even more extensively developed in *Albertosaurus* (see Parks 1928, fig. 1). The lateral surface of the flair is flat to slightly convex in *T. rex*. A keel is developed on the ventral margin of the flair for the M. serratus caudalis. In cross-section, the blade is oval to lanceolate (fig. 9.2E,F). The scapula is lateromedially thickest above the glenoid, where it is expanded into a narrow shelf on the medial side (figs. 9.2C, 9.3A). This shelf expands posteriorly onto the blade, where it forms a broad, roughly triangular surface for the M. subscapularis.

Coracoid (fig. 9.2). A complete coracoid is known for FMNH PR 2081 and a partial one for MOR 555. The coracoid is a large, oval, concave plate of bone that is pierced by the coracoid foramen. A large, deep chamber is present in the floor of the foramen (MOR 555). The lateral surface is demarcated into four areas, three of these separated by broad, low ridges, and the four by a prominent one. These areas probably indicate areas for the origin of the M. deltoideus major and minor and the two branches of the M. coracoideus dorsalis (fig. 9.1). Also on the lateral surface, dorsal and anterior to the glenoid, is a large distinct tubercle called by Ostrom (1974) the biceps tubercle (Walker [1990] denies that this is the same feature named by him in *Sphenosuchus*). One head of the biceps may have originated there, although usually this muscle has a more extensive origin (possibly the area for M. deltoideus major ventralis (fig. 9.1) is actually for one head of the M. biceps). The M. coracobrachialis ventralis occupies the lower portion of the coracoid medially. The anteroventral margin of the coracoid is thickened into a distinct lip laterally and medially, although it is best developed on the medial side.

The glenoid is developed on the scapula-coracoid suture (fig. 9.2). The coracoid portion of the glenoid is considerably larger in *Tyrannosaurus* than it is in *Gorgosaurus*. In sagittal section, the glenoid is longer and the anterior rim much less developed than in *Gorgosaurus* (fig. 9.3B). In cross-section, the glenoid is broader, slightly concave, and has a better developed medial wall than *Gorgosaurus*. The scapular portion of the glenoid is triangular in *Tyrannosaurus*, although consid-

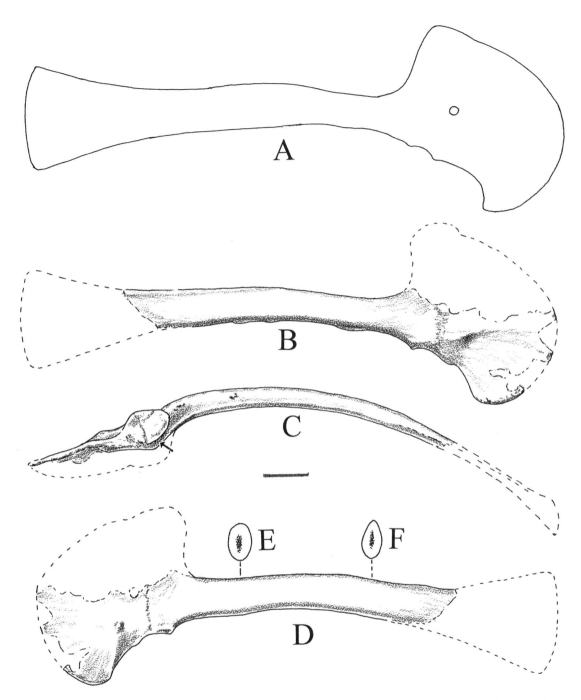

A

B

C

E   F

D

erably longer than wide. A thin wall occurs along the medial rim marking the edge of the cartilage cap.

The entire glenoid is parallelogram-shaped in *Tyrannosaurus* and subrectangular in *Albertosaurus*. This difference is due to the greater width of the scapular and coracoid across the glenoid in *Tyrannosaurus*. In addition, there is a notch on the lateral margin of the glenoid, along the scapula-coracoid suture in *Gorgosaurus*, but not *Tyranno-*

Figure 9.2. Scapula-coracoid of Tyrannosaurus rex: FMNH PR 2081 in (A) medial view (from a photograph); MOR 555 in (B) medial, (C) ventral (arrow denotes medial "shelf"), and (D) lateral views; (E) and (F) cross-sections. Scale: 10 cm.

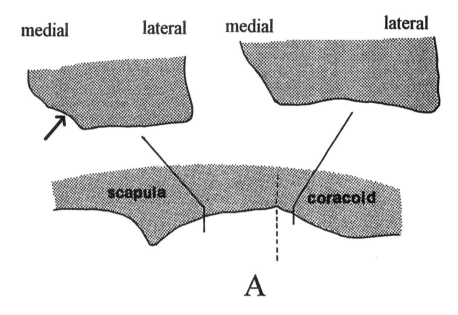

medial   lateral   medial   lateral

scapula   coracoid

A

medial   lateral   medial   lateral

scapula   coracoid

B

Figure 9.3. Comparative sections of the glenoid in T. rex *(A) and* Gorgosaurus *(B). Lower sections in (A) and (B) are sagittal sections comparing glenoid depth. Upper two sections in (A) and (B) are cross-sections. Note medial shelf (arrow) in (A) (upper left).*

saurus. In sagittal section, the glenoid of *Tyrannosaurus* is a shallow depression, rather than a deep depression as in *Gorgosaurus* (fig. 9.3A vs. 9.3B). In cross-section, the glenoid is almost flat in *Tyrannosaurus* and concave in *Gorgosaurus.*

Humerus. Two different morphs of humeri are known (figs. 9.4, 9.5) that seem to indicate sexual dimorphism, as also noted by Larson (1994). Humeri CMNH 9380 and FMNH PR 2081 have a pronounced medial curve proximally and are associated with robust skeletons (femur diameter/length = 0.41). The humerus of MOR 555 is almost a straight cylinder and is associated with a gracile skeleton (femur diameter/length = 0.37). The same two humeri morphs are also seen in *Tyrannosaurus bataar* (see Maleev 1974, figs. 33, 40B). The robust skeletons have been interpreted as females and the gracile skeletons as

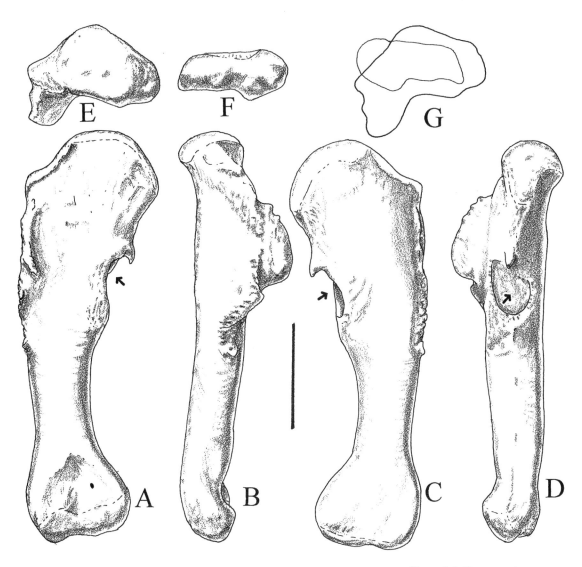

Figure 9.4. Humerus of Tyrannosaurus rex (FMNH PR 2081) in (A) anterior, (B) lateral, (C) posterior, (D) medial, (E) proximal, and (F) distal views. Note paleopathology (arrow) at the site of the medial head of the M. triceps humeralis. (G) compares proximal end (heavy line) with distal end (light line). Scale: 10 cm.

male (Carpenter 1990), a conclusion similar to that reached separately by Raath (1990) for *Syntarsus*. Cortical bone occupies 75% of the humeral diameter at midshaft.

The articulating surface of the humeral head faces dorsally and slightly posteriorly so that the head overhangs the shaft (fig. 9.4B,D). The margins of the articular surface, originally capped by the articular cartilage, are marked by a rim and tiny vertical rugosities. These features delineate the articular capsule and the position of the gleno-humeral and coracohumeral ligaments. In addition, scars around the head indicate the insertion points (fig. 9.6) for the M. subscapularis, M. scapulohumeralis, M. spinatus, M. deltoideus minor, and M. coraco-brachialis ventralis, and origins for the M. biceps and M. triceps humeralis (posterior head). On the humeral shaft, the deltopectoral crest is well developed and in lateral profile forms a broad-based triangle. The crest extends from just below the head to about one-third

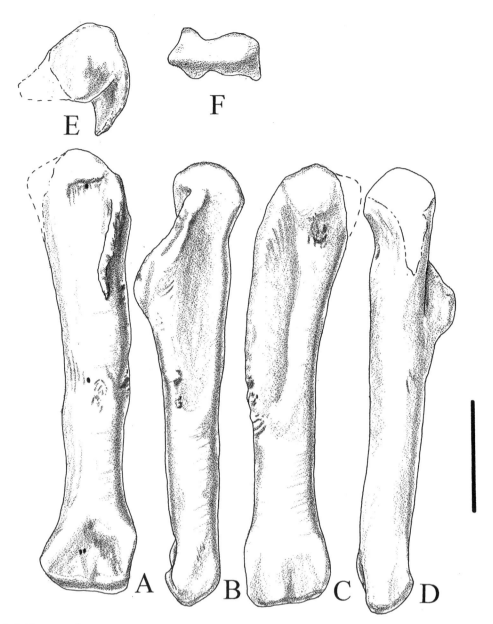

Figure 9.5. Humerus of Tyrannosaurus rex (MOR 555) in (A) anterior, (B) lateral, (C) posterior, (D) medial, (E) proximal, and (F) distal views. Scale: 10 cm.

of the humerus. Two heads of the M. pectoralis insert on the anterior and posterior sides of the crest. Elsewhere on the shaft are the insertions for the M. coracoideus dorsalis (combined heads), M. deltoideus major and M. latissimus dorsi, and the origins for M. triceps humeralis (medial head) and M. brachialis.

Distally, the radial condyle is larger than the ulnar condyle and both project anteroventrally. The articular surface of the epicondyles extend onto the anteroventral surface, however, much less so than in other theropods, such as *Allosaurus*. The small size of the epicondylar surfaces limits the amount of flexion and extension that could occur at the elbow (for further discussion of this implication see below). Proxi-

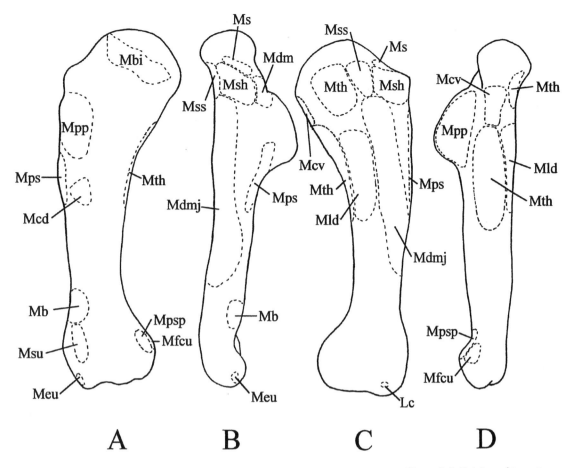

Figure 9.6. Origin and insertion sites on the humerus of Tyrannosaurus rex: (A) anterior, (B) lateral, (C) posterior, and (D) medial. See list of anatomical abbreviations for muscle and ligament names.

mal to the epicondyles, the radial fossa and ulnar fossa are partially separated by a ridge of bone; these fossa housed the synovial capsules of the elbow; on the posterior side, the olecranon fossa is practically nonexistent. The ulnar fossa is pierced by a single nutrient foramina in FMNH PR 2081 and a pair in MOR 555; this difference probably reflects individual variation, not sexual dimorphism. Sexual dimorphism, however, is reflected in the greater width across the epicondyles, with FMNH PR 2081 being about 12% wider, reflecting her greater robustness. The M. supinator, M. ectepicondylo-ulnaris, M. pronator superficilias et profundus, and M. flexor carpi ulnari originate around the distal end of the humerus. The origins for all extensors and flexors could not be found because some of them originate on the fascia covering the elbow.

There is a peculiar pathology on the medial side of the humerus of FMNH PR 2081. This pathology consists of an arclike, overhanging rim of exostosis and a spur located dorsally (fig. 9.4C,D). The surface of the humerus in visible within the pathology and is only slightly modified from its normal bone surface appearance. What little difference there is, consists of a small patch of irregular, hyperostosic bone beneath the overhang. The pathology occurs at the origin of the medial head of the M. triceps humeralis suggesting that the exostosis is in response to partial avulsion of the muscle due to abnormally high stress at the muscle-bone interface. The implications for this are discussed further below.

Ulna. The ulna is a stocky, thick-walled bone (fig. 9.6). In MOR 555 the cortical bone occupies 90% of the cross-sectional area of the ulnar shaft as measured 6 cm below the olecranon notch (comparable measurements are not available for FMNH PR 2081). The olecranon is very prominent, although it is more developed in FMNH PR 2081 than MOR 555. A very large scar is present on the posterior side for the M. triceps and for the fascia sheet (retinaculum) for several extensors and flexors (e.g., M. extensor metacarpi ulnaris) (fig. 9.8C,D). The olecranon notch is subtriangular in shape viewed dorsally and most of it slopes laterally towards the radial facet (fig. 9.7E). The notch is considerably wider in FMNH PR 2081 than in MOR 555, reflecting the difference in distal humeral width (fig. 9.7A vs. G). The radial notch is not very distinct (fig. 9.7E). Its posterior border, which corresponds to the Tuberculum ligamentum collateralis ventralis of birds (Baumel 1979), slopes laterally and does not contribute to the articular surface for the humeral ectocondyle, but lies beneath the intercondylar sulcus between the epi- and ectocondyles of the humerus. Instead, the tubercle supports a collateral ligament that extends to the posterolateral corner of the

Figure 9.7. Ulna of Tyrannosaurus rex: FMNH PR 2081 in (A) anterior, (B) lateral, (C) posterior, (D) medial, (E) proximal, and (F) distal views; MOR 555 in (G) anterior, (H) lateral, and (I) medial views. Radius of Tyrannosaurus rex: FMNH PR 2081 in (J) anterior, (K) lateral, (L) posterior, (M) medial, (N) proximal, and (O) distal views; MOR 555 in (P) anterior, (Q) lateral, and (R) medial views. Scale: 10 cm.

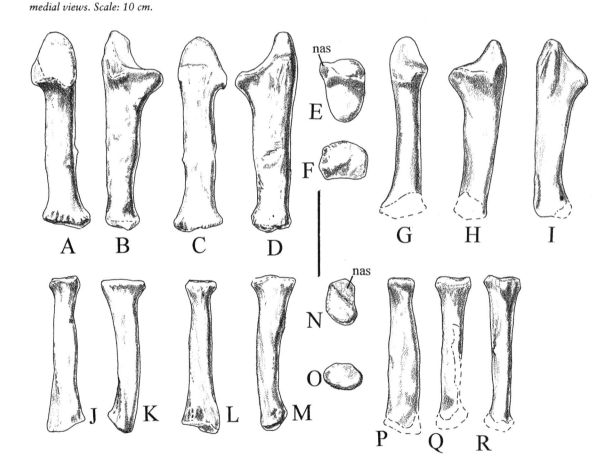

ectocondyle. Furthermore, the lateral overhang of this tubercle provides a gliding surface for the tendon of the M. flexor carpi ulnaris, as in birds. A shallow vertical groove on the lateral surface of the olecranon was occupied by the ulnar nerve. A prominent tubercle beneath the radial notch was probably for the insertion of the biceps. Insertion of the biceps on both the ulna and radius occurs by a divided tendon in birds (Chamberlain 1943). The ulnar shaft is compressed laterally and is longer anteroposteriorly to counter the stresses of the M. biceps (figs. 9.7, 9.8). Several extensors and flexors of the manus originate on the shaft (fig. 9.8). Distally, the ulna is wider than long with a low ridge extending obliquely across part of the articular face to a styloid process at the anteromedial corner for the ulnar collateral ligament (fig. 9.7F). The ulnar shaft near the humeral and carpal articular surfaces is rugose for the various ligaments binding the bones together. The ulna of FMNH PR 2081 is more robust than that of MOR 555.

Radius. The radius is a slender bone, subtriangular in cross-section (fig. 9.7). The surface of the radial head is D-shaped in dorsal view, with the straighter edge abutting against the ulna (fig. 9.7N). This flat surface prevents rotation of the radius, so that the human type of supination and pronation of the manus cannot occur. The result is that the manus faces medially, not ventrally, as is usually shown in life restorations by artists. A low, oblique ridge is present across the proximal surface of the radial head separating the articular surface for the ectocondyle of the humerus anterolaterally, from the nonarticular surface posteromedially. The small, triangular nonarticular surface lies opposite the Tuberculum ligamentum collateralis and also underlies the intercondylar sulcus between the distal condyles of the humerus. The shaft surface is very rugose just below the articular surface for the annular ligament that binds the bone to the ulna. The cortical bone occupies 82% of the cross-sectional area as measured 5 cm below the humeral articular surface. The radial shaft has a pronate twist along its length. On the medial surface, just below the radial head, is a small rugose area for the insertion of the M. biceps (fig. 9.8). This insertion is opposite that on the ulna for the M. biceps. The shaft of the radius has origins for several flexors and extensors for the manus (fig. 9.8). The distal end is wider than long (fig. 9.7O) and somewhat oval. A styloid process is developed at the anterolateral corner for the radial collateral ligament. The radius of FMNH PR 2081 is more robust than that of MOR 555.

Carpals. Two coossified carpals were found with MOR 555 (fig. 9.9A–C), but this may not represent the total number originally present (see Holtz 1994 for further discussion). Maleev (1974) reports the presence of a large, rectangular disclike intermedium in *Tyrannosaurus bataar,* whereas Barsbold (1983) reports the presence of four carpal elements in *T. bataar;* interestingly, the intermedium resembles that of *Allosaurus* (Gilmore 1920) and *Acrocanthosaurus* (Currie and Carpenter 2000), but not *Gorgosaurus* (Lambe 1917). In MOR 555 the two carpals probably represent the ulnare and radiale because a fragment of the distal end of the ulna(?) adheres to one of them, the ulnare(?). The presumed ulnare is slightly smaller than the radiale, a condition also

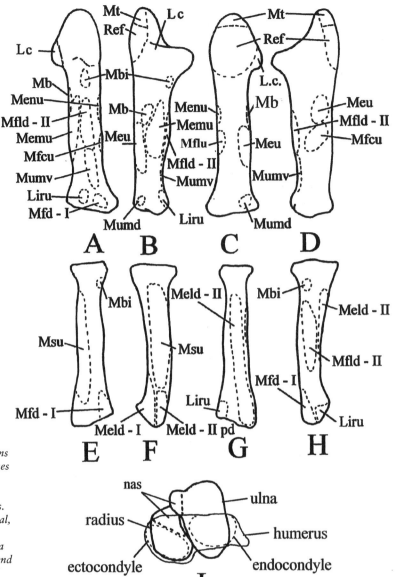

Figure 9.8. Origin and insertions sites on the lower forelimb bones of Tyrannosaurus rex. *Ulna in (A) anterior, (B) lateral, (C) posterior, and (D) medial views. Radius in (E) anterior, (F) lateral, (G) posterior, and (H) medial views. (I) proximal ends of ulna and radius (heavy line), distal end of the humerus (light line), and outline of the humeral epicondyles (dashed line). Note the nonarticular surface (nas) of the ulna and radius. See list of anatomical abbreviations for muscle and ligament names.*

reported for *Gorgosaurus* (Lambe 1917). Both carpals are small wedges that taper toward one another. There is a slight rim around the articular surfaces for the synovial membrane and for various ligaments connecting the bones to the ulna, radius, and metacarpals.

Metacarpal I. This bone is short, stocky, asymmetrical, and D-shaped in cross-section (fig. 9.9C–H). The proximal end is D-shaped in dorsal view (fig. 9G), is slightly concave for a carpal (part of the intermedium?), and slopes dorsomedially. The surface for articulation with metacarpal II is a depression, or synovial cavity, rimmed with rugosity for the collateral ligaments. The synovial cavity is deeper in FMNH PR 2081 than in MOR 555 and is probably due to individual

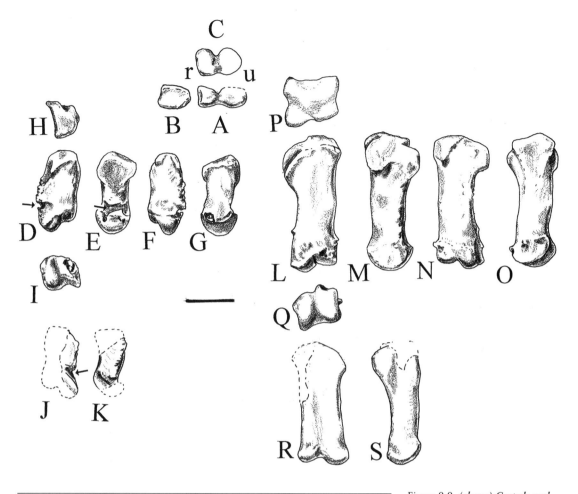

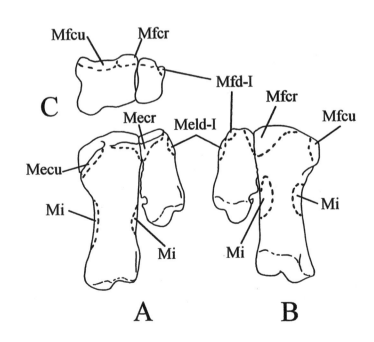

Figure 9.9. (above) Carpals and metacarpals of Tyrannosaurus rex. Carpals (MOR 555) in (A) anterior, (B) lateral, and (C) proximal views; r = radiale, u = ulnare. Metacarpal I: FMNH PR 2081 in (D) dorsal, (E) medial, (F) palmar, (G) lateral, (H) proximal, and (I) distal views; MOR 555 in (J) dorsal and (K) medial views (arrows show groove for radial artery to digit I). Metacarpal II: FMNH PR 2081 in (L) dorsal, (M) medial, (N) palmar, (O) lateral, (P) proximal, and (Q) distal views; MOR 555 in (R) dorsal and (S) medial views. Scale: 4 cm.

Figure 9.10. (left) Origin and insertion sites for the articulated metacarpal I and II of Tyrannosaurus rex in (A) dorsal, (B) palmar, and (C) proximal views.

Figure 9.11. (opposite page)
Figure 9.11. (opposite page)
Phalanges of Tyrannosaurus rex.
Phalanx I-1 (MOR 555) in
outline (upper portion missing)
in (A) dorsal, and (B) lateral
views. Phalanx I-2 (ungual)
(FMNH PR 2081) in (C) medial,
(D) dorsal, (E) lateral, and (F)
palmar views. Phalanx II-1:
FMNH PR 2081 in (G) dorsal,
(H) medial, (I) palmar, (J) lateral,
(K) proximal, and (L) distal
views; MOR 555 in (M) dorsal
and (N) medial views. Phalanx
II-2: FMNH PR 2081 in (O)
dorsal, (P) medial, (Q) palmar,
(R) lateral, (S) proximal, and (T)
distal views; MOR 555 in (U)
dorsal and (V) medial views.
Phalanx II-3 (FMNH PR 2081)
in (W) medial view. Scale: 4 cm.

variation. Ventral to the synovial cavity, just above the medial distal condyle, is a transverse groove (fig. 9.9C,D,I,J) that was probably for the radial artery. In humans, the radial artery extends along the anterior side of the forearm to the wrist, where it wraps around to the dorsal side of the thumb and passes into the palm between metacarpal I and II (Grant 1943; Gray and Goss 1973). This indirect passage to the palm prevents the M. opponeus pollicis from blocking blood flow when it flexes; an analogous situation must have also occurred in *T. rex* involving the M. adductor digiti I (M. adductor alulae of birds). The distal condyles of metacarpal I in *T. rex* are not equal, the external one being larger and extending further. A similar condition occurs in many dinosaurs (e.g., *Allosaurus, Plateosaurus;* see Gauthier 1986) and causes digit 1 to angle away from the axis of the manus. The biomechanical significance for this is discussed in further detail below. A lesion of secondary gout (due to uric acid buildup) is present on the external condyle of the metacarpal of FMNH PR 2081, as reported elsewhere (Rothschild et al. 1997). The damage is such that it affected the articular cartilage and part of the synovial capsule; leakage of synovial fluid may have occurred.

Metacarpal II. This metacarpal is almost twice as long as metacarpal I (fig. 9.9K–R). The proximal end is rectangular (fig. 9.9D) and also slopes dorsomedially. A rim denotes the edges of the articular cartilage and synovial capsule that once covered the joint surface. The external or lateral surface bears two facets at the corners for articulation with metacarpal I. Between these two facets is a shallow depression that is the counterpart to the synovial cavity of metacarpal I. On the medial side, opposite the metacarpal I facets, the corners of the articular surface overhang the shaft (fig. 9.9K,L). These tubercles are probably for insertion of the M. extensor carpi radialis and M. flexor carpi radialis. Between these two tubercles is a notch for metacarpal III, which is presumed to have been present, but is missing. The shaft is almost rectangular in cross-section and has a slight medial twist so that the lateral condyle is more palmar than the medial condyle (fig. 9.9P). The medial condyle is larger than the lateral condyle, and projects slightly lower as well (fig. 9.9K,M,Q). Scars, including a small tubercle for the collateral ligaments, are especially prominent in FMNH PR 2081. A rim around the articular surfaces denote the limits of the articular cartilage and synovial capsule. A gout lesion is present in FMNH PR 2081 as well (Rothschild et al. 1997).

Phalanx I-1. Only the ventral half of this phalanx is known (fig. 9.11A,B). It resembles that of *Gorgosaurus*, although the shaft does not show the degree of twist along its axis (the biomechanics for this are discussed in further detail below). The medial distal condyle is slightly larger than the lateral.

Phalanx I-2. This manal ungual (fig. 9.11C–F) is laterally compressed and the cotyles slightly asymmetrical in size; a rim demarcated the edge of the articular cartilage and synovial capsule of the joint. Ventrally the ungual surface is somewhat flat, and the entire surface slopes laterally. There is no "lip" extending above the articular cotyles for the insertion of the M. extensor digitorum. The lateral and medial

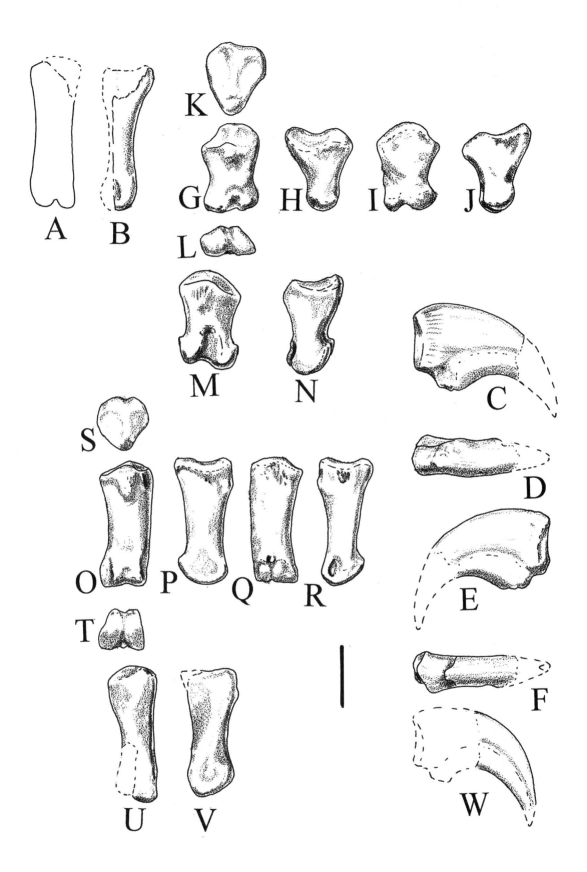

surfaces around the articular cotyles are rugose for the collateral ligaments. Ventrally, the tubercle for the surficial tendon of the M. flexor digitorum is prominent and inset from the articular cotyles, thus forming the characteristic ungual "step" or "heel." The palmar digital artery apparently crossed one side to the other dorsal to the synovial sheath encasing the M. flexor digitorum. On each side of the claw core, ungual arteries extended down each side of the claw core. The dorsal digital arteries apparently joined the ungual arteries at a Y-shaped junction on each side of the ungual, dorsal to the tubercle.

Phalanx II-1. This is a very unusual bone and could easily be mistaken for a pedal phalanx (fig. 9.11G–N). Its identification, however, is certain because an identical bone is present in *T. bataar* (Maleev 1974, fig. 36). The proximal end is triangular (fig. 9.11K) due to a very prominent tubercle for the M. extensor digitorum, which is the distal tendinous portion of the M. extensor metacarpi ulnaris. The M. interosseus dorsalis, which originated on the medial surface of metacarpal II, may also have partially inserted on the tubercle. This extensor tubercle shows variation in size, being more prominent in FMNH PR 2081. Ventrolaterally are scars for the insertion of M. interosseous and the deep tendons of M. flexor digitorum. The articular cotyles for reception of metacarpal II are asymmetrical in size, reflecting the difference in the distal articular condyles of that metacarpal; the cotyles are separated by a low ridge. Ventrally, there is a pair of small bumps below the cotyles for collateral ligaments and a rugose area between them for the tendon of the M. flexor carpi radialis. Distally, the articular condyles are asymmetrical, with the lateral condyle larger than the medial. Furthermore, the condyles face ventrally and not distally indicating that phalanx II-2 rotated palmar. Laterally, there are no collateral fossa for the collateral ligaments. Interestingly, phalanx II-1 is larger in MOR 555 than in FMNH PR 2081 (fig. 9.11 G,H vs. M,N), a reversal of the trend seen in the other forelimb bones.

Phalanx II-2. The articular cotyles are shallow, asymmetrical, and separated by a low ridge. The proximal end is triangular (fig. 9.11S) because of a prominent tubercle on the dorsal side for a tendinous slip of the M. extensor digitorum. Ventrolaterally are two small, raised scars for the collateral ligaments; a scar ventrally between them is for the insertion of the M. flexor digitorum. The shaft is slightly bent medially (Fig. 9.11O,U). Medially, on the ventral side, is a rugose arc in FMNH PR 2081 and a low tubercle in MOR 555; these are the terminal insertion sites for the deep tendons of the M. flexor digitorum. The distal articular condyles face ventrally more than dorsally. Dorsolaterally, this is a fossa for the collateral ligament (fig. 9.11RN); a shallow depression is present medially (fig. 9.11P,V). In FMNH PR 2081 there is a small piece of calcified Vincula brevia, the tendon that attaches to the M. flexor digitorum, ventrally between the distal condyles (fig. 9.11Q). The medial condyle in FMNH PR 2081 shows the degenerative condition of rheumatoid arthritis, which in this case, is connected to renal failure (Greenfield 1986), as evidenced by the gout lesions on the metacarpals.

Phalanx II-3. This ungual (fig. 9.11W) is tentatively identified as belonging to *T. rex* because it was found associated with FMNH PR 2081. It differs from the other ungual in being much more laterally compressed and in the more dorsally placed grooves for the ungual arteries.

Biomechanics

Having described the osteology and myology of the forelimb of *T. rex,* we now examine the forelimb biomechanically in order to answer Gould's question quoted at the beginning of the chapter. This analysis is done by first determining which of two mechanical systems the forelimb best represents, and then determining the power of the forelimb based on that system. The analysis compares the short forelimb of *T. rex* with those of *Deinonychus,* a long-forelimbed theropod, *Allosaurus,* an intermediate forelimbed theropod, and a human forelimb as a standard. Materials used include: *Deinonychus antirrhopus* coracoid (YPM 5236), scapula and humerus (AMNH 3015), ulna and radius (YPM 5220), and manus (YPM 5236); *Allosaurus fragilis* forelimb (DinoLabs uncataloged); *Tyrannosaurus rex* forelimb (MOR 555); and *Homo sapiens* (MOR uncataloged, commercially obtained, sex unknown). Preliminary results were previously presented by Smith and Carpenter (1990) and Carpenter and Smith (1995).

Mechanical Analysis. The forelimb is a third-class lever whose motive force (MF) is located between the resistive force (RF) and the axis of rotation (*motive force* as used here is the pull of a muscle on a bone, *resistive force* is the resistance of that bone to being moved, as when it is under load; see fig. 9.12). The motion of the limb may be viewed as one of two mechanical systems. One is a velocity-based system (VBS), typified by cursorial mammals, in which the insertion of limb muscles is close to the axis (allowing more rapid limb movement). The other is the force-based system (FBS), typified by fossorial mammals, in which the inserting limb muscles are farther from the axis (more power to the limb). Which of these two systems is in operation is determined by the ratio of the resistive force to the motive force. The resultant numerical value, the mechanical advantage (MA), is high in a VBS and is low in an FBS because MA decreases as the resistive force moves away from the axis of rotation. Although this dichotomous division is simplistic and in the real world there is actually a continuum between the two systems, recognition of only two systems makes analysis of the *T. rex* arm easier.

For the purpose of analysis, we follow Kreighbaum and Barthels (1985), who recognize that force (mass times acceleration) has four properties: (1) magnitude of force, (2) direction force is applied, (3) line of action (follows the direction of force), and (4) point of application (the attachment of the force that is being applied to the bone). If the line of action passes off-axis (eccentric force), a torque is produced resulting in a turning effect. The off-axis distance to the line of action is called the force arm. The motive force arm (MFA) and resistive force arm (RFA) is the distance from the axis of rotation to the line of action (MF or RF) and will always be perpendicular to the line of action. The equation

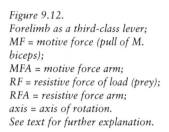

Figure 9.12.
Forelimb as a third-class lever;
MF = motive force (pull of M. biceps);
MFA = motive force arm;
RF = resistive force of load (prey);
RFA = resistive force arm;
axis = axis of rotation.
See text for further explanation.

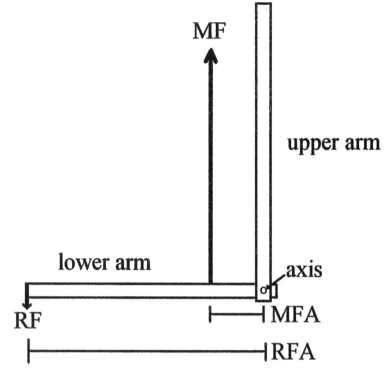

T = F (dL) is used to calculate torque, where T is torque, F is force, and dL is the perpendicular length of the force arm. Force is measured in newtons (N) and torque in newton-meters (Nm) or torque units. To calculate MA, either MF or RF must be known or assumed. For convenience, a MF of 100N was used, with the torque units for MF and RF the same to keep the lever arm static (i.e., the lower arm remains perpendicular to the upper arm). Data used in the analysis is presented in table 9.2, with limb lengths standardized to the shortest humerus (*Deinonychus*) in order to make comparisons equal. The following formulas were used:

MF × MFA = T
RF × RFA = T
RF/MF = MA

Using the data in table 9.2 for *Homo,*

100N × 0.25m = 2.5Nm
RF × 0.197m = 2.5Nm
RF = 2.5Nm/0.197m = 12.7
100N/12.7N = 7.9

An idealized force-based system would have an MA of 1, whereby one unit of motive force is needed to stabilize one unit of resistive force and an idealized velocity-based system would have an MA of 100. However, such values are not reached in nature because a multitude of forces are in operation on the system at any one time. In this study, the MAs range from 5.2 (*Tyrannosaurus*—FMNH PR 2081) to 21.7 (*Deinonychus*), with the highest value approaching that of a modern eagle,

TABLE 9.2

Data Used to Determine Mechanical Advantage (see fig. 9.12)
and Numerical Results; MFA and RFA standardized (see text).

	Homo sapiens	*Deinonychus*	*Allosaurus*	*Tyrannosaurus* FMNH PR 2081
MF	100N	100N	100N	100N
MFA	0.25m	0.010m	0.052m	0.042m
RFA	0.197m	0.215m	0.30m	0.22m
RF	12.7N	4.6N	17.3N	19.25N
MA	7.9	21.7	5.8	5.2

Aquila chrysaetos, with an MA of 25 (Smith unpubl.). Not surprisingly, the overall shape of the humerus in *Deinonychus* is similar to that of the eagle, suggesting similar stresses on the forelimb. The higher MA value in the eagle may be due to the glenoid being positioned more laterally so that the MF vector is more perpendicular to the scapulocoracoid. The high MA value indicates that the forelimb of *Deinonychus* is a velocity-based system (VBS), where reaching to grasp prey is more important than having the strength to hold the prey. The limb motion associated with grasping prey may be a precursor to the flight stroke in birds (see Gauthier and Padian 1985). *Homo sapiens* has an MA of 7.9. The close attachment of the MF to the axis, the resultant short MFA and long radius is a compromise between an FBS and a VBS. The relatively close insertion of the MF to the axis gives the radius a mechanical advantage in velocity. The MA for the *Allosaurus* is 5.8 and is therefore considered an FBS, where arm strength is important for holding prey. For *Tyrannosaurus* the lowest MA was 5.2. The scapula and coracoid are proportionately the largest in the study, indicating more surface area for the origin and insertion of various muscle groups. Clearly then, the forearm of *T. rex* is a force-based system, in which arm strength is more important than quickness of grasping.

Power Analysis. Having established that the forearm of *T. rex* is a forced-based system, we now examine the power of the forearm. To calculate the estimated force of a single muscle in a third-class lever, the cross-sectional area of the inserting tendon can be measured based on the size of the scar of insertion on the bone. There is a slight flaring of the tendon at its insertion, causing a slight increase in the size of the tendon at its insertion (personal observations), therefore the results for tendon diameter may be slightly overestimated. The tensile strength of tendon is known, allowing the maximum working range (MWR) and normal working range (NWR) to be calculated. The muscle we have identified as the major MF in our study is the M. biceps. In humans, the insertion is only on the humerus (radial tuberosity), whereas in *T. rex,* as in birds, there is a dual insertion on the radius and ulna (see above). We have assumed for the purpose of our study that MF is equally divided between these two insertions. Again, the human arm is used as a basis for comparison, although measurements are not standardized because absolute comparisons are required.

The surface area of the radial tuberosity of the human radius examined was 130 mm^2. The tensile strength of an average tendon is 100 N/mm^2 (Vogel, 1988). Tendons also have a built-in safety factor relative to the isometric contraction of ~3 (= MWR) and a normal working range is one-third of the safety factor (= NWR) (Biewener, pers. comm. to Smith 1987). Using the formula:

tendon tensile strength/area^2 × radial tuberosity area^2 =

estimated tendon tensile strength (approximates muscle strength) the result is:

100N/mm^2 × 130mm^2 = 13,000N

With a one-third safety factor:

13,000N/3 = 4,333N (= MWR),

and a normal working range of one-third of the safety factor:

4,333N/3 = 1,444N (= NWR)

Thus, the tensile strength of the human arm used in this study is 1,444N for the M. biceps. With MF = tensile strength; MFA = 0.04m; and RFA = 0.315m. Solving for RF using the formula:

MF × MFA = T

RF × RFA = T

4,333N × 0.04m = 173Nm

RF × 0.315m = 173Nm

RF = 550N (= MWR),

and

1,444N × 0.04m = 58N

RF × 0.315m = 58N

RF = 184N (= NWR)

Converting Newtons to kilograms (9.81 kg/N), the resistive force maximum working range (no safety factor) is 56 kg, and the normal working range (including safety factor) 18.75 kg for the skeleton used in the study (~180 cm tall).

An estimation of the size of the M. biceps can now be calculated. The strength per cross-sectional area of a muscle has been calculated to be between 4 kg/cm^2 and 8 kg/cm^2 in humans (Ikai and Fukunaga 1968) and 4 kg/cm^2 to 6 kg/cm^2 regardless of taxon (Schmidt-Nielsen 1975). An average strength of 5 kg/cm^2 was used to determine the cross-sectional area. The following formula is used:

MF(kg)/strength kg × cm^{-2} = cross-section cm^2

Using the previously determined MF of 147 kg (=1,444N) for the normal working range:

147kg/5kg × cm^{-2} = 29.4cm^2

Thus, the estimated cross-section of the human M. biceps in this study is 29 cm^2, or a diameter of 6.1 cm.

For *T. rex* (FMNH PR 2081), the area for the insertion of M. biceps on the radius is 132.7 mm^2 and on the ulna is 176.6 mm^2. Treating these as separate insertions and with MFA = 0.055m, and RFA = 0.29m:

tendon tensile strength for radius:

100N/mm^2 × 132.7mm^2 = 13,270N

MWR: 4423N × 0.055m = 243N

RF × 0.29m = 243N

RF = 838.8N = 85.5kg.

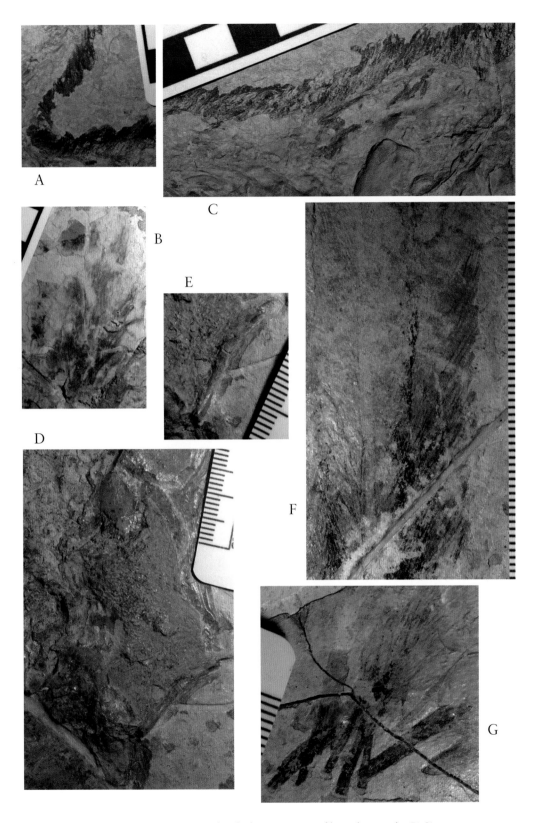

1. (A) *Sinosauropteryx* (NGMC 2123), detail of integumentary fibers along neck. (B) *Sinosauropteryx* (NGMC 2124), detail of tail feathers. (C) *Sinosauropteryx* (NGMC 2123), showing fibers along back and hips; compare to fig. 10.1A. (D) *Protarchaeopteryx* (NGMC 2125), showing end of right sternal plate (top) contacting superimposed right and left coracoids; the clavicle can be seen adjacent to the bottom border of the coracoid and is detailed in (E). (F) Tail feather of *Protarchaeopteryx* (NGMC 2125). (G) Plumulaceous feathers of *Protarchaeopteryx* (NGMC 2125). All scales in mm.

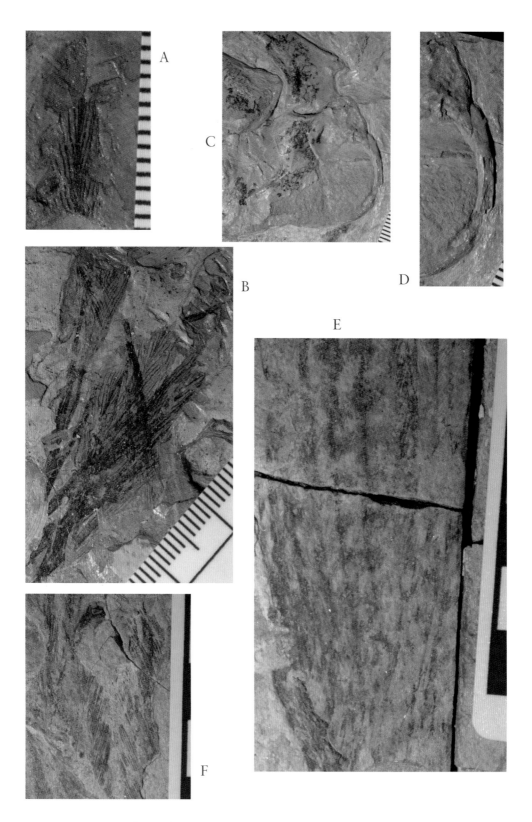

2. (A,B) Additional plumulaceous feathers of *Protarchaeopteryx* (NGMC 2125). (C) Glenoid ends of scapulae, superimposed right and left coracoids, and adjacent clavicle of *Caudipteryx* (NGMC 97-9-A); detail of the clavicle is shown in (D). (E) Tail feathers of *Caudipteryx* (NGMC 97-4-A). (F) Another more posterior caudal feather of the same specimen. All scales in mm.

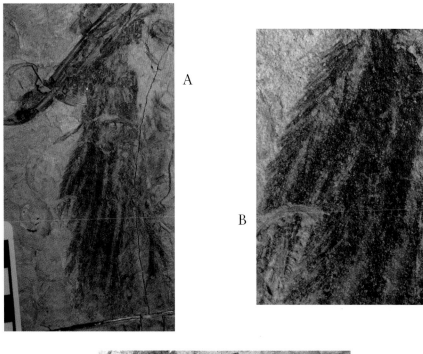

3. (A) feathers on the hand of *Caudipteryx* (NGMC 97-4-A), detailed in (B). (C) Fibrous integumentary structures along the anterior hemal arches of the same specimen, drawn in detail in fig. 10.1E. All scales in mm.

4. "Pursuit." A pair of *Troodon formosus* converge on a fleeing
Orodromeus makelai, separated from its herd in a collective
hunting alliance. Illustration © by John Bindon.

5. The feathered dinosaur *Protarchaeopteryx robusta*.
Sculpture © by Brian Cooley. [Two Views]

6. *Pachyrhinosaurus* from Alberta. Illustration © by Katherine I. Hargrove.

7. *Below: Centrosaurus* herd crossing a flooded river. Illustration © by Douglas Henderson.

8. A pair of *velociraptor* hunting. Illustration © by Douglas Henderson.

9. *Struthiomimus.* Illustration © by Douglas Henderson.

10. "The Old Guard." A male *Pachyrhinosaurus*-like
ceratopsian from the Grande Prairie Site, Alberta.
Sculpture © by Dan Lorusso. Photo by Michael Rizza.

11. *Above:* At an oxbow lake in the Late Cretaceous, Horseshoe Canyon Formation, an *Albertosaurus* pack attacks a herd of *Anchiceratops*, which wheel in defense. A group of *Hypacrosaurus* flees from the tumult, as do a flock of archaic water birds. Illustration © by Greg Paul.

12. *Hypacrosaurus altispinus.* Illustration © by John Sibbick.

13. *Oviraptor* and hatchlings.
Illustration © by Luis Rey.

14. *Below: Giganotosaurus*
vs. *Amargasaurus*.
Illustration © by Luis Rey.

15. A young *Diplodocus* is comforted by the presence of two adults.
Illustration © by Michael W. Skrepnick.

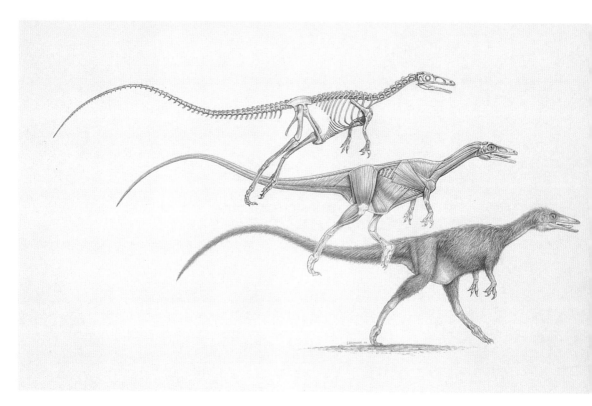

16. *Sinosauropteryx prima*. A three-stage drawing: skeleton, musculature, integument. Illustration © by Michael W. Skrepnick.

17. *Sinosauropteryx prima*, the first example of a dinosaur with a feather-like covering. Illustration © by Michael W. Skrepnick.

18. *Caudipteryx zoui*, a feathered dinosaur from China. Illustration © by Michael W. Skrepnick.

19. *Sinosauropteryx*. Illustration © by Jan Sovak.

20. *Albertosaurus* packhunt. Illustration © by Michael W. Skrepnick.

21. *Daspletosaurus torosus*, the large theropod from the Judith River Formation, charges across a mud flat trailing after its prey. Sculpture © by Michael Trcic.

22. *Tyrannosaurus rex*. Sculpture © by Michael Trcic.

NWR: 1474N × 0.055 = 81N
RF × 0.29m = 81N
RF = 279.6N = 28.5kg.
and ulna:
100Nmm² × 176.6mm² = 17,660N
   MWR: 5887N × 0.055m = 323.8N
RF × 0.29m = 323.8N
RF = 1116.5N = 113.8kg.
   NWR: 1962 × 0.055m = 107.9N
RF × 0.29m = 107.9N
RF = 371N = 37.9kg.

For the tendon, hence for the M. biceps of *T. rex*, MWR = 1955.3 (= 199kg.) and NWR = 650.6N (66.3kg). The diameter of M. biceps, based on the normal working range, is 9.2 cm.

Finally, the limb bones have unusually thick cortical bone. Curry and Alexander (1985) have examined the significance of this in modern animals and their results are applied here. Using their formulas:

ratio of marrow cavity radius/bone radius = K,

which is related to bone radius/cortical thickness (R/t).

Applying their formulas to *T. rex* (MOR 555), for the humerus K = 0.25, R/t = 1.34; radius K = 0.18, R/t = 1.22; ulna K = 0.1, R/t = 1.11. All these values indicate bone selected for ultimate strength or impact loading, rather than for static strength or stiffness.

Discussion

Our analysis of the osteology, myology, and biomechanics of the forelimb of *Tyrannosaurus rex* show that the arm was stoutly built and well muscled. Compared with a human, the M. biceps in *T. rex* was 3.5 times stronger per N of force in both the maximum and normal working ranges (other flexors, e.g., M. brachialis, were not considered and would certainly increase the arm strength). Such high strength suggests to us that the arms were not atrophied structures nor useless appendages, but rather had a definite function that we believe was to clutch struggling prey.

In support of this hypothesis, we note that the range of motion (ROM) for the forelimb from maximum extension to maximum flexion varies considerably among the taxa used in this study, as shown in table 9.3 (tendinous and muscle tissue restrictions of motion were not taken into account, except for *Homo;* these restrictions prevent the joints from achieving their maximum angle as determined from joint morphology only). ROM was determined from the rim for the articular cartilage on both surfaces of the joint. The range of maximum flexion and extension must be such that the rim for the articular cartilage did not pass into the joint (hyperflexion or hyperextension) (for further discussion, see Carpenter et al. 1994). The greatest ROM is the long arms of the human. Although, at 7.9, the mechanical advantage was rather low for the long lever arms, this seems to be a compromise between a velocity-based system and a force-based system. The second greatest range of motion is *Deinonychus,* which has a long forelimb, or long lever arms, and a very high MA value, typical of a VBS, for rapidly reaching out to grasp. The long cursorial running legs (tibia + metatarsals longer than femur) and long forelimbs indicate pursuit and grasping of prey. *Allosaurus* has a more restricted ROM, and proportionally shorter arms relative to hind limb length than *Deinonychus* or *Homo*. In addition, the MA of the moderately short lever arms is very low, indicative of a FBS in which clutching prey is more important than reaching out to grasp prey.

TABLE 9.3
Range of Movement for Taxa Used in This Study

Taxon	Shoulder-upper arm	Upper arm-lower arm (elbow)
Homo	360^0	165^0
Deinonychus	88^0	130^0
Allosaurus	65^0	62^0
Tyrannosaurus	40^0	45^0

Finally, the least amount of forearm motion is that of *T. rex* (fig. 9.13), which has short arms and the very low MA of an FBS. The limited ROM and short lever arm of the forelimb provided a very stable platform for the very powerful M. biceps. This indicates to us that the forelimbs were used to hold a struggling prey. In support of this interpretation, we note the pathology along the medial side of the humerus

Figure 9.13. Maximum range of motion of the forelimb of Tyrannosaurus rex in (A) lateral and (B) anterior views (see text for further information). Also shown is maximum extension and flexion of manus. Path of motion of the claw tips shown in dashed lines and dots. Angle of maximum movement in (A) is 25°. (C) is manus in maximum extension (light lines) and flexion (dark lines). Note that claws angle toward each other in flexion, ensuring they will not be pulled free.

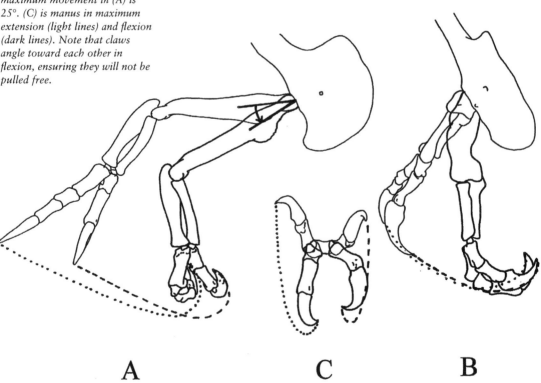

A   C   B

in FMNH PR 2081. The site of damage corresponds to the medial head of the M. triceps humeralis, which serves to adduct and extend the lower arm. As noted above, the pathology is characteristic of partial avulsion caused by abnormally high stress loads. Such loads might occur while clutching a large, struggling animal, such as an adult hadrosaur (see Carpenter in press) Indeed, the straight shaft of the humerus, as compared with that of *Allosaurus* (see Gilmore 1920), is precisely what is expected for maximum strength per unit mass (Bertram and Biewener, 1988). Such conditions occur where the bone must resist axial compression, as it would do in this case with the powerful M. biceps (see fig. 9.12). Furthermore, the very low K and R/t values for the humerus, ulna, and radius indicate bones selected for ultimate strength or impact loading. Finally, to ensure that the struggling prey not escape while the mouth is attempting to kill it, the two ungual claws point somewhat inward (fig. 9.13C) so that they do not slip out of the prey easily.

Conclusion

We conclude from our analysis of the forearms of *Tyrannosaurus rex* that they were used to hold prey while the head dispatched it. Such an interpretation is at odds with interpretation of *T. rex* as a scavenger because of the small size of the arms as advocated by Horner (1994), Horner and Lessem (1994), and Horner and Dobb (1997), but does support the fossil evidence presented elsewhere (Carpenter 2000).

We envision that *T. rex* stalked or ambushed prey, mostly subadult or young adult hadrosaurs (see Carpenter 2000). As with most extant predators, the mouth was used to grasp the prey. Then the short, powerful arms were used to grasp or clutch the prey against the body to prevent its escape while the teeth were disengaged and repeated bites made to kill the prey. We also believe that, as with most predators, *T. rex* was opportunistic and would have scavenged when given the opportunity.

Acknowledgments: We dedicate this chapter to Philip J. Currie in recognition for his work on theropods and for his many years of friendship. Tom Holtz reviewed an earlier version of this paper and we thank him for his input. We also thank several individuals for providing us with specimens or access to materials: Jack Horner (Museum of the Rockies), Peter Larson (Black Hills Institute), James Madsen (Dino-Labs), John Ostrom (Yale Peabody Museum), and Charlie McGovern (Stone Company). A complete cast of the bones of FMNH PR 2081 and of MOR 555 used in this study are housed at the Denver Museum of Natural History. This study was begun when FMNH PR 2081 was still housed at the Black Hills Institute and we thank Peter Larson for the loan of "Sue's" forelimb material.

References

Barsbold, R. 1983. [Carnivorous dinosaurs from the Cretaceous of Mongolia.] *Joint Soviet-Mongolian Palaeontological Expedition Transactions* 19: 1–117. (In Russian.)

Baumel, J. J. 1979. Arthrologia. In J. J. Baumel (ed.), *Nomina anatomica avium*, pp. 123–117. London: Academic Press.

Berge, J. C. V. 1979. Myologia. In J. J. Baumel (ed.), *Nomina anatomica avium*, pp. 175–219. London: Academic Press.

Bertram, J. E. A., and A. A. Biewener. 1988. Bone curvature: Sacrificing strength for load predictability. *Journal of Theoretical Biology* 131: 75–92.

Carpenter, K. 1990. Variation in *Tyrannosaurus rex*. In K. Carpenter and P. Currie (eds.), *Dinosaur Systematics: Approaches and Perspectives*, pp. 141–145. Cambridge: Cambridge University Press.

Carpenter, K. 2000. Evidence of predation in theropod dinosaurs. *Gaia: Revista de geociencias, Museu Nacional de Historia Natural* 15: 135–144.

Carpenter, K., and M. Smith 1995. Osteology and functional morphology of the forelimbs in tyrannosaurids as compared with other theropods (Dinosauria). *Journal of Vertebrate Paleontology, Abstracts with Program* 15 (3): 21A.

Carpenter, K., J. Madsen, and A. Lewis. 1994. Mounting of fossil vertebrate skeletons. In P. Leiggi and P. May, (eds.), *Vertebrate Paleontological Techniques*, pp. 285–322. Cambridge: Cambridge University Press.

Chamberlain, F. W. 1943. *Atlas of Avian Anatomy: Osteology, Arthrology, Myology*. East Lansing: Michigan State College.

Chure, D., and J. Madsen. 1996. On the presence of furculae in some non-maniraptorian theropods. *Journal of Vertebrate Paleontology* 16: 573–577.

Coombs, W. 1978. Forelimb muscles of the Ankylosauria (Reptilia: Ornithischia). *Journal of Paleontology* 52: 642–658.

Currie, P., and K. Carpenter. 2000. A new specimen of *Acrocanthosaurus atokensis* (Theropoda, Dinosauria), from the Lower Cretaceous Antlers Formation (Lower Cretaceous, Aptian) of Oklahoma, USA. *Geodiversitas* 22: 207–246.

Curry, J. D., and R. M. Alexander. 1985. The thickness of the wall of tubular bones. *Journal of Zoology* 206: 453–468.

Filla, J., and P. D. Redman. 1994. *Apatosaurus yahnahpin*: A preliminary description of a new species of diplodocid dinosaur from the Late Jurassic Morrison Formation of southern Wyoming, the first sauropod dinosaur found with a complete set of "belly" ribs. *Wyoming Geological Association Guidebook* 44: 159–178.

Galton, P. 1969. The pelvic musculature of the dinosaur *Hypsilophodon*. *Postilla* (Yale University) 131: 1–64

Gauthier, J. 1986. Saurischian monophyly and origin of birds. In K. Padian (ed.), The origin of birds and the evolution of flight. *Memoirs of the California Academy of Science* 8: 1–55

Gauthier, J., and K. Padian. 1985. Phylogenetic, functional, and aerodynamic analysis on the origin of birds and their flight. In M. K. Hecht, J. H. Ostrom, G. Viohl, and P. Wellnhofer (eds.), *The Beginnings of Birds*, pp. 185–197. Eichstätt: Freunde des Jura-Museums.

Gilmore, C. W. 1920. Osteology of the carnivorous Dinosauria in the United States National Museum, with special reference to the genera *Antrodemus* (*Allosaurus*) and *Ceratosaurus*. *Bulletin, U.S. National Museum* 110: 1–154.

Gould, S. 1980. *The Panda's Thumb*. New York: Norton.

Grant, J. C. 1943. *An Atlas of Anatomy*. 1: 1–214 Baltimore: Williams and Wilkins.

Gray, H., and C. M. Goss. 1973. *Anatomy of the Human Body*. Philadelphia: Lea and Feibiger.

Greenfield, G. B. 1986. *Radiology of Bone Disease*. Philadelphia: Lippincott.

Holtz, T. R., Jr. 1994. The phylogenetic position of the Tyrannosauridae: Implications for theropod systematics. *Journal of Paleontology* 68: 1100–1117.

Horner, J. R. 1994. Steak knives, beady eyes, and tiny arms (a portrait of *T. rex* as a scavenger). In G. D. Rosenberg and D. L. Wolberg (eds.), DinoFest. *Paleontological Society Special Publication* 7: 157–164.

Horner, J. R., and E. Dobb. 1997. *Dinosaur Lives*. New York City: HarperCollins.

Horner, J. R., and D. Lessem. 1994. *The Complete T. rex*. New York City: Simon and Schuster.

Huene, F. 1907–1908. Die Dinosaurier der europäischen Triasformation mit Berücksichtigung der äussereuropäischen Vorkommisse. *Geologische und paläeontologische Abhandlugen*, suppl. 1: 1–417

Ikai, M. and T. Fukunaga. 1961. Calculation of muscle strength per unit cross-section area of human muscle by means of ultrasonic measurement. *Internationale Zeitschrift für angewandte Physiologie einschlägige Arbeitsphysiologie* 26: 26–36.

Kreighbaum, E., and K. M. Barthels. 1985. *Biomechanics: A Qualitative Approach for Studying Human Movement*. Minneapolis: Burgess Publishing.

Lambe, L. 1917. The Cretaceous theropodous dinosaur *Gorgosaurus*. *Geological Survey of Canada Memoir* 100: 1–84.

Larson, P. L. 1994. *Tyrannosaurus* sex. In G. D. Rosenberg and D. L. Wolberg (eds.), DinoFest. *Paleontological Society Special Publication* 7: 139–155.

Lull, R. S., and N. Wright. 1942. Hadrosaurian dinosaurs of North America. *Geological Society of America Special Paper* 40: 1–242.

Makovicky, P., and P. J. Currie. 1998. The presence of a furcula in tyrannosaurid theropods, and its phylogenetic and functional implications. *Journal of Vertebrate Paleontology* 18: 143–149.

Maleev, E., 1974. [Giant carnosaurs of the family Tyrannosauridae.] *Joint Soviet-Mongolian Palaeontological Expedition Transactions* 1: 132–191. (In Russian.)

Mattison, R., and E. Griffin. 1989. Limb use and disuse in ratites and tyrannosaurids. *Journal of Vertebrate Paleontology, Abstracts* (suppl. to no. 3) 9: 32A.

Newman, B. 1970. Stance and gait in the flesh-eating dinosaur *Tyrannosaurus*. *Biological Journal of the Linnean Society* 2: 119–123.

Norman, D. B. 1986. On the anatomy of *Iguanodon atherfieldensis* (Ornithischia: Ornithopoda). *Bulletin du Institute Royal Science Naturelle de Belgique: Sciences de la Terre* 56: 281–372.

Osborn, H. 1906. *Tyrannosaurus*, Upper Cretaceous carnivorous dinosaur. *American Museum of Natural History Bulletin* 22: 281–296.

Ostrom, J. H. 1974. The pectoral girdle and forelimb function of *Deinonychus* (Reptilia: Saurischia): A correction. *Postilla* (Yale University) 165: 1–11.

Parks, W. A. 1928. *Albertosaurus arctunguis:* A new species of theropodous dinosaur from the Edmonton Formation of Alberta. *University of Toronto Studies*, Geological Series 25: 3–42.

Paul, G. 1986. The science and art of restoring the life appearance of dinosaurs and their relatives. In S. Cerkas and E. Olson (eds.), *Dinosaurs Past and Present* 2:5–49. Seattle: University of Washington Press.

Paul, G. 1988. *Predatory Dinosaurs of the World.* New York City: Simon and Schuster.

Raath, M. 1977. The anatomy of the Triassic theropod *Syntarsus rhodesiensis* (Saurischia: Podosauridae) and a consideration of its biology. Ph.D. thesis, Rhodes University (Rhodesia).

Raath, M. 1990. Morphological variation in small theropods and its meaning in systematics: Evidence from *Syntarsus rhodesiensis.* In K. Carpenter and P. J. Currie (eds.), *Dinosaur Systematics: Approaches and Perspectives,* pp. 91–105. New York: Cambridge University Press.

Romer, A. S. 1923. The pelvic musculature of saurischian dinosaurs. *Bulletin, American Museum of Natural History* 48: 605–617.

Rothschild, B. C., D. Tanke, and K. Carpenter. 1997. Tyrannosaurs suffered from gout. *Nature* 387: 357.

Russell, L. 1935. Musculature and function in the Ceratopsia. *National Museums of Canada Bulletin* 77: 39–48.

Schmidt-Nielsen, K. 1975. Scaling in biology: The consequences of size. *Journal of Experimental Biology* 194: 287–308.

Smith, M., and K. Carpenter. 1990. Forelimb biomechanics of *Tyrannosaurus rex. Journal of Vertebrate Paleontology, Abstracts* (suppl. to no. 3) 10: 43A.

Vogel, S. 1988. *Life's Devices.* Princeton: Princeton University Press.

Walker, A. D. 1990. A revision of *Sphenosuchus acutus* Haughton, a crocodylomorph reptile from the Elliot Formation (Late Triassic or Early Jurassic) of South Africa. *Philosophical Transactions of the Royal Society,* Biological Series 330: 1–120.

10. Feathered Dinosaurs and the Origin of Flight

Kevin Padian, Ji Qiang, and Ji Shu-an

Abstract

Feathers or featherlike integumentary structures in nonavian dinosaurs, unheard of before 1996, are now known from at least five and possibly more lineages of nonavian theropods. The size, structure, and roles of these integumentary features vary among the lineages in which they have been found. They begin in *Sinosauropteryx* as a dense, fine, short body covering, but in *Protarchaeopteryx* and *Caudipteryx* they attain a stronger central structure with more robust filaments that are gathered into several forms of featherlike organs. Most features of feathers seen in *Archaeopteryx* and living birds were present, though some are uncertain. The relatively short lengths of the feathers in *Protarchaeopteryx* and *Caudipteryx* contradict their function (and hence evolution) as aerodynamic organs per se. Consistent with this interpretation are the relatively short forelimbs, the absence of clear indications of the mechanical aptitude for a flight stroke, and the relatively primitive configurations of the pectoral girdles. Feathers and related structures did not evolve for flight; insulation is a possibility testable by further discoveries; aerodynamic functions such as thrust production have been proposed; behavioral functions such as camouflage, display, nesting, and species recognition are potentially testable, though indirectly, if features such as color patterns in more completely known plumages come to light in the future. Combining the hypotheses of insulation and behavior, the feathers on the arms at some evolutionary stage may have been at least partly selected so that adults could behaviorally modify the thermoregulation of eggs on the nest.

Introduction

At the 1996 meeting of the Society of Vertebrate Paleontology in New York City, Professor Chen Pei-ji of the Nanjing Institute of Geology and Paleontology astounded his colleagues by producing photographs of what was ostensibly a small theropod dinosaur with a dense fringe of fine filamentous integumentary structures surrounding its skull, neck, back, and tail. This fossil, like the others that we discuss here, came from the Sihetun area of Beipiao, Liaoning, China, in horizons of the Jehol Group (Lower Cretaceous). In the same year Ji and Ji had published a diagnosis in China of the counterpart to Chen's fossil, which they named *Sinosauropteryx* (Ji and Ji 1996; Chen et al. 1998). These publications drew more attention to these specimens than had been accorded to almost any fossils yet discovered in Asia.

Ever since John Ostrom proposed in 1973 that birds evolved from small carnivorous dinosaurs, it had been anticipated that some sort of featherlike covering would be found in the immediate relatives of birds among theropods. The Chinese discoveries appeared to many to confirm and vindicate Ostrom's hypothesis, and Chen et al.'s (1998) description of *Sinosauropteryx* included a detailed treatment of what its integumentary structures might signify for the origin and function of the first feathers (see also Unwin 1998).

But as spectacular as this discovery was, it was supplanted in the next year by the only logical follow-up: nonavian dinosaurs with true, if primitive, feathers. *Protarchaeopteryx* and *Caudipteryx,* described by Ji et al. (1998), both bear feathers (rectrices) with more or less symmetrical vanes and parallel barbs along still-elongated tails (Ji et al. 1998, figs. 3b, 8b). *Protarchaeopteryx* also bears remnants of feathers along the pelvis and proximal caudal vertebrae, which are reflexed over the ilium, as well as contourlike feathers near the pectoral region (Ji et al. 1998, fig. 3a). *Caudipteryx* bears similar feather remnants along the tail and in the pectoral region, and the fingers additionally bear distinct remiges that show faint traces of vanes, barbs, and shafts (Ji et al. 1998, fig. 8a). Both taxa bear tuftlike structures apparently composed of filaments gathered and perhaps cemented along their bases, similar to plumulaceous feathers (Ji et al. 1998, fig. 3a). These integumentary features are often found isolated at some distance from specific skeletal structures.

Since these discoveries, other taxa from Liaoning have been reported with filamentous integumentary structures like those seen in *Sinosauropteryx*. These include a therizinosaurid *Beipiaosaurus* (Xu, Tang, and Wang 1999), and a dromaeosaur (Xu, Wang, and Wu 1999). Similar structures have been reported on a Late Cretaceous alvarezsaurid (Schweitzer et al. 1999). The ornithomimid *Pelecanimimus* (Pérez-Moreno et al. 1994), from the Early Cretaceous of Spain, was originally thought to have integumentary structures somewhat like those of *Sinosauropteryx,* but these now appear to be remains of other kinds of soft parts (Briggs et al. 1997).

Originally suspected to be basal birds, more detailed cladistic analyses suggest that these Liaoning forms belong to quite separate lineages of coelurosaurian theropod dinosaurs. Chen et al. (1998) recognized that *Sinosauropteryx* is closely related to the basal coelurosaur *Compsognathus*. *Caudipteryx*, first described as the closest known taxon to Aves (*Archaeopteryx* and more derived birds; Ji et al. 1998), now appears to be more closely related to oviraptorosaurs (Sereno 1999; Witmer in press). *Protarchaeopteryx* was classified as a eumaniraptoran coelurosaur, in a trichotomy with Velociraptorinae and all taxa closer to birds than to the former two genera (Ji et al. 1998). A full cladistic analysis of these forms, including outgroups other than alvarezsaurids and velociraptorines, is still needed, because the phylogenetic position of alvarezsaurids is still not settled. Adding dromaeosaurs, alvarezsaurids, and therizinosaurs to oviraptorids and compsognathids comprises at least five separate lines of feathered nonavian coelurosaurs, and *Protarchaeopteryx* would represent a sixth if it turns out not to belong to one of the former lineages. In any event the origin of birds is thus constrained to the coelurosaurian theropods.

We focus on what the features of these feathered dinosaurs reveal about the origin of both feathers and flight itself. Discussions of the evolution of flight must center on the flight stroke, the sine qua non of powered flight, along with the skeletal structures that effect this stroke and an aerodynamically efficient wing (Padian 1985, 1987, 1995, in press). The necessary soft tissues and metabolic factors are largely lost to the paleobiologist. Photographs of the osteological and integumentary features are provided in plates 1–3 (see color insert); drawings of these and other structures are found in figures 10.1–10.3 in this chapter.

Institutional Abbreviations: NIGP, Nanjing Institute of Geology and Paleontology; NGMC, National Geological Museum of China; TMP, Tyrrell Museum of Palaeontology.

Sinosauropteryx

Ji and Ji (1996) erected the taxon on the basis of National Geological Museum of China NGMC (or GMV) 2123; the specimen described by Chen et al. (1998), Nanjing Institute of Geology and Paleontology NIGP 127586, is the counterpart to the same specimen. Chen et al. (1998) also described a second specimen, NIGP 127587. We base our discussions here on the holotype, plus NGMC 2124, a third complete skeleton.

The holotype is a small specimen, with a jaw length of about 70 mm. In death pose, the animal is collapsed on its right ventral side, so the skull is seen in left dorsolateral view, as is the ribcage. The hindlimbs are disarticulated. The ribs are splayed and flattened, as if they had been pressed obliquely into the substrate. The tail is seen in left lateral view.

The forelimbs are poorly preserved in this specimen, and are better seen in the NIGP counterpart than in the NGMC holotype part. Chen et al. (1998) noted that the ratio of humerus plus radius to femur plus tibia in the counterpart is approximately 29% in the holotype and 31% in NIGP 127586; it is 35% in NGMC 2124. The last specimen is larger than the other two: its humerus is 60 mm and its femur 108 mm long, compared to 20 mm and 53 mm, respectively, for the holotype and 35 mm and 86 mm for NIGP 127587. NGMC 2124 has a well-preserved

Figure 10.1. (opposite page) (A) right sternal plate of Protarchaeopteryx *(NGMC 2125), the corner of which can also be seen at the top margin of plate 1D. (B) furcula of* Confuciusornis *(TMP 98.14.1), anterior view. (C) right clavicle of* Protarchaeopteryx *(NGMC 2125); the dorsal and ventral ends, both broken, are shaded in, and the dark stippled region represents a pit created in the matrix that has not revealed additional bone. (D) left wrist of* Protarchaeopteryx *(NGMC 2125). (E) left wrist of* Caudipteryx *(NGMC 97-4-A). (F) right wrist of* Confuciusornis *(TMP 98.14.1). Abbreviations: R, radius; r, radiale; U, ulna; I-III, metacarpals; 1+2, 3, distal carpals. Scale: 5 mm.*

pectoral girdle. Its scapula is 55 mm long, 7 mm broad at its dorsal end, but narrowing to 4 mm for most of the midshaft. It widens abruptly as it approaches the glenoid fossa, which is poorly preserved, but appears to be oriented in the normal theropod direction (posteriorly). The coracoid transversely broadens to 21 mm but does not show the extended ventral process that would imply an articulation with the sternum. No sterna are preserved in the specimen, but there appears to be a thin remnant of a partial clavicle about halfway along the ventral border of one coracoid. It is less than 2 mm thick, so does not resemble the broad, fused, boomerang-shaped clavicles seen in many other tetanurans (allosaurids, tyrannosaurids, oviraptorids, etc.: reviews in Padian 1997; Padian and Chiappe 1998a,b). The poorly developed shoulder girdle and short forelimb suggest no characters particularly related to the evolution of flight (Padian 1985; Rayner 1988; Jenkins 1993).

The absence of skeletal characters related to flight places the unusual integumentary structures in an unexpected light. These structures have been generally described by Ji and Ji (1996) and Chen et al. (1998). In NGMC 2123, as in the other specimens, the fibers are finer and more densely packed than are the integumentary structures of *Protarchaeopteryx* and *Caudipteryx*. There are about 10 fibers per millimeter, as measured in areas where they appear to be sufficiently distinct (plate 1A). Featherlike structures at the end of the tail of NGMC 2123 are at least 5 cm long (plate 1B). All the caudal vertebrae bear filamentous structures but they are not well enough preserved to determine individual lengths or anatomical details. On the holotype specimen these fibers tend to be uniformly 8–10 mm long, but Chen et al. (1998) noted that they were much longer in the larger NIGP 127587, ranging from 13 mm posteriorly to 40 mm or longer along the caudal vertebrae.

On NGMC 2124, typical filaments stretch dorsally from the anterior cervicals through the anterior caudals and rarely along the more posterior caudals. The ilium is covered with black, slightly fibrous material that may represent remains of this integument, but filaments are not as clear, although some structures, horizontal and parallel to the backbone, are visible (plate 1C; fig. 10.1A). This integumentary pattern is locally discontinuous and resembles the integument of the type specimen of *Sinosauropteryx* in being very finely structured.

On the basis of these observations, we can dispense with the interpretations of the specimens by Geist et al. (1997), who opined from photographs of the holotype that the integumentary structures of *Sinosauropteryx* were merely frayed collagenous fibers of the midline that had been disturbed after death. In the first place, these integumentary structures do not occur only along the midline, as has been pointed out above, because the specimen is not seen in perfect lateral view. The dorsal midline of the skull is not contiguous with the profile of the specimen. In the second place, taphonomic factors have favored the preservation of soft-part structures along the perimeter of the fossils in these deposits, regardless of the local anatomy. But although sparsely distributed, these integumentary structures do not occur only along the perimeter of the specimen. In the type specimen they can be seen on the

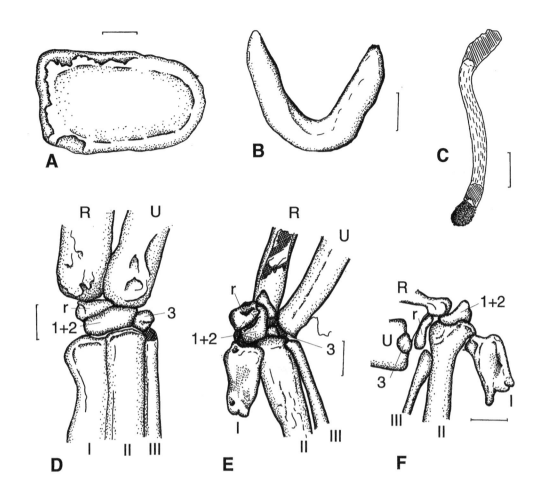

side of the skull, behind the right humerus, and in front of the right ulna (Chen et al. 1998). In NGMC 2124 integumentary remains are visible on the sides of the body and along the abdomen, near the knee joint, and several centimeters from the body. Chen et al. (1998) noted that the distance of the integumentary fibers from the bones could be expected to be proportional to the thickness of the soft tissue between the bones and skin. This appears true for the holotype, in which the peripheral halo of fibers is complemented by additional integumentary remains preserved in the region of the ilium (plate 1C; fig. 10.1A). The view that these structures are collagenous fibers from the body midline is simply indefensible.

Protarchaeopteryx

Still known only from the holotype (NGMC 2125; Ji et al. 1998), *Protarchaeopteryx* has a well-preserved skull and jaws about 70 mm long with sharp teeth that bear faint serrations. The humerus is 87 mm long and the forearm approximately 74 mm long; compared to the femur of 125 mm and tibia of 160 mm, the arm to leg ratio is approximately 57%.

The scapulae are incompletely preserved, and the coracoids overlie

Figure 10.2. (opposite page)
(A) drawing of left lateral
posterior dorsal region of
Sinosauropteryx (NGMC 2123),
as seen in plate 1C, showing four
posterior dorsal vertebrae and
ribs and proximal portion of left
femur. The ilium is indistinctly
preserved dorsal to the femur.
Well above the osteological
remains, and off center from
them, are the filamentous
integumentary fibers. (B) tail
feather of Protarchaeopteryx
(NGMC 2125), similar to that in
plate 1F. Note asymmetry of vane
angles; no pigment or structural
elements of the shaft can be
discerned. (C) isolated
plumulaceous feathers of
Protarchaeopteryx (NGMC
2125), similar to those shown in
plates 1G and 2A,B. (D)
posterior caudal region of
Caudipteryx (NGMC 97-4-A);
the dotted line at left represents a
break in the specimen, the
concentric arc segments represent
an ostracod shell, deep cross-
hatching represents unpreserved
bone. Horizontal lines dorsal to
the vertebrae represent pigmented
soft tissue; diagonal lines ventral
to the vertebrae represent the
bases of rectrices. (E) proximal
hemal arches of the caudal
vertebrae of Caudipteryx
(NGMC 97-4-A), showing
various fibrous and filamentous
integumentary structures. Scale: 5
mm.

each other, so that it is difficult to discern their outlines. They are disarticulated from the sterna (plate 1D). The left sternal plate (fig. 10.2A) is well preserved, slightly concave on its internal face, and approximately 25 by 15 mm, with rounded borders and corners. The right sternal plate underlies the coracoids. Ji et al. (1998, 753) stated, "The clavicles are fused into a broad, U-shaped furcula (interclavicular angle is about 60°) as in *Archaeopteryx, Confuciusornis* and many non-avian theropods." However, a different interpretation is possible: that this is only one clavicle, incompletely preserved. The preserved portion is about 22 mm long, with a sigmoid curvature (plate 1E; fig. 10.2C). The proximal 5 mm of its attachment to the right coracoid is visible only as an impression; from this point, the rodlike bone curves downward and inward toward the plane of the slab. As this arc relaxes, the distal end appears to curve into the slab, but preparation has revealed no further evidence of the bone. The shaft of the clavicle appears cylindrical, and is distinguished from the pectoral girdle bones by its grainy surface finish, suggesting incomplete growth (perhaps correlated with an absence of fusion). The diameter of the shaft is only slightly more than 2 mm and the bone was at least 30 mm long, counting impressions of unpreserved bone. Its aspect ratio is therefore high (AR = 15), much like the clavicles and the ribs (with which clavicles are often confused) in basal theropods, such as *Segisaurus* (Camp 1936). In contrast, the furcula in a specimen of *Confuciusornis* (fig. 10.2B; Tyrrell Museum of Palaeontology TMP 98.14.2) is 36 mm long but 4 mm broad (AR = 9), nearly twice as robust as in *Protarchaeopteryx*. Like the furculae of other basal (non-neornithine) tetanurans, that of *Confuciusornis* is boomerang shaped and ostensibly rather flat, not sigmoidal. Based on its shape, size, and aspect ratio, we suggest that the structure previously identified as a furcula in *Protarchaeopteryx* may be a single clavicle, incompletely preserved. It is not possible to determine whether its clavicles were fused into a furcula.

The humerus is 118% of the forearm in *Protarchaeopteryx*, negligibly different from the ratios in *Confuciusornis* (114–116%) and just longer than in *Archaeopteryx* (110%), but its arm is proportionally much shorter than the leg (57%) than in the latter two taxa (100% and 96%, respectively). On the basis of these ratios alone it is unlikely that any flight stroke motion would have been effective, although even with slight feathering the laterally extended forelimbs could have been effective in turning as well as in augmenting lift during a running leap (Caple et al. 1983). Burgers and Chiappe (1999) have suggested that even such a short arm would still be useful in generating thrust. Crucial to the generation of thrust in the flight stroke is the configuration of the wrist (Ostrom 1974, 1997; Padian, 1985; Vazquez, 1992, 1994), which in maniraptorans is typically semilunate (Ostrom 1974; Holtz 1996), although the possession of such a wrist does not by itself guarantee the ability to generate thrust. *Protarchaeopteryx* has three carpals: a radiale, a fusion of distal carpals 1 + 2, and a round distal carpal 3 (fig. 10.2D). The left wrist is better preserved; its radiale and dc 1 + 2 are flat and lozenge shaped, and the form of the latter does not appear especially semilunate. Distal carpal 3 is bordered by dc 1 + 2 and the second

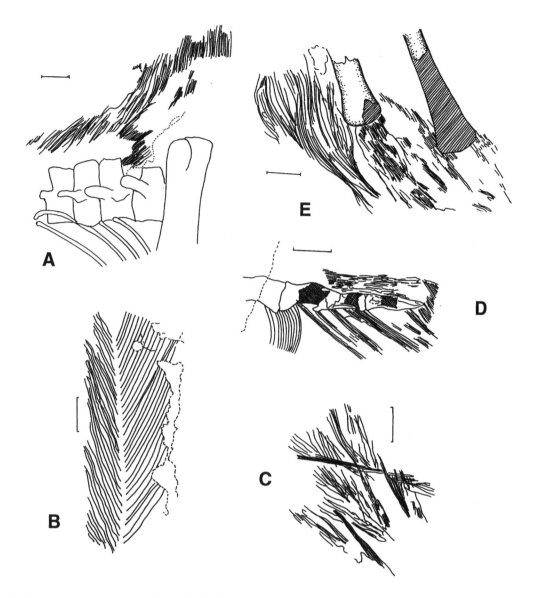

and third metacarpals, suggesting a limitation on the lateral rotation of the wrist. A break through the right wrist has distorted the carpals, and dc 3 is not preserved. As Ji et al. (1998) note, the hands are relatively longer than in any theropods other than *Archaeopteryx* and *Confuciusornis*.

Two kinds of feathers are preserved. The rectrices that radiate from the end of the tail (their anterior extent along the vertebral column cannot be determined) were at least 16 cm long; Ji et al. (1998) measured one at 132 mm from the closest caudal vertebra. The bases of the quills can be seen adjacent to the caudal vertebrae. The feathers clearly have a central rachis about 1 mm in diameter, seen as a light shadow with no distinct features. The vanes are symmetrical and up to 8 to 10 barbs occur per cm of shaft (plate 1F; fig. 10.1B); the regular spacing and straightness of the barbs have suggested the presence of barbules (Ji et al. 1998), but none can be seen clearly. Ji et al. (1998) noted at least 12 rectrices preserved, at least partially. The vanes are at least 5 mm wide on either side of the rachis, and individual barbs may exceed 15 mm in length.

The second kind of feather is plumulaceous. These feathers do not have a central rachis; rather, they consist of filaments about 30 mm long, of which half the length is free and half is gathered and apparently cemented into a proximal shaft of parallel-sided filaments (plates 1G, 2A,B; fig. 10.1C). The shafts are 1.3–1.5 mm in diameter. The filaments are finer and more densely gathered than the barbs of the rectrices. They are found in the chest area, near the femora and proximal caudals, and at the extreme upper left corner of the slab.

Caudipteryx

The type specimen is NGMC 97-4-A; the paratype is NGMC 97-9-A Ji et al. (1998). The skull length is 76 mm in the former and 79 mm in the latter specimens, and most bones are of approximately similar lengths where they can be measured. The humerus-to-forearm ratios are 144% and 122%, respectively, but a crack cleaves both forearms in the type specimen, so the former number may not be as reliable. The arm is 45% of the leg in the type specimen and 37% in the paratype, again suggesting a distortion based on the forearm of the type; the lower figure is more likely representative.

The paratype has the better preserved pectoral girdle (plate 2C,D). The posterior ends of the coracoids are obscured, so it is difficult to discern the sternal end of the coracoid and its shape. The scapula is 77 mm long and its midshaft diameter is 8 mm, broadening to 18 mm toward the coracoid, thus comparatively more robust than in *Sinosauropteryx*. The coracoid is 30 mm deep; it extends posteroventrally perhaps 4 to 5 cm, but this is difficult to ascertain. The glenoid fossa is relatively better preserved than in the other Liaoning forms; it is typically theropodan, oriented not laterally but posterolaterally. A clavicular fragment of the right side is preserved contiguous with the anterior border of the right coracoid; it is 15 mm long and 2 mm broad, and its distal end is incomplete. As in *Protarchaeopteryx*, this bone seems too slightly built to be part of a typical tetanuran furcula, and appears to retain the rodlike shape typical of basal theropods.

As Ji et al. (1998) noted, *Caudipteryx* has three distinct carpals that can be interpreted as the radiale, dc 1 + 2, and dc 3 (fig. 10.2E). They are very similar to the homologous bones in *Protarchaeopteryx*, though again the form of dc 1 + 2 is not as distinctly semilunate as in *Archaeopteryx* and various dromaeosaurs (Ostrom 1974); its surface is biconcave toward the radiale, against which it is closely appressed, as in dromaeosaurs. This situation is also similar to the condition in the therizinosauroid *Beipiaosaurus* (Xu, Tang, and Wang 1999, fig. 2d) and in *Confuciusornis* (fig. 10.2F).

Three kinds of feathers have been identified in *Caudipteryx* (Ji et al. 1998). The rectrices of the tail have been described by Ji et al. (1998). They noted ten complete and two partial rectrices, eleven attached to the left side of the tail, two attached to each side of the last five or six caudals, but not to more anterior ones. The length of the tail feathers exceeds 18 cm. As in *Protarchaeopteryx,* the feather shafts are seen only as light shadows suggestive of unpreserved structure (plate 2E,F). The distance between these shafts is 3–8 mm, but they may overlap extensively. The breadth of each vane ranges from 1.5 to 2 cm. The distal region of the tail preserves both a dorsal and ventral integumentary fringe, 10–15 mm long and typically filamentous. There are no distinct shafts in this region, but light spaces in their place (fig. 10.1D).

At least 14 remiges attached to the second digit of the type specimen have been identified (plate 3A,B). Ji et al. (1998) noted that in contrast to more derived birds, the distal remiges are shorter than the more proximal ones, ranging from 30 to 95 mm, with barbs 6.5 mm long. Like many of the tail feathers, they lack any evidence of a central shaft but a light shadow.

There are also isolated filamentous tufts, like the plumulaceous feathers seen in *Protarchaeopteryx,* in the pectoral region below the coracoids, around the hips, and at the base of the tail (plate 3C; fig. 10.1E). They are densely filamentous and about 2 cm long.

Discussion

Comparison of the Integumentary Structures of Sinosauropteryx *to Feathers*

The principal differences are, first, that the filaments of *Sinosauropteryx* are finer and denser by nearly an order of magnitude; and second, that there is no central shaft around which the filaments are organized (they do not appear to branch), nor any obvious secondary structures such as barbules. So what is the justification, if any, in regarding these integumentary structures as "protofeathers"—whatever that term may mean?

In the first place, these are integumentary and ostensibly therefore epidermal structures. Topologically, they are in the same position as feathers; morphologically, they share the same filamentous features; and compositionally, they appear to have similar keratinous structure, based on findings in *Shuvuuia* (Schweitzer et al. 1999). They therefore pass the tests of similarity that have been operational since Richard Owen synthesized the concept of homology in the 1840s. We cannot say if these structures grew from invaginated follicles like feathers do. We do not know how they developed; if they do not share all the developmental features of the feathers of living birds, it is because the latter structures are far more complex.

Furthermore, because these structures are the body covering, either feathers evolved from them, or they evolved from similar antecedent structures, or they are completely different and homologous only at a more remote epigenetic level. Although they are not feathers, we recognize feathers in other lineages of coelurosaurs, now including oviraptorosaurs, even though these do not perhaps have all the features and diversity of feathers in extant birds (for example, the presence of barbules in the Liaoning feathers has been inferred but has been difficult to confirm). They therefore pass the test of phylogenetic congruence.

Thus, in favor of the homology of these filamentous structures with feathers, we have the facts that (1) they appear in the same place as feathers, (2) they have at least some features of feathers (for example, they are based on thin, nonbranching, filamentous structures of high aspect ratio), and (3) they served at least some functions of feathers (they are de facto insulatory, and their colors would have either camouflaged them, advertised them, or offered a basis for species recognition). Perhaps beyond this, we have (4) that they fit at least a general expectation of what a simple precursor structure to a complex feather may have looked like. Histological and biochemical analyses provide additional information; already epidermal keratin signals of integumentary, featherlike structures in alvarezsaurids match most closely those of birds (Schweitzer et al. 1999). It is not important if these structures are not identical to those of living bird feathers, because there are so few structures with which feathers and these integumentary filaments can be compared.

It must be remembered that *Sinosauropteryx* was not the lineal ancestor of birds, so its features may not necessarily be strictly antecedent to those of birds. These integumentary structures could simply be a property of an evolutionary side branch in coelurosaurs. Against this, however, is the discovery of very similar features in therizinosaurids, alvarezsaurids, and dromaeosaurs, which suggests a more general distribution and hence a greater possibility that such things were indeed antecedent structures from which other integumentary expressions such as feathers evolved. How else can they be explained?

Possible Evolutionary Sequence from Protofeathers to True Feathers

The condition in *Sinosauropteryx* represents one starting point for basing hypotheses about the evolution of feathers on actual evidence, if the argument of the preceding section can be taken provisionally.

The filaments of *Sinosauropteryx* are distributed at about 10 to 15 per cm of parasagittal length. In *Caudipteryx* and *Protarchaeopteryx* the smaller, isolated plumulaceous feathers may represent such filaments gathered into tufts. The shafts of these feathers are of the same color as their terminal filaments, and the outsides of these shafts show parallel lines of the same breadth and direction as the terminal filaments. It appears possible that these shafts, at least, were formed by the consolidation of individual filaments. The shafts are approximately the same length as the terminal filaments, which do not branch and do not have central shafts (sometimes several filaments are gathered together, but they remain distinct in microscopic analysis: plates 1G, 2A,B; fig. 10.1C). We do not see clear evidence whether these shafts are solid or hollow.

At some point a central shaft, or rachis, evolved. It is fainter in the plumulaceous, downy feathers of living birds than in other feather types that have a more structural role in flight, display, or body covering. Shafts are present in the feathers of *Caudipteryx* and *Protarchaeopteryx,* but no direct evidence of them remains; their presence is inferred by the course of barbs from either side. The presence of a shaft alone provides evidence of a rudimentary feather. Vanes evolved either with this advance or afterward; we know of no feathers that are not organized into a two-dimensional form characterized by vanes, even though many feathers have loosely organized barbs at their bases that reflect the absence of barbules. These barbs may be homologous with the filaments of animals such as *Sinosauropteryx*.

The vanes of the feathers of *Caudipteryx* and *Protarchaeopteryx* have parallel barbs, which suggests the presence of barbules. Barbules had to evolve before feathers, as we know them in birds today, could be airworthy, that is, useful in an airfoil. If the neatly parallel barbs of the feathers of *Caudipteryx* and *Protarchaeopteryx* reflect the presence of barbules, then clearly barbules evolved before flight did, but why? A question that remains to be answered is whether some selective pressure in insulation or display, or another function, such as thrust generation (Burgers and Chiappe 1999), might have favored their evolution. The simplest explanation is that feathers evolved directly for flight, and that feathered nonavians have secondarily shortened the forelimbs and feathers. However, this scenario does not appear to be favored when tested against phylogenetic evidence (e.g., Sereno 1999).

Exaptive and Adaptive Roles of Feathers in Nonavian Theropods and Birds

If feathers did not evolve directly for flight, why did they evolve? Recurring to *Sinosauropteryx*, we find integumentary structures that would have had an insulatory function simply because they are so long and densely distributed all over the body. Therefore, this long-advanced function is upheld here. What is not yet clear is why an insulatory function was needed, and at what stage in development. Traditionally, advantages of thermoregulatory structures have been claimed for either juveniles or adults, and for either ectotherms or endotherms, but this may be a pair of false dichotomies. Neonate birds and mammals almost always have less integumentary insulation than adults do. But their growth rates and metabolic levels are usually higher than those of adults, because they are growing rapidly, and histological evidence shows that this was true of Mesozoic dinosaurs and crocodiles, as it is of living mammals and birds (Horner et al. in press). It is possible that feathers and fur first evolved as insulatory structures in neonates that were maintained and elaborated in adults for different purposes, but this hypothesis appears for the present untestable. On the other hand, many neonate birds and mammals take a long time to develop an insulatory covering and some (e.g., humans) never grow an effective coat.

The complementary question is why an insulatory function would be more useful in adults than in neonates, because adults have greater thermal inertia. Much depends on whether the organisms in question use their integumentary structures to retain heat or to help shed it; both functions are possible, and are used in birds. In all probability, behavioral thermoregulatory strategies preceded the evolution of such integumentary structures, which in turn allowed further behavioral and ecological possibilities. A further possibility is that at least some of these structures evolved in adults to help the young thermoregulate: several examples are now known of oviraptorids sitting on their nests as birds do today, using their forelimbs to cover their eggs (Clark et al. 1999). In these specimens, the forelimbs are partly folded, elbows drawn back, forearms and hands spread laterally. Given the distribution of feathering along the arms in *Caudipteryx* and *Protarchaeopteryx,* the latter regions covering the eggs may have provided the most insulatory effect. This may have had a strong adaptive role in the evolution of at least some feathers. If this is correct, then the role of insulation in the evolution of feathers may have been more important in adults for the purpose of the young than for either the adults or young in their own thermoregulation.

A behavioral role in display, camouflage, or species recognition has sometimes been advanced as the original function of feathers (e.g., Cowen and Lipps 1982). These suggestions, though not falsified, have been difficult to test. There is considerable evidence that many groups of dinosaurs were highly social (e.g., Horner 1999; Horner and Dobbs 1997). Variation in crests, horns, and frills among ornithischian dinosaur groups, with comparatively little postcranial variation, has suggested that these cranial adornments had primarily behavioral purposes (Vickaryous and Ryan 1997). Given that living birds are highly visual creatures it would seem plausible to draw the inference back phylogenetically through their theropod ancestors, especially inasmuch as it seems to hold in their more distant ornithischian relatives among the dinosaurs. But crocodiles, the closest living outgroup to dinosaurs, are not so highly visual, and so the test is not strongly reinforced (Witmer in press). Furthermore, neoceratopsians, hadrosaurs, and pachycephalosaurs must have evolved these complex structures independently, because their common ancestors among basal ornithischians do not share them (at least as far as hard parts go; behavior cannot be assessed).

The Liaoning deposits are the first to preserve such integumentary structures in nonavian theropods, even though these animals are known worldwide. But these deposits have far more theropods than any other sort of tetrapod; in fact, only the ceratopsian *Psittacosaurus* represents the other dinosaurs, and rumors of its integumentary structures are dubious because artifice is suspected. Among theropods, to some extent, we are required to argue mostly from negative evidence. Given the limited distribution of known integumentary structures, inferences are accordingly limited. There is a de facto insulatory function, and possible behavioral functions. Ostrom (1979) proposed that feathers evolved in part to trap prey, as a sort of insect net. He later abandoned this model, saying that it had accomplished its purpose heuristically, when Caple et al. (1983) showed that an animal leaping into the air, using small feathers as an insect net, would have upset its dynamic equilibrium and so have frustrated attempts at flight. But this demonstration does not invalidate the use of some feathers in some animals as insect nets; it only suggests that it would have been counterproductive to the evolution of flight. However, as we now know, feathers did not first evolve in birds, and they did not first evolve for flight. It is still possible, for example, that the feathers seen (and not seen) in some Liaoning taxa reflect annual ephemeral possession of feathers, perhaps during mating and brooding season.

As noted above, and in contrast to birds, the distal remiges in *Caudipteryx* are shorter than the more proximal ones (Ji et al. 1998). This feature is also found in *Archaeopteryx,* and Heilmann (1927, 100–105) surveyed its distribution, development, and use in living neonate birds. The hoatzin is perhaps the best known nestling with retarded distal primaries; when born, its claws are free, and it uses them and its beak to clamber through bushes. As Heilmann noted, if the distal primaries grew as quickly as the other feathers, the utility of the claws would be hampered. This has often been cited as evidence that bird ancestors were arboreal, but the hoatzin is neither a bird ancestor nor a basal bird. Furthermore, other birds such as the currasows, common fowl, turkeys, and megapodes have the same developmental sequence, yet their fingers are not separate and their claws not large and curved, nor used for climbing. As *Archaeopteryx* and *Caudipteryx* show, this feature in some living neonate birds is a reversion to a primitive character state. Because so much thrust is generated from these distal primaries, the wing cannot be effective in some types of flight until they are fully grown. *Archaeopteryx* shows, however, that the more proximal primaries and secondaries, when sufficiently long, are capable of sustaining flight.

Other Features Related to Flight and Their States in the Feathered Nonavian Dinosaurs

These other features relate mainly to the shoulder girdle and forelimb, though the tail and the structure of the bones, especially of the hind limb, are also of interest.

The flight stroke is the central focus of the evolution of flight, that is to say, active flapping flight (Padian 1985; Rayner 1988). An airfoil built for gliding may sustain the animal in the air, but in order to generate thrust, the forward component of flight, it must have an internal structure capable of providing integrity to the wing when it is deformed. The flight stroke, of course, deforms the airfoil even more than air currents do to the patagium of a passive glider. For this reason, the wings of flapping animals have structural elements, such as the feather shafts of birds, that help the wing to maintain an aerodynamically efficient profile.

The shoulder girdles of the Liaoning nonavian theropods are not generally well preserved. In none is the glenoid fossa well enough preserved to determine the range of movement possible for the humerus, although it appears to face posterolaterally in *Caudipteryx* (plate 2C). Jenkins (1993) pointed out that in nonavian theropods, as in other dinosaurs, the glenoid faces posterolaterally, in *Archaeopteryx* laterally, and in extant birds dorsolaterally. At least two nonavian coelurosaurs have been reconstructed with a lateral glenoid orientation (Novas and Puerta 1997; Norell and Makovicky 1997), but no integument was preserved with these specimens. At present, therefore, we do not understand the coevolution of feathers and shoulder elevation.

The form of the shoulder girdle itself, however, suggests some particulars about functional abilities related to the evolution of flight. In *Sinosauropteryx*, which did not have feathers and had short arms, the length of the scapula is more than ten times its breadth, and well over twice the depth of the coracoid. The remnant of the clavicle hugs the coracoid but is not broad or robust, suggesting the riblike form seen in basal theropods (Camp 1936). *Protarchaeopteryx* and *Caudipteryx*, as noted above, have similar clavicles of riblike form, not boomerang shaped as in many tetanurans (Padian 1997). Their pectoral girdles, however, are more robust than in *Sinosauropteryx*. *Protarchaeopteryx* and *Caudipteryx* have broad, ovoid sternal plates approximately as large as the coracoids. The coracoid seems to have a small ventral process in *Protarchaeopteryx* and *Caudipteryx*, but it is not clear whether this process articulates with a groove in the sternum, as seen in other maniraptorans (Norell and Makovicky 1997). The scapula is complete in the paratype of *Caudipteryx* (NGMC 97-9-A), and it is much broader than in *Sinosauropteryx*, flaring at its distal and proximal ends; a slight dorsal lip to the glenoid is indicated, and there is a faint acromion process.

Tetanuran theropods appear to have evolved features that were exaptive for the flight stroke, but did not first function in this capacity. The ossified sternal plates, the fusion of the clavicles into a broad furcula, and the lengthening of the arms and hands are found not only in birds, but in tetanurans that did not fly. Both the sternal plates and the furcula anchor muscles that in birds draw the forelimbs forward and medially, a major component of the flight stroke (Padian in press). In these nonavian theropods, the posterior component of the sternum was not as well developed as in birds, reflecting less capacity for retracting the forelimbs posteriorly, as birds do in flight. For these theropods it was not as important to draw back the forelimbs as it was to bring them forward and together, and the development of the sternal plates and furcula enabled this; the concomitant elongation of the prehensile hands and arms suggest a dedication to improvement of the predatory motions of the forelimb (Padian in press). The sternal plates of *Protarchaeopteryx* and *Caudipteryx* are comparable to those of *Archaeopteryx*, but the clavicles are not; nor are they as well developed as in allosaurs, tyrannosaurs, oviraptorosaurs, or dromaeosaurs. In contrast to the development of the feathers in the former two genera, the rudimentary condition of the clavicles suggests no particular adaptation toward the functions necessary to perform the flight stroke. (The furcula is absent in many ground-dwelling parrots, and is incompletely ossified even in some that fly actively.)

The configurations of the wrists in *Protarchaeopteryx* and *Caudipteryx* contrast morphologically and functionally. *Protarchaeopteryx* has a radiale of standard form, but its fused distal carpal 1 + 2 is not particularly semilunate (fig. 10.2D); moreover, distal carpal 3 articulates with it and metacarpals II and III, and would appear to limit the lateral rotation of the wrist necessary for an effective flight stroke (though some sideways rotation appears possible). The situation in *Caudipteryx* is complicated by the presence of supernumerary bones of indeterminate origin (fig. 10.2E), but is more like the condition in birds and dromaeosaurs. It has a large radiale and a small distal carpal 3. Distal carpal 1+2 is tightly articulated to the radiale and to the first two metacarpals; the same configuration is seen in *Confuciusornis* (fig. 10.2F) and in *Beipiaosaurus* (Xu, Tang, and Wang 1999, fig. 2d). In all these animals, as in dromaeosaurs and basal birds, the first metacarpal is exceptionally broad and robust, and the third is much less robust than the others.

The hindlimbs in each of the Liaoning nonavian theropods are robust and larger than the forelimbs. Taking an index of humerus + forearm + metacarpal II divided by femur + tibia + metatarsal III, the percentage of forelimb to hind limb is about 32% in the holotype of *Sinosauropteryx*, 55% in the holotype of *Protarchaeopteryx*, about 40% in the holotype of *Caudipteryx* and 34% in the paratype. In the Berlin specimen of *Archaeopteryx* the ratio is 94%, and two specimens of *Confuciusornis* yielded 95% and 100%. Consistent with the short feathers, none of these nonavian theropods appears to be close to achieving flight. On the other hand, their bone walls are remarkably thin. In most theropods the midshaft thickness of the bone walls is around 20% of the diameter (KP, unpub. data; 18%-23% is typical, depending on the section of bone). In table 10.1, the measurements also indicate that the thinning of the bone walls is not an adaptation for flight; rather, it evolved in theropods that had few or no other adaptations related to flight.

TABLE 10.1
Tibia Length Contrasted with Diameter
and Bone Wall Thickness (BWT) at Midshaft (mm)

Specimen	Tibia	BWT	Diam.	BWT/D
Sinosauropteryx NGMC 2123	60	0.5–1.0	6.4	8–15%
Sinosauropteryx NGMC 2124	147	1.5–1.65	10	15–16.5%
Caudipteryx NGMC 97–4–A	188	1.5	12–13	11.5–12.6%
Caudipteryx NGMC 97–9–A	190	<2.0	13	<15%
Protarchaeopteryx NGMC 2125	165	1.5–1.75	13–14	10.7–13.5%
Archaeopteryx (Berlin)	70	0.25–0.5	3.0–3.4	7.4–16%
Confuciusornis TMP 98.14.1	63	0.8–1.0	5	16–20%

Conclusions

The feathered nonavian dinosaurs of Liaoning are stunning in their preservation of feathers and other integumentary structures. These features occur in animals in which we could only have guessed that they would be present. And the diversity of coelurosaurian lineages that preserve such integumentary structures generally agrees with our expectations of the evolution of feathers according to phylogenetic hypotheses already postulated. That is, *Sinosauropteryx* is a basal compsognathid coelurosaur, and it has only rudimentary integumentary

Figure 10.3. A "gentleman's cladogram" representing a tree of the immediate relatives of birds among coelurosaurian theropods. This tree summarizes some of the current agreements and uncertainties of taxa most closely related to birds. The positions of troodontids and other taxa, such as tyrannosaurids, therizinosaurids, and Protarchaeopteryx, are mutable among various cladistic analyses. However, because filamentous integumentary structures are now known in Compsognathidae (Sinosauropteryx), Therizinosauroidea (Beipiaosaurus), and Dromaeosauridae (Sinornithosaurus), as well as alvarezsaurids (not pictured here, but variously proposed to be basal birds or

structures that, nevertheless, share some features with the rudimentary feathers seen in *Protarchaeopteryx* and *Caudipteryx*. These feathers, in turn, share some features with those of *Archaeopteryx* and the other true birds. But some other, more uncomfortable questions must also be asked. If feathers are found in oviraptorosaurs such as *Caudipteryx*, should they not also be present in troodontids if Arctometatarsalia is a valid taxon? If they are present in *Protarchaeopteryx* should we not expect to see them in dromaeosaurs, instead of merely the *Sinosauropteryx*-like integumentary structures reported to date? Either feathers evolved independently in several coelurosaurian lineages, or our understanding of coelurosaurian phylogeny needs revision (fig. 10.3).

Our analysis indicates that the feathered nonavian dinosaurs of Liaoning had few if any skeletal components with the equipment necessary for the avian flight stroke. It was already known that these

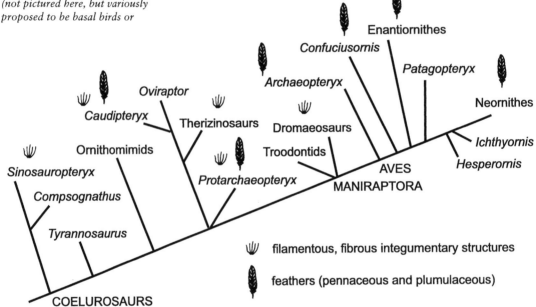

ornithomimid relatives), Caudipteryx, and Protarchaeopteryx, they can conservatively be considered present from basal Coelurosauria to Aves. Feathers are now known in Oviraptorosauria (Caudipteryx) and Protarchaeopteryx, as well as Aves, so they could also be found at least in Therizinosauroidea, Dromaeosauridae, and Troodontidae, if this phylogeny is correct. Based on several sources and kindly furnished by Dr. T. R. Holtz.

feathers must have evolved for purposes other than flight (Ji et al. 1998; Padian 1998). Behavioral mechanisms such as display, camouflage, and species recognition remain viable but poorly testable hypotheses. Thermoregulation is a given, but what sort of thermoregulation—shedding heat or gaining it? We suggest that this is a false dichotomy. Most animals that use their integument or other structures for thermoregulatory reasons do so not automatically, but flexibly: their behavior controls the use of the integument in shedding or gaining heat. We further suggest that feathers on the arms may have been at least partly selected at some stage for the thermoregulation of eggs in nests by their parents. And so, once again, we recur to behavior in several different respects as a primary potential motor of the evolution of feathers, which were exapted later for flight and other purposes.

Acknowledgments: We thank John Hutchinson, Luis Chiappe, and Tom Stidham for comments and readings of the manuscript, Tom Holtz for advice on current phylogenetic schemes, and the University of California Museum of Paleontology for support of the research for this chapter. Photographs were processed at the Scientific Visualization Center in the Valley Life Sciences Building at UC Berkeley. We are particularly grateful to Phil Currie for his hospitality and collaboration while this work was carried out, though he had no idea of its eventual purpose. We are indebted to Eva Koppelhus for her hospitality and logistic support. Above all, we congratulate Phil and celebrate his contributions to the paleontology of Canada and the world; no one could want a better colleague.

References

Briggs, D. E. G., P. R. Wilby, B. P. Pérez-Moreno, J. L. Sanz, and M. Fregenal-Martínez. 1997. The mineralization of dinosaur soft tissue in the Lower Cretaceous of Las Hoyas, Spain. *Journal of the Geological Society of London* 154: 587–588.

Burgers, P., and L. M. Chiappe. 1999. The wing of *Archaeopteryx* as a primary thrust generator. *Nature* 399: 60–62.

Camp, C. L. 1936. A new type of small bipedal dinosaur from the Navajo Sandstone of Arizona. *University of California Publications in the Geological Sciences* 24: 39–53.

Caple, G., R. P. Balda, and W. R. Willis. 1983. The physics of leaping animals and the evolution of preflight. *American Naturalist* 121: 455–476.

Chen P.-J., Dong Z.-M, and Zhen S.-N. 1998. An exceptionally well-preserved theropod dinosaur from the Yixian Formation of China. *Nature* 391: 147–152.

Clark, J. M., M. A. Norell, and L. M. Chiappe. 1999. An oviraptorid skeleton from the Late Cretaceous of Ukhaa Tolgod, Mongolia, preserved in an avianlike brooding position over an oviraptorid nest. *American Museum Novitates* 3265: 1–36.

Cowen, R., and J. Lipps. 1982. An adaptive scenario for the origin of birds and of flight in birds. In B. Mamet and M. J. Copeland (eds.), *Proceedings, Third American Paleontological Convention* 1: 109–112. Toronto: Business and Economic Service.

Geist, N. R., T. D. Jones, and J. A. Ruben. 1997. Implications of soft tissue preservation in the compsognathid dinosaur, *Sinosauropteryx*. *Journal of Vertebrate Paleontology* 7 (suppl. to no. 3): 48A.

Heilmann, G. 1927. *The Origin of Birds*. New York: Appleton.

Holtz, T. R., Jr. 1996. Phylogenetic taxonomy of the Coelurosauria (Dinosauria: Theropoda). *Journal of Paleontology* 70: 536–538.

Horner, J. R. 1999. Dinosaur reproduction and parenting. *Annual Reviews of Earth and Planetary Sciences* 28: 19–45.

Horner, J. R. and E. Dobb. 1997. *Dinosaur Lives: Unearthing an Evolutionary Saga*. New York: HarperCollins.

Horner, J. R., K. Padian, and A. de Ricqlés. In press. Comparative osteohistology of some embryonic and perinatal archosaurs: Phylogenetic and behavioral implications for dinosaurs. *Paleobiology*.

Jenkins, F. A., Jr. 1993. The evolution of the avian shoulder joint. *American Journal of Science* 293A: 253–267.

Ji Q., and Ji S.-a. 1996. [On the discovery of the earliest bird fossil in China and the origin of birds.] *Chinese Geology* 233: 30–33. (In Chinese.)

Ji Q., P. J. Currie, M. A. Norell, and Ji S.-a. 1998. Two feathered dinosaurs from northeastern China. *Nature* 393: 753–761.

Norell, M. A., and P. J. Makovicky. 1997. Important features of the dromaeosaur skeleton: Information from a new specimen. *American Museum Novitates* 3215: 1–28.

Novas, F. E., and P. F. Puerta. 1997. New evidence concerning avian origins from the late Cretaceous of Patagonia. *Nature* 387: 390–392.

Ostrom, J. H. 1973. The ancestry of birds. *Nature* 242: 136.

Ostrom, J. H. 1974. *Archaeopteryx* and the origin of flight. *Quarterly Review of Biology* 49: 27–47.

Ostrom, J. H. 1979. Bird flight: How did it begin? *American Scientist* 67 (1): 46–56.

Ostrom, J. H. 1997. How bird flight might have come about. In D. Wolberg and G. Rosenberg (eds.), *Dinofest International,* pp. 301–310. Philadelphia: Academy of Natural Sciences Press.

Padian, K. 1985. The origins and aerodynamics of flight in extinct vertebrates. *Palaeontology* 28: 423–433.

Padian, K. 1987. A comparative phylogenetic and functional approach to the origin of vertebrate flight. In B. Fenton, P. A. Racey. and J. M. V. Rayner (eds.), *Recent Advances in the Study of Bats,* pp. 3–22. Cambridge: Cambridge University Press.

Padian, K. 1995. Form and function: The evolution of a dialectic. In J. J. Thomason (ed.), *Functional Morphology and Vertebrate Paleontology,* pp. 264–277. Cambridge: Cambridge University Press.

Padian, K. 1997. Pectoral girdle. In P. J. Currie and K. Padian (eds.), *Encyclopedia of Dinosaurs,* pp. 530–536. San Diego: Academic Press.

Padian, K. 1998. When is a bird not a bird? *Nature* 393: 729–30.

Padian, K. In press. Stages in the evolution of bird flight: Beyond the arboreal-cursorial dichotomy. In J. A. Gauthier (ed.), *New Perspectives on the Origin and Early Evolution of Birds.* New Haven: Yale University Press.

Padian, K., and L. M. Chiappe. 1998a. The origin of birds and their flight. *Scientific American,* February 1998, 28–37.

Padian, K. and L. M. Chiappe. 1998b. The origin and early evolution of birds. *Biological Reviews* 73: 1–42.

Pérez-Moreno, B. P., J. L. Sanz, A. D. Buscalioni, J. L. Moratella, F. Ortega, and D. Rasskin-Guttman. 1994. A unique multitoothed ornithomimosaur dinosaur from the Lower Cretaceous of Spain. *Nature* 370: 363–367.

Rayner, J. M. V. 1988. The evolution of vertebrate flight. *Biological Journal of the Linnean Society* 34: 269–287.

Schweitzer, M. H., J. A. Watt, R. Avci, L. Knapp, L. Chiappe, M. Norell, and M. Marshall. 1999. Beta-keratin specific immunological reactivity in feather-like structures of the Cretaceous alvarezsaurid, *Shuvuuia deserti. Journal of Experimental Zoology (Molecular and Developmental Evolution)* 285: 146–157.

Sereno, P. C. 1999. The evolution of dinosaurs. *Science* 284: 2137–2147.

Unwin, D. M. 1998. Feathers, filaments, and theropod dinosaurs. *Nature* 392: 119–120.

Vazquez, R. 1992. Functional osteology of the avian wrist and the evolution of flapping flight. *Journal of Morphology* 211: 259–268.

Vazquez, R. 1994. The automating skeletal and muscular mechanisms of the avian wing. *Zoomorphology* 114: 59–71.

Vickaryous, M. K., and M. J. Ryan. 1997. Ornamentation. In P. J. Currie and K. Padian (eds.), *Encyclopedia of Dinosaurs,* pp. 488–493. San Diego: Academic Press.

Witmer, L. M. In press. The debate on avian ancestry. In Chiappe, L. M. and L. D. Witmer (eds.), *Mesozoic Birds: Above the Heads of Dinosaurs.* University of California Press.

Xu X., Tang Z.-I., and Wang X.-I. 1999. A therizinosaurid dinosaur with integumentary structures from China. *Nature* 399: 350–354.

Xu X., Wang X.-L., and Wu X. C. 1999. A dromaeosaurid dinosaur with a filamentous integument from the Yixian Formation of China. *Nature* 401: 262–266.

Section II.
Sauropods

11. New Titanosauriform (Sauropoda) from the Poison Strip Member of the Cedar Mountain Formation (Lower Cretaceous), Utah

Virginia Tidwell, Kenneth Carpenter, and Susanne Meyer

Abstract

A new titanosauriform sauropod from the Early Cretaceous of Utah is described, which displays an unusual mixture of primitive and derived characteristics. While many of these characters indicate an affiliation with the family Brachiosauridae, others hint at a more derived relationship. The specimen most closely resembles the Early Cretaceous brachiosaur *Cedarosaurus*, while displaying several autapomorphies which distinguish it from that genus: the centrum of the proximal caudal vertebra displays a convex anterior surface, whereas the posterior articulation is flat; neural arches and spines of the middle caudal vertebra are slightly inclined anteriorly and rest on amphiplatyan centra.

A preliminary review of recent Cretaceous sauropod discoveries highlights the growing variety of caudal centrum articulations. This variety illustrates the inadequacies of such traditional descriptive terminology as *amphicoelous* and *procoelous*, while emphasizing the need for more precise identifications of anterior and posterior articular face morphology.

Introduction

Sauropod discoveries in North America historically have occurred mainly in the Upper Jurassic Morrison Formation. This period is fairly well represented by abundant remains of such well-known taxa as *Camarasaurus*, *Diplodocus*, and *Apatosaurus*. The record of Cretaceous sauropods from this region, however, is much less complete. For many years *Pleurocoelus* and several specimens of the Maastrichtian titanosaur *Alamosaurus* comprised the only well-documented taxa. Scattered remains of other specimens were often assigned to either of these genera, although their fragmentary nature added little to our understanding of sauropod relationships.

Figure 11.1. (opposite page, top) Location of Tony's Bone Bed, type locality of Venenosaurus dicrocei. Inset, state of Utah showing location of the Cedar Mountain Formation. Dark irregular line is the distribution of the Cedar Mountain Formation.

Recent reports from Lower and Middle Cretaceous sediments have opened a window into this poorly known period (Britt and Stadtman 1996; Britt et al. 1997, 1998; Cifelli et al. 1997; Winkler et al. 1997). These preliminary reports, coupled with the complete description of the new Early Cretaceous brachiosaur *Cedarosaurus weiskopfae* (Tidwell et al. 1999), have begun to illuminate a wide diversity of taxa from the western interior of North America. This region is now the subject of intense study by several institutions, and additional specimens are coming to light on a regular basis.

In 1998 the Denver Museum of Natural History opened a small multitaxon quarry in the Cedar Mountain Formation in eastern Utah (fig. 11.1). To date this quarry has produced adult and juvenile ornithopod skeletons (DiCroce and Carpenter, chap. 13 of this volume), as well as a theropod and an adult and juvenile sauropod. All adult sauropod elements display a similar pattern of weathering or damage, and correspond in size to a small individual. These factors, coupled with an absence of duplicate elements, indicate that only one adult sauropod is represented in this quarry.

Figure 11.2. (opposite page, bottom) Stratigraphic column of the Cedar Mountain Formation in the vicinity of Tony's Bone Bed, showing stratigraphic level of the quarry.

Depositional Setting

The discovery was made in the Poison Strip Sandstone Member of the Cedar Mountain Formation, located in Grand County, east-central Utah. The basal member of the Cedar Mountain Formation is the Buckhorn Conglomerate (Kirkland et al. 1997). Four additional members established and named by Kirkland et al. (1997) are: the Yellow Cat Member, Poison Strip Sandstone, Ruby Ranch Member, and the Mussentuchit Member. The DMNH sauropod occurs in a small bone bed 3.75 m below the top of the Poison Strip Sandstone (fig. 11.2), which in the area of the bone bed is 11 m thick, and consists of medium- to coarse-grained sandstone and lenses of green mudstone. The bone bed lies immediately above a thin lens of gray carbonate, which is in turn overlain by a green, mottled paleosol. Carbonate growths are present on the bones from the quarry.

The Poison Strip Sandstone in the vicinity of Arches National Monument (Utah) is highly variable in thickness and is missing in some

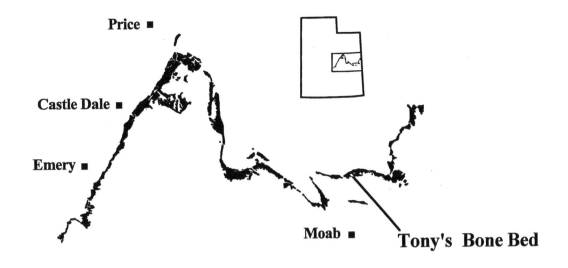

Price ■

Castle Dale ■

Emery ■

Moab ■

Tony's Bone Bed

10 km

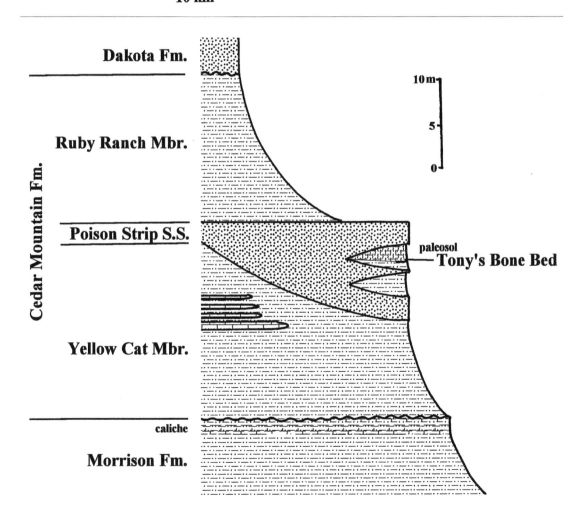

Dakota Fm.

Ruby Ranch Mbr.

Poison Strip S.S.

Cedar Mountain Fm.

paleosol
Tony's Bone Bed

10 m
5
0

Yellow Cat Mbr.

caliche

Morrison Fm.

areas. It represents a meandering river complex (see DiCroce and Carpenter, chap. 13 of this volume). The Poison Strip Sandstone was previously thought to correlate with the mid-Aptian Cloverly Formation (Kirkland et al. 1997), based on a supposed similarity in taxa. However, none of the classic dinosaurs of the Cloverly Formation occur, with the possible exception of *Sauropelta* (Carpenter et al. 1999). The previous report of the Cloverly iguanodontid *Tenontosaurus* in the Poison Strip Sandstone (Kirkland et al. 1997) is erroneous. The specimen is now known to have come from high in the Mussentuchit Member (Kirkland, pers. comm. 1999). A large deinonychid-type ischium found in the DMNH quarry may indicate the presence of *Utahraptor,* which is also known to occur in the College of Eastern Utah's Gaston Quarry and Brigham Young University's Dalton Wells Quarry (Kirkland et al. 1999). Both these localities are found in the stratigraphically lower Yellow Cat Member, suggesting a contemporaneous age for the DMNH quarry. Furthermore, the greenish mudstone lenses within the Poison Strip Sandstone are lithologically more similar to the mudstones of the Yellow Cat Member than of the overlying Ruby Ranch or Mussentuchit Members. While Kirkland et al. (1997) suggest a Barremian age for the Yellow Cat quarries, Britt finds little evidence to support this claim, preferring a more generalized Neocomian designation (B. Britt, pers. comm., 2000). A definitive conclusion must await additional study.

Institutional Abbreviations: DMNH, Denver Museum of Natural History; SMU, Southern Methodist University.

Systematics
Order: Sauropoda Marsh 1878
Titanosauriformes Salgado et al. 1997
Family: Incertae sedis
Venenosaurus gen. nov.

Holotype: DMNH 40932, a single specimen consisting of nine disarticulated caudal vertebrae, left scapula, right radius, left ulna, five metacarpals, four manus phalanges, right pubis, left and right ischium, three metatarsals, astragalus, chevrons, and ribs.

Horizon and Locality: Poison Strip Member, Cedar Mountain Formation (Lower Cretaceous); Grand County Utah, United States. Exact locality data is on file at DMNH.

Etymology: Venenos, "poison" (Latin); *saurus,* "reptile" (Greek). Named for the Poison Strip Member, Cedar Mountain Formation, from which the type specimen was collected.

Diagnosis: As for the species.

Venenosaurus dicrocei sp. nov.

Etymology: for Anthony DiCroce, who discovered the specimen.

Diagnosis: Proximal caudal centrum with convex anterior surface and a flat posterior surface; neural spines of middle caudals incline anteriorly, similar to *Cedarosaurus* and *Aeolosaurus,* centra amphyplatyan, as in *Brachiosaurus;* anterior and middle caudals with deep lateral fossae, shallow in *Brachiosaurus, Cedarosaurus,* and *Saltasaurus;* radius more slender than other taxa except *Cedarosaurus;* ulna with expanded medial wall, well-developed medial process, contrasting with *Camarasaurus* and *Brachiosaurus,* similar to *Cedarosaurus* and most titanosaurs; moderately developed olecranon process, ulna craniomedial process slightly concave, in contrast to *Camarasaurus* and *Brachiosaurus,* similar to titanosaurs; metacarpal I proximal end more slender anteroposteriorly than other sauropods; pubis longer than ischium; pubic articulation of ischium restricted to proximal half of the element, occupying much of the total length in *Andesaurus, Saltasaurus.*

Description

Measurements are given in tables 11.1 through 11. 3.

TABLE 11.1
Measurements of Caudal Vertebrae for DMNH 40932 (mm)

Caudal	Length	Anterior Centrum Width-Height	Posterior Centrum Width-Height	Total Height
Proximal	112	190–153	183–166	420
Anterior	80	185–165	175–155	—
Mid	102	140–112	118–99	362
Posterior	101	58–50	55–49	—

Caudal Vertebrae

A well-preserved proximal caudal vertebra, lacking only the prezygapophyses, displays a unique mixture of characters (fig. 11.3). The moderately elongate neural spine resembles that of *Camarasaurus* in length, although lacking the expanded apex and posterior inclination usually found in that genus. An elongated neural arch accentuates the apparent height of the neural spine. The postzygapophyses are small, complex, winglike structures that do not extend beyond the posterior edge of the centrum. The top half of each postzygapophysis faces dorsolaterally, whereas the bottom, articular half faces lateroventrally. They extend almost to the neural canal as a slight hyposphenal ridge. Prezygodiapophyseal laminae, as defined by Wilson (1999), extend ventrally from the broken prezygapophyses to the caudal ribs. A wide fossa is present on the anterior face of the lamina. The short, triangular caudal ribs are wide dorsally, with sloping, flattened sides that meet ventrally in a robust ridge.

The centrum of the proximal caudal differs from other known sauropods in that the articular surfaces are not biconcave, amphicoelous, or procoelous. Anteriorly it is convex, whereas the posterior surface is flat (fig. 11.3C). Yet, the development of the anterior convex-

TABLE 11.2
Measurements of Limbs and Pelvic Girdle for DMNH 40932 (mm)

Left scapula
    Length   1200
    Width of anterior plate   510 (incomplete)
    Width of posterior blade   292
    Least diameter   180
Radius
    Length   695
    Width, proximal end   155
    Width, distal end   144
    Least breadth   76
    Least circumference   235
Ulna
    Length   768
    Width, proximal end   265
    Width, distal end   121
    Least circumference   266
Pubis
    Preserved length   652
Left ischium
    Length   592
    Least width of shaft   70
Right ischium
    Length   581
    Length pubic articulation   231
    Width pubic articulation   155
    Least width of shaft   90

TABLE 11.3
Measurements for Metacarpals and Metatarsals (mm)

Metacarpal	I	II	III	IV	V
Greatest length	—	358	360	333	301
Proximal width	130	87	99	101	86
Least circum.	—	150	155	147	142

Metatarsal	I	II	IV(?)
Length	136	169	178
Prox. width	70	—	—
Prox. length	124	—	186
Dist. width	185	180	—
Least circum.	200	168	115

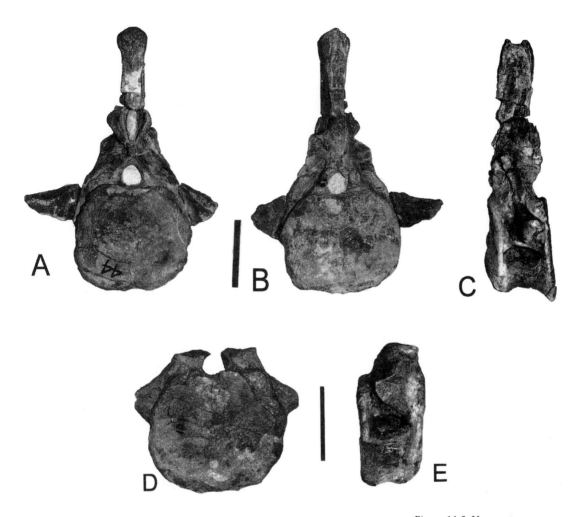

Figure 11.3. Venenosaurus dicrocei *(DMNH 40932). Proximal caudal vertebra in (A) posterior view, (B) anterior view, (C) left lateral view. Note convex anterior surface of centrum. Anterior centrum in (D) anterior view, (E) right lateral view. Note the deep lateral fossa on centrum. Scale: 10 cm.*

ity is slight, far less than that found in *Opisthocoelicaudia* (Borsuk-Bialynicka 1977). Ventrally, the centrum is wide and smooth, lacking the ventral keels or ridges found in *Diplodocus* and *Apatosaurus*. There is no evidence of chevron facets on either the anterior or posterior edge of the centrum, indicating that it is a proximal caudal. A second anterior caudal centrum is more typical of *Camarasaurus* or *Brachiosaurus* in that the articular surfaces are amphiplatyan, with moderately developed chevron facets (fig. 11.3D,E).

Seven middle and distal caudals also were recovered. All the middle caudal centra are short, in contrast to those of titanosaurs. *Andesaurus* (Calvo and Bonaparte 1991), *Malawisaurus* (Jacobs et al. 1993), *Aeolosaurus* (Salgado and Coria 1993), *Alamosaurus* (Gilmore 1946), and *Saltasaurus* (Powell 1992) all possess elongate centra in the middle caudals. None of the *Venenosaurus* centra show the lateral ridges that are often found in *Camarasaurus*. The articular surfaces are amphiplatyan and circular in outline. The wide ventral sides show no evidence of ventral keels or ridges; however, well-developed chevron facets are present on the posterior edges. All the caudal neural arches are located anteriorly on the centrum. This character is apparently shared by all Titanosauriformes. Despite considerable damage to the postzygapophyses, it is evident that none of these caudals posses a hyposphene.

Two of the middle caudals retain incomplete neural spines. These neural spines are angled anteriorly when the vertebrae are aligned so that the neural canal lies horizontally (fig. 11.4A,B). They resemble those of *Cedarosaurus* and the Late Cretaceous titanosaurs *Aeolosaurus* and *Gondwanatitan*. The rudimentary caudal ribs indicate a transitional position between anterior and middle caudals, approximately caudal 11 and 12. A single well-preserved distal caudal bears a neural

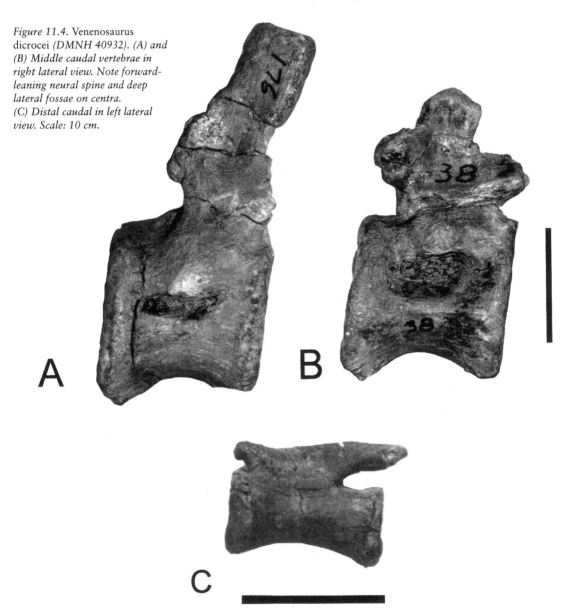

Figure 11.4. Venenosaurus dicrocei *(DMNH 40932). (A) and (B) Middle caudal vertebrae in right lateral view. Note forward-leaning neural spine and deep lateral fossae on centra. (C) Distal caudal in left lateral view. Scale: 10 cm.*

A

B

C

spine reduced to a posteriorly directed rod that extends to the posterior edge of the centrum (fig. 11.4C). Small chevron facets are present on the posterior border. There is no evidence of the well-defined lateral ridge that is found on the neural arch of *Cedarosaurus* distal vertebrae.

A unusual feature shared by many of the caudal vertebrae is the lateral fossa on the sides of the centra, ventral to the caudal ribs (figs. 11.3C, 11.4A,B). These fossae appear as deep depressions in the lateral walls, rather than as pneumatic openings into the interior of the centrum. There is no defined "lip," such as is often found in the caudal pleurocoels of *Diplodocus*. Moderately developed on the proximal caudal, the lateral fossae on the second centrum are deeper, measuring 23 mm and 33 mm on the right and left sides respectively. These are divided into two chambers by an internal ridge. Two *Venenosaurus* middle caudals display moderately developed lateral fossae beneath the rudimentary caudal ribs. Similar structures, although poorly developed, were also found on three of the anterior caudal centra of *Cedarosaurus*. Shallow lateral fossa are also found on anterior caudals of *Saltasaurus, Alamosaurus, Aeolosaurus, Gondwanatitan,* and *Malawisaurus* (Powell 1992; Gilmore 1946; Kellner and Azevedo 1999; Gomani 1999). However, none of these fossae are as deep nor appear to be subdivided into separate chambers as are the fossae in *Venenosaurus*. Only in *Venenosaurus* do these fossae continue into the middle caudal vertebrae. To date, the significance of lateral fossa in caudal vertebrae has not been adequately explored. As more Cretaceous sauropods are studied, it is hoped that the morphological and phylogenetic significance of these features will become clearer.

Forelimb

The left scapula is relatively short and robust (fig. 11.5A). Although the proximal plate is heavily weathered, it appears less expanded than that of *Brachiosaurus brancai* and *Cedarosaurus*. The ventral border is thick, tapering from a massive glenoid down to a thin distal plate. The distal end is not as expanded as that of *B. brancai* or *Camarasaurus*, but more closely resembles that of *Alamosaurus* (Gilmore 1946) and *Antarctosaurus* (Huene 1929).

All forelimb elements reflect the diminutive nature of this taxon (measurements are found in table 11.2).

The proximal end of the ulna is small in proportion to the length of the shaft (fig. 11.5D,E). The pronounced medial process resembles that found in *Pleurocoelus nanus* (Marsh 1888), *Cedarosaurus, Titanosaurus* (Jain and Bandyopadhyay 1997), and *Saltasaurus*. A moderate olecranon process contributes to the formation of a shallow concavity dorsal to the anteromedial process. A strongly developed concavity is cited by Upchurch (1995) as a character of the Titanosauroidea and represents the derived condition. The slight concavity found in *Venenosaurus* reflects a more intermediate stage. Along the anteromedial lower half of the shaft is found a broad groove that is bordered by two distinct ridges. This groove extends the length of the distal half of the shaft and corresponds to a ridge found on the lateral side of the lower portion of

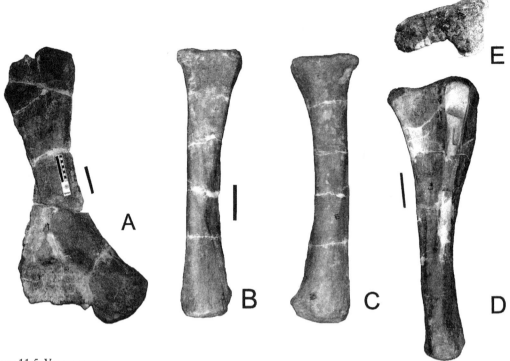

Figure 11.5. Venenosaurus
dicrocei *(DMNH 40932).*
Forelimb: left scapula (A) in
lateral view; left radius in
(B) medial and (C) lateral view;
left ulna in (D) anterior and
(E) dorsal view. Note the
expanded medial process and
moderately developed olecranon.
Scale: 10 cm.

the radius. The distal end of the ulna is quite expanded anteroposteriorly and is semicircular in outline.

Although the radius appears fairly robust, the maximum width of the proximal end is only 22% of the total length (fig. 11.5B, C). This element is more slender than similar elements of *Alamosaurus, Chubutisaurus* (Del Corro 1975), *Opisthocoelicaudia,* and *Saltasaurus,* as cited in Upchurch (1998). The ratio of length to least circumference produces a robustness ratio of 0.33, and the radius is more gracile than that of *Camarasaurus lewisi* (0.38) and *C. grandis* YPM 1901 (0.37) (McIntosh et al. 1996). It is however, less gracile than *Cedarosaurus* (0.31). The shaft is strongly bowed anteriorly. The ulnar rugosity is located near the distal end of the shaft, indicating the point of articulation between the radius and ulna. In this specimen the rugosity faces somewhat laterally, rather than posteriorly, as is common in sauropods. The distal end is subrectangular and the laterodistal corner is somewhat emarginate, possibly due to weathering. While it is difficult to distinguish generic differences in sauropod radii (Wilhite 1999), *Venenosaurus* differs from advanced titanosaurs in two aspects. The distal end lacks the lateral/medial expansion and ventrally directed medial process that are found in *Saltasaurus* and *Alamosaurus.* Overall, the radius most closely resembles that of *Brachiosaurus brancai.*

A nearly complete set of right metacarpals lacks only the shaft and distal end of metacarpal I (fig. 11.6A). All the elements are long and slender. The proximal end of metacarpal I is much more narrow anteroposteriorly than the transverse width. It resembles *Brachiosaurus* in tapering to a single lateral process, unlike *Camarasaurus,* which is a

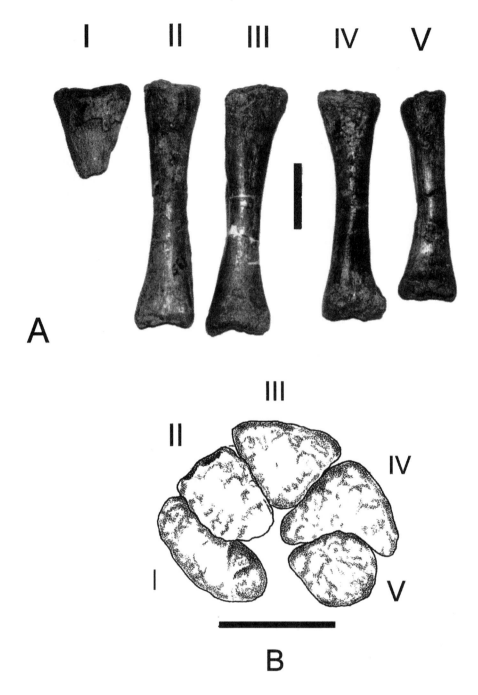

stout rectangle in dorsal view. It is however, more gracile than even *B. brancai* (Janensch 1961). A strong ridge extends ventrally down the posteromedial side. The proximal end of metacarpal II is nearly rectangular. Medially there is a broadly convex articulation with metacarpal I. Laterally a flat, broad, ventrally elongate surface articulates with metacarpal III. Metacarpal III is equal in length to metacarpal II. The proximal end of metacarpal III is a well-defined triangle whose lateral and medial surfaces are equal, similar to *B. brancai*. Proximally, metac-

Figure 11.6. Venenosaurus dicrocei (DMNH 40932). Right metacarpals in (A) anterior and (B) dorsal view. Scale: 10 cm.

arpal IV is triradiate with a robust anterolateral wall and a strong posteromedial wall. These are separated by a pronounced concavity that closely articulates with metacarpal V. The fifth metacarpal has a teardrop-shaped proximal end whose narrowest point is located posteromedially.

The distal ends of metacarpals II, III, and IV show clearly defined lateral and medial condyles extending posteriorly, just above the ventral edge. Distally, metacarpal V is flat, lacking distinct condyles. All the distal ends are rugose, rather than the smooth, rollerlike articular surfaces found on the metatarsals. The five metacarpals articulate proximally to form a tight arcade (fig. 11.6B). Pronounced ridges extending along the posterior side of each element at midshaft almost touch medially in the center of the columnar manus. Camarasaurs, brachiosaurs, and several advanced titanosaurs are known to possess slender, elongate metacarpals similar to those of *Venenosaurus*. The ratio of the length of metacarpal II to the radius in this specimen is 0.51, which is greater than *Camarasaurus* (0.43), equal to *Brachiosaurus* (0.51) (McIntosh 1990b), and less than *Aeolosaurus* (0.53) (Salgado et al. 1997).

Several manual phalanges were found in the quarry, and belong with the metacarpals, although no attempt has been made to assign their position. No distal unguals were recovered with this specimen.

Pelvic Girdle

Although no ilium was found, a badly eroded right pubis and two nearly complete ischia were recovered. Measurements are found in table 11.2. Significant portions of the proximal and distal ends of the pubis are missing, along with some of the ischial articular border (fig 11.7A,B). Most of the main body is complete, however, giving a general impression of this element. As preserved, the pubic foramen is open, rather than fully enclosed. While this is usually regarded an a juvenile characteristic, in this case it is likely due to weathering of the proximal end of the pubis. Just below the ischial articulation the pubis narrows significantly, although not as much as in *Brachiosaurus*. The anterior border is strongly curved proximal-distally in a manner that most closely resembles *Brachiosaurus*. The distal end appears to be moderately flared; however, erosion has obscured this area to some degree. Although the total length of the pubis is unknown, the preserved length of 65 cm is longer than either of the ischia.

The two well-preserved ischia are quite similar to those of *Brachiosaurus* in their proportions (fig 11.7 C–F). The prominent iliac peduncle is set significantly higher than the acetabulum border. The pubic articular surface is narrow anteroposteriorly. This articulation is confined to the proximal half of the element, rather than extending beyond midshaft, as in *Andesaurus* and *Saltasaurus*. Although the dorsal edge of the shaft is robust throughout its length, it also shows a significant thickening opposite the base of the pubic articular surface on both elements. The shaft is directed ventrally, as in *Brachiosaurus*. There is some rotation of the shaft to allow the two ischia to meet edge to edge along the symphysis. However, the unexpanded distal ends are directed

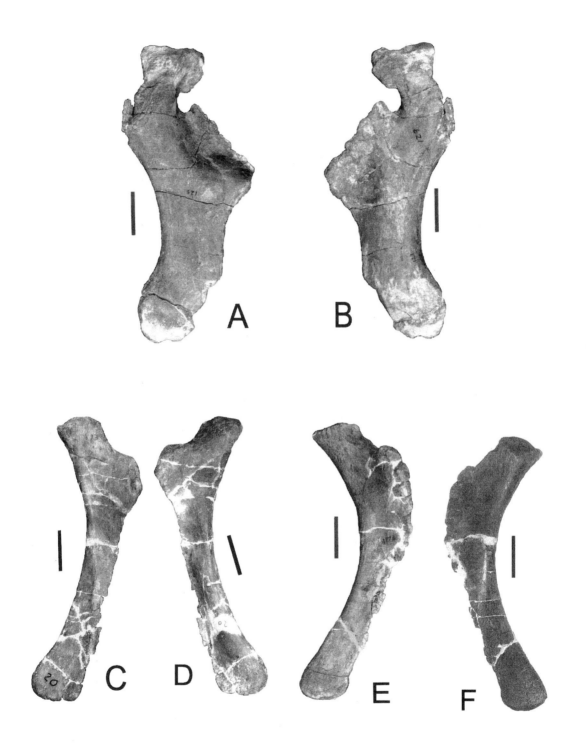

dorsolaterally rather than laterally, as in *Camarasaurus* and *Brachiosaurus*. This may be a result of taphonomic distortion because neither shaft appears to fully retain the natural shape.

In most sauropods the ischium is longer than the pubis, often by a considerable amount. This is not the case in *Opisthocoelicaudia, Andesaurus, Titanosaurus colberti,* and *Aeolosaurus* (Borsuk-Bialynicka 1977; Calvo and Bonaparte 1991; Jain and Bandyopadhyay 1997; Salgado and Coria 1993). In these taxa the total length of the ischium appears to be shortened, whereas the length of the pubic articulation has become elongate, extending well below midshaft. When measured from the ventral edge of the pubis-ischium articulation to the distal end, the pubis is longer than the ischium. Salgado et al. (1997) consider this to be characteristic of the clade Titanosauria, which they define as the most recent common ancestor of *Andesaurus delgadoi* and Titanosauridae and all its descendants. However, the condition is not known in *Saltasaurus* or *Alamosaurus* due to incomplete material. *Venenosaurus* displays a mixture of characters in having ischia that are somewhat shorter than the pubis in total length, yet the pubic articulation is restricted to the dorsal half of the element. In overall proportions, the pubis and ischium of *Venenosaurus* most closely resemble those of *Brachiosaurus*.

Three metatarsals were recovered (fig. 11.8). Metatarsal I is shortest and most robust. The proximal surface expands anteriorly upward and slopes down toward the medial side. The lateral side of the shaft is longer than the medial and lacks the constriction found medially. A slight depression occurs on the mid-lateral surface, slightly distal to the proximal articular surface, against which metatarsal II articulates. The distal end of metatarsal I expands into two condyles, the medial one larger than the lateral. Metatarsal II is longer and less stout than metatarsal I. The proximal surface is heavily eroded, yet still shows a very wide proximal end. The shaft is strongly constricted mediolaterally. Distally, despite damage to the anterior portion of the medial condyle, the two condyles are equal in size. A ridge between the lateral and posterior surfaces of the shaft extends upward from the lateral distal condyle. An additional metatarsal was also found with this specimen, whose proper position in the pes has not been determined. It is the longest and most slender of the three available metatarsals. The shaft is cylindrical distally and flattened anteromedially in the proximal portion. The proximal end is larger than the distal and concave in anterior view. The distal end is eroded, lacking portions of the lateral condyle. Whereas this element bears some resemblance to metatarsal IV, the proximal end is far more narrow anterior-posteriorly than in other specimens. There is also a prominent notch midway along the anterior side of the proximal end which gives this element a sinuous outline not found in other sauropod metatarsals. The tightly constricted shaft is reminiscent of metatarsal V; however, the proximal end is not as widely expanded as commonly found in that element. Although tentatively assigned to metatarsal IV(?), final determination of this element must be delayed until a complete set of metatarsals is recovered.

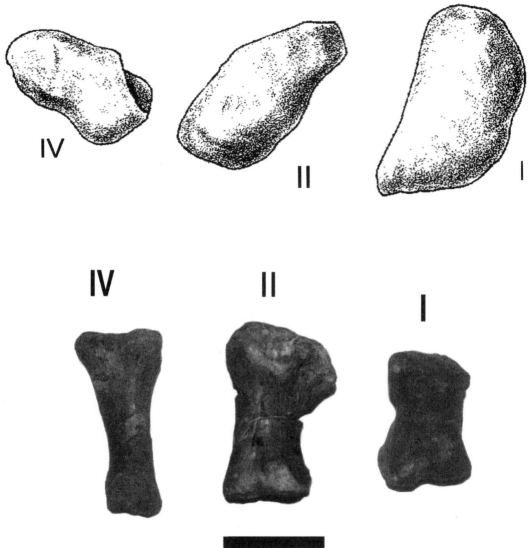

Figure 11.8. Venenosaurus dicrocei *(DMNH 40932).* *Metatarsals in anterior and dorsal view. Note very narrow distal condyle on metatarsal IV(?), and* *that metatarsal IV(?) is equal in* *length to metatarsal II.* *Scale: 10 cm.*

Dorsal Ribs

Numerous dorsal rib fragments were recovered, including a rib head from the right side of the body (fig. 11.9). It is fairly complete although the tuberculum is somewhat eroded, as is the junction of the capitulum and the tuberculum. There is a large foramen on the posterior surface. Measuring 40 mm, it lies near the base of the capitulum and is oval. This foramen leads proximally into a pneumatic cavity that occupies much of the interior of the capitulum. The size and position of this opening differs significantly from that of *B. altithorax* and *Malawisaurus* in being larger and more proximally positioned, as found in the Texas brachiosaur (Gomani et al. 1999). A similar rib foramen is illustrated as belonging to a specimen of *Apatosaurus* (formerly *Brontosaurus excelsus*) from Wyoming, in Marsh 1896 (fig. 11.10). Unfortunately, this particular element is not currently available for study. We cannot rule out the possibility that the rib is actually that of *Brachiosaurus*, also known from the Morrison Formation. However, this genus has not been reported from Wyoming to date. Wilson and Sereno (1998) regard "pneumatic dorsal ribs" as a titanosauriform synapomorphy, citing their occurrence in *Brachiosaurus, Euhelopus,* and titanosaurs. Alternatively, the illustration by Marsh raises the possibility that such "pneumatic" structures could occur more extensively throughout Sauropoda.

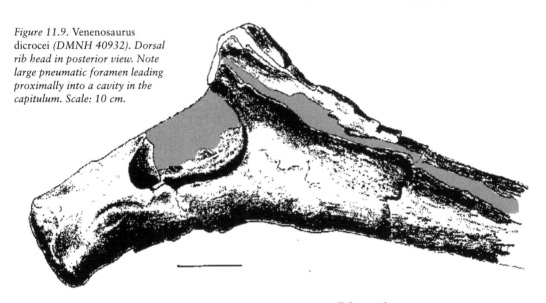

Figure 11.9. Venenosaurus dicrocei *(DMNH 40932). Dorsal rib head in posterior view. Note large pneumatic foramen leading proximally into a cavity in the capitulum. Scale: 10 cm.*

Discussion

Venenosaurus displays many characteristics that place it within the brachiosaur-titanosaur spectrum. This is an area of sauropod research that is undergoing intense scrutiny worldwide, driven by a wealth of newly excavated material. Consequently, several new cladistic analyses of brachiosaur-titanosaur lineages have been produced in the last few years (Salgado et al. 1997; Wilson and Sereno 1998; Upchurch 1998). All illuminate new areas of research and attempt to integrate the most recent taxa into newly defined clades. These efforts have made available a large amount of comparative information, and are much appreciated. However, they have also resulted in a maze of new character-based definitions that vary widely between the different analyses.

Salgado et al. (1997) define Titanosauriformes as the clade including the most recent common ancestor of *Brachiosaurus brancai, Chubutisaurus insignis,* Titanosauria, and all its descendants. Under this definition, *Venenosaurus* exhibits the following titanosauriform characteristics: (1) short caudal centra; (2) amphiplatyan anterior caudal vertebrae; (3) pubic articulation of the ischium restricted to the proximal half of the element; (4) neural arches positioned anteriorly in middle and posterior caudals; (5) ratio of the length of mcII/radius >0.50; (6) pneumatic rib heads; (7) it shares with *Andesaurus* and some

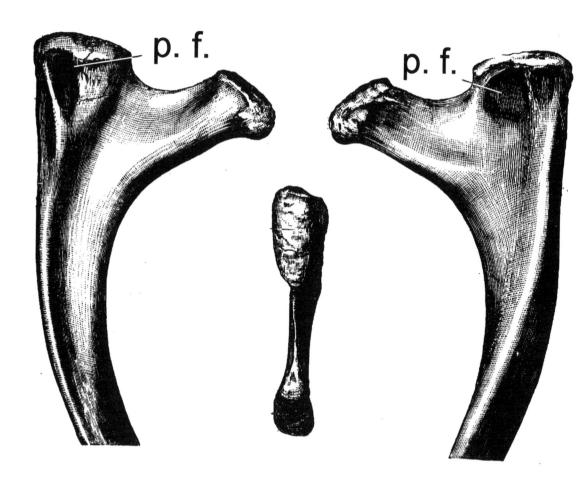

p. f.

p. f.

Figure 11.10. Dorsal rib head of Apatosaurus excelsus *as illustrated in Marsh's* Dinosaurs of North America. *P.F. denotes large "pneumatic foramen" located on the anterior and posterior sides of the tuberculum.*

members of Titanosauridae a pubis that is longer than the ischium; (8) anteriorly directed caudal neural spines similar to *Aeolosaurus;* (9) convex anterior surface of the proximal caudal centrum. The first three characters are plesiomorphic for sauropods. Characters 4, 5, and 6 are found in *Brachiosaurus* and suggest placement within the family Brachiosauridae, whereas the remaining features indicate the possibility of a closer relationship to more derived taxa. However, the last character is regarded as an autapomorphy for this genus.

Venenosaurus dicrocei appears to be a small adult sauropod, based on the complete fusion of neural arches and caudal ribs to the centra. However, no coracoid was found fused to the scapula and the pubic foramen is not closed. Whether these features reflect an immature condition or taphonomic processes is not known.

In comparing *Venenosaurus* with other sauropods, we find no distinctly camarasaurid characteristics. Rather, the distal end of the scapula is only moderately expanded, the radius and ulna are less robust, and the ratio of the length of mcII to the radius is significantly greater than in *Camarasaurus.*

Other genera more closely resembling *Venenosaurus* are *Brachiosaurus, Chubutisaurus, Pleurocoelus nanus,* the Texas *"Pleurocoelus"* SMU 61732, *Andesaurus,* and *Cedarosaurus.* All but *Brachiosaurus*

New Titanosauriform from the Cedar Mountain Formation • 155

are Early or Middle Cretaceous Titanosauriformes and exhibit one or more derived characteristics within that clade. For a brief review of these taxa, see Tidwell et al. 1999. *Venenosaurus* shares with all these taxa middle caudal centra which are amphiplatyan. However, none except *Cedarosaurus* exhibit the lateral fossa found on the anterior and middle caudals of *Venenosaurus*.

Brachiosaurus (Riggs 1904) differs from *Venenosaurus* in the proximal caudal, which is slightly amphicoelous, rather than anteriorly convex. *Brachiosaurus* also lacks forward-leaning neural spines in the anterior and middle caudals. The medial process of the ulna and the olecranon process are not as fully developed. Although the metacarpals resemble those of *Venenosaurus*, the proximal end of metacarpal I is much more robust in *B. brancai*. The pubis is tightly constricted below the ischial articulation, as in *Venenosaurus*, but is shorter than the ischium. Metatarsal I of *B. brancai* has a more rectangular distal surface. Metatarsal II is also more gracile and the proximal end is less widely flared than in *Venenosaurus*.

Pleurocoelus nanus (Marsh 1888) is composed of several disassociated juvenile specimens from Maryland that are poorly correlated with an undiagnostic holotype (Lull 1911). Unfortunately, the extremely immature nature of these specimens hinders meaningful comparisons with adult individuals. No proximal caudal is identifiable. The short caudal vertebrae consist of amphiplatyan centra and disassociated neural arches. None appear to be anteriorly inclined. The olecranon process is poorly developed but there is a well-developed medial process on the ulna.

Several articulated segments of caudal vertebrae from Texas (SMU 61732) have been referred to the genus *Pleurocoelus* and described by Langston (1974). Additional brachiosaurid specimens recently excavated by Southern Methodist University and the Fort Worth Museum of Science and Nature appear to belong to the same taxa (Winkler et al. 1997; Gomani et al. 1999). Neither the anterior nor the middle caudals possess anteriorly inclined neural spines. Instead, they range from vertically oriented to posteriorly inclined. Many of the caudal vertebrae show a well-developed hyposphene, a feature missing from *Venenosaurus*. The described Texas material includes a dorsal rib head with a pneumatic foramen similar in size and position to those of *Alamosaurus* and *Venenosaurus*. To date, no girdle or limb elements have been described from these specimens, so further comparison is not possible.

A number of undescribed elements, TMM 42488, from this locality were collected by Wann Langston and Jeff Pittman for the University of Texas in the 1980s (Langston pers. comm.). The prepared material includes a variety of elements, of which a scapula/corocoid, pubis, and ischium are shared by *Venenosaurus*. These elements show significant differences to the comparable bones in the Denver specimen.

An articulated hind limb from the Middle Cretaceous of Texas (Field Museum PR 977) has also been referred to *Pleurocoelus* (Langston 1974; Gallup 1989). It provides a useful comparison for the metatarsals of *Venenosaurus*. Metatarsal I has a deeper triangular depression instead of a slight depression on the mid-lateral shaft, equal-sized distal condyles instead of a larger medial condyle, and an extended lateral edge on the shaft. The curves along the medial side of the metatarsal II shaft are similar.

The basal titanosaur *Andesaurus* (Calvo and Bonaparte 1991) was recovered from Albian-Cenomanian sediments in Argentina. The proximal caudal is not known. Anterior caudal vertebrae are slightly procoelous, whereas the middle and distal caudals are amphicoelous. Middle caudal neural spines are markedly elongate anteroposteriorly, and are posteriorly inclined. The pubic articulation on the ischium is much more elongate and occupies a greater proportion of the total length than in *Venenosaurus*. Both taxa share the derived characteristic of a pubis that is longer than the ischium. The lower portion of the forelimb is unknown.

Cedarosaurus (Tidwell et al. 1999) is a brachiosaurid from the Yellow Cat Member of the Cedar Mountain Formation of Eastern Utah. It was found by a DMNH field crew at a significantly lower stratigraphic level than *Venenosaurus*. Although the proximal caudal is unknown, *Cedarosaurus* differs from *Venenosaurus* in a number of important areas. Anteriormost caudals display lateral fossa that are not as deep as in *Venenosaurus*. The centra display deeply concave anterior surfaces that contrast strongly with the amphiplatyan surfaces of *Venenosaurus*. Middle caudal neural spines are vertical to posteriorly inclined. These two taxa share radii that are more gracile than those of the other sauropods discussed here, although *Cedarosaurus* is the more slender of the two. On the radius, the proximal end is more narrow and drawn out into an elongate oval. The radial shaft also shows two very distinct, sharply defined ridges forming the edges of a concavity on the lower half of the bone. These features lead to a prominently raised ulnar rugosity. This is in sharp contrast with the single, poorly developed ridge and ulnar rugosity of *Venenosaurus*. The ulna of *Cedarosaurus* is more gracile, with a more widely expanded proximal end. There is also a prominent groove separating the lateral wall from the anconial process that extends well down the shaft. This groove is missing in *Venenosaurus*. The *Cedarosaurus* pubis is much less constricted below the ischial articulation. It also displays an unusual foramen located near the posterior, medial border that is not seen in *Venenosaurus*. Metatarsal I of *Cedarosaurus* is more elongate. The distal medial condyle rounds up to a greater extent forming a more prominent posteromedial process. The proximal surface is heavily eroded, but the undamaged posterior edge indicates less outward flare. Metatarsal II of *Cedarosaurus* is also more gracile than *Venenosaurus*. Along the shaft's anterolateral edge there is a rugose outward bulge. It is less stout, but also has equal distal condyles. A larger version of metatarsal IV(?) was recovered, although not figured, with Cedarosaurus.

The proximal caudal vertebra is very diagnostic for sauropods (Curtice 1996). Some Late Cretaceous titanosaurs have long been known to posses biconvex proximal caudals. This unusual morphology is found in *Alamosaurus* (Gilmore 1946), *Saltasaurus* (McIntosh 1990a), and *Neuquensaurus* (Huene 1929), reflecting a very derived condition. In *Titanosaurus colberti* the sixth sacral vertebra, identified as a fused caudosacral, is biconvex (Jain and Bandyopadhyay 1997). The plesiomorphic condition for sauropods is an amphiplatyan or amphicoelous proximal caudal, as found in most taxa that are more closely related to *Brachiosaurus* than to *Saltasaurus*. *Diplodocus* proximal caudals are mildly procoelous, as are several of the succeeding vertebrae. *Brachiosaurus altithorax* (Riggs 1904) and *Chubutisaurus* (Salgado 1993) retain the plesiomorphic state. In *Opisthocoelicaudia* (Borsuk-Bialynicka 1977) the proximal caudal is strongly opisthocoelus, having a large anterior ball coupled with a concave posterior surface. In *Pleurocoelus nanus, Andesaurus, Cedarosaurus,* and other Early Cretaceous taxa the condition is not well documented, because few specimens possess this element intact. The anterior convexity of *Venenosaurus,* coupled with a flat posterior surface, is unknown in any other sauropod. It may represent an intermediate state between the plesiomorphic condition found in most sauropods and the biconvex centra known in some titanosaurs.

In general, sauropod caudal neural spines are vertically oriented or are posteriorly inclined. These two morphologies are widespread throughout Eusauropoda (sensu Wilson and Sereno 1998), and represent the plesiomorphic state.

In several taxa the neural arch of anterior caudal vertebrae are anteriorly inclined. This feature is known from *Brachiosaurus, Cedarosaurus, Aeolosaurus, Gondwanatitan,* and *Lirainosaurus* (Sanz et al. 1999). These anteriorly inclined neural arches cause the anterior caudals to somewhat resemble parallelograms. The neural spines of *Brachiosaurus* and *Lirainosaurus* do not continue forward above the prezygapophyses, but are vertically oriented. Until the discovery of *Venenosaurus,* only *Aeolosaurus, Gondwanatitan,* and *Cedarosaurus* were known to possess anteriorly inclined caudal neural spines (fig. 11.11). In *Cedarosaurus* this feature is restricted to the first nine caudal vertebrae, and is coupled with centra having concave anterior articulations. By the anterior-middle caudal transition, as defined by the reduction of the caudal ribs, *Cedarosaurus* neural spines are posteriorly inclined, whereas those of *Venenosaurus* continue forward into the

Figure 11.11. Anterior caudal vertebrae possessing forward-leaning neural spines:
(A) Cedarosaurus,
(B) Aeolosaurus,
(C) Venenosaurus dicrocei.

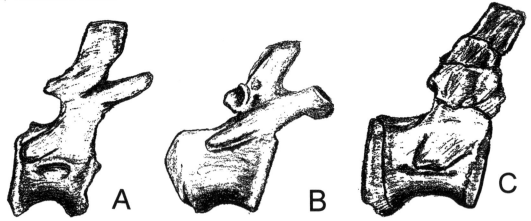

middle caudal region. In *Aeolosaurus* and *Gondwanatitan* a more pronounced forward-leaning neural spine is found in the anterior and middle caudal vertebrae. The centra of these caudals, like those of *Cedarosaurus*, possess concave anterior surfaces. In contrast, *Venenosaurus* has forward-leaning neural spines coupled with amphiplatyan caudal centra. This is a unique combination and represents an autapomorphy for this genus.

A Note on Caudal Centra Articulations

Historically, most Cretaceous sauropods have been associated with procoelous caudal vertebrae (Titanosauridae) or platycoelous caudals (Brachiosauridae). Here the term *procoelous* follows Romer (1956) as possessing a deep anterior concavity and prominent posterior convexity, whereas *platycoelous* centra are only slightly hollow at each end. The presence of procoelous caudal vertebrae has been strongly associated with titanosaurids beginning with the first discovery of *Titanosaurus* (Lydekker 1877) and continues with the recent description of two new Late Cretaceous titanosaurs (Jain and Bandyopadhyay 1997; Sanz et al. 1999). Platycoelous brachiosaurs include *Brachiosaurus*, *Pleurocoelus nanus*, and *Venenosaurus*. In all these taxa the caudal centra articulations display little variation throughout the length of the tail, and can be easily coded for cladistic analysis.

A number of recent discoveries of Early to Middle Cretaceous sauropods have been made in Africa, Thailand, and North and South America (Jacobs et al. 1993; Martin et al. 1994; Britt and Stadtman 1996; Salgado and Coria 1993; Tidwell et al. 1999). A broad range of morphology in caudal vertebrae has come to light as these taxa have been described. It is evident that the formerly adequate description of caudal vertebrae as platycoelous, amphiplatyan, or procoelous is no longer precise. In several of the newer taxa the caudal articulations change from one form to another depending on their position in the vertebral column. In *Andesaurus*, *Malawisaurus*, and the Texas brachiosaur, caudal articulation morphology changes from procoelous anterior caudals to platycoelous middle caudals. In other specimens the terms *procoelous* and *platycoelous* are proving to be too restrictive to adequately describe the caudal articulations that are present. *Cedarosaurus* anterior caudals are deeply concave anteriorly, which implies an incipient procoely. However, the posterior surfaces are flat, thus precluding the procoelous condition. This morphology of an anteriorly concave and posteriorly flat centrum is also found in some of the caudals of the Texas brachiosaur (Tidwell obser.). The terms *procoelous*, *platycoelous*, and *amphicoelous* all fail to inadequately describe this condition. When this inadequate terminology is incorporated into a cladistic analysis, it can lead to considerable confusion. Although a thorough reading of the original descriptive paper can help to clarify any misconceptions regarding a particular taxon, a more precise terminology is preferable. Unfortunately, these terms are commonly utilized in sauropod systematics with little clarification of the vertebral position, or extent of concavity present. Several authors have addressed the difficult situation which has begun to develop over the use of these generalized descriptions in cladistic analyses.

Sanz et al. (1999) note that not all the taxa reported to have procoelous centra display anteriorly concave and posteriorly convex centra, citing *Andesaurus* as an example. They recommend that only those vertebrae which exhibit a pronounced distal condyle be regarded as procoelous. They further restrict their focus to only the anterior caudals, leaving the middle and posterior caudals uncoded. Whereas this is an improvement over past practices, it does not provide the necessary flexibility to accurately detail the broad range of centrum articulations found in Titanosauriformes. In an extensive phylogenetic analysis of titanosaurs Salgado et al. (1997) divide character no. 23 (anterior caudals procoelous) into three states. The first reflects the moderate procoely found in *Andesaurus,* with shallow anterior faces and slightly convex posterior faces. State two describes anterior caudals as strongly procoelous, having "ball and socket" articular faces, with deeply excavated anterior face much like a "socket" and a prominent posterior ball. The third state includes the most derived titanosaurs and reflects middle and posterior caudals which are strongly procoelous. The thorough phylogenetic analysis of Upchurch (1998) also utilizes a three-part division to quantify the articulations in caudal centra. Character (C) 129 separates cranial caudals into amphicoelous/amphiplatyan and mildly/strongly procoelous varieties. C130 reflects whether they are amphicoelous–mildly procoelous or strongly procoelous, whereas C131 groups the middle and distal caudals into amphiplatyan or strongly procoelous sets.

These efforts are applauded, because they begin the process of more fully documenting the variety of caudal centra coming to light in Early Cretaceous taxa. These threefold divisions accurately measure differences in severity of procoely ranging from moderate to strong as it is found throughout known Titanosauriformes. Also, in differentiating between anterior and mid/posterior caudals they reflect the apparent posterior extension of procoelous characteristics along the vertebral column throughout the evolution of this clade. Table 11.4 illustrates how the divisions utilized by these authors correctly document the shift in *Malawisaurus* from strongly procoelous anterior caudals to platycoelous middle and posterior caudals (Jacobs et al. 1993). The transition in *Andesaurus* from mildly procoelous to amphiplatyan vertebrae is also recorded.

However, both Upchurch and Salgado retain the term *procoelous caudal vertebrae,* utilizing the paired characteristics of anterior concavity and posterior convexity. This term cannot accurately describe anterior caudal vertebra found in *Cedarosaurus* and the Texas "*Pleurocoelus*" SMU 61732 (see table 11.4). Many of the anterior caudals of *Cedarosaurus* show strongly concave anterior surfaces coupled with flat posterior surfaces, a condition also found in some of the caudals in the Texas brachiosaur (fig. 11.12). The new combinations of anterior and posterior surfaces found in these taxa illustrate the need for additional refinements in the coding of caudal vertebra in phylogenetic ana-

Figure 11.12. (opposite page) Anterior caudal vertebrae illustrating (A) the "procoelous" condition, Malawisaurus, *and (B) the "procoelous/distoplatyan" condition,* Cedarosaurus. *Note concave anterior surface on both centra, coupled with a pronounced posterior ball on* Malawisaurus, *and a flat posterior surface on* Cedarosaurus.

TABLE 11.4

Anterior–Posterior Articulations of Some Titanosauriform Caudal Centra
M = moderately developed; S = strongly developed

0-flat / 1-concave / 2-convex	Anterior Caudals ant. face / pos. face						Mid Caudals ant. face / pos. face						Posterior Caudals ant. face / pos. face					
	0	1	2	0	1	2	0	1	2	0	1	2	0	1	2	0	1	2
Sonorasaurus	-	-	-	-	-	-	-	-	M	-	M	-	-	-	-	-	-	-
Brachiosaurus	-	M	-	-	M	-	-	M	-	-	M	-	-	M	-	-	M	-
Pleurocoelus nanus	M	-	-	M	-	-	M	-	-	M	-	-	M	-	-	M	-	-
Venenosaurus	M	-	-	M	-	-	M	-	-	M	-	-	M	-	-	M	-	-
Chubutisaurus	M	-	-	M	-	-	M	-	-	M	-	-	-	-	-	-	-	-
Cedarosaurus	-	S	-	M	-	-	-	S	-	M	-	-	M	-	-	M	-	-
SMU 17298	-	M	-	-	-	M	-	M	-	M	-	-	-	M	-	-	M	-
Andesaurus	-	M	-	-	-	M	M	-	-	M	-	-	-	-	-	-	-	-
Malawisaurus	-	S	-	-	-	S	-	M	-	-	M	-	-	M	-	-	M	-
Aeolosaurus	-	S	-	-	-	S	-	S	-	-	-	S	-	S	-	-	-	S
Alamosaurus	-	S	-	-	-	S	-	S	-	-	-	S	-	S	-	-	-	S
Saltasaurus	-	S	-	-	-	S	-	S	-	-	-	S	-	S	-	-	-	S
Titanosaurus indicus	-	S	-	-	-	S	-	S	-	-	-	S	-	S	-	-	-	S

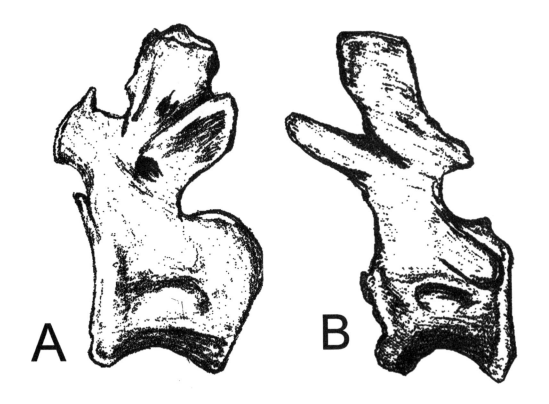

lyses. New descriptive combinations such as *procoelous/distoplatyan* could more accurately reflect the centrum articulations in *Cedarosaurus* and the Texas brachiosaur, and their use is strongly encouraged. We expect to see increasing amounts of variation within Early Cretaceous sauropods as more specimens are described in the future. Only as these variations are accurately represented in future cladistic analyses will a clearer picture of Cretaceous sauropod relationships emerge.

Conclusion

The unique morphology in the *Venenosaurus* proximal caudal has not been reported in any other sauropod. The slightly convex anterior surface suggests an incipient biconvex state, reminiscent of advanced titanosaurs. However, it is coupled with a flat posterior surface, a plesiomorphic character. This feature is probably a precursor to the fully derived biconvex condition. A similar mixture of characteristics is also found in the middle caudals, which combine a forward-leaning neural spine with amphiplatyan centra. These two autapomorphies set this specimen apart from all previously described sauropods, requiring the designation of *Venenosaurus* as a new genus. In addition, the combination of brachiosaur and titanosaur characters found in the proximal caudal, middle caudals, and pelvic elements reinforce the close relationships of these groups. Other Early Cretaceous titanosauriforms recently excavated by Southern Methodist University, University of Texas, Denver Museum of Natural History, and several other institutions, also display differing combinations of brachiosaur and titanosaur features. When fully described, these taxa will have significant ramifications for sauropod phylogenetic studies.

Acknowledgments: We are grateful for access to several specimens for comparative purposes. Our thanks to Timothy Rowe and Chris Bell at University of Texas Memorial Museum; Louis Jacobs, Dale Winkler, and Elizabeth Gomani at Southern Methodist University; and Robert Purdy, United States Museum of Natural History.

Informative discussions regarding Early Cretaceous sauropods were most helpful from John McIntosh, Ray Wilhite, Brian Curtice, Brooks Britt, and Leonard Salgado. Our thanks to Jim Mead, John McIntosh, Brooks Britt, and Paul Upchurch for their insightful comments regarding early versions of this manuscript.

The holotype of *Venenosaurus* was collected under Bureau of Land Management Excavation Permit no. UT-EX-99-005. It was discovered by Denver Museum of Natural History volunteer Tony DiCroce. Its excavation and speedy preparation were only made possible through the help of many museum volunteers.

A translation of Powell (1992) was made at DMNH by N. Ecker, and a partial translation of Janensch 1961 was made by Carla Henebrey and J. Peterson. Translations of Calvo and Bonaparte 1991 and Salgado et al. 1995 were made by J. Wilson and obtained courtesy of the Polyglot Paleontologist website: (http://www.uhmc.sunysb.edu/anatomicalsci/paleo). Drawings of the metacarpals, metatarsals, and dorsal rib were made by Judy Peterson.

References

Borsuk-Bialynicka, M. 1977. A new camarasaurid sauropod *Opisthocoe-licaudia skarzynskii,* gen. n. sp. n. from the Upper Cretaceous of Mongolia. *Palaeontologica polonica* 37: 5–64.

Britt, B. B., and K. L. Stadtman. 1996. The Early Cretaceous Dalton Wells dinosaur fauna and the earliest North American titanosaurid sauropod. *Journal of Vertebrate Paleontology Abstracts,* 16 (suppl. to no. 3): 24A.

Britt, B. B., K. L. Stadtman, R. D. Scheetz, and J. S. McIntosh. 1997. Camarasaurid and titanosaurid sauropods from the Early Cretaceous Dalton Well Quarry (Cedar Mountain Formation), Utah. *Journal of Vertebrate Paleontology Abstracts* 17 (suppl. to no. 3): 34A.

Britt, B. B., R. D. Scheetz, J. S. McIntosh, and K. L. Stadtman. 1998. Osteological characters of an Early Cretaceous titanosaurid sauropod dinosaur from the Cedar Mountain Formation of Utah. *Journal of Vertebrate Paleontology Abstract* 18 (suppl to no. 3): 29A.

Calvo, J. O., and J. F. Bonaparte. 1991. *Andesaurus delgadoi* gen. et sp. nov. (Saurischia-Sauropoda), Dinosaurio Titanosauridae de la Formación Rio Limay (Albiano-Cenonaniano), Neuquén, Argentina. *Ameghiniana* 28 (3–4): 303–310.

Carpenter, K., J. Kirkland, D. Burge, and J. Bird. 1999. Ankylosaurs (Dinosauria: Ornithischia) of the Cedar Mountain Formation, Utah, and their stratigraphic distribution. In D. Gillette (ed.), *Vertebrate Paleontology in Utah.* Utah Geological Survey Miscellaneous Publication 99–1: 244–251.

Cifelli, R., J. Gardner, L. Nydam, and D. Brinkman. 1997. Additions to the vertebrate fauna of the Antlers Formation (Lower Cretaceous), Southeastern Oklahoma. *Oklahoma Geology Notes* 57 (4): 124–131.

Curtice, B. 1996. Codex of diplodocid caudal vertebrae from the Dry Mesa dinosaur quarry. Master's thesis, Brigham Young University.

Del Corro, G. 1975. Un nuevo sauropodo del Cretacico *Chubutisaurus insignis* gen. et sp. nov. (Saurischia-Chubutisauridae nov.) del Cretacico Superior (Chubutiano), Chubut, Argentina. *Actas del Congreso de Paleontología y Biostratigrafía,* 2: 229–240.

Gallup, M. R. 1989. Functional morphology of the hindfoot of the Texas sauropod *Pleurocoelus* sp. indet. *Geological Society of America, Special Paper* 238: 71–74.

Gilmore, C. W. 1946. Reptilian fauna of the North Horn Formation of Central Utah. *U.S. Geological Survey Professional Paper* 210–C: 1–52.

Gomani, E. M. 1999. Sauropod caudal vertebrae from Malawi, Africa. In Y. Tomida, T. H. Rich, and P. Vickers-Rich (eds.), *Proceedings of the Second Gondwanan Dinosaur Symposium,* pp. 235–248. *National Science Museum Monographs,* no. 15, Tokyo.

Gomani, E. M., L. L. Jacobs, and D. A. Winkler. 1999. Comparison of the African titanosaurian *Malawisaurus* with a North American Early Cretaceous sauropod. In Y. Tomida, T. H. Rich, and P. Vickers-Rich (eds.), *Proceedings of the Second Gondwanan Dinosaur Symposium,* pp. 223–233, *National Science Museum Monographs,* no. 15, Tokyo.

Huene, F. 1929. Los Saurisquios y Ornitisquios del Cretáceo argentino. *Anales Museo La Plata,* series 2, 3: 1–196.

Jacobs, L., D. Winkler, and E. Gomani. 1993. New material of an Early Cretaceous titanosaurid sauropod dinosaur from Malawi. *Palaeontology* 36 (3): 523–534.

Jain, S. L., and S. Bandyopadhyay. 1997. New titanosaurid (Dinosauria: Sauropoda) from the Late Cretaceous of Central India. *Journal of Vertebrate Paleontology* 17 (1): 114–136.

Janensch, W. 1961. Die Gliedmassen und Gliedmassengürtel der Sauropoden der Tendaguru-Schichten. *Palaeontographica* (suppl. 7) 3: 177–235.

Kellner, A. W., and S. A. de Azevedo. 1999. A new sauropod dinosaur (Titanosauria) from the Late Cretaceous of Brazil. In Y. Tomida, T. H. Rich, and P. Vickers-Rich (eds.), *Proceedings of the Second Gondwanan Dinosaur Symposium*, pp. 111–142. *National Science Museum Monographs*, no. 15. Tokyo.

Kirkland, J., B. Britt, D. Burge, K. Carpenter, R. Cifelli, F. Decourten, J. Eaton, S. Hasiotis, and T. Lawton. 1997. Lower to Middle Cretaceous dinosaur faunas of the Central Colorado Plateau: A key to understanding 35 million years of tectonics, sedimentology, evolution, and biogeography. *Brigham Young University Geology Studies* 42 (2): 69–103.

Kirkland, J., R. Cifelli, B. Britt, D. Burge, F. DeCourten, J. Eaton, and J. M. Parrish. 1999. Distribution of vertebrate faunas in the Cedar Mountain Formation, east-central Utah. In D. Gillette (ed.), *Vertebrate Paleontology in Utah*. Utah Geological Survey Miscellaneous Publication 99–1:244–251.

Langston, W., Jr. 1974. Nonmammalian Comanchean tetrapods. *Geoscience and Man* 8: 77–102.

Lull, R. S. 1911. Systematic paleontology of the Lower Cretaceous deposits of Maryland: Vertebrata. *Maryland Geological Survey. Lower Cretaceous,* pp. 181–211.

Lydekker, R. 1877. Notices of new and other Vertebrata from Indian Tertiary and Secondary Rocks. *Records of the Geological Survey of India* 10: 30–43.

Marsh, O. C. 1888. Notice of a new genus of Sauropoda and other new dinosaurs from the Potomac Formation. *American Journal of Science,* series 3, 35: 89–92.

Marsh, O. C. 1896. The dinosaurs of North America. *U.S. Geological Survey Annual Report for 1894–95* 16: 133–244.

Martin, V., E. Buffetaut, and V. Suttethorn. 1994. A new genus of sauropod dinosaur from the Sao Khua Formation (Late Jurassic to Early Cretaceous) of Northwestern Thailand. *Comptes rendus de l'Academie des Sciences de Paris* 319 (11): 1085–1092.

McIntosh, J. 1990a. Sauropoda. In D. Weishampel, P. Dodson, H. Osmólska (eds.), *The Dinosauria,* pp. 345–401. Berkeley: University of California Press.

McIntosh, J. 1990b. Species determination in sauropod dinosaurs. In K. Carpenter and P. J. Currie (eds.), *Dinosaur Systematics: Approaches and Perspectives,* pp. 53–69. Cambridge: Cambridge University Press.

McIntosh, J. S., W. E. Miller, K. L. Stadtman, and D. D. Gillette. 1996. The Osteology of *Camarasaurus lewisi* (Jensen 1988). *Brigham Young University Geology Studies* 41: 73–115.

Powell, J. E. 1992. Osteología de *Saltasaurus loricatus* (Sauropoda—Titanosauridae) del Cretacico Superior del noroeste Argentino. In J. L. Sanz and A. D. Buscalioni (Coords), *Los dinosaurios y su entorno biótico,* pp. 165–230. Actas 2 Carso de Paleontologia en Cuenca, Instituto "Juan de Valdes." Ayuntamiento de Cuenca.

Riggs, E. S. 1904. Structure and relationships of opisthocoelian dinosaurs.

Part 2. The Brachiosauridae. *Publications of the Field Columbian Museum, Geological Series.* 2 (6): 229–247.

Romer, A. S. 1956. *Osteology of the Reptiles.* Chicago: University of Chicago Press.

Salgado, L. 1993. Comments on *Chubutisaurus insignis* Del Corro (Saurischia, Sauropoda). *Ameghiniana* 30 (3): 265–270.

Salgado, L., and R. A. Coria. 1993. El genero *Aeolosaurus* (Sauropoda, Titanosauridae) en la Formación Allen (Campaniano-Maastrichtiano) de la provincia de Río Negro, Argentina. *Ameghiniana* 30 (2): 119–128.

Salgado, L., R. Coria, and J. Calvo. 1997. Evolution of titanosaurid sauropods. I: Phylogenetic analysis based on the postcranial evidence. *Ameghiniana* 34 (1): 3–32.

Sanz, J. L., J. E. Powell, J. L. Loeuff, R. Martinez, and X. P. Suberbiola. 1999. Sauropod remains from the Upper Cretaceous of Lano (north-central Spain): Titanosaur phylogenetic relationships. *Estudios del Museo Ciencias Naturales de Alava* 14: 235–255.

Tidwell, V., K. Carpenter, and B. Brooks. 1999. New sauropod from the Lower Cretaceous of Utah, USA. *Oryctos* 2: 21–37.

Upchurch, P. 1995. The evolutionary history of sauropod dinosaurs. *Philosophical Transactions of the Royal Society of London,* series B, 349: 365–390.

Upchurch, P. 1998. The phylogenetic relationships of sauropod dinosaurs. *Zoological Journal of the Linnean Society* 124: 43–103.

Wilhite, R. 1999. Ontogenetic variation in the appendicular skeleton of the genus *Camarasaurus.* M.S. thesis, Brigham Young University.

Wilson, J. A. 1999. A nomenclature for vertebral laminae in sauropods and other saurischian dinosaurs. *Journal of Vertebrate Paleontology* 19:639–653.

Wilson, J., and P. Sereno. 1998. Early evolution and higher-level phylogeny of sauropod dinosaurs. *Journal of Vertebrate Paleontology, Memoir 5*: 1–68.

Winkler, D., L. Jacobs, and P. A. Murry. 1997. Jones Ranch: An Early Cretaceous sauropod bone-bed in Texas. *Journal of Vertebrate Paleontology Abstracts* 17 (3): 85A.

12. Gastroliths from the Lower Cretaceous Sauropod *Cedarosaurus weiskopfae*

Frank Sanders, Kim Manley, and Kenneth Carpenter

Abstract

A set of 115 clasts, ranging in size and mass from 0.04 cc to 270 cc and 0.1 gm to 715 gm, has been collected in association with the brachiosaurid *Cedarosaurus weiskopfae* in the Cedar Mountain Formation. The clasts were partially matrix supported, and some were supported on edge. There were a number of clast-to-clast and clast-to-bone contacts. The clasts are most parsimoniously interpreted as gastroliths, making this the first set discovered in situ in this formation. The gastrolith surfaces are mostly polished. Tight spatial distribution, partial matrix support, and some instances of on-edge orientation indicate that they were deposited while contained within carcass soft tissue. Low-energy depositional conditions and apparent initial containment within soft tissue indicate that the set is complete. Surface characteristics and distributions of shape, mass, volume, and composition have been determined. More than half the clasts are less than 10 cc in volume. Drab colors indicate low selectivity by the sauropod for this character. Most of the gastroliths are chert or quartzite, but some are sandstones and siltstones. High surface reflectance values for the majority of the gastroliths is consistent with the results of previous studies on other bona fide gastroliths.

Introduction

Behavioral characteristics of extinct fauna are generally inaccessible through the fossil record. An exception is the habit of clast swallowing; collections of clasts may be preserved in situ with fossilized skeletons of animals that carried gastroliths. The Denver Museum of Natural History has collected a complete set of gastroliths from the gut region of a nearly complete specimen of the Cretaceous brachiosaurid *Cedarosaurus weiskopfae* (Tidwell et al. 1999). This is the first documented discovery of in situ gastroliths in the Cedar Mountain Formation, and one of the few in situ discoveries in any sauropod specimen (table 12.1).

TABLE 12.1
**Documented Occurrences of Gastroliths in Direct Association
with Mesozoic Fauna**

Taxon	Period	Documented Occurrences
Prosauropods	Triassic	Bond 1955; Raath 1974: *Massospondylus* (at least four specimens)
Sauropods	Jurassic	Cannon 1906, discovered by O. C. Marsh party in Colorado, 1877: *Apatosaurus* specimen
Sauropods	Jurassic	Janensch 1929: *Barosaurus* specimen
Sauropods	Jurassic	Gillette 1990: *Seismosaurus* specimen
Sauropods	Jurassic	Dantas et al. 1998: *Lourinhasaurus* specimen
Sauropods	Cretaceous	Calvo 1994: aff. *Rebbachisaurus* (two specimens)
Sauropods	Cretaceous	Sanders and Carpenter 1998: *Cedarosaurus* specimen
Hadrosaurids	Cretaceous	Brown 1907: possible occurrence with a hadrosaurid specimen
Psittacosaurs	Cretaceous	Brown 1941; Bird 1985; Xu 1997: *Psittacosaurus* (two specimens)
Ankylosaurs	Cretaceous	Carpenter 1997: *Panoplosaurus* specimen
Theropods	Cretaceous	Ji et al. 1999: occurrence in two specimens (holotype and paratype) of feathered theropod *Caudipteryx*
Ornithomimids	Cretaceous	Kobayashi et al. 1999: occurrence in twelve articulated skeletons
Plesiosaurs	Cretaceous	Williston 1903; Williston 1904; Brown 1904; Riggs 1939; Welles and Bump 1949; Shuler 1950; Darby and Ojakangas 1980; Chatterjee and Small 1989; Martin 1994: frequent occurrence of gastroliths in elasmosaur and pliosaur plesiosaur specimens

Occurrence of Gastroliths in Mesozoic Taxa

Gastroliths have been collected in situ with a variety of Mesozoic reptilian taxa (table 12.1). These include the prosauropod *Massospondylus;* a number of sauropod genera (*Apatosaurus, Barosaurus, Seismosaurus, Lourinhasaurus,* aff. *Rebbachisaurus,* and *Cedarosaurus*); the psittacosaur *Psittacosaurus;* the ankylosaur *Panoplosaurus;* the holotype and paratype of the feathered theropod *Caudipteryx;* a dozen ornithomimid skeletons; and frequent discoveries with specimens of elasmosaur and pliosaur plesiosaurs.

Complete sets of gastroliths are found in association with many elasmosaur plesiosaurs (Williston 1903, 1904; Brown 1904; Riggs 1939; Welles and Bump 1949; Shuler 1950; Darby and Ojakangas 1980; Chatterjee and Small 1989; Martin 1994). They were probably used for buoyancy control in these taxa (Taylor 1993, 1994). Gastroliths used as digestive aids have been commonly found in association with specimens of the recently extinct herbivorous moa (Burrows et al. 1981).

Gastroliths associated with dinosaur skeletons, in contrast, are relatively rare. Documented associations with nonsauropod Dinosauria include three possible gastroliths found near a hadrosaurid's forelegs (Brown 1907) and two unequivocal discoveries associated with psittacosaurs (Brown 1941; Bird 1985; Xu 1997). Carpenter (1997) mentions gastroliths associated with the ankylosaur *Panoplosaurus.* Gastroliths have been found with specimens of the prosauropod *Massospondylus* (Bond 1955; Raath 1974). Recently, gastroliths have been reported from two specimens (holotype and paratype) of the feathered theropod *Caudipteryx* (Ji et al. 1999). Cannon (1906) documented sauropodan gastroliths. Wieland (1906, 1907) claimed stegosaurid gastroliths, but Brown (1907) indicated that the Wieland clasts were not associated with the stegosaur bones in question.

Only a few occurrences of gastroliths in association with sauropods have been described. Most of them are Jurassic (Cannon 1906; Janensch 1929; Gillette 1990; Dantas et al. 1998). Calvo (1994) describes a gastrolith set from a Cretaceous sauropod, but notes that many supposed dinosaur gastroliths have not been associated with skeletons (Schaffner 1938; Stauffer 1945; Greene 1956; Sperry 1957). Calvo questions whether such putative gastroliths can generally be inferred to be bona fide. Stokes (1987) recognizes the lack of evidence for gastroliths among many sauropods, including the lack of gastroliths with sauropods in the Cedar Mountain Formation as of that date. He argues, however, for a gastrolith origin for polished clasts commonly found in that formation. He suggests that clast swallowing may have occurred on a recycling basis among some sauropod genera in the Early Cretaceous of the North American western interior, and that the gastroliths may have been preferentially selected on the basis of bright coloration. The discovery of gastroliths in association with *Cedarosaurus* demonstrates that at least one early Cretaceous sauropod genus in western North America utilized gastroliths, though not selectively for color.

Occurrence in *Cedarosaurus*

The Denver Museum of Natural History (DMNH) collected a set of 115 smooth-to-polished clasts (fig. 12.1) from within the skeleton of *Cedarosaurus weiskopfae* in the Cedar Mountain Formation of Grand County, Utah. All but three of the clasts were located within an area of approximately 0.5 m × 0.5 m × 0.25 m, for a volume of about 0.06 m³. The pocket of clasts occurred immediately above a sternal plate, where they left impressions, and about 30 to 50 cm anterior to the left femoral condyle, about 30 cm ventral to the vertebral column, and adjacent to a pubis and ilium (fig. 12.2); thus, the clasts were located in the gut region of the sauropod. Gastrolith physical statistics are listed (table 12.2). Three additional polished clasts were found elsewhere within the skeleton (fig. 12.2). No other clasts occurred in the quarry (i.e., within a volume of about 11 m³). The quarry measured approximately 6 m wide, cut into a steeply sloping hillside. Taking the cross section of the quarry as a right triangle with legs 2.5 m deep by 1.5 m high, the approximate quarry volume was 0.5 · (2.5 m · 1.5 m) · 6 m . 11 m³. The clasts were partially matrix supported in a hard, uniform mudstone. Some disk-shaped clasts were supported on edge (fig. 12.3). The clasts in the main pocket were distributed in three dimensions (fig. 12.4) and there were a number of clast-to-clast contacts, as well as some clast-to-bone contacts. The occurrence of this compact, isolated pocket of polished, matrix-supported clasts within the sauropod's abdominal region is most parsimoniously interpreted as their being gastroliths. Their compact spatial distribution, instances of on-edge orientation, and three-dimensional distribution indicate that they were contained within carcass soft tissue at the time of burial. Because the gastroliths were found in a single pocket deep within the quarry, the set is believed to be complete.

Depositional Setting

Cedarosaurus weiskopfae was collected in the Yellow Cat Member of the Cedar Mountain Formation in eastern Utah. The Yellow Cat unconformably overlies the Brushy Basin Member of the Morrison Formation and underlies the Poison Strip Sandstone of the Cedar Mountain Formation (Kirkland et al. 1997). The skeleton lay in a hard, maroon mudstone, lacking inclusions or lenses of other materials. The depositional environment was low energy, consistent with a floodplain. Such an environment and semiarticulation of the *Cedarosaurus* skeleton demonstrate that fluvial transport of the clasts to the interior of the skeleton would have been unlikely. This conclusion is supported by the lack of inclusions in the surrounding matrix, the on-edge orientation of some of the clasts, and the clast-to-bone contacts that occur. The skeletal position indicates that the animal's carcass came to rest on the belly; this explains why the gastroliths were discovered in situ.

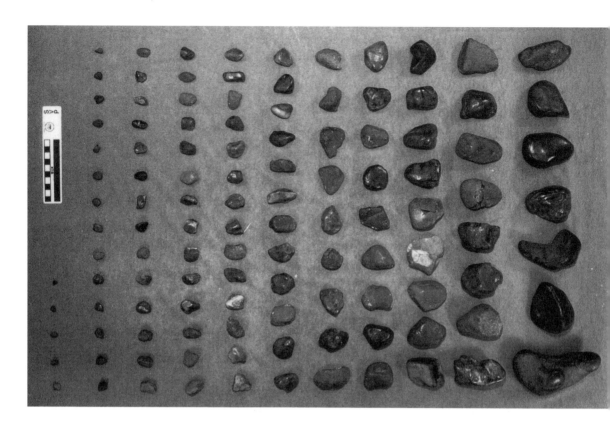

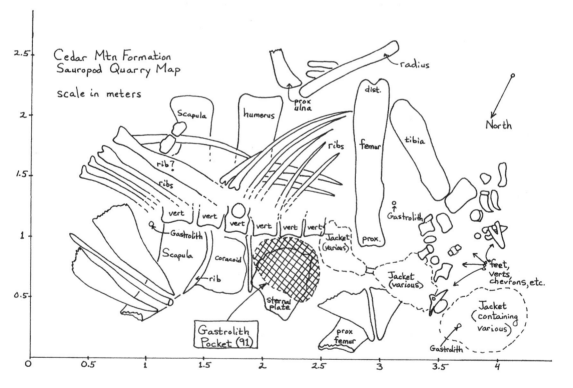

Cedar Mtn Formation
Sauropod Quarry Map

scale in meters

Figure 12.1. (opposite page above) Gastroliths found in situ with Cedarosaurus weiskopfae.

Figure 12.2. (opposite page below) Cedarosaurus *quarry diagram. Volume of gastrolith pocket is approximately 0.06 m³ (0.5 m × 0.5 m × 0.25 m).*

Figure 12.3. (above) In situ Cedarosaurus *gastrolith assemblage, exhibiting three-dimensional distribution (white arrows) and on-edge configurations (black arrows).*

TABLE 12.2
Basic Physical Statistics of *Cedarosaurus* Gastroliths

Gastrolith Physical Parameter	Parameter Value
Total mass	7.00 kg
Total volume	2703 cc
Total surface area	~ 4410 cm²

Figure 12.4. Cedarosaurus gastrolith assemblage with larger clasts replaced in their approximate original positions.

Figure 12.5. (opposite page above) Cedarosaurus *gastrolith size distribution.*

Figure 12.6. (opposite page below) Cedarosaurus *gastrolith shape distribution. Axes after Krumbein (1941).*

Description

The volume range of the individual gastroliths is 0.04 cc to 270 cc. More than half the clasts (67 of 115) are less than 10 cc in volume. Masses range from 0.1 gm to 715 gm. The size distribution is skewed toward small sizes (fig. 12.5). The largest clast measures 16.5 × 6.8 × 5.7 cm.

Each gastrolith was measured along three orthogonal, but not necessarily intersecting, axes (see Krumbein 1941). The major, intermediate, and short axes—labeled a, b, and c, respectively—are used to determine the shapes of the clasts by taking the ratios b/a and c/b and plotting them as ordered pairs (fig. 12.6). The shape distribution tends toward spherical; only 7% of the gastroliths are highly irregular in shape (table 12.3). However, some of the largest clasts are the most irregular (ellipsoidal).

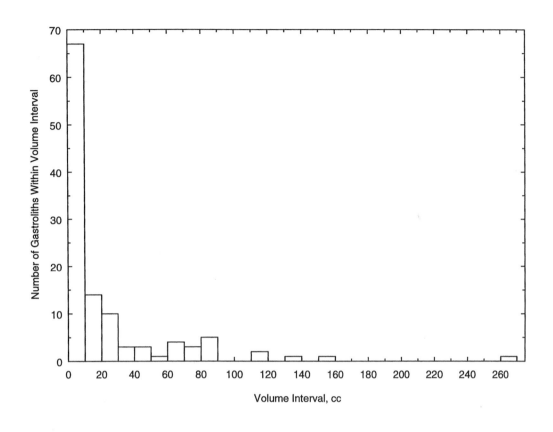

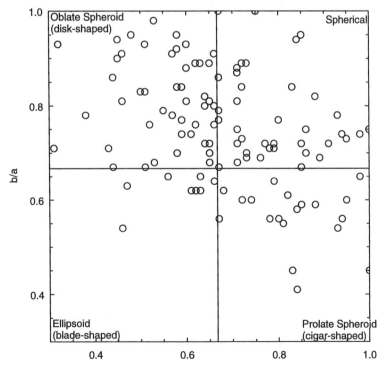

Gastroliths from the Lower Cretaceous Sauropod *Cedarosaurus weiskopfae* • 173

TABLE 12.3

TABLE 12.3
Shape Distribution of *Cedarosaurus* Gastroliths, Rounded to Nearest Percent

Gastrolith Shape	Percent
Oblate spheroid (disk-shaped)	43
Spheroid	34
Prolate spheroid (rod-shaped)	16
Ellipsoidal (blade-shaped)	7

Figure 12.7. (opposite page above) Cedarosaurus gastrolith *sphericity as a function of size.*

Figure 12.8. (opposite page below) Cedarosaurus gastrolith *roundness as a function of size.*

The smallest gastroliths exhibit the largest range of sphericity (fig. 12.7; see appendix 12.1, formula 1). At the largest sizes, the sphericity tends to the median value for the entire set, 0.68. A notable exception is the largest gastrolith of the set, which has a notably low sphericity of 0.48. Roundness at the largest sizes also tends to the set's median value, between 0.5 and 0.6 (fig. 12.8). Surface area was computed for the gastrolith set. This parameter might be expected to be important if gastroliths assisted the maceration of gut contents (as summarized by Farlow 1987). Surface areas were determined by categorizing each clast as either oblate spheroid, prolate spheroid, spheroid, or ellipsoid, based on the quadrant occupied by in figure 12.6 by each clast (see appendix 12.1, formulas 2–5).

Accuracy of computed surface areas was checked by measuring ten clasts, ranging from the largest to the smallest, directly with the latex method (see appendix 12.1, formula 5). Although individual clasts varied by as much as 10% in computed versus measured surface area, the variations in the computed areas varied equally above and below the measured surface areas, and the cumulative difference between measured and computed surface areas for the ten clasts was 1.5%.

Of the total gastrolith surface area of 4410 cm² (table 12.2), 51% (2267 cm²) is provided by the largest 17% of the clasts (20 largest specimens). If ball-mill maceration occurred and surface area was significant, then the presence of the large clasts may have been especially important. Two gastroliths exhibited notable shape characteristics. They are among the few ellipsoids (that is, they are among the most irregularly shaped) and have the longest physical dimensions (16.5 cm and 11.9 cm) of the clasts. They are the largest and fourth-largest of the set by mass and volume (715 gm, 270 cc; 320 gm, 115 cc, respectively). Their highly irregular shapes, indicated by their low sphericity values (0.48 and 0.51, respectively), set them apart from the other gastroliths. Swallowing these clasts may have presented a challenge; their presence practically rules out the possibility of accidental ingestion of clasts, and suggests either that irregular shapes were attractive to *Cedarosaurus,* or that the sauropod was not selective for shape. The irregularity of their shapes gives them high surface area compared to their size.

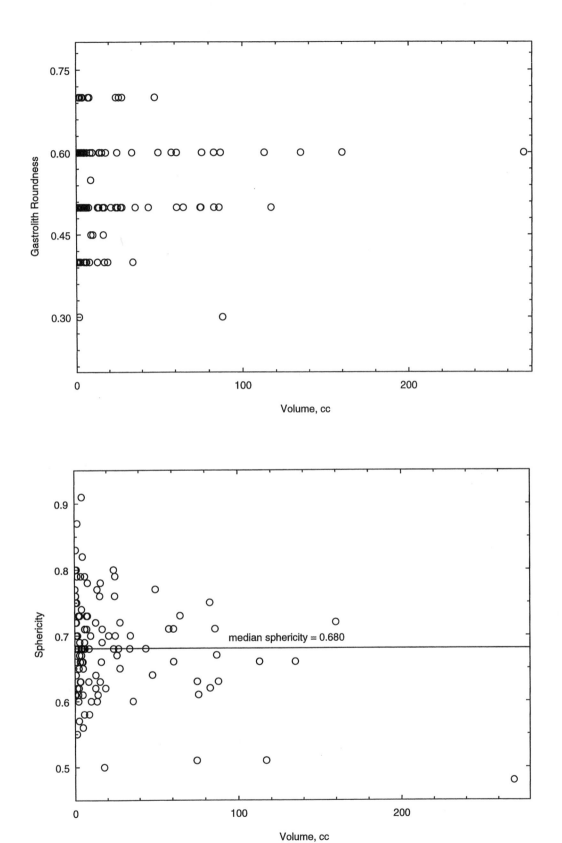

The gastroliths are composed of cherts, sandstones, siltstones, and quartzites in the ratios given in table 12.4. Some cherts are banded and some contain fossils. Pitting is common on several. The sandstones are moderately cemented, poorly sorted, and medium grained to well sorted and fine grained. They are relatively fragile; some were fractured while being jostled against other gastroliths in collection bags during transport (they were subsequently repaired). The siltstones represent only a small fraction of the total, but are the most enigmatic. They tend to be small; the largest is 33.6 gm and 15.6 cc. They are even more fragile than the sandstones. As with the sandstones, some of the siltstones were fractured in collection bags during transport. None of the gastroliths have a soapy feel, making that popular notion for distinguishing gastroliths unreliable.

TABLE 12.4
Composition of *Cedarosaurus* Gastroliths, Percent by Number, Mass, and Volume

Composition	Percent by Number	Percent by Mass	Percent by Volume
chert	62	65	64
sandstone and siltstone	31	29	29
quartzite	7	6	7

Gastrolith colors range from blue-gray and purplish red to dark brown or nearly black. The drab coloration of most of the clasts in the set argues against selection on the basis of coloration, contrary to the suggestion by Stokes (1987).

Surface polish measurements, using laser light-scattering (Manley 1991) were made on 83 of the 115 gastroliths (Of the 115 clasts in the complete set, 95 had been extracted from field jackets at the time the measurements were performed. Twelve of the 95 were too small to measure or had surface problems that prevented accurate data collection, resulting in a set of 83 that were measured). Three to ten measurements were made on each of two different locations on all but eight of the samples. For those eight, the size or surface characteristics precluded more than one location being measured. All measurements were converted to reflectance values by determining the peak intensity reflectance value at 90°, from which the average of the shoulder values of the light-scattering curve at ±30° was subtracted. The resulting number represents the difference in reflected specular light (peak value) and diffuse light (shoulder values), and is therefore an indication of the degree of surface polish. This number becomes the reflectance value (RV) when expressed as a percentage of the peak value. Comparisons between clasts can thus be made. Where two locations were measured for a gastrolith, their reflectance values were averaged.

In a previous study of populations of gastroliths and clasts from beaches and streams (Manley 1993), RVs greater than 35% appear to

TABLE 12.5
**Number of Occurrences of Gastroliths within Reflectance Value Ranges,
from 83 *Cedarosaurus* Gastroliths**

0–9%	10–19%	20–29%	30–39%	40–49%	50–59%	60–69%	70–79%	80–89%	90–100%
2	7	9	9	9	9	11	14	11	2

fairly well separate the gastrolith from nongastrolith populations, and only 8% of the beach clasts had RVs between 50% and 80%. Thus RVs above 50% are reasonably diagnostic for gastroliths.

The RVs for the *Cedarosaurus* gastroliths are shown in table 12.5. The RV range for all locations is 0–99%. The range using averages for two locations per gastrolith, where available, is 6–95.5%. Forty-seven of the gastroliths (57%) have RVs above 50%, while 62 (75%) have RVs greater than 35%. Of the gastroliths with RVs above 80%, all are cherts.

Complete polish measurements were somewhat hampered by the presence on many gastroliths of a metallic coating (possibly hematite) that probably originated from the iron-rich mudstone. Parallel scratches are visible in many places on this coating due to expansion and contraction of the mudstone around the inflexible gastroliths.

Conclusion

At least one North American Lower Cretaceous sauropod genus, *Cedarosaurus*, utilized gastroliths composed variously of chert, sandstone, siltstone, and quartzite. The high RVs of the clast surfaces are consistent with those of bona fide gastroliths in previous studies (Manley 1991, 1993) and probably indicate long residence in the animal's gut. Drab color and wide variation in the shape of the gastroliths indicate a low degree of selectivity for such factors by the sauropod.

The relative fragility of the sandstone and siltstone clasts is puzzling. Their smooth surfaces indicate long residence inside the gut. If they were as fragile inside the living animal as they are at present, the hypothesis that gastroliths rolled or tumbled within the gut (Bryan 1931; Gillette 1994, 1995) would tend to be supported. It is also possible, however, that the postdepositional environment has weakened these clasts chemically by dissolution of cement. In that case, the ball-mill maceration model for gastroliths (summarized by Farlow 1987) may be applicable to the *Cedarosaurus* clasts. The low sphericity of the twenty largest clasts (which translates into a high ratio of surface area to volume for the largest clasts) would also be consistent with a grinding or crushing model for gastrolith function in the animal's gut.

Acknowledgments: We thank the many volunteers for their many years of assistance in the field, and especially in the excavation of *Cedarosaurus weiskopfae*. Thanks to Darren Tanke for review comments, and to the Los Alamos National Laboratories for use of the laser scattering equipment.

References

Bird, R. T. 1985. *Bones for Barnum Brown, Adventures of a Dinosaur Hunter.* Fort Worth: Texas Christian University Press.

Bond, G. 1955. A note on dinosaur remains from the Forest Sandstone (upper Karroo). *Arnoldia* 2 (20): 795–800.

Brown, B. 1904. Stomach stones and food of plesiosaurs. *Science,* n.s., 20 (501): 184–185.

Brown, B. 1907. Gastroliths. *Science,* n.s., 25 (636): 392.

Brown, B. 1941. The last dinosaurs. *Natural History* 48: 290–295.

Bryan, K. 1931. Wind-worn stones or ventifacts: A discussion and bibliography. In *Natural Resources Council Report on Sedimentation 1929– 1930,* pp. 29–50. Washington D.C.: National Academy of Science.

Burrows, C. J., B. McCulloch, and M. Trotter. 1981. The diet of moas based on gizzard contents samples from Pyramid Valley, North Canterbury, and Scaifes Lagoon, Lake Wanaka, Otago. *Records of the Canterbury Museum* 9 (6): 309–336.

Calvo, J. O. 1994. Gastroliths in sauropod dinosaurs. *Gaia* 10: 205–208.

Cannon, G. L. 1906. Sauropodan gastroliths. *Science,* n.s., 24 (604): 116.

Carpenter, K. 1997. Ankylosaurs. In J. O. Farlow and M. K. Brett-Surman (eds.), *The Complete Dinosaur,* pp. 307–316. Bloomington: Indiana University Press.

Chatterjee, S., and B. Small. 1989. New plesiosaurs from the Upper Cretaceous of Antarctica. In J. A. Crane (ed.), Origins and Evolution of the Antarctic Biota. *Geological Society (London) Special Publication* 47:197–215.

Dantas, P., C. Freitas, T. Azevedo, A. G. de Carvalho, D. Santos, F. Ortego, V. Santos, J. L. Sanz, C. M. da Silva, and M. Cachão. 1998. Estudo dos gastrólitos do dinossáurio *Lourinhasaurus* do Jurássico superior portugués. *Communicaçãoes do Instituto Geológico e Mineiro, V Congresso Nacional de Geologia (Lisbon)* 84 (1): 87–90.

Darby, D. G., and R. W. Ojakangas. 1980. Gastroliths from an Upper Cretaceous plesiosaur. *Journal of Paleontology* 54 (3): 548–556.

Farlow, J. O. 1987. Speculations about the diet and digestive physiology of herbivorous dinosaurs. *Paleobiology* 13 (1): 60–72.

Gillette, D. 1990. Gastroliths of a sauropod dinosaur from New Mexico. *Journal of Vertebrate Paleontology* 10 (suppl. 3): 24A.

Gillette, D. 1994. *Seismosaurus.* New York: Columbia University Press.

Gillette, D. 1995. True grit. *Natural History* 104 (6): 41–43.

Greene, W. D. 1956. Dinosaur gizzard stones, Wyoming. *Mineralogist* 24: 51–55.

Janensch, W. 1929. Magensteine bei Sauropoden der Tendaguru-Schichten. *Paleontographica* 7 (1): 134–144.

Ji, Q., P. J. Currie, M. A. Norell, and S.-A. Ji. 1999. Two feathered dinosaurs from northeastern China. *Nature* 393: 753–761.

Kirkland, J., B. Britt, D. Burge, K. Carpenter, R. Cifelli, F. Decourten, J. Eaton, S. Hasiotis, and T. Lawton. 1997. Lower to Middle Cretaceous dinosaur faunas of the Central Colorado Plateau: A key to understanding 35 million years of tectonics, sedimentology, evolution, and biogeography. *Brigham Young University Geology Studies* 42 (2): 69–103.

Kobayashi, Y., J.-C. Lu, Z.-M. Dong, R. Barsbold, Y. Azuma, and Y. Tomida. 1999. Herbivorous diet in an ornithomimid dinosaur. *Nature* 402: 480–81.

Krumbein, W. C. 1941. Measurement and geological significance of shape and roundness of sedimentary particles. *Journal of Sedimentary Petrology* 11 (2): 64–72.

Manley, K. 1991. Two techniques for measuring surface polish as applied to gastroliths. *Ichnos* 1: 313–316.

Manley, K. 1993. Surface polish measurements from bona fide and suspected sauropod gastroliths, wave and stream transported clasts. *Ichnos* 2: 167–169.

Martin, J. E. 1994. Gastric residues in marine reptiles from the Late Cretaceous Pierre Shale in South Dakota: Their bearing on extinction. *Journal of Vertebrate Paleontology* 14 (3): 36A.

Raath, M. A. 1974. Fossil vertebrate studies in Rhodesia: Further evidence of gastroliths in prosauropod dinosaurs. *Arnoldia* 7 (5): 1–7.

Riggs, E. S. 1939. A specimen of *Elasmosaurus serpentinus*. *Field Columbian Museum of Natural History,* Geological Series 6 (25): 385–391.

Sanders, F. H., and K. Carpenter. 1998. Gastroliths from a Camarasaurid in the Cedar Mountain Formation. *Journal of Vertebrate Paleontology* 18 (3): 74A.

Schaffner, D. C. 1938. Gastroliths in the Lower Dakota of northern Kansas. *Kansas Academy of Sciences Transactions* 41: 225–226.

Shuler, E. W. 1950. A new elasmosaur from the Eagle Ford Shale of Texas. *Southern Methodist University,* Fondren Science Series 1 (part 2): 1–32.

Sperry, G. 1957. Collecting gizzard stones in Utah. *Desert Magazine,* July, pp. 4–5.

Stauffer, R. C. 1945. Gastroliths from Minnesota. *American Journal of Science* 243 (6): 336–340.

Stokes, W. L. 1987. Dinosaur gastroliths revisited. *Journal of Paleontology* 61 (6): 1242–1246.

Taylor, M. A. 1993. Stomach stones for feeding or buoyancy? The occurrence and function of gastroliths in marine tetrapods. *Philosophical Transactions of the Royal Society (London), B* 341: 163–175.

Taylor, M. A. 1994. Stone, bone, or blubber? Buoyancy control strategies in aquatic tetrapods. In L. Maddock and L. M. V. Rayner (eds.), *Mechanics and Physiology of Animal Swimming,* pp. 151–161. Cambridge: Cambridge University Press.

Tidwell, V., K. Carpenter, and W. Brooks. 1999. New sauropod from the Lower Cretaceous of Utah, USA. *Oryctos* 2: 21–37.

Welles, S. P., and J. D. Bump. 1949. *Alzadasaurus pembertoni,* a new elasmosaur from the Upper Cretaceous of South Dakota. *Journal of Paleontology* 23 (5): 521–535.

Wieland, G. R. 1906. Dinosaurian gastroliths. *Science,* n.s., 23 (595): 819–821.

Wieland, G. R. 1907. Gastroliths. *Science,* n.s., 25 (628): 66–67.

Williston, S. W. 1903. North American plesiosaurs, part 1. *Field Columbian Museum,* Geological Series, 73: 75–77.

Williston, S. W. 1904. The stomach stones of the plesiosaurs. *Science,* n.s., 20: 565.

Xu, X. 1997. A new psittacosaur (*Psittacosaurus mazongshanensis* sp. nov.) from Mazongshan area, Gansu province, China. In Z.-M. Dong (ed.), *Sino-Japanese Silk Road Dinosaur Expedition,* pp. 48–67. Institute of Paleontology and Paleoanthropology Academia Sinica. Beijing: China Ocean Press.

APPENDIX 12.1.
Formulas Used with the Study of Gastroliths

Formula 1: Sphericity is defined as the ratio of two diameters,

$$\left(\frac{d_1}{d_2}\right), \text{ where } d_1 = \sqrt[3]{\frac{6V}{\Pi}} \text{ and } d_2 = \text{longest linear dimension of the clast.}$$

d_1 is the diameter of a sphere having the same volume, V, as the clast.

Formula 2: Surface area (S) of the oblate spheroids was computed as

$$S_{os} = \frac{\pi a^2}{2} + \frac{\pi c^2}{4e} \ln\left(\frac{1+e}{1-e}\right) \text{ where } a = \text{length of the major axis,}$$

$$c = \text{length of the minor axis, and } e = \sqrt{\frac{a^2 - c^2}{a}}.$$

Formula 3: Surface area of the prolate spheroids was computed as

$$S_{ps} = \frac{\pi c^2}{2} + \frac{\pi ac}{2e} \sin^{-1}(e).$$

Formula 4: Surface area of the spheroids was computed as $S_{sphere} = \pi a^2 \psi^2$.

Formula 5: Because there is no closed-form mathematical solution for the surface area of an ellipsoid, the surface areas of the clasts that fall within the ellipsoid region of figure 12.6 were directly measured by coating the clasts with latex rubber, cutting the latex off the clasts, and then flattening the latex sheets on a piece of finely ruled graph paper and counting the number of ruled squares within the outline of the latex.

Section III.
Ornithischians

13. New Ornithopod from the Cedar Mountain Formation (Lower Cretaceous) of Eastern Utah

Tony DiCroce and Kenneth Carpenter

Abstract

The first occurrence of an iguanodontian from the Lower Cretaceous Poison Strip Member of the Cedar Mountain Formation, Utah, is described and named. This new taxon is represented by a well-preserved ilium, femora, tibiae, and vertebrae, as well as other material. The femora are typical for an ornithopod, but the ilium has a short, horizontal postacetabular process that is functionally an antitrochanter.

Introduction

Late Cretaceous Ornithopoda are among the most diverse groups of dinosaurs in North America and Europe (see table 13.1); however, except for *Iguanodon* and *Hypsilophodon* from Europe, they are poorly known from the Early Cretaceous. Galton and Jensen (1979) emphasized that the poor representation of Lower Cretaceous ornithopods may reflect inadequate collecting from strata of appropriate age. In North America, the most numerous ornithopod from Lower Cretaceous is *Tenontosaurus tilletti* from the Aptian-Albian Cloverly Formation of Montana and Wyoming (Ostrom 1970). Montana alone has produced over 80 specimens since the first discovery in 1903 by a field crew from the American Museum of Natural History (Forster 1990). A more primitive species, *T. dossi*, is known from the Aptian Twin Mountain Formation in Texas (Winkler et al. 1997).

TABLE 13.1
Early to Middle Cretaceous Ornithopoda
(in part from Sues and Norman 1990)

Taxon	Stratum/Age/Location	Reference
	Cedar Mountain, Utah	
Iguanodontidae		
Eolambia caroljonesa	Albian / Mussentuchit Member	Kirkland 1998
Planicoxa venenica	Barremian-Aptian / Poison Strip Member	DiCroce and Carpenter, this volume
Iguanodon ottingeri	Barremian-Aptian / Yellow Cat Member	Galton 1979
	Other North America	
Hypsilophodontidae		
Hypsilophodon wielandi	Aptian / Lakota Formation, South Dakota	Galton 1979
Zephyrosaurus schaffi	Aptian-Albian / Cloverly Formation	Sues 1980
Iguanodontidae		
Tenontosaurus tilletti	Aptian / Cloverly Formation, Montana	Ostrom 1970
T. dossi	Aptian / Twin Mountains Formation, Texas	Winkler et al. 1997
Iguanodon lakotaensis	Barremian / Lakota Formation, South Dakota	Weishampel and Bjork 1989
"Camptosaurus" depressus	Barremian / Lakota Formation, South Dakota	Gilmore 1909
	Other Countries	
Hypsilophodontidae		
Fulgurotherium australe	Albian / Griman Creek Formation, Australia	Huene 1932
Valdosaurus canaliculatus	Barremian / Wealden Isle of Wight, West Sussex, England	Galton 1975
V. nigeriensis	El Rhaz Formation / Agadez, Niger	Taguet 1976
Altascopcosaurus loadsi	Aptian-Albian / Otway Group, Victoria, Australia	Rich and Rich 1989
Hypsilophodon foxii	Barremian-Aptian / Wealden Marls, Isle of Wight, England	Huxley 1869
	Las Zabacheras Beds, Teruel, Spain	
Leaellynasaura amicagraphica	Aptian-Albian / Otaway Group, Victoria, Australia	Rich and Rich 1989
"Iguanodontian"		
Probactrosaurus gobiensis	Aptian-Albian / Dashuigou Fm., China	Rozhestvensky 1966
Iguanodontidae		
Iguanodon anglicus	Valanginian-Barremian / Wealden, East Sussex, England	Holl 1829
I. atherfieldensis	Valanginian-Aptian / Wealden, West Sussex, England	Hooley 1925
*		
Muttaburrasaurus langdoni	Albian / Griman Creek Fm., Queensland, Australia	Bartholomai and Molnar 1981
Ouranosaurus nigeriensis	Aptian / El Rhaz Formation, Agadez, Niger	Taquet 1976

* See addendum.

184 • Tony DiCroce and Kenneth Carpenter

Galton and Jensen (1979) described two species from the Lower Cretaceous of North America that they referred to the European genera *Hypsilophodon* and *Iguanodon*. The *Iguanodon*, named *I. ottingeri,* is significant because it was collected from the Cedar Mountain Formation, in eastern Utah. Originally collected in 1968, *I. ottingeri* was at first believed to have been recovered from the Brushy Basin Member of the Morrison Formation. The specimen is now correctly established as being from the Yellow Cat Member of the Cedar Mountain Formation (Kirkland et al. 1997). Galton and Jensen (1997) concluded that the presence of *Iguanodon* and *Hypsilophodon* in North America suggested a land connection between North America and Europe during the Early Cretaceous. Another species of *Iguanodon, I. lakotaensis,* is known from the Lower Cretaceous Lakota Formation, South Dakota (Weishampel and Bjork 1989). Unfortunately only a few elements were recovered: a nearly complete skull, a partial mandible, and two very incomplete vertebrae (a caudal and a dorsal). At the time of its discovery *I. lakotaensis* was considered by Weishampel and Bjork (1989) the first indisputable record of *Iguanodon* from North America. More recently, *Eolambia caroljonesa* has been described from high in the Cedar Mountain Formation. Originally described as a primitive crestless lambeosaurine hadrosaurid (Kirkland 1998), this taxon is now considered a *Probactrosaurus*-grade iguanodontian (Head 1999).

Yet another recent discovery was made in the Cedar Mountain Formation in the spring of 1998 by a team of volunteers from the Denver Museum of Natural History, (DMNH). The discovery was made in the Poison Strip Sandstone Member of the Cedar Mountain Formation in eastern Utah. The quarry site (fig. 13.1) was named Tony's Bone Bed; its exact location is on file at the DMNH, Denver,

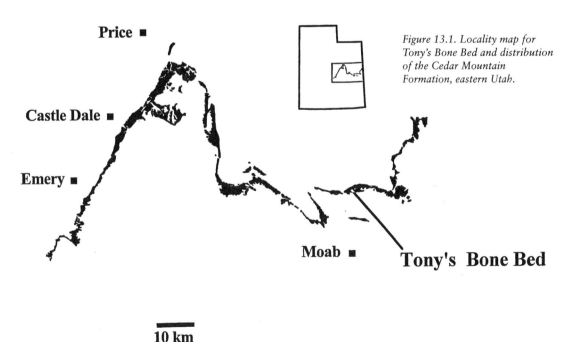

Figure 13.1. Locality map for Tony's Bone Bed and distribution of the Cedar Mountain Formation, eastern Utah.

New Ornithopod from the Cedar Mountain Formation • 185

Colorado. The associated fauna includes an unidentified juvenile ornithopod, an adult and juvenile of a new sauropod (Tidwell et al., chap. 11 of this volume), and a theropod, (possibly *Utahraptor ostrommaysorum)*. Previous discoveries of dinosaurs from the Poison Strip Member are limited to the nodosaurid ankylosaur *?Sauropelta* and a few isolated theropod and sauropod bones (Kirkland et al. 1997; Carpenter et al. 1999).

Geological and Taphonomic Setting

Tony's Bone Bed is located 3.75 m below the top of the Poison Strip Sandstone Member (fig. 13.2) of the Cedar Mountain Formation. At the nearby stratotype locality, the Poison Strip Sandstone is 5 m thick and composed of fine- to medium-grained sandstone with suspended black, gray, and white chert pebbles. It is trough-crossbedded with pale greenish mudstone partings toward the top (Kirkland et al. 1997). However, near Tony's Bone Bed, the Poison Strip is 11 m thick, is composed of medium- to coarse-grained sandstone, and contains lenses of greenish mudstone. Elsewhere in the vicinity of Arches National Monument, the Poison Strip Sandstone is highly variable in thickness and is missing in some places. It may represent a meandering river complex.

Figure 13.2. Stratigraphic column of the Cedar Mountain Formation in the vicinity of Tony's Bone Bed. The bone bed occurs at the base of a paleosol, near the middle of the Poison Strip Sandstone.

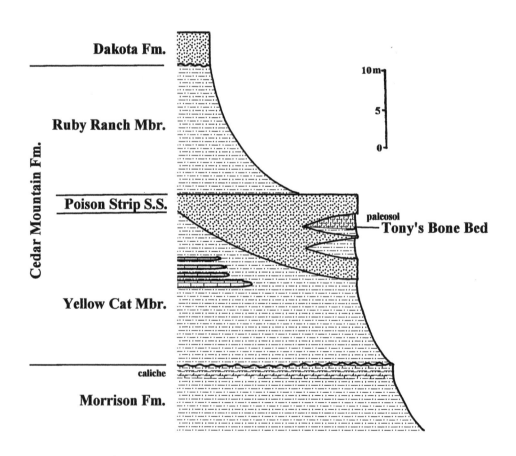

Previous best age estimates of the Poison Strip Sandstone placed it as mid-Aptian (Kirkland et al. 1997). However, none of the typical dinosaurs of the Cloverly Formation occur, with the possible exception of *Sauropelta* (Carpenter et al. 1999). The previous report of *Tenontosaurus* (Kirkland et al. 1997) is erroneous and the specimen is now known to have come from high in the Mussentuchit Member (Kirkland, pers. comm. 1999). The possible presence of *Utahraptor* (based on a large deinonychid-type ischium) suggests the fauna may be closer to that of the underlying Yellow Cat Member (see Kirkland et al. 1999 for faunal list). This interpretation is supported by the sauropod from the quarry (see Tidwell et al., chap. 11 of this volume), which is closer to those from the Dalton Wells Quarry in the Yellow Cat Member. Furthermore, the greenish mudstone lenses within the Poison Strip Sandstone are lithologically more similar to the mudstones of the Yellow Cat Member than of the overlying Ruby Ranch or Mussentuchit Members. All the evidence then supports an older age for the Poison Strip Sandstone than originally thought. Most likely, this age is Barremian, the same as the Yellow Cat Member (Kirkland et al. 1997).

The bones occur at the interface between gray lacustrine limestone and overlying greenish mudstone. Some of the bones extend down into the limestone, but most occur within the lower few centimeters of mudstone. Evidence of pedogenesis (e.g., red-green mottling, calcareous nodules, possible rhizoliths) occurs in the upper part of the mudstone, suggesting paleosol development superimposed on a former lacustrine setting.

The bone quality is highly variable. Some of the material shows weathering prior to burial and other mechanical damage, possibly due to trampling. However, the majority of the fossils are in very good condition and several vertebrae show delicate lamina. No articulated bones were found, indicating passage of time between death and burial. No tooth marks are present on the bones, although it is possible that some of the damage to the ends of the bones may have been caused by scavenging. None of the bones shows transport abrasion, although the long bones do show weak bimodal orientation at right angles. Thus, the most likely interpretation for the accumulation of the bone bed is death of several different individuals over a span of time, possibly during an interval of low water associated with the dry season.

Systematics
Order: Ornithischia Seeley 1888
Suborder: Ornithopoda Marsh 1871
Family: Iguanodontidae?
Planicoxa n.g.

Etymology: plani "flat" or "level"; *coxa,* a feminine noun for "hip" (Latin). *Planicoxa* refers to the flat appearance of the ilium, the defining characteristic.

Holotype: DMNH 42504, left ilium.

Type Locality: Tony's Bone Bed, Poison Strip Member of the Cedar Mountain Formation, Utah.

Paratypes: One cervical neural arch, DMNH 42511; seven dorsal vertebral arches, DMNH 42516, DMNH 42518, DMNH 42519, DMNH 42520, DMNH 42521, DMNH 42522, and DMNH 42524; three dorsal centra, DMNH 42513, DMNH 42515, and DMNH 42525; three dorsal rib fragments, DMNH 42523, DMNH 42526, and DMNH 42527; one sacral vertebra, DMNH 42510; two caudal centra, DMNH 42514 and DMNH 42517; left humerus (proximal end), DMNH 42508; left ulna, DMNH 42507; left femur, DMNH 42505; right femur, DMNH 40917; two right tibiae, DMNH 40914 and DMNH 40918; left tibia (distal end), DMNH 42506; left metatarsal II, DMNH 42509; and pedal phalanx, DMNH 42512.

Diagnosis: As for the species.

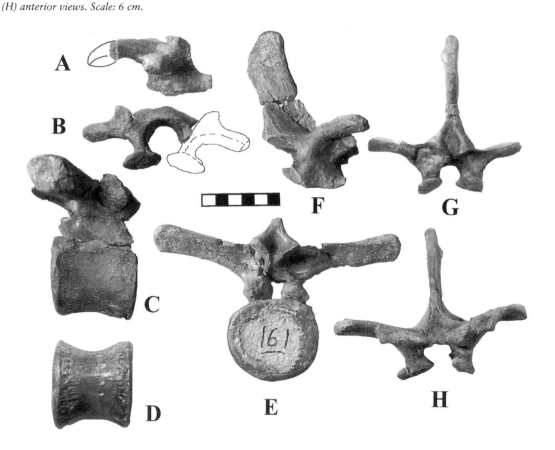

Figure 13.3. Planicoxa venenica: DMNH 42511 cervical arch in (A) right lateral and (B) anterior views; DMNH 42518 and DMNH 42513 anterior dorsal arch and centrum in (C) right lateral, (D) ventral, and (E) proximal views; DMNH 42518 posterior dorsal arch in (F) right lateral, (G) proximal (note deep triangular fossa), and (H) anterior views. Scale: 6 cm.

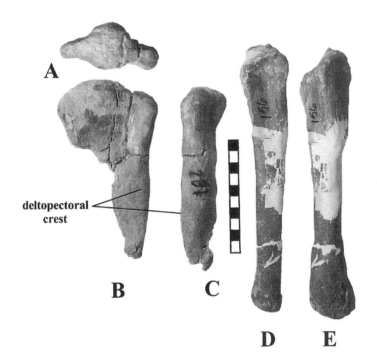

deltopectoral
crest

Figure 13.4. Planicoxa venenica:
DMNH 42508 left humerus in
(A) proximal, (B) anterior, and
(C) medial views (note well-
developed hemispherical head on
anterior surface); DMNH 42507
left ulna in (D) anterior and
(E) medial views. Scale: 10 cm.

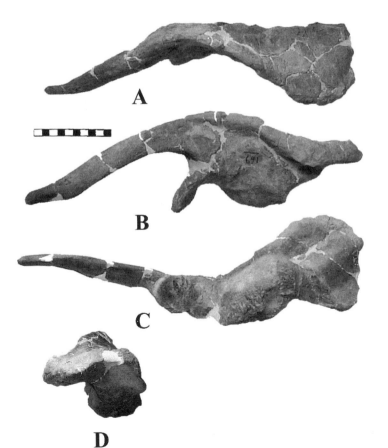

Figure 13.5. Planicoxa venenica
(DMNH 42504) holotype: left
ilium in (A) dorsal, (B) lateral,
(C) ventral, and (D) posterior
views. Scale: 10 cm.

New Ornithopod from the Cedar Mountain Formation • 189

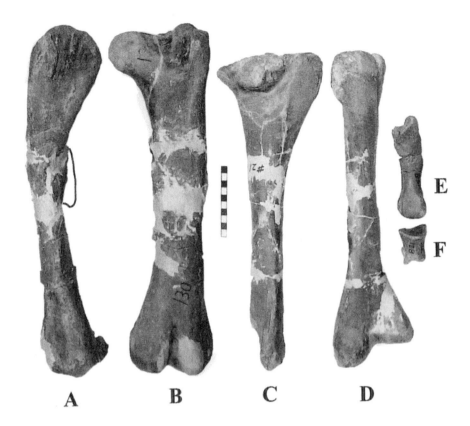

A   B   C   D

E

F

Figure 13.6. Planicoxa venenica:
DMNH 42505 left femur in
(A) lateral and (B) anterior views;
DMNH 40914 right tibia in
(C) caudal and (D) lateral views;
DMNH 42509 left metatarsal II
in (E) anterior view; DMNH
42512 phalanx in (F) anterior
view. Scale: 10 cm.

P. venenica, n. sp.
Figs. 13.3–13.6

Etymology: venenica, "poison" (Latin), in reference to the Poison Strip
Sandstone Member of the Cedar Mountain Formation, where the
discovery was made.

Diagnosis: Caudals with paired ventral ridges connecting anterior and
posterior chevron facets, humeral head extending onto anterior
surface, postacetabular blade of ilium short and horizontal.

Description

Vertebrae

Cervical Vertebra: The cervical neural arch (fig. 13.3A,B) lacks a
neural spine, a feature described for *Camptosaurus* (Gilmore 1909).
The surface that would have been occupied by the spine is rugose for the
nuchal ligament. This feature is considered a synapomorphy of the
Iguanodontidae (Sereno 1999).

Dorsal Vertebrae: The neural arches are detached from their respective centra (fig. 13.3), indicating the immaturity of the individuals.
The articular surfaces are round and slightly amphiplatyan as compared to *I. lakotaensis,* where they are heart-shaped and moderately
opisthocelous (Weishampel and Bjork 1989). The anterior centrum is
longer than tall and is constricted laterally (fig. 13.3C,D). Ventrally it

has a low, rounded ridge extending anteroposteriorly. The ventral anterior and posterior rims are rugose, as they are in the posterior dorsals, in a manner similar to that seen in *Thescelosaurus*. The posterior dorsal centrum is the same length as the anterior dorsal centrum, but about 25% larger in diameter. All centra are slightly constricted laterally and have a slightly developed ridge ventrally. A nutrient foramen occurs on each side of the ridge. The neural arches are characterized by short, wide neural spines. The parapophyses and diapophysis are located on the transverse process and are widely separated to accommodate a double-headed rib. The prezygapophyses are missing in all but two specimens. Where present, they are very short, rounded, and not very prominent, unlike *T. tilletti* (Forster 1990). There is a triangular fossa between the postzygapophysis that appears to be deepest in the posteriormost neural arches. Because all the arches in the quarry were widely separated from their centra, it is difficult to accurately reconstruct their sequence.

Sacral Vertebra: The single sacral centrum is not fused to its neural arch. Compared to *T. tilletti*, it appears to be either number 2 or 3, and has a greatly expanded cranial surface and a small rounded caudal surface.

Caudal Vertebra: The caudal vertebrae are represented by one proximal and one distal centrum with no fused neural arches. The caudals are characterized by a pair of low ridges that connect the anterior and posterior chevron facets along the ventral. In contrast to *I. lakotaensis* the articulated surfaces are just slightly amphicoelous, with oval articular surfaces, and do not retain the strong laterally placed horizontal ridge retained in *I. lakotaensis*.

Forelimb

Humerus: The humerus (Fig. 13.4A–C) is represented by a 16 cm fragment long, of the proximal end, including most of the deltopectoral crest (see table 13.2). The hemispherical head is well developed on the

TABLE 13.2
Measurements of *Planicoxa venenica* (cm)

Element	Length	Proximal Width	Distal Width	Head to 4th Trochanter	Minimum Circumference
Humerus, r.	—	9	—		
Ulna, l.	23	55	4.5		
Femur, r.	52	15	14	23.5	21
Femur, l.	44.5	15	13.5	23*	20
Tibia, l.	NA	—	12.5		
Tibia, r.	44.5	15.5	12.5		
Tibia, r.	48	16	12		

*Estimated.

posterior surface and continues onto the anterior surface as well, a condition also seen in *Thescelosaurus* and *Iguanodon*. In contrast, the head is restricted to the posterior side of the humerus in *Tenontosaurus* and *Camptosaurus*. The deltopectoral crest is small and gracile, as opposed to the larger, more massive one characteristic of *Tenontosaurus*.

Ulna: The ulna (fig. 13.4D,E) is straight shafted, with an expanded distal end. The olecranon process is relatively large, whereas it is moderately developed in *Hypsilophodon* (Galton 1974b). The radial notch is slightly concave and shallow. The ulna is slender, as in *Hypsilophodon*, straight as *Camptosaurus*, but more slender than *Thescelosaurus* (Galton 1974a).

Pelvis

Ilium: The ilium is well preserved, although some minor fracturing has occurred during burial (fig. 13.5). The most distinctive feature of the ilium is the horizontally directed postacetabular process (fig. 13.5A,D). A somewhat similar condition exists in *Thescelosaurus*, although not to the degree seen here (Carpenter, obs.). The lateral edge of the process is thickened into an antitrochanter, just dorsal and posterior to the ischiac peduncle. An antitrochanter is absent in *Thescelosaurus* (Galton 1974a) and *Tenontosaurus* (Ostrom 1970). The brevis shelf is triangular and is located ventromedially of the short postacetabular process. The dorsal border continues to the preacetabular process in a smooth unbroken surface, with no evidence of the sharp concave dorsal margin characteristic of *Tenontosaurus* (Ostrom 1970). The preacetabular blade curves ventrally similar to *Tenontosaurus* as described by Ostrom (1970). The blade, however, is proportionally much more slender and longer than *Tenontosaurus*, and compares much closer to *I. atherfieldensis*. The length of the preacetabular blade is approximately 23 cm, which is 50% of the total ilium length; in *Tenontosaurus* the blade is about 35% (Forster 1990). In dorsal view, the preacetabular blade is divergent, although the anteriormost tip curves medially to project forward. The anterior surface of the pubic peduncle is thick and wide. Laterally, the peduncle is long and thin and compares with *Camptosaurus* (Gilmore 1909). In cross-section, the peduncle is triangular, with the anterolateral surface for the origin of the M. ilio-femoralis internus. The ischial peduncle is very stout and, with the long thin pubic peduncle, forms a very deep concave acetabulum.

Hind Limb

Femora: The shaft of both femora (fig. 13.6) are damaged, but were clearly bowed as in many small to moderate-size ornithopods. The crista trochanteris is at or slightly above the level of the femoral head. The greater trochanter is very broad and fan-shaped. It is separated from the anterolaterally placed anterior trochanter (lesser trochanter by some authors) by a long narrow cleft. This character is retained in both basal and higher iguanodontians (Coria and Salgado 1996). The anterior trochanter appears to be much shorter than *Tenontosaurus* and is more posterolaterally placed; it is not as laterally placed as in *Iguanodon*. The fourth trochanter medially positioned along the shaft has an extremely long deep groove, which is one-fourth the length of the shaft, and is incomplete in the femur. Distally, the intercondylar groove is very deep on the anterior side. Unfortunately, the bone is damaged on each side of the groove, so it is not known if the groove was partially closed or narrowed.

Tibiae: Both tibiae (fig. 13.6) have a round straight shaft with very wide rugose proximal and distal ends. The proximal end has a distinct lateral condyle and a very narrow cnemial crest. The long axis of the distal end of both tibiae is about perpendicular to the proximal long axis, and thus differs remarkably from *Tenontosaurus,* where the long axes are separated about 65% (Forster 1990). The malleoli are separated by a deep flexor groove and the lateral malleolus extends more distally from the medial malleolus when compared to *Tenontosaurus.*

Metatarsal: The proximal end is slightly damaged from crushing. In anterior view, the metatarsal (fig. 13.6) is slightly bowed toward the medial direction. The shaft has a flat posterior surface, a rounded anterior surface, and it terminates in a knobby distal end.

Phalanx: The phalanx (fig. 13.6) does not differ markedly from other ornithopod phalanges. It is proximally concave where it fits against the metatarsal. It also has well-defined flexor and extensor grooves.

Discussion

The discovery of *Planicoxa* adds to the dinosaur diversity of the Cedar Mountain Formation, and to the diversity of Early Cretaceous ornithopods. The absence of a neural spine on the cervical arch of *Planicoxa* is an Euornithopoda synapomorphy (Sereno 1998, 1999). However, unlike other euornithopods, the postacetabular process of the ilium is anteroposteriorly short and horizontal. This horizontally flattened process is an autapomorphy that has not been described before in euornithopods. Functionally, it acts as an antitrochanter and shows an expanded scar for the M. ilio-femoralis internus. Along the dorsal rim is the origin of the M. ilio-tibialis lateralis. Ventrally, the overhanging postacetabular blade is divided into two areas. The most anterolateral of these corresponds to the origin of the M. ilio-fibularis as identified by Galton (1969) for *Hypsilophodon.* The more postero-medial site is the ventral aspect of the brevis shelf, which traditionally has been identified as the site for the M. caudi-femoralis brevis (e.g., Romer 1927; Galton 1969), after the crocodile (Romer 1923). However, based on avian anatomy (e.g., Drushel and Caplan 1991), the two muscle origins may be the M. ilio-fibularis pars cranialis for the antero-lateral scar and the M. ilio-fibularis et caudalis for the posteromedial one. This hypothesis, however, needs to be examined in more detail.

Conclusion

The discovery of *Planicoxa venenica* adds significant new information to the Barremian-Albian fauna of the Cedar Mountain Formation. *P. venenica* represents a new genus and new species of iguanodontian from the Poison Strip Member with characteristics such as the horizontally flattened postacetabular blade and the loss of cervical neural spine.

Acknowledgments: We would first like to thank the entire staff of the of the Earth Science Department at the Denver Museum of Natural History. Their commitment to the Paleontology Certification Program and to the volunteers that participate has allowed fieldwork and research to be performed by several individual amateurs from many different backgrounds. All the volunteers share one thing in common, their love for dinosaurs and their discovery. Second, we would like to thank everyone that assisted at the quarry. Their help made this discovery possible. The excavation of Tony's Bone Bed was made under Bureau of Land Management Excavation Permit no. UT-EX-99-005.

We are especially thankful to Dale Winkler for his review and valuable comments that greatly improved this work.

References

Bartholomai, A., and R. E. Molnar. 1981. *Muttaburrasaurus,* a new iguanodontid (Ornithischia: Ornithopoda) dinosaur from the Lower Cretaceous of Queensland. *Memoir of the Queensland Museum* 20: 319–349.

Carpenter, K., J. Kirkland, D. Burge, and J. Bird. 1999. Ankylosaurs (Dinosauria: Ornithischia) of the Cedar Mountain Formation, Utah, and their stratigraphic distribution. In D. Gillette (ed.), Vertebrate Paleontology in Utah. *Utah Geological Survey Miscellaneous Publication* 99–1: 244–251.

Coria, R. A., and L. Salgado. 1996. A basal iguanodontian (Ornithischia: Ornithopoda) from the Late Cretaceous of South America. *Journal of Vertebrate Paleontology* 16: 445–457.

Drushel, R. F., and A. I. Caplan. 1991. Three-dimensional reconstruction and cross-sectional anatomy of the thigh musculature of the developing chick embryo *(Gallus gallus). Journal of Morphology* 208: 293–309.

Forster, C. A. 1990. The postcranial skeleton of the ornithopod dinosaur *Tenontosaurus tilletti. Journal of Vertebrate Paleontology* 10: 273–294.

Galton, P. M. 1969. The pelvic musculature of the dinosaur *Hypsilophodon* (Reptilia. Ornithischia). *Postilla* (Peabody Museum) 131: 1–64.

Galton, P. M. 1974a. Notes on *Thescelosaurus,* a conservative ornithopod dinosaur from the Upper Cretaceous of North America, with comments on ornithopod classification. *Journal of Paleontology* 48: 1048–1067.

Galton, P. M. 1974b. The ornithischian dinosaur *Hypsilophodon* from the Wealden of the Isle of Wight. *Bulletin of the British Museum (Natural History), Geology* 25: 1–152.

Galton, P. M. 1975. English hypsilophodontid dinosaurs (Reptilia: Ornithischia). *Palaeontology* 18: 741–752.

Galton, P. M., and J. A. Jensen. 1979. Remains of ornithopod dinosaurs from the Lower Cretaceous of North America. *Brigham Young University Geology Studies* 25: 1–10.

Gilmore, C. W. 1909. Osteology of the Jurassic reptile *Camptosaurus,* with a revision of two new species. *Proceedings of the United States National Museum* 36: 197–332.

Head, J. J. 1999. Reassessment of the systematic position of *Eolambia caroljonesa* (Dinosauria, Iguanodontia) and the North American iguanodontian record. *Journal of Vertebrate Paleontology, Abstracts with Programs* 19 (3): 50A.

Holl, F. 1829. *Handbuch der Petrefaktenkunde.* Part 1. Quedlinburg.

Hooley, R. W. 1925. On the skeleton of *Iguanodon atherfieldensis* sp. nov., from the Wealden shales of Atherfield (Isle of Wight). *Quarterly Journal of the Geological Society of London* 81: 1–61.

Huene, F. von 1932. Die fossile Reptil-Ordnung Saurischia, ihre Entwicklung und Geschichte. *Monograph für Geologie und Paläeontologie,* 1st ser., 4: 1–361.

Huxley, T. H. 1869. On *Hypsilophodon,* a new genus of Dinosauria. *Abstract and Proceedings of the Geological Society London* 204: 3–4.

Kirkland, J. I. 1998. A new hadrosaurid from the upper Cedar Mountain Formation (Albian-Cenomanian Cretaceous) of eastern Utah: The oldest known hadrosaurid (Lambeosaurine). In S. G. Lucas, J. I. Kirkland, and J. W. Estep (eds.), Lower and Middle Cretaceous Terrestrial Ecosystems. *New Mexico Museum of Natural History and Science Bulletin* 14: 283–295.

Kirkland, J. I., B. Britt, D. Burge, K. Carpenter, R. Cifelli, F. Decourten, J. Eaton, S. Hasiotis, and T. Lawton. 1997. Lower to Middle Cretaceous dinosaur faunas of the Central Colorado Plateau: A key to understanding 35 million years of tectonics, sedimentology, evolution, and biogeography. *Brigham Young University Geology Studies* 42. (2): 69–103.

Kirkland, J. I., R. Cifelli, B. Britt, D. Burge, F. DeCourten, J. Eaton, and J. Parrish. 1999. Distribution of vertebrate faunas in the Cedar Mountain Formation, east-central Utah. In D. Gillette (ed.), Vertebrate Paleontology in Utah. *Utah Geological Survey Miscellaneous Publication* 99–1: 201–217.

Ostrom, J. H. 1970 Stratigraphy and paleontology of the Cloverly Formation (Lower Cretaceous) of the Bighorn Basin area, Wyoming and Montana. *Bulletin of the Peabody Museum of Natural History* 35: 1–234.

Rich, T. H. V., and Rich, P. V. 1989. Polar dinosaurs and biotas of the Early Cretaceous of southeastern Australia. *National Geographic Research* 5: 15–53.

Romer, A. S. 1923. Crocodilian pelvic muscles and their avian and reptilian homologies. *American Museum of Natural History Bulletin* 48: 533–552.

Romer, A. S. 1927. The pelvic musculature of ornithischian dinosaurs. *Acta Zoologica* 8: 226–275.

Rozhestvensky, A. K. 1966. New iguanodonts from Central Asia. *International Geology Review* 9556–9566.

Sereno, P. C. 1998. A rationale for phylogenetic definitions, with application to the higher-level taxonomy of Dinosauria. *Neues Jahrbuch für Mineralogie, Geologie, und Paläontologie* 210: 41–83.

Sereno, P. C. 1999. The evolution of dinosaurs. *Science* 284: 2137–2147

Sues, H.-D. 1980. Anatomy and relationships of a new hypsilophodontid dinosaur from the Lower Cretaceous of North America. *Palaeontographica* A 169: 51–72.

Sues, H.-D., and D. B. Norman. 1990. Hypsilophodontidae, *Tenontosaurus,* Dryosauridae. In D. B. Weishampel, P. Dodson, and H. Osmólska (eds.), *The Dinosauria,* pp. 498–509. Berkeley: University of California Press.

Taquet, P. 1976. Géologie et paléontologie du gisement de Gadoufaoua (Aptien du Niger). *Cahiers de paléontologie* (Paris, CNRS).

Weishampel, D. B., and P. R. Bjork. 1989. The first indisputable remains of *Iguanodon* (Ornithischia: Ornithopoda) from North America: *Iguanodon lakotaensis* n. sp. *Journal of Vertebrate Paleontology* 9: 56–66.

Winkler, D. A., P. A. Murry, and L. L. Jacobs. 1997. A new species of *Tenontosaurus* (Dinosauria: Ornithopoda) from the Early Cretaceous of Texas. *Journal of Vertebrate Paleontology* 17 (2): 330–348.

Addendum to Table 13.1

Under *Other Countries,* Iguanodontidae:

Iguanodon bernissartensis	Valanginian-Hauterivian/Wealden, Bernissart, Belgium	Casier 1960, Norman 1980

Addendum to References

Casier, E. 1960. Les Iguanodons de Bernissart. L'Institut Royal des Sciences Naturelles de Belgique. Goemaere: Brussels, 134 p.

Norman, D. B. 1980. On the Ornithischian Dinosaur Iguanodon Bernissartensis from the Lower Cretaceous of Bernissart (Belgium). L'Institut Royal des Sciences Naturelles de Belgique Mémoir no. 178, 103 p.

14. A Baby Ornithopod from the Morrison Formation of Garden Park, Colorado

KATHLEEN BRILL
AND KENNETH CARPENTER

Abstract

Several blocks of maroon sandstone contain a partial skeleton of a small ornithopod from the Morrison Formation of Garden Park, Colorado. The skeleton consists of several long bones, including a femur, humerus, and partial tibia and fibula, a portion of the articulated vertebral column, two phalanges, a few ribs, and paired frontals with a postorbital. The vertebrae consist of five posterior cervical, at least five dorsal, four sacral, and two caudal vertebrae that are clearly defined.

The scarcity of diagnostic features makes it difficult to identify the skeleton. However, the ratio of the lengths of the tibia and femur, the absence of a pit near the fourth trochanter on the femur, and the shape of the proximal end of the humerus indicates that the specimen may be *Othnielia rex*.

The individual was most likely a juvenile because the bones are small, the neural arches are not fused to the centra, and the ends of the long bones are spongy and incompletely formed. The femur length is ~78.5 mm, which is approximately one-third the adult femur length. The straight humeral shaft and pronounced emargination of the frontal above the orbit differ from the adult skeleton and are probably juvenile features.

Introduction

The partial skeleton of a small ornithopod was found in the Morrison Formation in Garden Park, Colorado (fig. 14.1). The skeleton (DMNH 21716) consists of several long bones, including a left humerus, a right femur, a partial right tibia and fibula, a portion of the articulated vertebral column, two phalanges, several ribs, and paired frontals with a left postorbital. The proximity, orientation, and sizes of the of the bones indicate that they represent one individual. The bones are very small, making this one of the youngest articulated Morrison ornithopods found to date.

The skeleton was found in five maroon sandstone blocks. The blocks had split along the skeleton, forming part and counterpart slabs. At least one of the matching slabs, which must contain the anterior side of the femur and humerus and most of the dorsal vertebrae, was not recovered. The bones were so poorly preserved that they were removed with hydrochloric acid, leaving molds. Casts were then made of the

Figure 14.1. Area map of Garden Park, Colorado, showing where the DMNH 21716 was found.

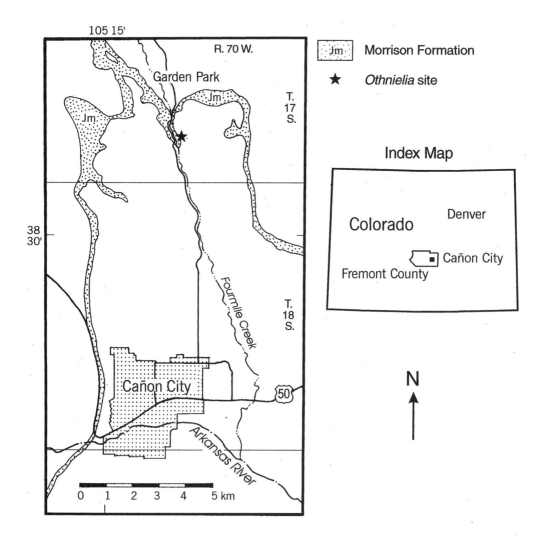

molds with gray-tinted latex. The molds are fine enough that many details of the original bone features are seen in the casts. Pertinent measurements of the bones discussed are found in table 14.1.

TABLE 14.1
Measurements of Postcranial Skeleton of DMNH 21716 (mm)

Element	Measurement
Femur length	79
Femur width, distal end	16.8
Femur—minimal distance from proximal end to distal edge of 4th trochanter	34
Tibia length (estimated)	> 85
Tibia width distal end	12
Humerus length (estimated)	53
Humerus width, proximal end	12
Last cervical vertebra, centrum length	8.8
Last cervical vertebra, transverse process width	13.9
Dorsal vertebra, centrum length	8.3
1st sacral vertebra, centrum length	9.9
2nd sacral vertebra, centrum length	12.6

Description

The cranium elements consist of articulated frontal and left postorbital bones seen in ventral view (fig. 14.2). The emargination of the two frontals at the lateral margins is very pronounced. The anterior width of the frontals is about half the posterior width, a feature also seen in juvenile *Orodromeus* (Scheetz 1999). Sutures at the anterior end and between the frontals as well as the sulcus for the olfactory tract are clearly seen. The margin of the postorbital overlaps the frontal. The "neck" of the frontal process of the postorbital narrows considerably before flaring to form the jugal and squamosal processes. The rim of the orbit is distinctly seen as a ridge that curves medially on the frontals and merges into the jugal process of the postorbital.

The axial skeleton consists of five posterior cervical vertebrae, at least five anterior dorsals, four sacral, and two caudal vertebrae that are clearly defined (figs. 14.3, 14.5, 14.8). All except the caudals are articulated. The linear orientation of the vertebrae and the presence of fragments of the neural arches of additional dorsal vertebrae indicates that the entire series may have been originally preserved.

The lateral sides of the centra of the cervical vertebrae have a distinctive "pinched-in" appearance and the ventral sides have a sharp keel, features that are also present in other hypsilophodontids (Galton and Jensen 1973). One transverse process is articulated with each of the

Figure 14.2. Cast of paired frontal and left postorbital bones in ventral view. Scale: 10 mm.

Figure 14.3. Casts of the cervical and dorsal vertebrae. Views A and B are part and counterpart slabs of the same articulated cervical and dorsal vertebrae. The arrows point to cervical vertebra 9. Juvenile features shown are sutures between the transverse processes and centra, and separation of the neural arches from the centra. Scale: 10 mm.

Figure 14.4. Cast of the left humerus. Scale: 10 mm.

Figure 14.5. Cast of the right femur. S1 and S2 indicate sacral vertebrae 1 and 2. R indicates the sacral rib. Scale: 10 mm.

A Baby Ornithopod from the Morrison Formation • 201

Figure 14.6. Cast of the right tibia and fibula. View A shows the shafts and view B shows the distal end of the tibia. Scale: 10 mm.

cervicals. There is a distinct line between each process and centrum, showing that they are incompletely fused. The transverse processes narrow toward the tips. A prezygapophysis is visible on the last cervical. The neural arches are present but not attached to the centra.

The neural arches have separated from each of the dorsal centra. The dorsal centra are more rounded ventrally than the cervicals and compressed in the center. At least two rib fragments are present, but not articulated with the vertebrae.

Details of the centra of the first two sacral vertebrae are more visible than in the last two. The centra are widest at the anterior and posterior faces. The second sacral is longer than the first. Both have facets for an intervertebral sacral rib. One sacral rib is present but is not articulated with the vertebrae. The posterior sacral vertebrae are only faintly visible. The caudal vertebrae have a groove on the ventral surface and a small facet for a chevron.

The left humerus (fig. 14. 4) is sloped on the proximal end lateral to the spongy head, and is arced significantly at the low deltopectoral crest. The distal shaft is very straight. Details of the distal end cannot be seen because it is missing.

The right femur (fig. 14.5) was preserved nearly intact. It has a deep cleft between the anterior and greater trochanters. The femur also has an incompletely formed head, probably because it had not yet ossified. The fourth trochanter is distinctively protuberant. There is a shallow depression on the medial side of the femur that merges into the base of the fourth trochanter. There is no pit at the base of the fourth trochanter, as occurs in *Dryosaurus* (Galton 1983). The presence or absence of an anterior intercondylar groove cannot be determined because only the posterior side of the bone is presented. The proximal ends of both the tibia and fibula (fig. 14.6) are missing. The distal end of the tibia is bladelike and triangular in shape. The distal end of the fibula is missing. The first and second phalanges of the third toe of the left (?) pes are present.

The elements that make up this specimen are shown in figure 14.7.

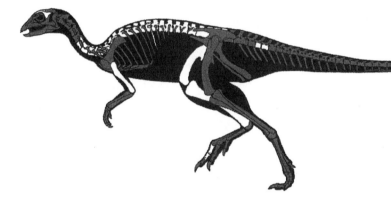

Figure 14.7. Skeleton of a hypsilophodontid dinosaur. DMNH 21716 elements are shown in white. Modified from Gregory Paul (1996).

Discussion

Several small ornithopods are found in the Morrison formation of the Garden Park area, including *Camptosaurus* spp., *Othnielia rex*, *Dryosaurus altus*, and "*Nanosaurus agilis*." "*N. agilis*" is fragmentary, and may be a juvenile of some other taxon. The remaining ornithopods are well described. The ratio of the lengths of the tibia and femur, the absence of a pit at the base of the fourth trochanter of the femur, and the shape of the proximal end of the humerus make it reasonable to refer this specimen to the hypsilophodontid *Othnielia rex* (Galton 1983).

DMNH 21716 has many juvenile characteristics. The heads of the femur and humerus are small and bladelike, the bone ends appear very spongy, and some bones have a fibrous surface texture (fig. 14.8). Sutures and pronounced emargination of the anterior end, which differs from older individuals, are very prominent on the frontal bones (fig. 14.2). Transverse processes appear to be incompletely fused to the

Figure 14.8. Ontogenetic features: view A shows the fibrous surface texture of the vertebrae with the arrow pointing to an especially clear example. View B shows the spongy nature of the proximal end of the femur (arrow). Scale: 10 mm.

centra of the cervical vertebrae and neural arches are separated from all the vertebrae (fig. 14.3). The distal shaft of the humerus is straight, which has been identified as a ontogenetic feature in other ornithopods (Carpenter 1994).

An adult *Othnielia rex* (YPM 1882) femur is 244 mm long with total body length of approximately 2300 mm (Galton 1983). Based on the length of the femur and length of the posterior cervical vertebrae, DMNH 21716 was about one third the length of an adult, or approximately 750 mm. The age of this animal cannot be inferred from its size because no eggs or hatchlings for *Othnielia* have been found.

Acknowledgments: We thank Dr. Mike Williams for the discovery of "Mike's Baby" during field work by the Denver Museum of Natural History in the early and middle 1990s. Review comments by Jack Horner are appreciated and helped improve the manuscript.

References

Carpenter, K. 1994. Baby *Dryosaurus* from the Upper Jurassic Morrison Formation of Dinosaur National Monument. In K. Carpenter, K. F. Hirsch, and J. R. Horner (eds.), *Dinosaur Eggs and Babies*, pp. 296–297. Cambridge: Cambridge University Press.

Galton, P. 1983. The cranial anatomy of *Dryosaurus*, a medially dinosaur from the Upper Jurassic of North America and East Africa with a review of hypsilophodontids from the Upper Jurassic of North America. *Geologica et Palaeontologica* 17: 207–243.

Galton, P., and J. Jensen. 1973. Skeleton of a hypsilophodontid dinosaur (*Nanosaurus(?) rex*) from the Upper Jurassic of Utah. *Brigham Young University Geology Studies* 20: 137–157.

Paul, G. 1996. *The Complete Illustrated Guide to Dinosaur Skeletons.* Tokyo: Gakken Mook.

Scheetz, R. D. 1999. Osteology of *Orodromeus makelai* and the phylogeny of basal Ornithopod dinosaurs. Ph.D. dissertation, Montana State University, Bozeman.

15. Evidence of Hatchling- and Nestling-Size Hadrosaurs (Reptilia: Ornithischia) from Dinosaur Provincial Park (Dinosaur Park Formation: Campanian), Alberta

DARREN H. TANKE AND
M. K. BRETT-SURMAN

Abstract

The occurrences of dinosaur eggshell and neonate-size hadrosaur skeletal material from the Late Cretaceous (Campanian) of Dinosaur Provincial Park, Alberta, is reviewed and some skeletal specimens in TMP collections are described. Eggshell fragments occur rarely and only in two microfossil sites dominated by invertebrate shells. This factor may be related to the calcium in the invertebrate shells acting as a buffer to the acidic water conditions of the time. Many of the hadrosaur skeletal elements show little or no stream abrasion, suggesting they originated from areas near the Park, if not from the Park itself. This new material supports recent suggestions that hadrosaurs did not nest only in upland areas, but nested in lowland environments as well.

Introduction

More than two decades of fieldwork conducted by staff of the Royal Tyrrell Museum of Palaeontology in Dinosaur Provincial Park, Alberta, Canada (fig. 15.1), has resulted in the recovery of many thousands of vertebrate fossils, including those of hatchling or extremely young individuals. On the basis of teeth or skeletal material (especially the latter), neonate-size *Champsosaurus,* crocodilians, turtles, and dinosaurs, including ornithomimids, small theropods (*Saurornithoslestes* and *Troodon*), tyrannosaurs, possible ankylosaurs, centrosaurine ceratopsians, and especially hadrosaurs, are now documented. Occurrences of non-nested, neonate hadrosaurs are rarely reported in the

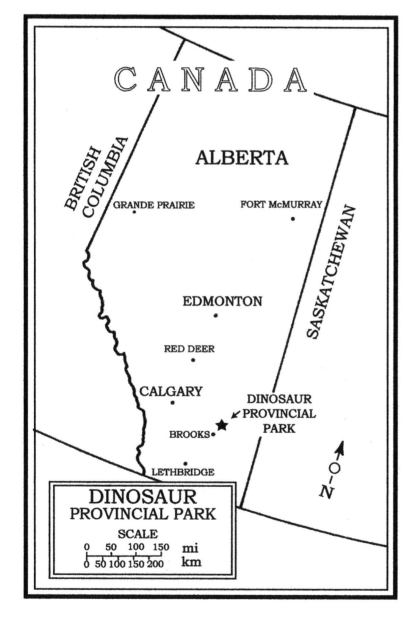

Figure 15.1. Map of Alberta, Canada, showing the location of Dinosaur Provincial Park.

literature (see Carpenter and Alf 1994; Carpenter 1999 for reviews). Numerous disarticulated skeletal remains of neonate-size hadrosaurs from DPP are now known. Some specimens were derived from extremely small and possibly embryonic individuals.

Hatchling- or nestling-size (< 1.5 m total length) hadrosaur skeletal remains have always been poorly represented in the Upper Cretaceous lowland deposits of western North America and elsewhere. This rarity, and the presence of eggs, nests, or nesting horizons preserved in upland facies, originally led previous workers (Sternberg 1955; Horner 1982, 1984) to consider hadrosaur breeding and nesting habits to have occurred in more upland regions. More recent research has indicated hadrosaurs nested in lowlands as well (Carpenter 1982, 1992; Fiorillo 1987). However, neonate-size hadrosaur specimens have only occasionally been previously reported from the lowland facies in Dinosaur Provincial Park (DPP). Sternberg (1955) described several isolated jaws (NMC 619 and NMC 8525), and what is still the best single known specimen from a hatchling-size individual, a partial skull of an unidentified hadrosaurine (NMC 8917). Russell (1967) mentioned isolated remains of hadrosaurs and saurischian dinosaurs from the Late Cretaceous of Alberta that came from individuals "no larger than a full-grown turkey." Dodson (1983, 98) commented briefly on the occurrences of juvenile hadrosaur material recovered from vertebrate microfossil localities. Brinkman (1986) noted occurrences of dinosaur eggshell in the Park. These short and infrequent notations give the impression that fossil material referable to hatchling- and nestling-size hadrosaurs in DPP is extremely rare. This, however, is not true. The apparent rarity of remains is due to fossil collecting biases against smaller specimens or simply not recognizing them in the field.

Recently, the view that hadrosaurs nested predominantly in upland environments has come to be challenged. Fiorillo (1987, 1989), described juvenile hadrosaur material from a lowland deposit in central Montana. Carpenter (1992, 1999) described and figured very small hadrosaur footprints from the Blackhawk Formation of Utah. Most recently, Clouse (1995) reported on extensive hadrosaur nesting grounds and embryonic remains from possible lowland facies near Havre, Montana. Fieldwork at microvertebrate sites and bone beds by the Royal Tyrrell Museum at DPP has resulted in the discovery of several thousand taxonomically diverse specimens. Among these samples occur material from hatchling- to nestling-size hadrosaurs. While such specimens at first appear to be relatively uncommon, an experienced collector can usually find several such specimens per day (Tanke, field observation). The 1992 field season marked the first year a concerted effort was made to find and collect neonate-size hadrosaur bones and this effort was successful, with 43 specimens collected. These finds mostly consist of edentulous dentary fragments, limb bones of varying completeness, pedal elements, and centra, although other elements are represented. Some of the material shows little or no transport abrasion, supporting the hypothesis of nesting by hadrosaurs within or near the Park boundaries, and also confirming Carpenter's (1982) and Fiorillo's (1987, 1989) hypotheses for a lowland nesting behavior in these dinosaurs.

Institutional Abbreviations: DPP, Dinosaur Provincial Park, Alberta; NMC, National Museum of Canada (now Canadian Museum of Nature), Ottawa; TMP, Royal Tyrrell Museum of Palaeontology, Drumheller, Alberta.

Description

Eggshell

Dinosaur eggshell in the Dinosaur Park Formation is known from only two small localities within DPP. Both localities are vertebrate microfossil sites or bone beds (BB), containing countless fragmented remains of pisidiid clams, rare unionid clams (Brinkman 1986; Brinkman et al. 1987), and rare gastropods (Eberth 1990). The presence of many invertebrate shells within these sites [BB 31 (Quarry 156) and BB 98] apparently released calcium carbonate into the acidic water, which acted as a buffering agent to raise the local water pH to levels conducive to nondissolution of eggshell (Carpenter 1982, 1987, 1999).

The eggshell pieces are fragmentary, with no specimens exceeding 1 cm in greatest dimension; they have a pebbled surface texture. They are similar to dinosaur eggshell from Montana described by Jepsen (1931) and Sahni (1972) from the Judith River Formation, and from the Two Medicine Formation (Horner 1999). Referring the DPP eggshell to hadrosaurs is somewhat problematic because a single egg can have different types of surface texture on different parts of the egg. However, the DPP eggshell is similar to that of *Maiasaura,* a hadrosaurid best known from Montana, and eggshell from the Devil's Coulee hadrosaur nesting locality (Currie 1988) in southern Alberta. While it is beyond the scope of this chapter to report on the eggshell histologically, in gross appearance some of the eggshell is hadrosaurian (Zelenitsky, pers. comm. 1999).

Bones

A complete listing of DPP embryonic and neonate hadrosaur bone material housed at TMP is listed in appendix 15.1. Measurements of some of the more complete material are given in appendix 15.2.

Dentaries are well represented in the collection. Many of the dentary fragments bear fresh erosional damage resulting in loss of the coronoid process. It is likely that these specimens were originally buried entire, such as those described by Sternberg (1955). A combination of factors, such as the small size of the specimens and the high erosion rates in DPP, make it difficult to find complete dentaries and other neonate bones. Few specimens preserve the entire series of grooves for tooth emplacements. As noted by Sternberg (1955) and Dodson (1983), most dentaries preserve only about 10 tooth files. TMP neonate hadrosaur material from DPP exhibit 12 tooth files in the maxilla, and 11 tooth files in the dentary (fig. 15.2a; TMP96.12.12).

Limb bones have the same morphology and general proportions as those found in adult animals, and often show predepositional erosion or concave articular ends. This condition is no doubt due to the cartilaginous cap on the ends of these bones (Horner and Weishampel 1988).

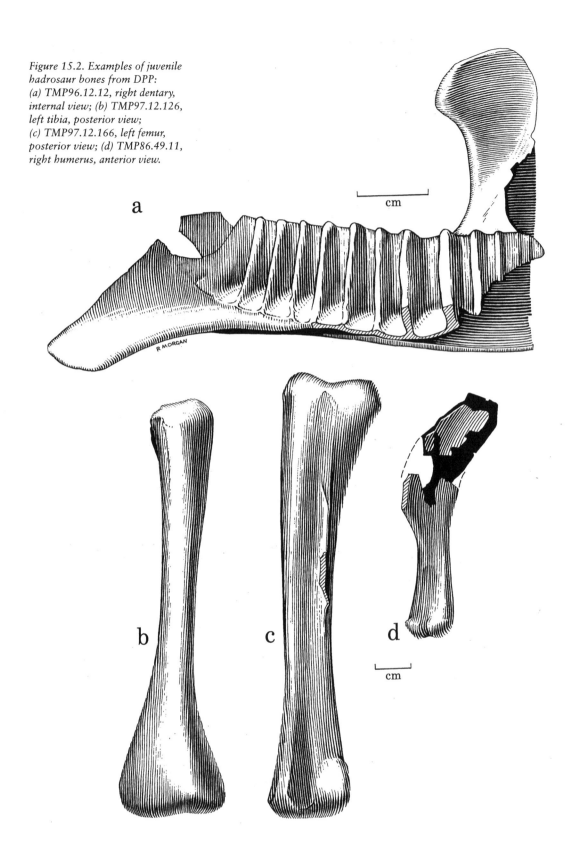

Figure 15.2. Examples of juvenile hadrosaur bones from DPP: (a) TMP96.12.12, right dentary, internal view; (b) TMP97.12.126, left tibia, posterior view; (c) TMP97.12.166, left femur, posterior view; (d) TMP86.49.11, right humerus, anterior view.

One left humerus (fig. 15.2d; TMP86.49.11) is typical of hadrosaurines, with a deltopectoral crest extending below the midpoint of the shaft. The external texture of the deltopectoral crest (4 mm thick) shows the early formation of the compacta (<0.5 mm thick). The entire length of the humerus is 70 mm.

A left femur (fig. 15.2c; TMP97.12.166) has the apex of the fourth trochanter at the midpoint of the shaft. The base of the trochanter, however, extends cranially almost one-quarter the length of the shaft, a feature not seen in many adults of Campanian age. The shape of the fourth trochanter is primitively an asymmetric triangle that is derived from the pendant hook in the iguanodont condition (Brett-Surman 1989). Over geologic time, the trochanter became an isosceles triangle, a condition seen in many crown group genera of hadrosaurids (Brett-Surman 1989). The change in trochanter shape, from juveniles with a trochanteric base extending cranially, to a trochanter with a restricted base only at the midpoint of the femoral shaft, is a heterochronic feature. The neck between the head and the greater trochanter shows the beginning of development. The anterior trochanter is offset from the greater trochanter. The channel between the distal articular condyles is already formed, even though the condylar surfaces are still not fully ossified. Closure of the channel is seen only in adults, and thus is a growth feature, not taxonomic.

An isolated left tibia (fig. 15.2b; TMP97.12.126) is more developed distally than proximally. The proximal condyles are slightly developed, with the median condyle larger than the lateral condyle and the cnemial crest. The distal end of the tibia has the articular facet for the astragalus formed, mostly on the caudal surface. The craniolateral surface of the outer condyle already has the vertical striations on the compacta where the distal end of the fibula articulates. Fusion of the astragalus and calcaneum to the tibia and fibula is rarely seen, and then only in very old adults.

The pelvis has features in common with those of an adult, indicating that it forms very early in ontogeny. The prepubic portion of one pubis (TMP91.36.367) has a neck (50 mm) that is longer than the blade (34 mm), as is typical of hadrosaurines. The postpubis is as long as the prepubis, a condition of neonates (Brett-Surman 1989). Most of the bone texture is woven, with the compacta on the external surface just beginning to form.

Vertebrae are represented by centra only. Like the limb bones, these show proportions comparable to adult-size animals. Embryonic *Hypacrosaurus* material from Devil's Coulee shows that the neural arches were well developed, with all processes present, but are not fused to the centra. It is likely that tiny neural arches in DPP were lost prior to burial, destroyed by contemporary erosion, are overlooked, or not recognized.

Discussion

Hatchling hadrosaurs had limb bones with poorly developed (highly cartilaginous) articular ends (Horner and Weishampel 1988). These bones would not be expected to last long in an active silt and sand-laden river system, especially in water with a higher pH caused by the acids produced by the breakdown and release of tannins into the ecosystem by the abundant local coniferous vegetation (K. Aulenback, pers. comm. 2000). In fact, some specimens do show what appears to be transport wear with ends or edges rounded to varying degrees, indicating that some of the neonate bones may have traveled unknown distances.

Sternberg (1955) suggested that juvenile carcasses may have washed down from upland nesting grounds. While this may be true in part, such carcasses surviving the ravages of crocodilians, turtles, fish, and other scavengers before arriving in DPP makes this scenario seem doubtful. More likely, those bones showing transport abrasion had been moved only short distances from the distal upland areas or from lowland nesting sites.

The presence of commonly unabraded, near perfect neonate and possible embryonic hadrosaur bones and sharp-edged hadrosaur eggshell fragments in situ indicate nesting sites must have occurred nearby the area of deposition. These findings confirm hypotheses that hadrosaurs nested in lowland environments (Carpenter 1982, 1992; Fiorillo 1987, 1989). Because hatchling hadrosaur bone occurs in bone beds and microfossil sites throughout the stratigraphic section of the Dinosaur Park Formation within DPP, it is clear that hadrosaurs nested in the area during the 2.5 million years of deposition (Eberth 1990).

Bone beds 23, 28, 47, and 50 have yielded unusually high numbers of neonate hadrosaur bones relative to other bone beds in DPP (Tanke, field observation). Several small outcrops within BB 50 are particularly rich and produce up to a dozen baby hadrosaur specimens annually (the exact GPS coordinates and photographs of these sites are on file in TMP collections). Dinosaur eggshell has not been found in these bone beds. Perhaps these higher rates of neonate bone are related to close proximity to an active hadrosaur nesting ground (Horner 1994). Certainly the uneroded nature of many of the neonate bones would indicate minimal stream transport. Collecting biases or collecting intensities in one particular site over others does not appear to be related to the abundance of bone (Tanke, field observation).

Conclusions

Hadrosaurs apparently nested in both lowland and upland environments, dependent on their diet, soil conditions, habits, competition with other hadrosaur or dinosaurian genera, or other unknown factors. It has not been established whether a particular genus nested in a single type of environment (lowland/upland), or could have nested in both. Rare hadrosaurids in DPP (such as the lambeosaurine *Parasaurolophus* and hadrosaurine *Brachylophosaurus*) may be carcasses derived from migrating individuals. Or, they originated from upland feeding or nesting grounds, but more work in upland environment depositional systems will need to test this hypothesis. Presently, little is known of the hadrosaurs that lived in the upland and more northern Campanian-aged facies in Alberta.

Acknowledgments: We have benefited from discussions with Phil Currie, Darla Zelenitsky, and Kevin Aulenback. We thank Phil for his unwavering support over the years and offer our congratulations on the occasion of his 25th anniversary of paleontological activities in Alberta and the world. We thank Kenneth Carpenter and Patty Ralrick for helping review and critique the manuscript. P. Ralrick also assisted in data recording. A final word of thanks to the overworked TMP collections staff. All figures were prepared by Rod Morgan (Calgary, Alberta).

References

Brett-Surman, M. K. 1989. Revision of the Hadrosauridae (Reptilia: Ornithischia) and their evolution during the Campanian and Maastrichtian. Ph.D. diss., George Washington University.

Brinkman, D. 1986. Microvertebrate sites: Progress and prospects. In B. G. Naylor (ed.), *Dinosaur Systematics Symposium, Field Trip Guidebook to Dinosaur Provincial Park,* pp. 24–37. Drumheller: Tyrrell Museum of Palaeontology.

Brinkman, D., D. A. Eberth, and P. A. Johnston. 1987. Bonebed 31: Palaeoecology of the Upper Cretaceous Judith River Formation at Dinosaur Provincial Park, Alberta, Canada. In D. A. Eberth (ed.), Fourth Symposium on Mesozoic Terrestrial Ecosystems, Field Trip "A" Guidebook. *Tyrrell Museum of Palaeontology, Occasional Paper* 3, pp. 12–13.

Carpenter, K. 1982. Baby dinosaurs from the Late Cretaceous Lance and Hell Creek Formations and a description of a new species of theropod. *Contributions to Geology* (University of Wyoming) 20:123–134.

Carpenter, K. 1987. Potential for fossilization in Late Cretaceous–Early Tertiary swamp environments. *Geological Society of America, Abstracts with Programs* 19 (5): 264.

Carpenter, K. 1992. Behavior of hadrosaurs as interpreted from footprints in the "Mesaverde" Group (Campanian) of Colorado, Utah, and Wyoming. *Contributions to Geology* (University of Wyoming) 29 (2): 81–96.

Carpenter, K. 1999. *Eggs, Nests, and Baby Dinosaurs.* Bloomington: Indiana University Press.

Carpenter, K., and K. Alf. 1994. Global distribution of dinosaur eggs, nests, and babies. In K. Carpenter, K. F. Hirsch, and J. R. Horner (eds.), *Dinosaur Eggs and Babies,* pp. 15–30. Cambridge: Cambridge University Press.

Clouse, V. R. 1995. Paleogeography of an extensive dinosaur nesting horizon in the Judith River Formation of north-central Montana. *Geological Society of America, Abstracts with Programs* 27 (4): 6.

Currie, P. J. 1988. The discovery of dinosaur eggs at Devil's Coulee. *Alberta* 1 (1): 3–10.

Dodson, P. 1983. A faunal review of the Judith River (Oldman) Formation, Dinosaur Provincial Park, Alberta. *Mosasaur* 1: 89–118.

Eberth, D. A. 1990. Stratigraphy and sedimentology of vertebrate microfossil sites in uppermost Judith River Formation (Campanian), Dinosaur Provincial Park, Alberta, Canada. *Palaeogeography, Palaeoclimatology, Palaeoecology* 78: 1–36.

Fiorillo, A. R. 1987. Significance of juvenile dinosaurs from Careless Creek Quarry (Judith River Formation), Wheatland County, Montana. In P. J. Currie and E. H. Koster (eds.), *Fourth Symposium on Mesozoic Terrestrial Ecosystems: Short Papers,* pp. 89–95. Tyrrell Museum of Palaeontology, Occasional Paper 3.

Fiorillo, A. R. 1989. The vertebrate fauna from the Judith River Formation (Late Cretaceous) of Wheatland and Golden Valley Counties, Montana. *Mosasaur* 4: 127–142.

Horner, J. R. 1982. Evidence of colonial nesting and "site fidelity" among ornithischian dinosaurs. *Nature* 297: 675–676.

Horner, J. R. 1984. The nesting behavior of dinosaurs. *Scientific American* 250 (4): 130–137.

Horner, J. R. 1994. Comparative taphonomy of some dinosaur and extant bird colonial nesting grounds. In K. Carpenter, K. F. Hirsch and J. R. Horner (eds.), *Dinosaur Eggs and Babies,* pp. 116–123. Cambridge: Cambridge University Press.

Horner, J. R. 1999. Egg clutches and embryos of two hadrosaurian dinosaurs. *Journal of Vertebrate Paleontology* 19 (4): 607–611.

Horner, J. R., and D. B. Weishampel. 1988. A comparative embryological study of two ornithischian dinosaurs. *Nature* 332: 256–257.

Jepsen, G. L. 1931. Egg shells sixty million years old. *Discovery* 12: 180–183.

Russell, L. S. 1967. Reply to J. M. Cys 1967. The inability of dinosaurs to hibernate as a possible key factor in their extinction. *Journal of Paleontology* 41 (1): 266–267.

Sahni, A. 1972. The vertebrate fauna of the Judith River Formation, Montana. *Bulletin of the American Museum of Natural History* 147: 321–412.

Sternberg, C. M. 1955. A juvenile hadrosaur from the Oldman formation of Alberta. *National Museum of Canada, Bulletin* 136: 120–122.

Embryonic, Neonate, and Small Juvenile Hadrosaur Specimens from Dinosaur Provincial Park (Dinosaur Park Fm.; Campanian), Alberta. Specimens housed in Royal Tyrrell Museum of Palaeontology (Drumheller, Alberta) collections. All specimen numbers are preceded by the acronym TMP. List compiled and updated by Darren Tanke (TMP) and Patty Ralrick; listing accurate up to March 31, 1999.

Partial skull: 82.4.2 (cast of NMC 8917 described by Sternberg 1955).

Basisphenoid: 92.36.1047; 98.92.152.

Jugal: 92.36.428 (not neonate, but juvenile).

Maxilla: 80.16.1826 (fragment with 4 teeth); 81.22.6 (nearly complete, with 7 teeth); 82.16.177 (with 7 teeth; not neonate, but small juvenile); 86.36.265 (fragment); 88.36.4 (right maxilla with teeth; not neonate, but small juvenile); 88.36.10; 94.12.327 (fragment); 89.50.50; 95.134.4 (fragment).

Surangular: 97.12.125.

Ceratohyal: 98.93.165 (nearly complete; not neonate, but small juvenile).

Right dentaries: 67.17.4; 67.20.232 (complete); 81.16.414; 81.41.131 (fragment); 82.20.197; 82.20.202; 82.20.472; 83.67.33; 86.78.57; 87.36.386; 89.36.414; 89.50.154; 91.36.57; 93.36.662; 93.40.19; 94.12.542; 95.12.165; 96.12.9; 96.12.12 (complete); 96.12.158; 98.12.19 (complete); 98.93.146.

Left dentaries: 79.8.396; 79.8.588; 79.14.420; 80.13.47; 80.16.1260 (fragment); 81.19.274; 82.16.61; 82.16.259; 85.36.173; 86.36.6 (fragment); 87.48.95 (fragment); 90.50.14; 91.36.821; 92.36.12; 92.36.721; 92.36.724 (complete); 92.50.183; 93.36.69; 94.12.46; 98.68.96; 98.93.153 (complete).

Undifferentiated dentary fragments: 67.20.4; 92.36.127; 94.12.627 (not neonate, but small juvenile).

Unidentified jaw fragments: 94.12.904.

Surangular: 79.8.254 (not neonate, but small juvenile).

Teeth: 79.8.639; 90.36.3; 90.78.4 (3 teeth); 93.36.71; 94.12.552 and 94.12.579 (both not neonate, but small juvenile); 95.127.19.

Cervical vertebrae:

Centra: 96.12.166 (not neonate; but small juvenile); 96.12.171 (from cervico-dorsal transition region; not neonate, but small juvenile); 97.12.210 (not neonate; but small juvenile).

Diapophysis: 94.12.426 (diapophysis? Tiny = embryonic?).

Dorsal vertebrae:

Centra: 79.8.412 (not neonate, but small juvenile); 85.63.65; 91.36.206; 92.36.584; 95.127.14; 98.93.15; 98.93.161 (not neonate, but small juvenile).

Neural arch: 82.31.1 (not neonate, but small juvenile).

Sacral centra: 92.36.152 (nonhadrosaurian?); 92.50.142; 94.172.137 (embryonic?); 98.93.50 (not neonate, but small juvenile).

Caudal centra: 80.16.290 (medial); 80.16.1248 (proximal); 81.20.51 (proximal); 82.31.82 (medial? eroded); 83.36.112 (medial); 85.97.51 (proximal); 86.77.25 (medial); 90.50.177 (medial); 90.50.204 (medial); 92.36.166 (medial); 92.36.339 (proximal); 92.36.1171 (proximal); 94.12.431 (medial); 94.12.724 (proximal; not neonate, but small juvenile); 94.12.727 (proximal); 94.12.980 (medial; fused neural arch base); 94.112.63 (proximal); 94.112.74 (proximal; not neonate, but small juvenile); 95.12.58 (medial); 97.12.169 (proximal); 98.93.16 (proximal); 98.93.24 (proximal). 95.12.159 (2 unassociated centra, 1 posterior dorsal, and 1 proximal caudal).

Coracoid: 92.36.470 (nearly complete-left).

Right humerus: 86.49.11 (complete); 92.36.963; 93.36.386 (midshaft region—embryonic?); 94.12.676 (midshaft); 94.12.757 (shaft); 97.12.167 (complete).

Left humerus: 84.163.26 (?ceratopsian); 85.36.164 (shaft); 85.59.55 (shaft); 86.78.82 (shaft); 87.72.24

(midshaft section—embryonic?); 92.36.138 (complete); 92.36.371 (bears small theropod toothmarks on shaft); 92.36.1001 (distal half).

Undifferentiated humerus: 87.36.358 (distal end); 92.36.472 (distal end); 92.36.1064 (distal end); 97.12.154 (distal end in 3 pieces).

Ulna: 81.16.373 (complete); 91.36.600 (proximal end); 92.36.982 (shaft); 93.36.3 (nearly complete-in two pieces); 93.36.123 (?ulna shaft); 95.405.46 (?proximal end fragment); 96.12.168 (proximal end); 98.93.33 (shaft).

Metacarpal: 94.12.668 (?half metacarpal); 97.12.156 (complete).

Manual phalanges: 95.12.123 (manual phalanx or carpal; not neonate, but small juvenile).

Rib: 80.16.634 (left dorsal); 80.16.1343 (right dorsal; both not neonate, but small juvenile); 94.12.860 (right dorsal); 94.12.933 (incomplete right dorsal rib in 3 pieces).

Right ilium: 94.12.700.

Pubis: 91.36.367 (complete).

Ischium: 98.93.137 (complete; not neonate, but small juvenile).

Right femur: 73.8.360; 81.16.372 (embryonic?); 90.36.412; 92.36.112; 92.36.426 (in 2 pieces); 92.36.921 (complete); 92.40.4 (proximal end); 94.12.427 (nearly complete; in 2 pieces); 96.12.172 (complete); 96.12.175 (complete); 97.12.166 (complete); 97.12.173 (complete).

Left femur: 89.36.173 (complete); 89.36.415 (complete); 92.36.240 (proximal and distal ends); 92.36.600; 92.36.920; 92.36.1069 (shaft and distal end; not neonate, but small juvenile); 94.12.483 (proximal half); 98.93.132 (complete?).

Undifferentiated femur fragments: 87.36.375 (distal end); 92.36.130 (4 associated fragments); 92.36.471 (distal end); 92.36.1070 (distal end); 93.150.3 (shaft); 94.12.491 (distal end fragment); 94.12.492 (shaft fragment); 94.12.742 (distal end region); 95.405.50 (shaft fragment); 95.405.51 (distal fragment); 96.12.157 (distal end); 96.12.170 (distal end); 98.93.26 (distal end region).

Right tibia: 80.16.818 (distal end); 91.36.547 (distal end); 92.36.536 (complete); 92.36.732 (distal end); 97.12.216 (complete).

Left tibia: 67.20.339 (not neonate, but small juvenile); 84.67.60 (complete); 85.36.138 (complete); 87.67.60 (complete); 89.36.113; 91.36.783 (not neonate, but small juvenile); 92.36.585; 94.12.425 (distal end; non-hadrosaur?); 94.12.835 (partial tibia in 2 pieces); 94.12.956 (distal half); 94.45.8 (complete?); 97.12.126 (complete); 97.12.197.

Undifferentiated tibia shafts: 91.36.733 (proximal end-embryonic?); 92.36.165 (2 pieces); 92.36.804; 92.36.1048; 94.12.822; 94.12.911 (?tibia shaft, embryonic?); 95.405.38; 97.12.185; 98.93.25.

Undifferentiated tibia shaft fragments: 92.36.125; 94.12.485; 94.12.486; 94.12.493; 94.45.9; 94.45.10; 96.12.160; 96.12.163; 97.12.177; 98.93.32; 98.93.60; 98.93.61.

Fibula: 92.36.457 (distal half; possibly distal ischium?); 94.12.821 (eroded and fragmented slender limb bone—?fibula).

Metatarsals: 80.8.189 (distal end); 80.16.1760 (distal end); 84.163.59 (complete); 85.59.38 (complete); 85.59.274 (metatarsal in 2 pieces); 86.77.72 (nearly complete; in 2 pieces); 86.78.34; 92.36.922 (proximal end; not neonate, but small juvenile); 93.110.44 (not neonate, but small juvenile); 94.12.422; 94.12.424 (midshaft region); 94.12.939 (midshaft region); 94.45.11; 96.12.169 (complete left Mt. III); 96.12.432 (distal end; Mt. III); 97.12.168; 98.93.151.

Pedal phalanges: 80.16.963; 84.67.54; 87.36.358 (ungual missing tip); 92.36.121; 96.12.164 (embryonic?); 97.12.155.

Unidentified specimens: 92.36.490 (small bone shaft-radius?, ulna?); 93.36.73 (fragment); 94.12.810 (vial of 13 neonate hadrosaur bone fragments); 95.405.11 (bone shaft).

Unnumbered specimens for destructive histological analysis: metatarsal, distal end; small bone shaft (?fibula).

Measurements of Selected Juvenile Hadrosaur Material in TMP Collections (mm)

Phalanges are pedal elements. Abbreviations: D.= Digit; Lt.= Left; Mid.= Middle; Mt.= Metatarsal; p.= phalanx; Prox.= Proximal; Rt.= Right.

Specimen Number	Element	Length	Prox. Width	Prox. Height	Distal Width	Distal Height
	Cranial					
98.92.152	Basisphenoid	14	Max. Width 13		Condyle Width 12	
	Postcrania					
92.36.470	Lt. coracoid		7.5	14		
97.12.167	Rt. humerus	118			26	
92.36.138	Lt. humerus	73	18+		15.9	
87.36.358	Humerus				16.3	
92.36.1064	Humerus				17	
97.12.154	Humerus				17	
81.16.373	Lt. ulna	107	21		15.5	
91.36.600	Ulna		18			
94.170.251	Ulna	77.5				
96.12.168	Ulna		16.5			
97.12.156	Metacarpal	41.2	7.9			
96.12.164	Rt. D. IV, p. I.	10.3	9.5	7.2	8	
84.67.54	D. III, p. I	13.2	13	8	10	
97.12.155	D. III, p. I	18			12.5	8.9
92.36.121	?Lt. D. IV, p. I.	16.3	14.5	10		
98.93.137	Ischium	265	62.8		11	15
90.36.412	Rt. femur	119+				
92.36.921	Rt. femur	97.2				
96.12.172	Rt. femur	155	37		31	
89.36.173	Lt. femur	132	28		23	
89.36.415	Lt. femur	113.5	24.5		20	
92.36.240	Lt. femur		29.5		24	
96.12.175	Lt. femur	125	28		21	
97.12.153	Lt. femur	125			22	
81.16.372	Femur		13			
81.16.375	Femur				23	
92.36.1069	Femur				31	
92.40.4	Femur		16.2			
97.12.197	Rt. tibia				24	
97.12.216	Rt. tibia	143	37		39.5	
84.67.60	Lt. tibia, tooth-marked?	108	28			

Specimen Number	Element	Length	Prox. Width	Prox. Height	Distal Width	Distal Height
85.36.138	Lt. tibia	115.5	28.5		27+	
94.12.956	Lt. tibia				20	
94.45.8	Lt. tibia	124+			33	
92.36.536	Tibia	110				
96.12.169	Lt. Mt. III	46		13	11	
84.163.59	Rt. Mt. II	48.9	15	15.3	14	11.8
85.59.279	Rt. Mt. II		12.8	15	15.4	13.9
80.16.1760	Lt. Mt. II				13	11
85.59.38	Lt. Mt. II	39.5	10.2	11.2	10.2	9.8
86.77.72	Lt. Mt. II			13	12.8	11
80.8.189	Mt. III				16	14.5
86.78.34	Mt. III				13.8	
	Vertebral Centra		Max. Width		Max. Height	
79.8.412	Dorsal	19	18		20	
85.63.65	Dorsal	6.2	7.1		7.9	
86.77.25	Dorsal	6.5	8		6.5	
91.36.206	Dorsal	15	14.5		15.2	
92.36.584	Dorsal	18.3	15.5		20	
92.50.5	Dorsal	13	12		13.3	
95.127.14	Dorsal	5	7.8		6.5	
96.12.159	Dorsal	11	10		10	
96.12.431	Dorsal	6.7	10		8	
98.93.15	Dorsal	10	16		15	
92.50.142	Sacral	13.9	20		14.2	
94.172.137	Sacral	5	8		5.8	
90.50.177	Mid. caudal	5	8.2		8	
92.36.166	Mid. caudal	11.3	11.2		10.9	
94.12.446	Mid. caudal	8.1	13		11	
94.12.980	Mid. caudal	5.5	9.5		7.5	
80.16.1248	Prox. caudal	7	12.5		12+	
81.20.51	Prox. caudal	10.5	16		15.3	
85.97.51	Prox. caudal	4.5	7.9		8.9	
90.50.204	Prox. caudal	5.9	9.7		7.2	
92.36.339	Prox. caudal	11.4	17		17	
92.36.1171	Prox. caudal	13	21		20.5	
97.12.169	Prox. caudal	4.1	9		8	
98.93.16	Prox. caudal	13	23.5		24.9	
98.93.24	Prox. caudal	12	27		25	

16. Taphonomy and Paleoenvironment of a Hadrosaur (Dinosauria) from the Matanuska Formation (Turonian) in South-Central Alaska

ANNE D. PASCH AND KEVIN C. MAY

Abstract

The discovery of postcranial elements of a hadrosaur in the Talkeetna Mountains 150 kilometers northeast of Anchorage is: (1) the first known occurrence of a hadrosaur in south-central Alaska; (2) a new high-latitude locality for dinosaurs; (3) middle Turonian in age, making it one of the few well-dated early hadrosaurs in the world; (4) one of only four vertebrate fossils known from the Wrangellia composite terrane; and (5) the first association of dinosaur bones in Alaska that can be attributed to a single individual. A closely associated assemblage of marine invertebrates provides a reliable age and indicates an outer-shelf or upper bathyal depositional environment. The bone surfaces indicate that the postcranial skeleton was damaged by marine scavengers after transport. Scavenged bones occur in soft mudstone matrix whereas bones with intact surfaces occur in indurated calcareous concretions. The spatial relationships of the concretions within the bone quarry may be the result of the distribution of fragments of flesh from the disintegrating carcass. This occurrence of a hadrosaur in Alaska provides a geographic link between early hadrosaurs of Asia and North America.

Introduction

The record for dinosaurs in Alaska south of the Brooks Range is very poor. It consists of Upper Jurassic theropod and ornithopod tracks that have never been described on the Alaska Peninsula (Weishampel 1990), a partial skull of *Edmontonia* from a Campanian/Maastrichtian unit in the Talkeetna Mountains (Gangloff 1995), and the Talkeetna Mountains hadrosaur (TMH) (Pasch and May 1995). Evidence for pre-Campanian hadrosaurs is very rare, consisting of occurrences in Asia, Europe, and North America (Brazzatti and Calligaris 1995; Currie and Eberth 1993; Head 1998; 1999; Horne 1994; Kirkland 1994; Pasch and May 1997; Rozhdestvensky 1968; Weishampel and Horner 1990). Many are fragmentary or poorly constrained as to age.

Figure 16.1. (opposite page) Map of Alaska showing location of the Talkeetna Mountains hadrosaur (TMH) quarry in south-central Alaska and location in borrow pit.

The TMH specimen was discovered in a quarry being excavated for road material in 1994. Over 70 skeletal elements of this specimen, along with a diverse assemblage of marine invertebrates were collected during the fall of 1994 and the summer of 1996. Ammonites indicate a Turonian age for the strata. According to Horner's (1979) checklist of Upper Cretaceous dinosaurs from marine sediments in North America, the TMH is typical of dinosaurs found in marine settings because it is a hadrosaur, is apparently a juvenile, and is associated with shark teeth, ammonites, and other mollusks. The hadrosaur elements are cataloged at the University of Alaska Museum (AK) in Fairbanks (accession no. 2000 P-02).

Location and Geologic Setting

The quarry is situated in the Talkeetna Mountains in south-central Alaska, approximately 150 kilometers northeast of Anchorage, near the Glenn Highway. Its elevation is 950 meters (3119 feet) at approximately 61° 52.93' N and 147° 20.28' W (fig. 16.1).

The quarry is situated in a borrow pit along the Glenn Highway in the Matanuska Formation on the southeast limb of the nose in the displaced portion of an anticline (Grantz 1961a,b). The bone-bearing unit consists of an easily weathered, dark gray, marine mudstone, containing highly indurated calcareous concretions and finely disseminated pyrite crystals. Horizontal laminae, ripples, and evidence of bioturbation are faintly visible only on wet fresh surfaces. The unit has been subject to postdepositional deformation as indicated by the joint sets, faults, secondary deposition of calcite, and degree of induration.

Age of the Bone-Bearing Unit

Foraminifera and a well-preserved collection of fossil mollusks from the quarry provides a secure Middle Turonian age and a marine setting for the bone-bearing unit (table 16.1). Will P. Elder (U.S. Geological Survey) identified seven species of ammonites, six species of bivalves, and two gastropods of a Middle Turonian assemblage. The presence of the ammonite, *Muramotoceras*, strongly suggests the age is Middle Turonian, because this genus is known only from two species that occur in Middle Turonian sequences. This is the first noted occurrence of this unusual heteromorph outside Japan. The ammonite genus

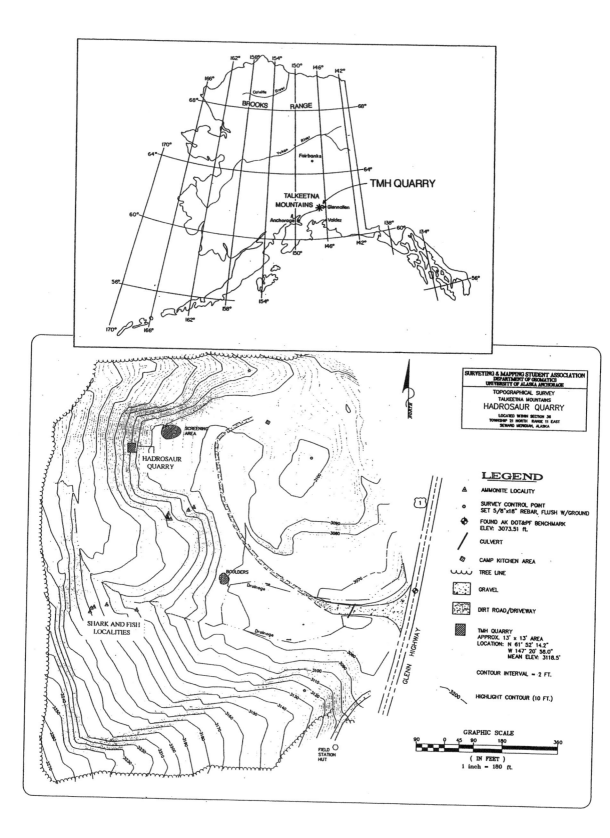

Eubostrychoceras is known from Japan, Germany, and Madagascar (fig. 16.2). *E. japonicum* is Turonian and probably Middle Turonian (Matsumoto 1977). The inoceramid bivalves have a worldwide distribution and are used as Cretaceous guide fossils from the Albian through the Maastrichtian (Thiede and Dinkelman 1977). Other fossils include

TABLE 16.1
Flora and Fauna Associated
with the Talkeetna Mountains Hadrosaur (TMH)

VERTEBRATA	"Mako-type" shark teeth* Fish teeth, jaw fragments, scales*
CEPHALOPODA (Elder, personal communication)	*Eubostrychoceras* cf. *japonicum*** *Gaudryceras* aff. *G. denseplicatum* *Mesopuzosia* cf. *M. indopacifica* *Muramotoceras* aff. *M. yezoense*** *Sciponoceras* sp. *Tetragonites* aff. *T. glabrus* *Yezoites puerculus* (*Otoscaphites teshioensis*)
PELECYPODA (BIVALVIA) (Elder, personal communication)	*Inoceramus* aff. *I. cuvieri* *Inoceramus* aff. *I. hobetsensis** *Inoceramus* aff. *I. mamatensis** *Inoceramus* aff. *I. teshioensis* *Acila (Truncacila)* sp.* *Nucula* sp.*
GASTROPODA (Elder, personal communication)	*Biplica* sp. (or similar opisthobranch)* Naticid*
SCAPHOPODA	*Dentalium* sp.*
CNIDARIA	Small solitary hexacoral (*Platycyathus?*)*
PORIFERA (Larson, personal communication)	Sponge spicule fragment

PROTISTA
 (Larson, personal communication)
	Radiolarians*
	 Foraminifera: planktic and benthic forms*
	     Marginotruncana cf. *sigali*, *Vaginulina* sp.
	     Bathysiphon spp., *Guttulina* sp.
	     Dentalina sp., *Gyroidinoides* sp.
	     Gavellinella cf. *velascoensis?*
	     Haplophragmoides spp.

PYRRHOPHYTA (Larson and Reid, pers. comm.)	Dinoflagellates
TRACHEOPHYTA (Reid and Pasch 1999)	Palynomorphs: Lycopodophyta (1 sp.), Pteridophyta (69 sp.),* Ginkgophyta, Cycadophyta, Pinophyta (9 sp.), Anthophyta (5 sp.)* Wood fragments
ICHNOFOSSILS	*Planolites* sp. (?)* Calcareous worm tubes* *Teredolites* sp.*

* First occurrence in Matanuska Formation.
** First occurrence in North America.

teleost fish teeth and jaw fragments, scales, shark teeth, scaphopods, a solitary hexacoral, foraminifera, palynomorphs, trace fossils, *Teredolites,* and wood fragments (table 16.1). Both the lithology and the invertebrates of the bone-bearing unit strongly suggest that the quarry section belongs to the lower portion of C-1, an informal stratigraphic unit of Turonian age in the lower half of the Matanuska Formation as defined by Jones and Grantz (1967) and Member 4 (Turonian) as defined by Jones (1963).

*Figure 16.2. Heteromorph ammonite (*Eubostrychoceras *sp.) recovered from the TMH quarry site.*

Hadrosaur Skeletal Material from the Talkeetna Mountains

Over 70 elements of the postcranial skeleton have been recovered. They were concentrated in a four-square-meter area containing a large concretion nearly a meter in length. Some bone elements required little preparation and some remain fully or partially encased in calcareous mudstone concretions. Axial and appendicular elements include articulated and isolated vertebrae and portions of all four limbs, as well as a large concretion, which has been partially prepared, containing pelvic elements. To date, 2 scapulae, a coracoid fragment, 2 humeri, 2 ulnae, 1 radius, portions of both femora, a tibia, fibula, astragalus, metacarpal, 4 metatarsals, and 15 pedal phalanges from the appendicular skeleton have been identified, along with rib fragments, 23 caudal centra, 2 chevrons, and a few centimeters of ossified tendon from the axial skeleton. They are shown diagrammatically in figure 16.3. All elements are closely associated and some are articulated. No elements are duplicated and the identified bones all fall within a narrow size range, suggesting they represent a single individual. Preliminary comparisons with other specimens suggest the animal was a juvenile approximately 3 meters long.

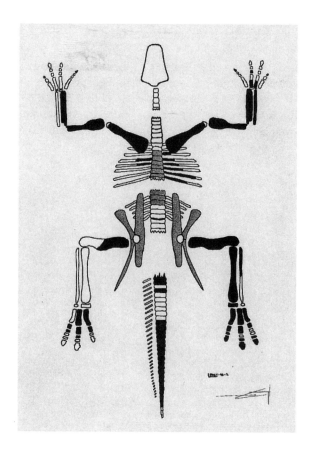

Figure 16.3. Schematic of TMH elements recovered to date. Positive identification in black, tentative identification in gray.

Systematics

Dinosauria Owen 1842
Ornithischia Seely 1888
Ornithopoda Marsh 1871
Hadrosauridae Cope 1869

Identification to the family level is based on three nearly complete right pedal phalanges (II-1, III-1, IV-1), which were compared with material at the University of Alaska Museum in Fairbanks and the Royal Tyrrell Museum of Palaeontology in Drumheller, Alberta. It is not known whether this individual is a hadrosaurid or lambeosaurid. However, the position of the deltopectoral crest on the humerus (fig. 16.4) and the elongation of the caudal centra (fig. 16.5) are very different from those of *Edmontosaurus* (table 16.2).

Paleoecologic Context

Deposition in a middle- to outer-shelf or upper bathyal environment below wave base is demonstrated by the invertebrate assemblage, which is dominated by ammonites and inoceramid bivalves (table 16.1). The thin-shelled heteromorphic ammonites were probably inhabitants of the outer shelf (36–183 m) (Tasch 1973). Inoceramids are

TABLE 16.2
Comparison of Humeri and Caudal Centra Dimensions of the Talkeetna
Mountain Hadrosaur (TMH) with *Edmontosaurus* from the Prince Creek
Formation (Maastrichtian), North Slope, Alaska

Abbreviations:
L = length   CW = condyle width
DLP = deltopectoral crest length   H = height
DLPW = deltopectoral crest width   W = width

	Humeri Dimensions (cm)			
	L	DLP	DLPW	CW
Edmontosaurus	25.00	14.00	6.00	5.00
	30.00	16.50	6.50	6.20
	28.50	15.00	7.50	8.00
	22.50	12.00	5.50	6.00
	23.00	12.50	5.75	6.25
	20.50	10.50	4.50	5.00
	23.50	12.00	5.70	5.90
	21.00	11.50	5.50	5.00
	23.00	11.50	6.50	6.50
	24.50	13.00	6.75	6.50
	22.00	11.00	5.50	6.00
	31.50	16.50	8.50	8.00
TMH	25.00	12.00	5.50	6.00
	25.00	N/A	N/A	6.00

	Caudal Centra Dimensions (cm)				
	L	W	H	L/W	L/H
Edmontosaurus	2.00	3.60	3.20	0.56	0.62
	2.00	3.30	2.90	0.61	0.69
	1.65	3.20	2.60	0.52	0.63
	1.60	3.10	2.50	0.52	0.64
	1.30	2.50	1.90	0.52	0.68
TMH	4.00	3.60	3.10	1.11	1.29
	4.10	3.25	3.00	1.26	1.37
	3.80	3.10	2.70	1.23	1.41
	3.80	3.00	2.70	1.27	1.41
	3.40	2.30	2.00	1.48	1.70

thought to have inhabited a wide range of depths, but seem to be confined to the upper bathyal and neritic environments close to continental or island margins (Thiede and Dinkelman 1977). The lack of heavy-shelled, shallow-water pelecypods also suggests an outer neritic zone or deeper water location for the strata at the quarry (Jones 1963).

The density of the invertebrates suggests an environment where organisms were either very rare or arrived only after death. The preservation suggests rapid burial. The shells lack signs of postmortem biological activity such as borings or encrustations. They show no signs of abrasion, and broken surfaces are fresh (fig. 16.6). Some are nearly whole and undeformed whereas others are fragmented, crushed, and

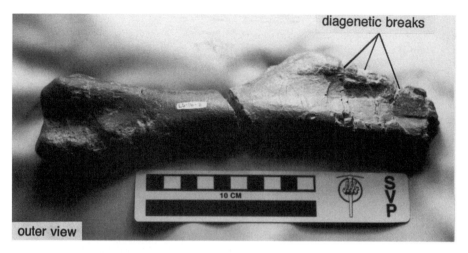

outer view

diagenetic breaks

10 CM

SVP

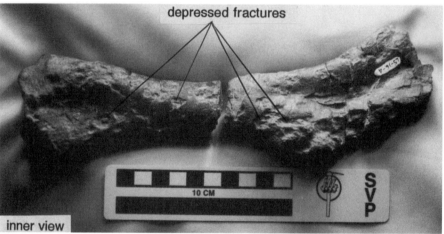

inner view

depressed fractures

10 CM

SVP

Figure 16.4. Left humerus (AK-485-V-03) with depressed fractures. Maximum width = 54.2 mm, length = 235.0 mm.

greatly compressed. The orientation of the larger planar valves (up to 20 cm in diameter) in the quarry was always parallel to bedding. The lack of abrasion and the recovery of fragile heteromorph ammonites suggest that the invertebrates could not have been reworked. The proximity of the hadrosaur bones to each other implies that they were not disturbed a great deal by scavengers. The occurrence of pyrite in the mudstone suggests that bacterial degeneration of soft tissue had occurred resulting in a sulfide-rich environment (Hogler 1994).

Whether or not the fossil assemblage represents a community of organisms, which can be used for the reconstruction of specific paleoecologic conditions, is an open question. The muddy substrate may have been unstable, subject to submarine slides and slumps. Elder (pers. comm.) states that transportation of delicate and complete shells in this type of environment is very common. The organisms, whether transported or not, show some ecological affinities to each other. Ecological interpretation is confounded by the possibility that ammonite shells can float long distances after death. This may be true for pelagic genera with normal planispiral shells such as *Mesopuzosia*, however, it may not be

lateral view   caudal view   ventral view

Edmontosaurus

TMH

Figure 16.5. (above) Centra of caudal vertebrae (AK-4885-V-07 and AK-485-V-06). Comparison with Edmontosaurus. (Illustration by Lee Post).

Figure 16.6. Coiled ammonite, Mesopuzosia sp. Specimen is highly compressed but surfaces are intact.

true for the heteromorphs. Seilacher and Labarbera (1995) suggest that the septum closing off the living chamber of heteromorph ammonites was not calcified and that it decomposed with other soft parts, thus limiting the drift time of the shell. A benthic mode of life, which has been suggested for heteromorphs of this type, would also have placed limits on the distance of transport after death. Matsumoto (1977) suggests that *Eubostrychoceras,* with its open coiling, was not adapted for rapid swimming but for a benthic life style, and may even have been partly embedded in the substrate. The spinose flared ribs of the shell may have been used to stabilize the animal as it sat on the bottom.

The most abundant mollusks in the borrow pit are inoceramids, an extinct group of bivalves thought to be related to modern oysters. They were benthic, with large relatively flat shells typical of species living on soft, muddy substrates. They are characterized by large robust valves with lengths which can exceed 27 mm and thicknesses of 2–3 mm. The shells have multiple ligamental pits, which provided anchorage for threadlike ligaments that attached them to the substrate. Inoceramids are common constituents of dark, gray calcareous, laminated mudstones, which indicate reducing conditions below the sediment-water interface (Thiede and Dinkelman 1977). They were probably filter feeders living below wave base and may have harbored chemosynthetic symbionts to supplement their diet (MacLeod and Hoppe 1992).

Nucula, a small primitive bivalve represented by several specimens, is a ubiquitous genus of an infaunal detritus feeder often found in organic muds (Tasch 1973). It is an important component of ancient and modern deep-water communities. It is indicative of a low-diversity assemblage in a soft, water-saturated substrate, rich in organic matter, with abundant hydrogen sulfide, somewhat depleted in oxygen. Nine typical extant, deep-water species live below bottom waters which have temperatures from 2.3° to 9.2° C (Kauffman 1976).

The abundance of *Bathysiphon* sp. and spherical radiolarians indicate bathyal to outer neritic paleodepths (Larson, pers. comm.). Whether transported or not, the heteromorphs, inoceramids, nuculids, and protists all indicate that the TMH was buried at a paleodepth greater than 35 m.

Taphonomy

Deposition in an outer-shelf depositional environment would imply that the TMH carcass had bloated with gases and floated to a marine environment before it sank (Martill 1991). Because no elements of the skull were found, the head must have been detached before the carcass sank. The body of the animal came to rest on its left side with all four limbs extended to the east, somewhat parallel with each other. After deposition, some ribs and distal elements were detached from the carcass, but all except a few caudal centra were located within the four-square-meter quadrant containing the largest concretion (fig. 16.7). Approximately 20 concretions were excavated from an eight-square-meter quadrant. They are of two types: those that are bone bearing, which range from 25 to 110 cm in diameter (fig. 16.8), and those that are devoid of bone material with a nearly uniform diameter of 20 to 30 cm. The highly indurated nature of the concretions made preparation difficult. However, all bone surfaces exposed from the concretionary matrix are smooth and lacked depressed fractures. These include the right scapula and pedal elements. The concentration of the concretions within the quarry is relatively high in comparison with the rest of the outcrop. Approximately 20% of the bones were surrounded by these highly indurated calcareous concretions.

The remaining 80% of the prepared elements were removed from the poorly cemented mudstone matrix. Without exception, the elements that are not encased in concretions are characterized by surfaces with numerous closely spaced conical depressions (depressed fractures). These depressed fractures occur on two or more sides of the damaged elements. They are subround to oval in planar view and in places coalesce into irregularly shaped depressions. In cross-section they are U-shaped or conical and many have displaced cortical bone fragments forming an irregular surface on the bottom of the depressions. Those distinct enough to be measured range from 2.12 to 5.81 mm in diameter and from 1.64 to 3.62 mm in depth (table 16.3). This

Figure 16.7. TMH quarry map showing distribution of bone-bearing and barren concretions.

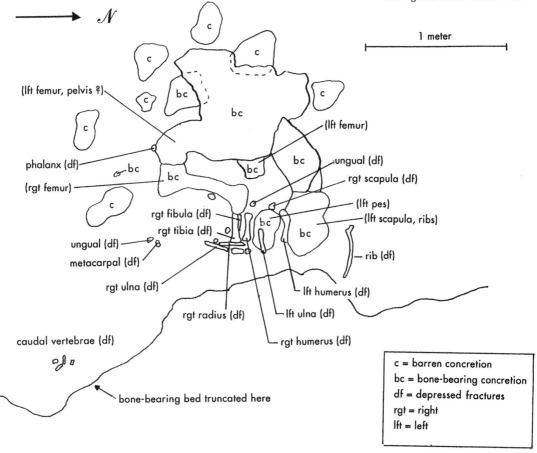

Figure 16.8. (above) Bone-bearing concretion containing elements of right pes (metatarsal and phalanges) with partially prepared bones showing no evidence of scavenging (no depressed fractures) (AK-485-V-09).

Figure 16.9. (opposite page above) Right (upper) (AK-485-V-01) and left (lower) (AK 485-V-02) ulnae. Both were excavated from mudstone and both show depressed fractures indicative of scavenging. Right ulna: maximum width = 39.0 mm, length = 195.0 mm.

Figure 16.10. (opposite page below) Dorsal view of two unguals with depressed fractures. On left, phalanx III, digit II, left pes (AK-485-V-05): maximum width = 31.3 mm, length = 62.6 mm; on right, phalanx V, digit IV, left pes (AK-485-V-04): maximum width = 28.26 mm, length = 49.8 mm.

morphology occurs on both ulnae (fig. 16.9), a rib, both humeri, a metatarsal, two unguals (fig. 16.10), and on the right tibia and fibula. Tooth-damaged dinosaur bone can be recognized by such distinctive markings (Chin 1997). Comparison of these bones with those of other vertebrates, known to have been damaged by predators and scavengers, led to the conclusion that the depressed fractures are bite marks. The depressions do not have the perfect symmetry of gastropod drill marks or the geometry of sponge borings.

TABLE 16.3

Comparison of Depressed Fractures on Talkeetna Mountain Hadrosaur (TMH) Bone Elements with Dimensions of Mosasaur Teeth

	TMH Fractures mm	Mosasaur Teeth mm
Diameter	(at bone surface) (n = 28)	(2.5 mm from tip) (n = 11)
Mean	3.80	4.54
Range	2.12–5.81	3.70–5.77
Depth of fractures	(n = 7)	
Mean	2.59	–
Range	1.64—3.62	–

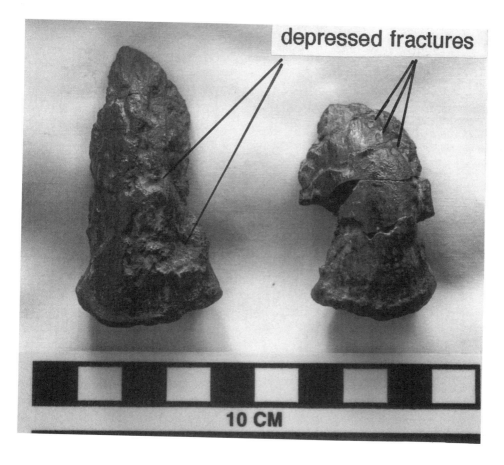

depressed fractures

Figure 16.11. (above) Shark tooth
recovered from the TMH site.
Scale in mm.

Figure 16.12. Teleost fish teeth in
jaw fragment recovered from the
TMH site. Scale in mm.

Figure 16.13. (below) Dentary
and maxillary teeth of the
mosasaur, Tylosaurus proriger,
from the Upper Cretaceous of
Kansas. The apexes of these teeth
are conical and a close match for
the depressed fractures in the
TMH elements. Scale in cm.

Numerous shark and teleost fish teeth were recovered from the outcrop. They have been discounted as those of the probable scavenger because they are smaller than, and do not match the shape of, the depressed fractures. The shark teeth are bladelike and average about 1.5 cm from base to apex (fig. 16.11). Three mm from the tip they are less than 2.5 mm wide. The teleost fish teeth are approximately 2 mm long and less than 2 mm wide at the base (fig. 16.12). Many of the depressed fractures do closely match the size, shape, and spacing of the dentary, maxillary, and pterygoid teeth of the mosasaur, *Tylosaurus proriger* (fig. 16.13). Diameters measured 2.5 mm from the tip range from 3.70 to 5.77 mm, near the size range of the depressed fractures. Therefore, a good candidate for the primary scavenger is a marine reptile with similar teeth.

The fact that the carcass was able to float to an outer-shelf/upper bathyal environment indicates that its thoracic and abdominal cavities remained intact during transport. If the damage had been caused by a terrestrial predator, it is likely that the thoracic and abdominal cavities would have been punctured, thus preventing the carcass from bloating and floating. Therefore, the damage was most likely inflicted by a marine scavenger and not a terrestrial predator. Evidence that the body cavity may been opened in the marine setting is provided by several disarticulated ribs.

The arrangement of the bone-bearing concretions suggests that the distribution of the bite marks (depressed fractures) was controlled by the distribution of flesh. The bones of the lower extremities have the least amount of flesh and therefore sustained the most damage. The scavenger was unable to get its mouth around the more robust parts of the upper extremities and axial skeleton. These were left relatively intact and the flesh reacted with the substrate to form concretions (Berner 1968). Bones pulled free from the carcass were scavenged and buried in mud, whereas those within the carcass were protected by flesh until buried.

The distribution of concretions represents the distribution of pieces of tissue, some with bone and some without. The bone-bearing concretions contain closely associated and articulated elements, which remained attached to the carcass. The barren concretions represent chunks of flesh torn from the carcass. In both cases, the decay of the flesh created micro-geochemical environments conducive to the formation of the concretions (Davis 1992; Berner 1968).

The stages of deterioration of large reptiles on the sea floor have never been documented for a single individual, but Hogler (1994) described three types of benthic communities that have been associated with saurian fossils. The extent to which they modified the carcass is an indication of the amount of oxygen present and how long it was exposed to decomposition. In anoxic stagnant water the primary organisms of destruction were anaerobic bacteria. Where some oxygen was present, multicellular organisms were associated with marine reptile remains that commonly showed evidence of having been scavenged, bored, or gnawed. Skeletal elements remaining after soft tissues were removed were modified by encrusters and borers. The bare bones provided niches and hiding places for nestling and cryptic creatures.

A Hadrosaur (Dinosauria) from the Matanuska Formation • 233

The organisms associated with the TMH suggest the second type of Hogler's benthic communities was present at the time of burial. If the TMH was host only to bacterial mats, the irregular surfaces on the bones would be difficult to explain. Bacterial mats leave smooth surfaces reminiscent of soft tissue outlines (Hogler 1994). No excavations of boring organisms were found. No evidence for encrusters such as serpulid worms or encrusting bivalves, or for any nestling, cryptic creatures, was found. The scattering of distal elements, gnawed bone surfaces, and the associated invertebrates indicate that the TMH skeleton was subject to decomposition and scavenging organisms before burial but was buried before all soft tissue was removed (Hogler 1994).

Summary

The Talkeetna Mountains hadrosaur is an important addition to the rich fossil record of duck-billed dinosaurs, and represents the first individual of this group to be found in south-central Alaska, as well as one of the earliest hadrosaurids known in the world. It has the potential to contribute to our understanding of the timing and direction of the spread of this group of ornithopods. The distribution of concretions in and surrounding the skeleton suggest they represent fragments of flesh, which created a geochemical environment different from that of the surrounding mud. The depressed fractures on bones surrounded by soft matrix represent bite marks of a scavenger, possibly a mosasaur with blunt, conical teeth. The carcass was partially consumed prior to burial on a muddy shelf at a minimum depth of 35 meters. Burial occurred at a time when the skeleton was still partially enclosed in flesh.

Acknowledgments: We are indebted to MB Construction for permission to conduct the excavation in their quarry and for donating all specimens to the state of Alaska. We thank John Larson and Will Elder for identification of the invertebrates. Special thanks to Kenneth Carpenter for his provocative and extremely helpful suggestions and to the reviewers, Roland Gangloff and Darren Tanke. As newcomers to the excitement of vertebrate paleontology, we thank Phil Currie for his encouragement and the inspiration his enthusiastic pursuit of the understanding of vertebrates fossils has provided. This work was supported in part by grants from the Dinosaur Society, the Eagle River Parks and Recreation Board, the Joe Kapella Memorial Fund, the Edward and Anna Range Schmidt Charitable Trust, the Chugach Gem and Mineral Society, Alaska Geology, Inc., and the Alaska Museum of Natural History.

References

Berner, R. A. 1968. Calcium carbonate concretions formed by the decomposition of organic matter. *Science* 159: 195–197.

Brazzatti, T., and R. Calligaris. 1995. Studio preliminare di reperti ossei di dinosauri del Carso Triestino. *Atti del Museo Civico di Storia Naturale di Trieste* 46: 221–226.

Chin, K. 1997. What did dinosaurs eat? Coprolites and other direct evidence of dinosaur diets. In J. O. Farlow and M. K. Brett-Surman (eds.),

The Complete Dinosaur, pp. 371–382. Bloomington: Indiana University Press.

Currie, P. J., and D. A. Eberth. 1993. Paleontology, sedimentology, and palaeoecology of the Iren Dabasu Formation (Upper Cretaceous), Inner Mongolia, People's Republic of China. *Cretaceous Research* 14: 127–144.

Davis, R. A. 1992. *Depositional Systems.* Englewood Cliffs, N.J.: Prentice-Hall.

Gangloff, R. A. 1995. *Edmontonia* sp., the first record of an ankylosaur from Alaska. *Journal of Vertebrate Paleontology* 15: 195–200.

Grantz, A. 1961a. Geologic map and cross-sections of the Anchorage (D-2) quadrangle and northeasternmost part of the Anchorage (D-3) quadrangle, Alaska. *U.S. Geological Survey Miscellaneous Geological Investigations Map I-342.*

Grantz, A. 1961b. Geologic map of the north two-thirds of Anchorage (D-1) quadrangle, Alaska. *U.S. Geological Survey Miscellaneous Geological Investigations Map I-343.*

Head, J. J. 1998. A new species of basal hadrosaurid (Dinosauria, Ornithischia) from the Cenomanian of Texas. *Journal of Vertebrate Paleontology* 18: 718–738.

Head, J. J. 1999. Reassessment of the systematic position of *Eolambia caroljonesa* (Dinosauria, Iguanodontia) and the North American Iguanodontian record. *Journal of Vertebrate Paleontology, Abstracts* (suppl. to no. 3) 19 (3): 50A.

Hogler, J. A. 1994. Speculations on the role of marine reptile deadfalls in Mesozoic deep-sea paleoecology. *Palaios* 9: 42–47.

Horne, G. S. 1994. A Mid-Cretaceous ornithopod from central Honduras. *Journal of Vertebrate Paleontology* 14: 147–150.

Horner, J. R., 1979. Upper Cretaceous dinosaurs from the Bearpaw Shale (marine) of south-central Montana with a checklist of Upper Cretaceous dinosaur remains from Marine sediments in North America. *Journal of Paleontology* 53: 566–577.

Jones, D. L. 1963. Upper Cretaceous (Campanian and Maastrichtian) ammonites from southern Alaska. *U.S. Geological Survey Professional Paper* 432.

Jones, D. L., and A. Grantz. 1967. Cretaceous ammonites from the lower part of the Matanuska Formation, southern Alaska. *U.S. Geological Survey Professional Paper* 547.

Kauffman, E. G. 1976. Deep-sea Cretaceous macrofossils: Hole 317A, Manihiki Plateau. In E. D. Jackson et al., *Initial Reports of the Deep Sea Drilling Project,* 33: 503–535. Washington, D.C.: Government Printing Office.

Kirkland, J. T. 1994. A large primitive hadrosaur from the Lower Cretaceous of Utah. *Journal of Vertebrate Paleontology, Abstracts (suppl. to no. 3)* 14 (3): 32A.

MacLeod, K. G., and K. A. Hoppe. 1992. Evidence that inoceramid bivalves were benthic and harbored chemosynthetic symbionts. *Geology* 20: 117–120.

Martill, D. M. 1991. Bones as stones: The contribution of vertebrate remains to the lithologic record. In S. K. Donovan (ed.), *The Processes of Fossilization,* pp. 270–292. New York: Columbia University Press.

Matsumoto, T. 1977. Some heteromorph ammonites from the Cretaceous of Hokkaido. *Kyushu University Memoirs of the Faculty of Science,* series D, Geology 23 (3): 303–366.

Pasch, A. D., and K. C. May. 1995. The significance of a new hadrosaur (Hadrosauridae) from the Matanuska formation (Cretaceous) in south-central Alaska. *Journal of Vertebrate Paleontology, Abstracts (suppl. to no. 3)* 15 (3): 48A.

Pasch, A. D., and K. C. May. 1997. First occurrence of a hadrosaur (Dinosauria) from the Matanuska Formation (Turonian) in the Talkeetna Mountains of south-central Alaska. In J. G. Clough and F. Larson (eds.), *Short Notes on Alaska Geology, 1997*, pp. 99–109. Alaska Department of Natural Resources.

Reid, S. L., and A. D. Pasch. 1999. The significance of a Turonian spore and pollen flora from the Matanuska Formation, Talkeetna Mountains, Alaska. *American Association of Stratigraphic Palynologists, Program and Abstracts, 32nd Annual Meeting*, p. 34.

Rozhdestvensky, A. K. 1968. Hadrosauridae of Kazakhstan. "Nauka" publishing House, Moscow. (Translation from Russian.) In L. P. Tatarinov et al. (eds.), *Upper Paleozoic and Mesozoic Amphibians and Reptiles*, pp. 1–120. Moscow: Akademia Nauk.

Seilacher, A., and M. Labarbera. 1995. Ammonites as cartesian divers. *Palaios* 10: 493–506.

Tasch, P. 1973. *Paleobiology of the Invertebrates*. New York: Wiley.

Thiede, J., and M. G. Dinkelman. 1977. Occurrence of *Inoceramus* remains in late Mesozoic pelagic and hemipelagic sediments. In P. R. Supko et al., *Initial Reports of the Deep Sea Drilling Project* 29: 899–910. Washington, D.C.: Government Printing Office.

Weishampel, D. B. 1990. Dinosaurian distribution. In D. B. Weishampel, P. Dodson, and H. Osmólska (eds.), *The Dinosauria*, pp. 63–139. Berkeley: University of California Press.

Weishampel, D. B., and J. R. Horner. 1990. Hadrosauridae. In D. B. Weishampel, P. Dodson, and H. Osmólska (eds.), *The Dinosauria*, pp. 534–561. Berkeley: University of California Press.

17. Primitive Armored Dinosaur from the Lufeng Basin, China

Dong Zhiming

Abstract

A small armored ornithischian is referred to the Scelidosauridae (Thyreophora) as a new genus and species. The material consists of an incomplete right lower jaw with teeth and cranial fragments. It was collected from the Dark Red Beds of the Lower Lufeng Formation (Early Jurassic), Lufeng Basin, Yunnan.

Introduction

Since 1938 many dinosaur remains have been discovered in the Lufeng Basin, China. The basin has been visited and surveyed by many Chinese and foreign paleontologists. So far, seven genera and 12 species of dinosaurs have been reported from these Upper Triassic and Lower Jurassic fossiliferous beds (Bien 1941; Sun et al. 1985). The taxa include the prosauropods *Lufengosaurus, Yunnanosaurus,* and *Anchisaurus,* the theropods *Lukousaurus* and *Sinosaurus,* the ornithopods *Dianchungosaurus* and *Tawasaurus,* and the primitive stegosaur *Tatisaurus* (Young 1982; Simmons 1965; Sun and Cui 1986; Sun et al. 1985; Dong 1990).

Recently, I examined and prepared dinosaurian fossils collected by M. N. Bien from the Lufeng Basin in 1938 and 1939. Among the specimens, a new small scelidosaur was discovered, and is named as a new genus and species below. The specimen was found in the Dark Red

Beds of the Lower Lufeng Formation, and includes a nearly complete lower jaw and several fragments of the skull.

The earliest reported ankylosaur, *Scelidosaurus,* is from the Early Jurassic (Liassic stage) of England (Owen 1863; Romer 1968). Ankylosaurians were a major group of ornithischians with broad, low bodies, but are most easily identified by their dermal armor. They originated as small scelidosauridlike animals sometime before the Early Jurassic. The new taxon provides evidence for ankylosaurs in the Early Jurassic (Liassic).

Systematic Paleontology
Order: Ornithischia Seeley 1888
Sudorder: Thyreophora Nopcsa 1915
Family: Scelidosauridae Cope 1869
Genus: *Bienosaurus* gen. nov.

Etymology: *Bieno-,* for M. N. Bien, who collected the holotype; *saurus,* "lizard" (Greek).
Locality and Horizon: Dark Red Beds, Lower Lufeng Formation, Lower Jurassic; Lufeng Basin, Yunnan Province, China.

B. lufengensis sp. nov.

Etymology: for the Lufeng Basin, where the holotype was found.
Holotype: IVPP V 9612; A nearly complete right lower jaw, a fragmentary frontal, and other cranial fragments.
Diagnosis: Small armored ornithischian dinosaur; predentary short, wide; frontal thick, with small bony scutes fused to dorsal surface; dentary wide, as typical of ankylosaurs, with ornamented lateral surface; teeth small, leaf-shaped, with symmetrical crown and developed cingulum.
Description: The right mandibular ramus measures 60 mm as preserved (figs. 17.1, 17.2). A fragment of the predentary is attached to the anterior end of the dentary. A shallow notch is present across the predentary-dentary junction. The predentary is small and wide. The lateral surface of the dentary is convex. The dentary is a relatively thick bone, as is typical of ankylosaurs, with an ornamented lateral surface. The anterior end of the dentary is relatively narrow, and from the sixth tooth backward, the transverse diameter becomes wider. The lateral surface of the dentary bears finely ornamented depressions, whereas the medial side has a large open Meckelian canal. On the posterior end of the dentary, an incomplete coronoid is preserved. Based on the length of the mandible, the estimated length of the skull is 65–70 mm.

Eleven teeth or roots are preserved in the dentary and two large alveoli posterior to the last tooth. Of the first two teeth, only the roots remain, although replacement teeth can be seen within the alveoli. The tooth row is curved, as in ankylosaurs. Anterior teeth are relatively small and become larger posteriorly. They are leaf-shaped with symmetrical crowns, as in other ankylosaurs (fig. 17.2C). Both the lateral and medial surfaces of the crown are covered with a thin layer of

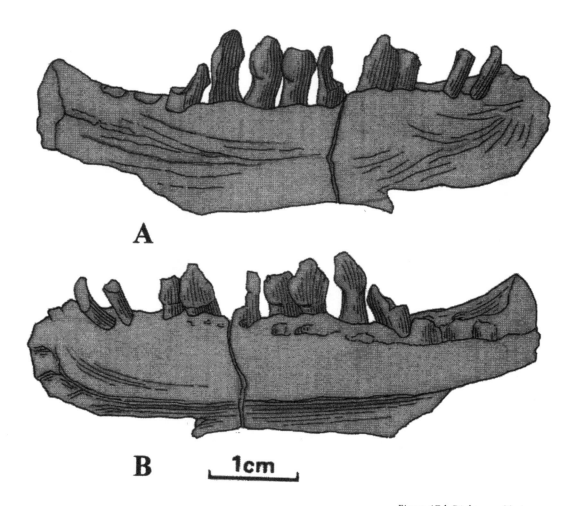

A

B   1cm

Figure 17.1. Right mandibular
ramus of Bienosaurus lufengensis
gen. et sp. nov. IVPP V. 9612 in
(A) lateral and (B) medial views.

enamel. The anterior and posterior edges of the crown carry small denticles. The central ridge is never prominent and there is a slight swelling at the base of the crown.

Pieces of the fragmentary skull were obtained from the slab containing the lower jaw and are believed to have come from the same individual. The largest pieces are identified as the right frontal (fig 17.2D) and supraorbital (fig. 17.2E). The supraorbital is thick, with small, bony scutes fused to its dorsal surface. The other fragments possibly represent the maxilla and pterygoids (fig. 17.2F)

Comparison and Discussion

Based on features of the predentary and teeth, V 9612 is clearly referred to as a small ornithischian. Three other ornithischians—*Dianchungosaurus*, *Tawasaurus*, and *Tatisaurus*—have been reported from the Lower Lufeng Formation (although *Tawasaurus* may represent a hatchling prosauropod; Sereno 1991).

The dentary of the *Bienosaurus* is ornamented on both lateral and medial surfaces. There is also fusion of small scutes to the frontal and

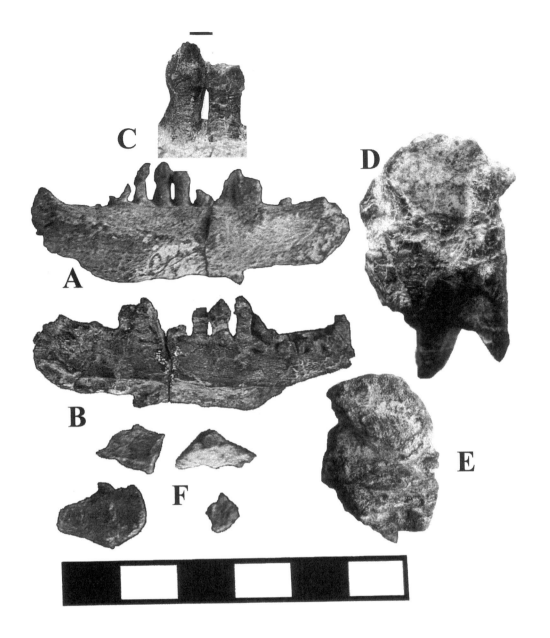

Figure 17.2. Bienosaurus lufengensis *gen. et sp. nov. IVPP V. 9612.* Right mandibular ramus in (A) lateral and (B) medial views; (C) close-up of two teeth; (D) frontal; (E) supraorbital?, and (F) other cranial fragments, possibly of a maxilla and pterygoid. Scale in (C): 2 mm; larger scale: cm.

supraorbital bones. These features do not occur in primitive ornithopods (including fabrosaurids, hypsilophodontids, and heterodontosaurids) or primitive stegosaurs (*Tatisaurus, Huayangosaurus*), but do occur in ankylosaurs. The teeth of *Bienosaurus* show characters of both ankylosaurs and stegosaurs. However, its dentary is thick, with a strongly curved tooth row, as in ankylosaurs.

The ankylosaurs have been divided into two families: Ankylosauridae and Nodosauridae (Coombs 1978; Maryanska 1977). In 1968, *Scelidosaurus* was referred to the Ankylosauria by Romer (1968). Coombs et al. (1990) recognized small armored ornithischians, including *Scutellosaurus, Scelidosaurus,* and possibly *Tatisaurus,* as basal thyreophorans. Recently, the Scelidosauridae was considered to be a sister group to the Stegosauria and the Ankylosauria by Olshevsky (1991). I remove the Scelidosauridae to Ankylosauria as a basal family. This family includes several small armored ornithischians (*Emausaurus, Lusitanosaurus, Scutellosaurus,* and *Scelidosaurus*).

Scutellosaurus is a small creature from the Kayenta Formation of the Lower Jurassic of Arizona. In *Scelidosaurus* the mandible is straight, dorsoventrally slender, and without ornamentation on the lateral side. *Scelidosaurus,* from the lower Jurassic of Britain, is relatively large and has tooth crowns that are mesiodistally expanded, and broadly triangular. The buccal aspect of the crown is relatively flat. *Bienosaurus* differs in all these features. Because the new form does not include armor plates, as in ankylosaurs and stegosaurs, it is best regarded as neither stegosaur nor ankylosaur, but as a primitive thyreophoran.

Acknowledgments: The author is especially grateful to Dr. Philip J. Currie of the Royal Tyrrell Museum of Palaeontology and T. Osmólska of the Palaeozoological Institute of the Polish Academy of Sciences for reading this manuscript and providing kind advice.

References

Bien, M. N. 1941. Discovery of Triassic saurischian and primitive mammalian remains at Lufeng, Yunnan. *Bulletin of the Geological Society of China* 20: 225–234.

Coombs, W. P. 1978. The families of the ornithischian dinosaur order Ankylosauria. *Palaeontology* 21: 143–170

Coombs, W. P., D. B. Weishampel, and L. M. Witmer. 1990. Basal Thyreophora. In D. Weishampel, P. Dodson, and H. Osmólska (eds.). *The Dinosauria,* pp. 427–434. Berkeley: University of California Press.

Dong, Z. 1990. Stegosaurs of Asia. In K. Carpenter and P. Currie (eds.), *Dinosaur Systematics: Approaches and Perspectives,* pp. 255–268. New York: Cambridge University Press.

Maryanska, T. 1977. Ankylosauridae (Dinosauria) from Mongolia. *Palaeontologica Polonica* 37: 85–152.

Olshevsky, G. 1991. A revision of the Parainfraclass Archosauria Cope, 1869, excluding the advanced Crocodylia. *Mesozoic Meanderings* 2: 1–196.

Owen, R. 1863. Monographs on the British fossil Reptilia from the Oolite Formations. Part 2, *Scelidosaurus harrisonii* and *Pliosaurus grandis. Palaeontological Society Monograph* 14: 1–26.

Romer, A. S. 1968. *Notes and Comments on Vertebrate Paleontology.* Chicago: University of Chicago Press.

Sereno, P. C. 1991. *Lesothosaurus,* Fabrosaurids, and the early evolution of Ornithischia. *Journal of Vertebrate Paleontology* 11 (2): 168–197.

Simmons, D. L. 1965. The non-therapsid reptiles of the Lufeng Basin, Yunnan, China. *Fieldiana, Geology* 15: 1–93.

Sun, A. L., and K. H. Cui. 1986. A brief introduction to the Lower Lufeng saurischian fauna (Lower Jurassic: Lufeng and Yunnan, People's Republic of China). In K. Padian (ed.), *The Beginning of the Age of Dinosaurs: Faunal Change across the Triassic-Jurassic Boundary,* pp. 275–278. New York: Cambridge University Press.

Sun A., G. Cui, Y. Li, and X. Wu. 1985. A verified list of Lufeng saurischian fauna. *Vertebrata PalAsiatica* 23: 1–12.

Young, C. C. 1982. On a new genus of dinosaur from Lufeng, Yunnan. In C. C. Young (ed.), *Collected Papers of Yang Zhong Jian,* pp. 31–35. Beijing: Science Press.

18. A *Montanoceratops cerorhynchus* (Dinosauria: Ceratopsia) Braincase from the Horseshoe Canyon Formation of Alberta

PETER J. MAKOVICKY

Abstract

An isolated braincase, collected by an American Museum of Natural History expedition in 1910, represents the first nonceratopsid neoceratopsian remains from the Horseshoe Canyon Formation of southern Alberta. This braincase shares a number of anatomical features with two other North American nonceratopsid neoceratopsians, *Montanoceratops cerorhynchus* and *Leptoceratops gracilis,* including a vertical supraoccipital, a deep notch between the base of each basipterygoid process and the body of the basisphenoid, and posteriorly everted basioccipital tubera. Other characters, such as the presence of a midline depression on the supraoccipital, and a narrow notch, rather than a tab, separating the basioccipital tubera, distinguish it from *Leptoceratops gracilis,* but allow referral to *Montanoceratops cerorhynchus.* The Horseshoe Canyon Formation is temporally equivalent to the St. Mary River Formation, which has yielded two other specimens of *Montanoceratops cerorhynchus,* and the braincase represents a paleogeographic range extension of this taxon. The large number of braincase characters shared by *Montanoceratops cerorhynchus* and *Leptoceratops gracilis* stands in contrast to recent phylogenetic work, which has posited these two North American taxa as occupying very different positions within ceratopsian phylogeny. A reevaluation of basal ceratopsian relationships incorporating new data from the holotype of *Montanoceratops cerorhynchus,* as well as the braincase described here, nests the latter taxon in a clade with *Leptoceratops gracilis* and *Udanoceratops tschizhovi,* exclusive of other ceratopsian taxa.

Introduction

Basal neoceratopsian dinosaurs have composed a relatively inconspicuous constituent of the known Late Cretaceous faunas of western North America. Our current knowledge of basal neoceratopsian diversity from this region and age is largely restricted to the two taxa *Leptoceratops gracilis* and *Montanoceratops cerorhynchus* from the Scollard Formation (Brown 1914; Sternberg 1951) and Lancian beds of Wyoming (Ostrom 1978), and the St. Mary River Formation (Brown and Schlaikjer 1942), respectively. Other, but more fragmentary, non-ceratopsid neoceratopsian remains from western North America include a fragmentary skeleton from the Early Cretaceous of Idaho (Weishampel et al. 1990), an isolated braincase from Horseshoe Canyon beds of the Red Deer River Valley (Lull 1933; Forster 1990), isolated teeth from the Milk River Formation of southernmost Alberta (Baszio 1997), an isolated mandible referred to *Leptoceratops* sp. from the Campanian Dinosaur Park Formation (Ryan and Currie 1998), and two very fragmentary skeletons from the Two Medicine Formation referred to *Leptoceratops* sp. by Gilmore (1939), and suggested to be distinct from *Leptoceratops gracilis* by Sternberg (1951) and Ostrom (1978). Recently, Chinnery and Trexler (1999) reported on bone bed material that represents a new species of *Leptoceratops* from the Two Medicine Formation, and which could be conspecific with Gilmore's (1939) material. A surge of discoveries of new basal neoceratopsian taxa and specimens in recent years (Nessov 1989, 1995; Dong and Azuma 1997; Chinnery and Weishampel 1998; Wolfe and Kirkland 1998; Ryan and Currie 1998; Zhao et al. 1999) suggests a more complicated picture for basal ceratopsian diversity, biogeography, and biostratigraphy than previously portrayed.

More data, especially from North America, is required to address these issues, and the isolated braincase mentioned above offers the opportunity of adding to our knowledge of both the anatomy and distribution of North American basal neoceratopsians. This specimen was collected in 1910 by a field party from the American Museum of Natural History under the leadership of Barnum Brown, but its sole mention in the literature consists of inclusion in a list of ceratopsian material in Lull's (1933) monograph, and comparative descriptive comments in a doctoral dissertation (Forster 1990).

Specimen: AMNH 5244, partial neoceratopsian braincase preserving the supraoccipital, both opisthotic-exoccipitals, the basioccipital, basisphenoid, both laterosphenoids, and parts of the fused parietals.

Horizon and Locality Data: Horseshoe Canyon Formation (previously lower part of Edmonton Formation), 53.4 m (175 ft.) above river level; east bank, Red Deer River, 3.5 miles below Tolman Ferry. Field photographs taken by Barnum Brown suggest that AMNH 5244 was collected close to the AMNH 5245 (*"Ankylosaurus"*) quarry. This ankylosaurid specimen has later been referred to *Euoplocephalus* sp.

Institutional Abbreviations: AMNH, Dept. of Vertebrate Paleontology, American Museum of Natural History, New York; CMGP, Central Museum of Geological Prospecting, St. Petersburg, Russia; IVPP, Institute of Vertebrate Paleontology and Paleoanthropology, Beijing; MOR, Museum of the Rockies, Bozeman, Montana; NMC, Canadian Museum of Nature, Ottawa; PIN, Paleontologicheskii Institut, Russian Academy of Sciences, Moscow; ZPAL, Zaklad Paleobiologii, Polish Academy of Sciences, Warsaw.

Description

AMNH 5244 is well preserved but displays oblique distortion toward the left side of the specimen, with the greatest degree of distortion present at the front of the braincase (fig. 18.1A). Most of the sutures between individual skull bones are obliterated by fusion, suggesting that the specimen was either a mature individual or very close to maturity. The braincase is relatively short and deep, as is characteristic for neoceratopsians.

The supraoccipital is tall, as witnessed by the remains of its suture with the overlying parietals in occipital view. It probably formed a small part of the foramen magnum border, but the extent of this participation, if present, is difficult to judge because of the complete fusion between the exoccipitals and supraoccipital. According to Chinnery and Wesihampel (1998), the supraoccipital does not participate in the foramen magnum in a juvenile *Montanoceratops cerorhynchus* braincase (MOR 542), but the borders of the element are difficult to determine in the specimen, which also displays strong mediolateral crushing that distorted the shape of the foramen magnum. Supraoccipital participation in the foramen magnum is present in *Protoceratops andrewsi* (Brown and Schlaikjer 1940), *Leptoceratops gracilis* (Sternberg 1951), and *Bagaceratops rozhdestvenskyi* (Maryanska and Osmólska 1975), but absent in ceratopsids (Brown and Schlaikjer 1940). Dorsal to the foramen magnum, the supraoccipital bears two low ridges, which meet dorsally to form a low, midline ridge of bone that buttresses the supraoccipital-parietal suture.

Unlike most ceratopsian taxa, for which the supraoccipital is known, a tear-drop-shaped parasagittal depression separates the two ridges on the lower half of the supraoccipital. The preserved part of the supraoccipital of *Montanoceratops cerorhynchus* (MOR 542) displays two faint ridges extending dorsally from the foramen magnum border toward the supraoccipital-parietal suture, similar to the ridges than define the midline depression on the supraoccipital of AMNH 5244. Shallower depressions are present on either side of the midline ridge, as in ceratopsid supraoccipitals, where this ridge continues on to the exoccipitals dorsal to the foramen magnum. A tall midline ridge is present on the supraoccipital of *Psittacosaurus mongoliensis* (AMNH

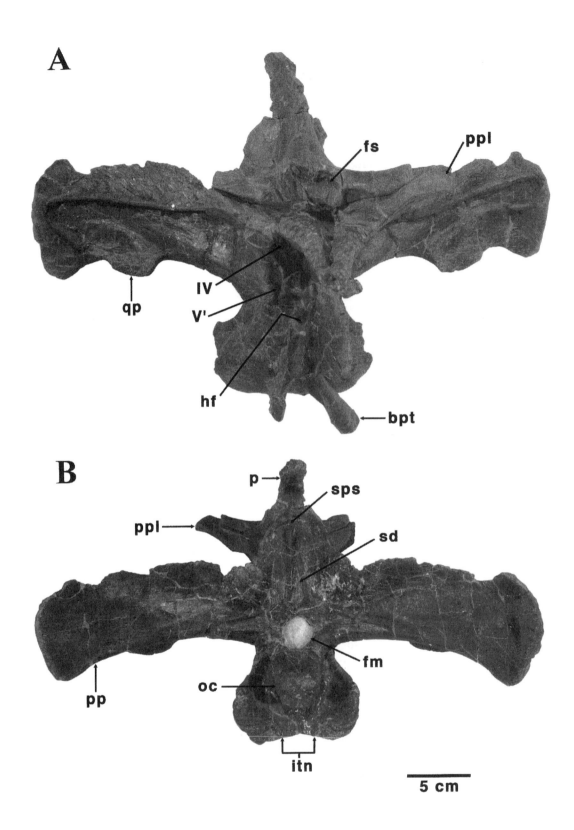

6254), and a lower ridge is present in *Leptoceratops gracilis* (NMC 8888), some specimens of *Protoceratops andrewsi* (AMNH 6429) and in ceratopsids (Dodson and Currie 1990, fig. 29.5), but no neoceratopsian other than *Montanoceratops cerorhynchus* is known to have two ridges that converge dorsally and define a depression between them.

The posterior face of the supraoccipital is oriented in the same plane as the basioccipital tubera, a condition also seen in *Leptoceratops gracilis* (NMC 8888), but not in most other basal neoceratopsians, including *Archaeoceratops oshimai* (IVPP V 11114), *Protoceratops andrewsi* (AMNH 6466), and *Bagaceratops rozhdestvenskyi* (ZPAL MgD I/129), as well as psittacosaurs (AMNH 6254) and pachycephalosaurs, where the supraoccipital is anterodorsally inclined relative to the posterior face of the basioccipital. The supraoccipital of the only known braincase of *Montanoceratops cerorhynchus* (MOR 542) is partly broken and displaced (Chinnery and Weishampel 1998). The condition in ceratopsids is difficult to evaluate because the supraoccipital is displaced from the foramen magnum by the enlarged exoccipitals. In *Triceratops horridus* (AMNH 970), *Centrosaurus apertus* (AMNH 5239), and *Anchiceratops ornatus* (AMNH 5251) the supraoccipital appears to be oriented in nearly the same plane as the basioccipital plate.

The foramen magnum is circular and has a smaller diameter than the occipital condyle (fig. 18.1B). Robust ridges of bone extend laterally from the foramen magnum along the ventral edge of each paroccipital process. A shallow depression is present above this ridge near the base of the paroccipital process, just dorsolateral to the foramen magnum. The occipital condyle is spherical, and complete fusion between its constituent bones prevents determination of the proportional contribution of the exoccipitals. In immature specimens of other basal neoceratopsians, such as *Montanoceratops cerorhynchus* (MOR 542) and *Leptoceratops gracilis* (NMC 8888), the exoccipitals meet below the foramen magnum and contribute about one-third of the occipital condyle. The condyle of AMNH 5244 does not have a distinct, constricted neck, as seen in the occipital condyle of ceratopsids. Unlike *Leptoceratops gracilis* (NMC 8889) and *Psittacosaurus mongoliensis* (AMNH 5254), there is no shallow depression or pit at the base of the condylar neck. Ventrolateral to the condyle, the exoccipital is perforated by three foramina for cranial nerves IX–XII in a triangular configuration, as in other nonceratopsid neoceratopsians (Brown and Schlaikjer 1940, 1942). The basioccipital tubera are widely flared in a transverse plane below the occipital condyle.

The tubera are rounded ventrolaterally, and are circumscribed anteriorly by a deep groove that curves around onto the posterior surface of the basioccipital near the midline. The smooth area on the midline between the emergence of the groove on either side of the occiput is narrow (approximately one-third of maximum transverse width of basal tubera) and indented (fig. 18.1B), as in a partial braincase of a juvenile specimen of *Montanoceratops cerorhynchus* (MOR 542). This condition is unlike that in *Leptoceratops gracilis* (NMC 8889), where this smooth area is proportionately wider and forms a ventral convexity between the tuberal grooves. This midline convexity bears a number

Figure 18.1. (opposite page) AMNH 5244 Montanoceratops cerorhynchus braincase in (A) anterior, and posterior (B) views. Abbreviations: IV, exit for trochlear nerve; V, exit for ophthalmic branch of trigeminal nerve; bpt, basipterygoid process; fm, foramen magnum; fs, sutural surface for frontal; hf, hypophysial fossa; itn, intertuberal notch; oc, occipital condyle; p, parietal; pp, paroccipital process; ppl, postorbital process of laterosphenoid; qp, process for reception of pterygoid wing of quadrate; sd, supraoccipital depression; sps, supraoccipital-parietal suture.

of low ridges in *Leptoceratops gracilis* (NMC 8889). Another similarity between the juvenile *Montanoceratops cerorhynchus* braincase and AMNH 5244 is the posteroventral eversion of the tubera, which forms a lip or shelf under the occipital condyle (Chinnery and Weishampel 1998). An everted tuberal rim is present in *Leptoceratops gracilis* (NMC 8888), but is absent in *Protoceratops andrewsi* (AMNH 6429; 6429; 6466) and in ceratopsids, such as *Centrosaurus apertus* (AMNH 5239) and *Triceratops horridus* (AMNH 970).

A prominent otosphenoidal crest descends ventrally from the anterior border of the facial nerve exit and overlaps the anteroventral face of the basisphenoid, where it forms the lateral wall of the carotid canal. There is no gap or foramen on the ventral midline between the basisphenoid and basioccipital, as seen in a juvenile braincase of *Bagaceratops rozhdestvenskyi* (ZPAL MgD I/133). A small foramen is also present on the edge of the basisphenoid midline at its suture with the basioccipital in a small specimen of *Leptoceratops gracilis* (NMC 8888) and in a basicranium referred to *Microceratops gobiensis* (Bohlin 1953). This feature is variably present in some specimens of *Protoceratops andrewsi* (AMNH 6466) and *Triceratops horridus* (Forster 1990) and may be subject to individual and perhaps ontogenetic variation. The foramen has been compared to the median Eustachian system of crocodilians (Maryanska and Osmólska 1975), but there is no evidence that it connects to the middle ear in ceratopsians. A basicranium referred to *Asiaceratops salsopaludalis* (CMGP 496/12457) has a midline notch along the ventral edge of the basisphenoid, but it is filled by a triangular process of the basioccipital.

Anteriorly, the base of the basisphenoid rises steeply from the basal tubera. A large foramen for passage of the internal carotid artery is present on both sides of the basisphenoid, and breakage in the hypophysial region reveals a cross-section of these canals at their level of entry into the hypophyseal fossa. The basipterygoid processes project ventrally and curve posteriorly from the braincase, just anterior to the carotid foramina. This curvature of the basipterygoid processes is proposed as an autapomorphy of *Montanoceratops cerorhynchus* (Chinnery and Weishampel 1998). As in *Montanoceratops cerorhynchus* (MOR 542) and *Leptoceratops gracilis* (NMC 8889), the base of each basipterygoid process is separated from the posterior edge of the basisphenoid by a deep notch. This character differs from *Protoceratops andrewsi* (AMNH 6425), *Bagaceratops rozhdestvenskyi* (AMNH-MAE unnumbered specimen), and ceratopsids (Dodson and Currie 1990, fig. 29.5), where the basipterygoid processes are located near the posterior border of the basisphenoid, close to its suture with the basioccipital tubera.

Anterior to the hypophyseal fossa, the cultriform process forms a thin lamina, which is broken along its rostral margin (fig. 18.2). The laterosphenoids flare widely dorsal to this lamina, and bear robust postorbital processes. These extend slightly beyond the laterally directed frontal process of the parietal, and brace the postorbital beneath the frontal-postorbital suture (Forster 1990). The laterosphenoids do not meet dorsal to the brain cavity, as in ceratopsids, and there is a rhomboid gap between the anteroventral edge of the laterosphenoids and the anterior edge of the parietal (Forster 1990). An orbitosphenoid may have been fused to the laterosphenoid, as suggested by the full anterior closure of the braincase anterior and dorsal to the hypophyseal fossa and medial to the optic nerve exits (Dodson and Currie 1990).

Each paroccipital process is relatively wide and flares at the rounded distal end. Distal to the ridge of bone extending from the foramen magnum, the posterior surface of the paroccipital process is flat and smooth. A robust ridge of bone buttresses the anterior face of the paroccipital process from the opisthotic region of the braincase. Comparison with the immature *Montanoceratops cerorhynchus* (MOR 542) braincase suggests that this ridge is formed by the opishtotic, whereas the exoccipital forms the majority of the paroccipital process (Chinnery and Weishampel 1998). The ventral face of this ridge forms a deep overhanging groove that leads into the middle ear. At midlength, the anterior surface of the paroccipital process bears a large suboval articular facet for reception of the quadrate wing of the pterygoid. Its surface is rugose and marked with striae. The ventral border of this facet forms a broad, ventrally convex process that extends markedly below the ventral border of the paroccipital process (fig. 18.1A). A sutural surface is

Figure 18.2. Close-up of right lateral view of AMNH 5244 Montanoceratops cerorhynchus braincase showing cranial nerve foramina. Abbreviations: I, exit for olfactory nerves; II, exit for optic nerve; III, exit for occulomotor nerve; IV, exit for trochlear nerve; V', exit for ophthalmic branch of trigeminal nerve; V'', exit for maxillary and mandibular branches of trigeminal nerve; VII, exit for facial nerve; X, foramen for conduit of vagus nerve through metotic strut; hf, hypophysial fossa; oc. occipital condyle.

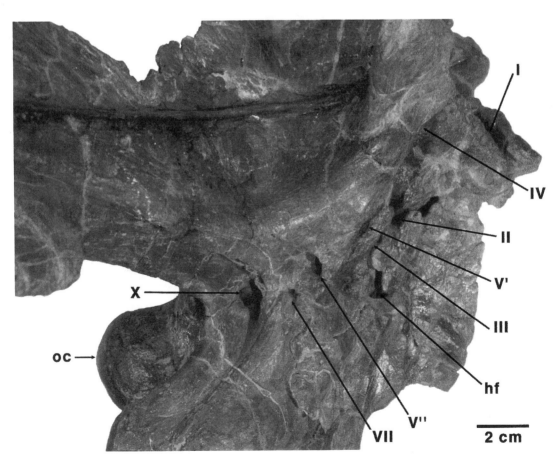

also preserved on the distal part of the paroccipital process in a juvenile specimen of *Montanoceratops cerorhynchus* (MOR 542), but it is less developed than in AMNH 5244, possibly due to the immature nature of the former specimen. A process buttressing the pterygoid is also present on the anterior face of the paroccipital process of an adult specimen of *Leptoceratops gracilis* (NMC 8889), but it is relatively small and is not visible below the paroccipital process in posterior view.

The middle ear region forms a deep recess at the base of the paroccipital process. A narrow groove runs from the ventral apex of the middle ear opening to the circumtuberal groove on the basioccipital tubera (fig. 18.2). The crista interfenestralis is partially preserved in the right ear, and appears to subdivide the opening into a larger fenestra ovalis and a smaller fenestra pseudorotunda.

Dorsal to the middle ear region, the opisthotic region is shallowly concave and is confluent with the supratemporal fossa of the parietal. The parietals are imperceptibly fused to each other, but sutures are still visible between the anterolateral process of the parietals and the postor-bital process of the laterosphenoids, and in posterior view at the apex of the parietal-supraoccipital suture. The anterior sutural surface for reception of the frontals is open (fig. 18.1A), and there is no indication of a frontoparietal depression as in *Protoceratops andrewsi* (Brown and Schlaikjer 1940). The preserved frontal fragments of the holotype specimen of *Montanoceratops cerorhynchus* (Brown and Schlaikjer 1942, AMNH 5464) indicate that the skull roof was relatively wide and flat, as in *Leptoceratops gracilis* (Sternberg 1951). The beginning of a sagittal crest is present along the preserved portion of the parietals just beyond the frontal sutural region.

Cranial Nerves

Exits for cranial nerves II–XII are preserved on AMNH 5244 (fig. 18.2). Nerve exits are better preserved on the right side, those of the left having been partly obscured by the postmortem distortion of the brain-case. The olfactory nerves (CN-I) exited the AMNH 5244 braincase dorsal to the fused anterior tips of the laterosphenoids and beneath the frontals as stated by Forster (1990). The exit for the optic nerve (CN-II) is very large and situated posterior and ventral to a midline ossifi-cation at the front of the braincase, which probably constitutes the orbitosphenoid. Directly ventral to this foramen, and just dorsal to the hypophyseal fossa, is a smaller exit for the oculomotor nerve (CN-III) (fig. 18.2). The trochlear nerve (CN-IV) exit is the most dorsal and lateral of the preserved cranial nerve exits and lies far from the midline at the base of the postorbital process of the laterosphenoid (figs. 18.1A, 18.2). The exit for CN-IV indicated by Maryanska and Osmólska (1975) for *Bagaceratops rozhdestvenskyi* (ZPAL MgD I/133) is more likely for the optic nerve (CN-II). The ventromedially open configur-ation of the foramen is probably due to lack of ossification in this juvenile specimen.

The trigeminal (CN-V) has two separate exits for the ophthalmic and the maxillary and mandibular branches, respectively (fig. 18.2), as described for other ceratopsians (Hay 1909; Brown and Schlaikjer

1940; Forster 1996). An anteriorly facing foramen that is continuous with a deep groove along the base of the laterosphenoid marks the course of the ophthalmic branch (CN-V$_1$). The maxillary and mandibular nerves (CN-V$_2$ and V$_3$) exited the braincase through a larger laterally facing foramen anterior to and level with the middle ear. The posterior border of this foramen is formed by the prootic in the juvenile *Montanoceratops cerorhynchus* braincase (Chinnery and Weishampel 1998). This opening corresponds to a large anteroventral foramen enclosed between the prootic and laterosphenoid of the juvenile *Bagaceratops rozhdestvenskyi* (ZPAL MgD I/133) and originally misidentified as for the oculomotor nerve (CN-III) (Maryanska and Osmólska 1975). The actual position of the oculomotor exit in ZPAL MgD I/133 is closer to the midline of the skull; it is confluent with the opening for CN-II due to a lack of ossification or breakage. The supposed presphenoid of *Bagaceratops rozhdestvenskyi* is here reinterpreted as a broken piece of the laterosphenoid. Ventral to the ophthalmic nerve exit and just anterior to the middle ear of AMNH 5244, the laterosphenoid is pierced by a small foramen for the facial nerve (CN-VII), as in *Montanoceratops cerorhynchus* (MOR 542) and *Protoceratops andrewsi* (AMNH 6466). A pit situated posterior to the hypophyseal fossa and directly ventral to the ophthalmic nerve foramen may mark the exit of the abducent nerve (CN-VI), although preservation prevents positive determination.

The exits for the vestibulocochlear nerve (CN-VIII) are obscured by the matrix still filling the brain cavity. The glossopharyngeal (CN-IX) and vagus (CN-X) nerves exited the braincase through the metotic fissure, but the vagus (X) nerve turned posteriorly and passed through the metotic strut to emerge through the most lateral of the three exoccipital foramina (Forster 1996). A small embayment on the anterior face of the paroccipital process, lateral and adjacent to the fenestra pseudorotunda, probably marks the point where the vagus nerve passed into the metotic strut to exit on the occipital face of the skull (fig. 18.2). A foramen passing through the metotic strut is present in similar position on the posterior wall of the middle ear chamber in the juvenile specimen of *Bagaceratops rozhdestvenskyi* (ZPAL MgD I/133) and an immature specimen of *Leptoceratops gracilis* (NMC 8887). The two medial exoccipital foramina mark the course of the accessory (XI) and hypoglossal (XII) nerves. The hypoglossal foramen is the larger of the two and ovate in shape.

Discussion

The Horseshoe Canyon Formation has yielded a rich dinosaur fauna of early Maastrichtian age (Eberth 1997), but thus far no basal neoceratopsian specimens have been reported. The AMNH 5244 braincase represents the first such discovery, and in conjunction with other recent discoveries, underscores that nonceratopsid neoceratopsian dinosaurs are present in most Late Cretaceous terrestrial sedimentary units of Alberta and Montana, albeit in small numbers (see Ryan and Currie 1998, table 1).

The AMNH 5244 braincase lacks synapomorphies of ceratopsids such as reduction of the number of hypoglossal exits from three to two, fusion of the laterosphenoids dorsal to the brain cavity, and very deep paroccipital processes. Instead, it is morphologically similar to the braincase anatomy of the two North American basal neoceratopsians *Leptoceratops gracilis* and *Montanoceratops cerorhynchus* in several features. Both of these taxa have the lateral edges of the basioccipital tubera strongly everted, a feature unique among ceratopsians. Both also display a deep notch between the posteroventral rim of the basisphenoid and each of the basipterygoid processes, although this character is also present in a braincase referred to *Asiaceratops salsopaludalis* (CMGP 496/12457). The preserved part of the supraoccipital in *Montanoceratops cerorhynchus* (MOR 542) displays a faint midline depression on the supraoccipital as in AMNH 5244, a derived feature absent in other neoceratopsians. Furthermore, AMNH 5244 has posteriorly curved basipterygoid processes, a purported autapomorphy of *Montanoceratops cerorhynchus* (Chinnery and Weishampel 1998), although the condition is unknown for *Leptoceratops gracilis* due to lack of preservation or postburial deformation of the specimens. Another character in which AMNH 5244 differs from *Leptoceratops gracilis* concerns the proportions and morphology of the basioccipital tubera. In *Leptoceratops gracilis* (NMC 8889) these are widely separated by a ventrally convex tab, whereas in AMNH 5244 the basioccipital tubera are enlarged and separated by a proportionately narrower, concave notch, as in *Montanoceratops cerorhynchus* (MOR 542). In summary, the Horseshoe Canyon Formation braincase shares a number of derived characters such a supraoccipital midline depression and curved basipterygoid process, as well as other similarities, with *Montanoceratops cerorhynchus,* and is referred to that taxon.

The Horseshoe Canyon Formation is temporally equivalent to the St. Mary River Formation of southernmost Alberta and northern Montana (Eberth 1997; Ryan and Currie 1998), which has yielded the type and referred specimens of *Montanoceratops cerorhynchus* (Brown and Schlaikjer 1942; Chinnery and Weishampel 1998). Thus AMNH 5244 is approximately coeval with known specimens of *Montanoceratops cerorhynchus* and indicates a broader geographical range for this taxon than previously recognized. An undescribed specimen referred to *Montanoceratops cerorhynchus* (Ryan and Currie 1998) from the Foothills area of southern Alberta is thought to be from latest Campanian beds that are a little older than the Horseshoe Canyon Formation.

Recent phylogenetic hypotheses for Ceratopsia have portrayed the two well-known North American basal neoceratopsian taxa as occupying different phylogenetic positions with *Montanoceratops cerorhynchus* as the sister group to the Ceratopsidae (Sereno 1984, 1986; Forster 1990; Chinnery and Weishampel 1998) or a Ceratopsidae + *Turanoceratops* grouping (Sereno 1997; 1999), and *Leptoceratops gracilis* in a more basal position. A study by Dodson and Currie (1990) instead placed these two taxa within a monophyletic Protoceratopsidae that encompasses all nonceratopsid neoceratopsians, but *Leptocera-*

tops gracilis was posited as the most primitive protoceratopsid, while *Montanoceratops cerorhynchus* was apical within the group. All these phylogenetic hypotheses are at odds with the large number of potential synapomorphies in the braincase shared by these two taxa exclusive of other ceratopsians. Reevaluation of the holotype of *Montanoceratops cerorhynchus* (AMNH 5464) following removal of the preserved skull parts from the reconstructed plaster skull (Brown and Schlaikjer 1942, fig. 1), shows that a number of supposed synapomorphies linking this taxon to ceratopsids are in fact based on misidentifications (Makovicky in prep). These circumstances cast doubt on current hypotheses regarding the affinities of *Montanoceratops cerorhynchus,* and warrant reexamination of basal neoceratopsian relationships.

A preliminary analysis of basal neoceratopsian relationships was undertaken, incorporating the new braincase character data described above, as well as new information based on the reidentification of parts of the *Montanoceratops cerorhynchus* holotype. A matrix of 98 characters (80 binary, 18 multistate) was compiled for 12 ingroup ceratopsian taxa (see fig. 18.3, app. 18.2) and three outgroups. *Hypsilophodon foxii, Stegoceras validum,* and one of the basalmost ceratopsians *Psittacosaurus mongoliensis* were used as outgroups with trees rooted on *Hypsilophodon foxii.* The ingroup encompasses most well-known basal neoceratopsian taxa, including the enigmatic *Chaoyangsaurus youngi,* suggested to be a neoceratopsian by Sereno (1999), and the two ceratopsids *Triceratops horridus* and *Centrosaurus apertus. Turanoceratops tardabilis* (Nessov 1995) was excluded because it lacks diagnostic features, and the identity of many of the referred elements, for example a misidentified ornithopod braincase, is questionable. *Breviceratops kozlowskii* (Kurzanov 1991) is here regarded as synonymous with *Bagaceratops rozhdestvenskyi* (V. Alifanov, pers. comm., pers. obs.). All analyses were run using the Branch and Bound algorithm of PAUP*4.2 Beta, with branches collapsed when maximum length = 0. Polymorphic characters were treated as polymorphisms, because characters are scored at species level. The analysis yielded a single tree, which is shown in figure 18.3. Thirteen multistate characters were ordered based on primary homology considerations. Reanalysis of the data with all characters treated as unordered yielded the same topology, and only caused a decrease of one step in Bremer support at some internal nodes. The characters and data matrix are presented in appendices 18.1 and 18.2, respectively, and unambiguous synapomorphies of the various clades are listed in appendix 18.3.

The phylogenetic hypothesis found here places *Montanoceratops cerorhynchus* as the sister taxon to a *Leptoceratops gracilis–Udanoceratops tschizhovi,* rather than as a sister taxon to the Ceratopsidae. This clade of medium-size, robust neoceratopsians is characterized by, among other synapomorphies, a specialized vertical-notch dentition, in which the maxillary teeth wear a horizontal shelf into an expansion on the labial face of the dentary teeth. One of the braincase characters uniting this clade is the everted form of the basioccipital tubera (unknown for *Udanoceratops tschizhovi*). The stem-based taxon Lepto-

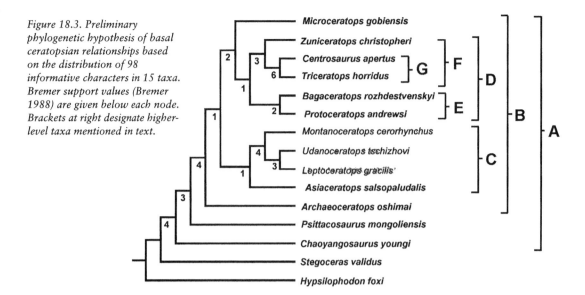

Figure 18.3. Preliminary phylogenetic hypothesis of basal ceratopsian relationships based on the distribution of 98 informative characters in 15 taxa. Bremer support values (Bremer 1988) are given below each node. Brackets at right designate higher-level taxa mentioned in text.

ceratopsidae is erected here to accommodate this clade, and is defined as all taxa closer to *Leptoceratops gracilis* than to *Triceratops horridus*. Weak support is found for *Asiaceratops salsopaludalis* as occupying a basal position within Leptoceratopsidae, and the group is diagnosed by possession of a deep pit at the front of the dentary for reception of the rostral. The validity of the latter species is questionable (Dodson and Currie 1990), but excluding it from analysis does not alter the relationships between other taxa.

A monophyletic Mongolian clade containing *Protoceratops andrewsi* and *Bagaceratops rozhdestvenskyi* (Protoceratopsidae [Sereno 1998]) is recovered as the sister taxon to a Ceratopsidae-*Zuniceratops christopheri* group (Ceratopsoidea [Sereno 1998]) and this large clade conforms to Sereno's (1998) definition of Coronosauria. Protoceratopsidae is united by four unambiguous characters, including the possession of an anterior prong on the quadratojugal, a posteriorly bifurcated splenial, and an anterior extension of the surangular on the lateral face of the mandible. Ceratopsoidea is diagnosed by possession of orbital horns, vertical occlusion between teeth, and a curved ischium. The only unambiguous synapomorphy uniting the coronosaur clade is the possession of hooflike pedal unguals, but many ambiguous characters, such as the presence of a nasal horn (incipient in *Protoceratops andrewsi*), an epijugal capping the end of the jugal horn, and a bladelike postpubis, support this clade under some optimizations (DELTRAN). The poorly known *Microceratops gobiensis* is the sister taxon to Coronosauria, and its fragmentary nature accounts for much of the ambiguity in the distribution of synapomorphies in this part of the tree. The *Microceratops gobiensis*–coronosaur clade is characterized by frill fenestration, divergent temporal bars, and a postquadratic process on the squamosal.

Archaeoceratops oshimai is here found to be the most primitive neoceratopsian taxon, whereas *Chaoyangsaurus youngi* lies outside the *Psittacosaurus mongoliensis*–Neoceratopsia clade, at the base of Ceratopsia, as suggested by Zhao et al. (1999), but in contrast to Sereno's (1999) findings. Neoceratopsian monophyly is well supported by 13 unambiguous synapomorphies, including possession of an epijugal, expansion of the parietal frill, a mediolaterally expanded quadratojugal, a keeled rostral, and the presence of a primary ridge on at least the maxillary teeth. Ceratopsia is diagnosed by a suite of characters such as a large, triangular head, presence of a rostral, and the basioccipital forming the basal tubera. Possession of jugal horns and a coronoid process situated lateral to the tooth row are two of the synapomorphies that unite the *Psittacosaurus mongoliensis*–neoceratopsian grouping to the exclusion of *Chaoyangsaurus youngi*.

The phylogenetic hypothesis presented here is generally similar to that of Sereno (1999), with the exception of the positions of *Chaoyangsaurus youngi, Montanoceratops cerorhynchus,* and *Microceratops gobiensis.* It differs more substantially from hypotheses published by Chinnery and Weishampel (1998) and Sereno (1986) and is least congruent with the hypothesis of a monophyletic Protoceratopsidae encompassing all nonceratopsid neoceratopsians, as suggested by Dodson and Currie (1990). A minimum of three dispersal events is required to accommodate the distribution of neoceratopsian taxa between North American and Asia according to the current phylogeny. An Asian to North American dispersal is required for the *Zuniceratops christopheri*–ceratopsid clade, but the presence of *Montanoceratops cerorhynchus* and *Leptoceratops gracilis* in North America can be accounted for ambiguously by either a separate dispersal event for each of these taxa, or a single dispersal event for their ancestor and a reverse dispersal for *Udanoceratops tschizhovi.* Similar patterns of endemism punctuated by dispersal of both singular taxa and larger clades between Asian and North American faunas occur in oviraptorosaur (Makovicky and Sues 1998; Norell et al. in press) and troodontid (Norell et al. 2000) theropods, and among multituberculate mammals (Rougier et al. 1997).

Acknowledgments: I dedicate this paper to Phil Currie, who first introduced me to research in dinosaur paleontology, and whose advice and friendship helped stoke my passion for this field. Under Phil's expert guidance, I began my work on theropod dinosaurs, a group that has consumed most of my time since. It is therefore very gratifying to dedicate my first nontheropod paper to Phil. Thanks also to Darren Tanke for initiating this project and for inviting my contribution, and to K. Carpenter and an anonymous reviewer for suggesting editorial and substantive improvements. I am grateful to Drs. Vladimir Alifanov (PIN), Aleksander Averianov (Zoological Institute, St. Petersburg), B. Chinnery (Johns Hopkins), Halszka Osmólska (ZPAL), Kieran Shepherd (NMC), Dong Zhiming, Xu Xing, and X. Zhao (IVPP) for arranging access to specimens in their care, and for their hospitality during my visits. A big "thank you" is due to Jack Horner for having the skull parts of the *Montanoceratops cerorhynchus* holotype removed from their positions in an "artistic" skull reconstruction, which helped clear up a lot of questions. This work was funded by the Danish Research Academy, a Theodore Roosevelt Memorial Fund fellowship, and a Dinosaur International Society fellowship.

References

Baszio, S. 1997. Systematic paleontology of isolated dinosaur teeth from the latest Cretaceous of south Alberta, Canada. *Courier Forschungsinstitut Senckenberg* 196: 33–77.

Bohlin, B. 1953. *Fossil Reptiles from Mongolia and Kansu.* Reports from the Scientific Expedition to the Northwestern Provinces of China under the Leadership of Dr. Sven Hedin, Sino-Swedish Expedition, Publication 37. Stockholm: Statens Etnografiska Museum.

Bremer, K. The limits of amino acid sequence data in angiosperm phylogenetic reconstruction. *Evolution* 42: 795–803.

Brown, B. 1914. *Leptoceratops,* a new genus of Ceratopsia from the Edmonton Cretaceous. *Bulletin of the American Museum of Natural History.* 33 (36): 567–580.

Brown, B., and E. M. Schlaikjer. 1940. The structure and relationships of *Protoceratops. Annals of the New York Academy of Sciences* 40: 133–266.

Brown, B., and E. M. Schlaikjer. 1942. The skeleton of *Leptoceratops* with the description of a new species. *American Museum Novitates* 1169: 1–15.

Chinnery, B. J., and D. Trexler. 1999. The first bonebed occurrence of a basal ceratopsian, with new information on the skull morphology of *Leptoceratops. Journal of Vertebrate Paleontology* 19 (suppl. to 3): 38A.

Chinnery, B. J., and D. B. Weishampel. 1998. *Montanoceratops cerorhynchus* (Dinosauria: Ceratopsia) and relationships among basal neoceratopsians. *Journal of Vertebrate Paleontology* 18 (3): 569–585.

Dodson, P., and P. J. Currie. 1990. Neoceratopsia. In D. B. Weishampel, P. Dodson, and H. Osmólska (eds.), *The Dinosauria,* pp. 593–618. Berkeley: University of California Press.

Dong Z. M., and Y. Azuma. 1997. On a primitive neoceratopsian from the Early Cretaceous. In Z. M. Dong (ed.), *Sino-Japanese Silk Road Dinosaur Expedition,* pp. 68–89. Beijing: China Ocean Press.

Eberth, D. A. 1997. Edmonton Group. In P. J. Currie and K. Padian (eds.), *Encyclopedia of Dinosaurs,* pp. 199–204. San Diego: Academic Press.

Forster, C. A. 1990. The cranial morphology and systematics of *Triceratops,* with a preliminary analysis of ceratopsian phylogeny. Ph.D. diss., University of Pennsylvania, Philadelphia.

Forster, C. A. 1996. New information on the skull of *Triceratops. Journal of Vertebrate Paleontology* 16: 246–258.

Gilmore, C. W. 1939. Ceratopsian dinosaurs from the Two Medicine Formation, Upper Cretaceous of Montana. *Proceedings U.S. National Museum* 87 (3066): 1–18.

Hay, O. P. 1909. On the skull and the brain of *Triceratops* with notes on the brain-cases of *Iguanodon* and *Megalosaurus. Proceedings U.S. National Museum* 36 (1660): 95–108.

Kurzanov, S. M. 1991. A new Late Cretaceous protoceratopsid genus from Mongolia. *Paleontological Journal* 1990: 85–91.

Lull, R. S. 1933. A revision of the Ceratopsia or horned dinosaurs. *Peabody Museum of Natural History Memoir* 3: 1–175.

Makovicky, P. J., and H.-D. Sues. 1998. Anatomy and phylogenetic relationships of the theropod dinosaur *Microvenator celer* from the Lower Cretaceous of Montana. *American Museum Novitates* 3240: 1–27.

Maryanska, T., and H. Osmólska. 1975. Protoceratopsidae (Dinosauria) of Asia. *Palaeontologica polonica* 33: 133–182.

Nessov, L. A. 1989. Ceratopsian dinosaurs and crocodiles of the middle Mesozoic of Asia. In T. N. Bogdanova and L. I. Kozhatsky (eds.), *Theoretical and Applied Aspects of Modern Paleontology,* pp. 142–149. Leningrad: Nauka.

Nessov, L. A. 1995. [*Dinosaurs of Northern Eurasia: New Data about Assemblages, Ecology, and Paleobiogeography.*] St. Petersburg: University of Saint Petersburg. (In Russian.)

Norell, M. A., J. M. Clark, and P. J. Makovicky. In press. Relationships among Maniraptora: Problems and prospects. In J. A. Gauthier (ed.), *New Perspectives on the Origin and Evolution of Birds: Proceedings of the International Symposium in Honor of John H. Ostrom. Bulletin of the Peabody Museum of Natural History.*

Norell, M. A., P. J. Makovicky, and J. M. Clark. 2000. A new troodontid theropod from Ukhaa Tolgod, Mongolia. *Journal of Vertebrate Paleontology* 20: 1–5.

Ostrom, J. H. 1978. *Leptoceratops gracilis* from the "Lance" Formation of Wyoming. *Journal of Paleontology* 52: 697–704.

Rougier, G. W., M. J. Novacek, and D. Dashzeveg. 1997. A new multituberculate from the Late Cretaceous locality Ukhaa Tolgod, Mongolia: Considerations on multituberculate interrelationships. *American Museum Novitates* 3191: 1–26.

Ryan, M. J., and P. J. Currie. 1998. First report of protoceratopsians (Neoceratopsia) from the Late Cretaceous Judith River Group, Alberta, Canada. *Canadian Journal of Earth Sciences* 35: 820–826.

Sereno, P. C. 1984. The phylogeny of the Ornithischia: A reappraisal. In W. Reif and F. Westphal (eds.), *Third Symposium on Mesozoic Terrestrial Ecosystems,* pp. 219–226. Tübingen: Attempto Verlag.

Sereno, P. C. 1986. Phylogeny of the bird-hipped dinosaurs (order Ornithischia). *National Geographic Research* 2: 234–256.

Sereno, P. C. 1997. The origin and evolution of dinosaurs. *Annual Review of Earth and Planetary Sciences* 25: 435–489.

Sereno, P. C. 1998. A rationale for phylogenetic definitions, with application to the higher-level taxonomy of Dinosauria. *Neues Jahrbuch für Geologie und Paläontologie* 210: 41–83.

Sereno, P. C. 1999. The evolution of dinosaurs. *Science* 284: 2137–2147.

Sternberg, C. M. 1951. Complete skeleton of *Leptoceratops gracilis* Brown from the Upper Edmonton Member on Red Deer River, Alberta. *National Museum of Canada Bulletin, Annual Report* (1949–50) 123: 225–255.

Weishampel, D. B., A. McCrady, and W. A. Akersten. 1990. New ornithischian dinosaur material from the Wayan Formation (Lower Cretaceous) of eastern Idaho. *Journal of Vertebrate Paleontology* 10 (suppl. to 3): 48A.

Wolfe, D. G., and J. I. Kirkland. 1998. *Zuniceratops christopheri* n. gen and n. sp., a ceratopsian dinosaur from the Moreno Hill Formation (Cretaceous, Turonian) of west-central New Mexico. In S. G. Lucas, J. I. Kirkland, and J. W. Estep (eds.), *Lower and Middle Cretaceous*

Terrestrial Ecosystems, pp. 303–317. *New Mexico Museum of Natural History and Science, Bulletin* 14.

Zhao, X., Z. Cheng, and X. Xu. 1999. The earliest ceratopsian from the Tuchengzi Formation of Liaoning, China. *Journal of Vertebrate Paleontology* 19: 681–691.

APPENDIX 18.1.
List of Characters

(1) Head size small relative to body (0) or large relative to body (1).

(2) Head shape in dorsal view: elongate, ovoid (0) or triangular, wide over jugals (1).

(3) Rostral bone forming beak absent (0) or present (1).

(4) Rostral ventral process absent (0) or present (1).

(5) Anterior face of rostral round, convex (0), or sharply keeled (1).

(6) Premaxillary palatal region flat in ventral view (0) or vaulted dorsally (1).

(7) Relative height of premaxilla (snout) to orbital region low (0) or deep (1).

(8) Premaxillary-maxillary suture situated posterior to convex buccal process at front of upper jaw (0) or extends through process (1).

(9) Nares position close to buccal margin (0) or dorsal, away from buccal margin (1).

(10) Depression on premaxilla anteroventral to naris absent (0) or present (1).

(11) Nasal horn absent (0), small (1), or large (2). Ordered.

(12) Nares width less than 10% of skull length (0) or more than 10% of skull length (1).

(13) Position of choana on palate: anterior to maxillary tooth row (0) or level with maxillary tooth row (1).

(14) Maxillae from opposite sides separated by vomers at anterior border of the internal choanae (0) or maxillae contact each other anterior to choanae (1).

(15) Dentigerous margin of maxilla straight (0) or ventrally convex (1).

(16) Antorbital fossa reduced (0) or large and rounded (1).

(17) Palpebral free, articulating with lacrimal (0) or fused to orbital margin (1).

(18) Orbit diameter more than 20% of skull length (0) or less than 20% skull length (1).

(19) Epijugal ossification absent (0) or present (1).

(20) Epijugal position on jugal: along dorsal edge of horn [epijugal trapezoidal] (0) or capping end of horn [epijugal conical] (1).

(21) Jugal-lacrimal contact reduced (0) or expanded (1).

(22) Jugal horns absent (0), present and laterally directed (1), or present and ventrally directed (2). Ordered.

(23) Orbital horns absent (0) or present (1).

(24) Postorbital inverted and L-shaped (0) or triangular and platelike (1).

(25) Postorbital with dorsal part rounded and overhanging lateral edge of supratemporal fenestra (0) or with concave dorsal shelf bordering supratemporal fenestra (1).

(26) Lower temporal opening with postorbital participation in margin (0), postorbital excluded from margin (1), or jugal-squamosal contact very wide and postorbital situated far from fenestra (2).

(27) Lower temporal fenestra width more than 10% of skull length (0) or less than 10% of skull length (1).

(28) Squamosal subtriangular in lateral view (0) or T-shaped, with postquadratic process (1).

(29) Temporal bars of squamosals parallel (0) or posteriorly divergent (1).

(30) Quadratojugal mediolaterally flattened (0), transversely expanded and triangular in

coronal section (1), or triangular in coronal section, but with slender anterior prong articulating with jugal (2).

(31) Quadrate shaft anteriorly convex in lateral view (0) or straight (1).

(32) Elongate parasagittal process of the palatine absent (0) or present (1).

(33) Ectopterygoid exposed in palatal view (0) or reduced and concealed in palatal view (1).

(34) Ectopterygoid contacts jugal (0) or ectopterygoid reduced and restricted to contact with maxilla (1).

(35) Ventral ridge on mandibular process of pterygoid defining Eustachian canal absent (0) or present (1).

(36) Pterygoid-maxilla contact at posterior end of tooth row absent (0) or present (1).

(37) Parieto-frontal contact flat (0), depressed (1), or invaginated by fontanelle (2). Ordered.

(38) Parieto-squamosal frill absent (0), or parietal frill less than 70% of basal length of skull (1), or more than 70% of basal length (2). Ordered.

(39) Frill solid (0) or fenestrated near posterior margin (1).

(40) Epoccipital ossifications/frill scallops absent (0) or present (1).

(41) Basioccipital participates in foramen magnum (0) or basioccipital is excluded from foramen magnum and exoccipitals form less than one-third of condyle (1) or exoccipitals form about half or more of occipital condyle (2). Ordered.

(42) Basioccipital excluded from basal tubera by basisphenoid and limited to occipital midline (0) or basioccipital tubera (1).

(43) Basipterygoid process orientation anterior (0), ventral (1), or posteroventral (2). Ordered.

(44) Basal tubera flat, in plane with basioccipital plate (0) or everted posterolaterally, forming lip beneath occipital condyle (1).

(45) Notch between posteroventral edge of basisphenoid and base of basipterygoid process deep (0) or notch shallow and base of basipterygoid process close to basioccipital tubera (1).

(46) Exoccipital with three exits for cranial nerves X–XII near occipital condyle (0) or with two exits (1).

(47) Exoccipital-quadrate separated by ventral flange of squamosal (0) or in contact (1).

(48) Paroccipital processes deep (height^3 <1/2> length) (0) or significantly narrower (1)

(49) Supraoccipital participates in dorsal margin of foramen magnum (0) or excluded from foramen magnum by exoccipitals (1).

(50) Supraoccipital anteriorly inclined relative to basioccipital (0) or in same plane as posterior face of basioccipital (1).

(51) Supraoccipital shape tall, triangular (0), wider than tall, trapezoid (1), or square (2).

(52) Predentary less than two-thirds of dentary length (0) or equal to or more than two-thirds of dentary length (1).

(53) Predentary buccal margin sharp (0) or with a rounded, beveled edge (1) or with grooved, triturating edge (2).

(54) Large pit at anterior end of dentary absent (0) or present (1).

(55) Diastema between predentary and first dentary tooth absent (0) or present (1).

(56) Ventral margin of dentary curved (0) or straight (1) in lateral view.

(57) Contact between dentary and prearticular absent (0) or present (1).

(58) Distal end of coronoid process rounded (0) or with anterior expansion (1).

(59) Coronoid process positioned close to main axis of dentary and posterior to tooth row (0), set lateral to tooth row, and end of tooth row covered by anterior part of coronoid process (1), or tooth row level with posterior edge of coronoid process (2). Ordered.

(60) Surangular without distinct lateral ridge or shelf overhanging angular (0) or shelf/ridge present (1).

(61) Angular-surangular-dentary contact triradiate (0) or surangular with long ventral process overlapping angular, and dentary-surangular and angular-surangular sutures form acute angle on lateral face of mandible (1).

(62) Posterior end of splenial simple or with shallow dent (0) or with bifid overlap of angular (1).

(63) Three or more teeth in premaxilla (0), two caniniform teeth in premaxilla (1), or premaxilla edentulous (2).

(64) All teeth with single roots (0), some cheek teeth with double roots (1), or all cheek teeth with double roots (2). Ordered.

(65) Cheek teeth spaced (0) or closely apressed with determinate eruption and replacement pattern (1).

(66) Teeth occlude at an oblique angle (0), at a vertical angle (1), or at a vertical angle but dentary teeth have a horizontal shelf on the labial face (2).

(67) Teeth without median primary ridge (0), only maxillary teeth with primary ridge (1), or both maxillary and dentary teeth with primary ridge (2). Ordered.

(68) Base of primary ridge confluent with the cingulum on maxillary teeth (0) or base of primary ridge set back from cingulum, which forms a continuous ridge at the crown base (1).

(69) Tooth row double, with only one replacement tooth present at a time (0) or batterylike with multiple (3) rows of replacement teeth (1).

(70) Both lingual and buccal sides of teeth covered with enamel (0) or enamel restricted to lateral side of maxillary and medial side of dentary teeth (1).

(71) Dentary tooth crowns with continuous, smooth root-crown transition (0) or bulbous expansion at root-crown transition on labial side of tooth (1).

(72) Number of alveoli in dentary less than 20 (0) or more than 20 (1).

(73) Cheek teeth with cylindrical roots (0) or roots with anterior and posterior grooves along root (1).

(74) Tooth crowns radiate or pennate in lateral view (0), maxillary crowns ovate in lateral view (1), or both maxillary and dentary teeth ovate in lateral view (2).

(75) Atlas intercentrum semicircular (0) or disc-shaped (1).

(76) Atlas intercentrum not fused to odontoid (0) or fused (1).

(77) Atlas neurapophyses free (0) or fused to intercentrum/odontoid (1).

(78) Axial neural spine low (0), tall and hatchet-shaped (1), or elongate and posteriorly inclined (2).

(79) Syncervical absent (0), partially fused [centra but not arches] (1), or completely coossified (2). Ordered.

(80) Dorsal vertebrae with flat articular zygapophyses (0) or tongue and groove articulations on zygapophyses (1).

(81) Number of sacrals: five (0), six (1), seven (2), or eight or more (3). Ordered.

(82) Outline of sacrum defines rectangle or hourglass in dorsal view (0) or oval in dorsal view (1).

(83) Caudal neural spines short and inclined (0) or tall and straight (1).

(84) Distal chevrons with lobate expanded shape (0) or rodlike (1).

(85) Clavicles absent (0) or present and robust (1).

(86) Scapula distinctly curved in sagittal view (0) or relatively flat (1).

(87) Scapular blade at acute angle relative to glenoid (0) or almost perpendicular to glenoid (1).

(88) Olecranon process relatively small (0) or enlarged (one-third of ulnar length) (1).

(89) More than two distal carpals (0) or less than two distal carpals (1).

(90) Manus much smaller than pes (0) or closer to pes in size (1).

(91) Shaft of postpubis round (0) or mediolaterally flattened, bladelike (1) in cross section.

(92) Postpubis long and ventrally oriented (0) or short and posteriorly directed (1).

(93) Prepubis short and rod-shaped (0) or long and flared at anterior end (1).

(94) Ischial shaft straight (0) or curved, posterodorsally convex (1).

(95) Femoral fourth trochanter large and pendant (0) or reduced (1).

(96) Tibio-femoral ratio more than one (0) or less than one (1).

(97) Foot gracile with long, constricted metatarsus, elongate phalanges (0) or short and uncompressed, stubby phalanges (1).

(98) Pedal unguals pointed (0) or moderately rounded, hooflike (1).

Characters were compiled from both the literature and from direct comparison of specimens of all ingroup and outgroup taxa except for *Zuniceratops christopheri*. Main literature sources: Brown and Schlaikjer 1940; Maryanska and Osmólska 1975; Sereno 1984, 1986, 1999; Dodson and Currie 1990; Forster 1990, 1996; and Chinnery and Weishampel 1998.

APPENDIX 18.2.
Data Matrix

Archaeoceratops oshimai

11101 11110 00??0 10010 0101? 100?1 1???? ??1?? 11101 ?0100 101?0 1?011 ??100 010??
?0?10 00?01 000?? ????? ??00? ?10

Asiaceratops salsopaludalis

????? 1?110 ????0 1???? ????? ????? ????? ????? 11?00 ????? ???10 1???0 ??200 0?001
0011/2? ????? ??1?? ?1??? ????? ??0

Bagaceratops rozhdestvenskyi

11111 11110 1?010 10011 01010 10112 11001 10/1?10 11101 10100 11101 11011 11200 02000
0012? ????? ????? ????? ?0??? ???

Chaoyangsaurus youngi

11100 1?0?0 ????0 ??00– ?0??? ?0??0 ????0 0???? 01?0? ????? ?0010 1?000 ??100 00001
00000 0000? ????? ????? ????? ???

Centrosaurus apertus

11111 11001 21110 01111 12110 21111 1?111 12211 21101 21011 20201 1112? 00211 12010
01121 11220 41011 01111 11111 101

Leptoceratops gracilis

11111 11010 00011 10010 01011 00001 100?1 10100 11?10 11101 01110 01010 00200 22101
10120 10010 10110 11000 00000 010

Microceratops gobiensis

11??? ????? ??0?? ??0?? ?1??? ??111 1???? ???10 ?1??? ????? ????? 1?011 ???00 0200?
00??? ????1 ??1?? 1???? ?0??0 010

Montanoceratops cerorhynchus

1???? ?1?1? ??0?0 10?10 ?1011 ??0?1 1???? ?0??? 11210 1??01 0???? 1101? ?0?00 2210?
10120 11111 20110 11?00 00000 010

Protoceratops andrewsi

11111 11110 10010 10011 01010 10112 11001 10/1210 11101 10100 11101 01011 11100 02000 00120
11111 20110 11000 10000 011

Psittacosaurus mongoliensis

11100 11–10 00110 0000– 01000 00000 00100 10000 01000 10100 10000 10010 0?200 00001
00000 00000 0000? 11000 00000 010

Stegoceras validum

000– 00–00 00100 0100– 100– 0100? 0?002 000–0 00000 00000 00?00 1000? ?0000 00001
0000? ?0??1 0???? 000?0 ???01 010

Triceratops horridus

11111 11001 21110 01111 12110 21111 10111 12201 21101 21011 20201 1112? 00211 12010
011?1 11220 41011 01111 11111 101

Udanoceratops tschizhovi

1?111 11–10 0?0?1 ???10 ?1??? ????1 1???? ????? ????? ????? ???1? 0?010 00200 22100
1012? ????0 201?? 110?? ????0 ???

Zuniceratops christopheri

????? ????? ????0 ????? ?11?? ????? ????? ????? ?1??? ????? ????? 1?0?? ???0? 12000
?01?? ????? ??0?? ????? ???1? ???

Hypsilophodon foxii

000– 00–00 00000 1000– 00000 00000 00000 000–0 00000 2?100 00000 11000 00000 00001
00000 00000 0000? 00000 00000 010

APPENDIX 18.3.
List of Unambiguous Synapomorphies

The list includes Ceratopsia and clades within it. The first number refers to the character list (appendix 18.1) and the digit in parentheses refers to the state derived for the given node.

Ceratopsia: 1 (1), 2 (1), 3 (1), 6 (1), 42 (1), 63 (1).

Psittacosaurus mongoliensis + Neoceratopsia: 22 (1), 36 (1), 59 (1).

Neoceratopsia: 5 (1), 16 (1), 19 (1), 24 (1), 30 (1), 31 (1), 38 (1), 41 (1), 43 (1), 53 (1), 67 (1), 74 (1), 80 (1).

Coronosauria + Leptoceratopsidae: 4 (1), 67 (2), 74 (2), 76 (1), 79 (1), 81 (2), 83 (1).

Leptoceratopsidae: 54 (1).

M. cerorhynchus + (*U. tschizhovi* + *L. gracilis*): 44 (1), 66 (2), 68 (1), 71 (1).

(*U. tschizhovi* + *L. gracilis*): 15 (1), 56 (0), 80 (0).

M. gobiensis + Coronosauria: 28 (1), 29 (1), 39 (1).

Coronosauria: 98 (1).

Protoceratopsidae: 30 (2), 32 (1), 61 (1), 62 (1).

Ceratopsoidea: 23 (1), 66 (1), 83 (0), 94 (1).

Ceratopsidae: 58 (1), 64 (1), 69 (1), 72 (1).

19. Speculations on the Socioecology of Ceratopsid Dinosaurs (Ornithischia: Neoceratopsia)

Scott D. Sampson

Abstract

Behavior-morphology correlations in extant vertebrates can be integrated with paleontological and taphonomic data to enhance speculations about the socioecology of ceratopsid dinosaurs. Ceratopsids possess a variety of species-specific cranial specializations, principally horns and frills, that likely functioned as signals both to recognize and to compete for mates. Multiple occurrences of low-diversity mass death assemblages suggest that ceratopsids formed gregarious associations for at least a portion of the year. The co-occurrence of large body sizes and highly derived shearing dentitions indicates that ceratopsids exploited poor-quality fodder, and current evidence of paleohabitat suggests that food resources were patchily distributed. Sexual dimorphism, if present, appears to have been minimal and restricted largely to variations in these same mating signals. These mating signals (e.g., horns and frills) were subject to delayed growth, at least among centrosaurines. Agonistic encounters involved display and combat, and infrequently resulted in injury. These various lines of evidence, when compared with extant analogues, suggest that (at least some) ceratopsid species periodically formed large, nonterritorial, hierarchically structured, mixed-sex aggregations, or "herds." Predator defense may have been a principal factor dictating herd size.

Introduction

Ceratopsids comprise a diverse radiation of Late Cretaceous large-bodied ornithischian herbivores. Although relatively conservative in the postcranium, ceratopsids exhibit a broad array of skull types distinguished largely by differing morphologies of the dermal skull roof—in particular, nasal and supraorbital ornamentations and adorned parietosquamosal frills. Ceratopsid skulls also possess derived dentitions and jaw morphologies adapted for vertical shearing, likely for processing tough, fibrous vegetation. Despite this remarkable degree of novelty, horned dinosaurs exhibit a number of evolutionary trends common to various clades of Cenozoic mammalian herbivores; namely, they constitute a geologically brief evolutionary radiation in which the evolution of large body size is associated with the development of highly derived feeding structures and varied hornlike organs (Geist 1974, 1978, 1987; Sampson 1999).

There has been considerable speculation about the function of ceratopsid horns and frills, as well as the possible social behaviors of these horned dinosaurs. Some authors (e.g., Currie and Dodson 1984; Currie 1989) have postulated the existence of socially complex "herds" among ceratopsids, largely based on the multiple occurrence of low-diversity bone beds. Alternatively, Lehman (1997) proposed that ceratopsids lacked elaborate social structures, suggesting instead that the evidence is more consistent with "infestations" akin to those formed by tortoises and crocodilians. Below, I argue that current evidence is most consistent with the view that (at least some) ceratopsid species formed mixed-sex, nonterritorial, hierarchically structured "herds." This argument is based on a combination of fossil evidence—morphological and taphonomic—together with behavior-morphology correlations derived from studies of extant vertebrate analogues.

Socioecological Correlates in Extant Vertebrates

Numerous studies of living vertebrates demonstrate strong interrelationships among various biological parameters (e.g., Lack 1968; Orians 1969; Schaffer and Reed 1972; Selander 1972; Trivers 1972; Jarman 1974, 1983; Emlen and Oring 1977; Carothers 1984; Demment and Van Soest 1985; Janis 1990). Jarman (1974, 1983) described correlations between a number of ecological variables among ungulates, including feeding style, body size, group size, home range, antipredator behavior, growth pattern, and social organization. Jarman (1974) created five ecological categories that outline a general trend from small-bodied, territorial, monomorphic, monogamous species with highly selective diets, to large-bodied, gregarious, often nonterritorial, dimorphic, and polygynous forms with more generalized diets. He hypothesized a causal chain from dispersion of resources, to dispersion of females, to male mating strategies and social organization. That is, the primary or ultimate factors determining social organization are resource exploitation and the dispersion of females, closely associated with body size and physiological adaptations.

Jarman (1983) argued that sexual dimorphism is integrally related to the pattern of growth, which in turn can be related to ecological strategies. Bimaturism, distinct differences between males and females at sexual maturity (Wiley 1974), is a common theme among vertebrates. Bimaturism is thought to evolve, at least in some cases, because females select mates with tangible survival qualities, which in turn may lead to selection in males for survival signals. Prolongation of body growth and/or an increased rate of growth for males are seen in a variety of mammals, birds, and reptiles, particularly large-bodied forms (Geist 1968, 1978; Jarman 1983). By prolonging body growth and/or growth of weapons, species produce systems of signals enabling older, larger males with fully mature horns and hornlike structures to inhibit the reproductive output of younger males. Adolescent males often do not enter the breeding pool but rather focus on survival until they can successfully compete for mates. Bimaturism is one result of this process, and delayed maturation of males relative to females is necessary for the evolution of extreme polygyny and dimorphism (Jarman 1983). Thus, for example, male African buffalo (*Syncerus*) become sexually mature at 2.5 to 3.0 years of age but are not fully armed until 6.0 or 7.0 years and do not enter the breeding pool until about 8.0 years of age (Estes 1974). Similarly, in bighorn sheep (*Ovis*), adolescence (i.e., the time between sexual maturity and the attainment of adult morphology) begins at end of the second year and lasts from four to six years, with rams entering the rut only after full development of horns (Geist 1968). Bimaturism (but not necessarily neoteny) in relation to dominance hierarchies is also seen in marine mammals (Godsell 1991), birds (Selander 1965; Wiley 1974), and lizards (Iverson 1979; Fitch and Henderson 1977a,b). In all these examples, males, but not females, typically delay breeding activities until several years after sexual maturity.

In numerous vertebrates, the sexes show differential growth and dimorphism (heteromorphy). Sexual dimorphism tends to be least in small-bodied forms, greatest in medium-sized forms, and reduced in large-bodied forms, particularly gregarious taxa inhabiting open environments (Walther 1966; Estes 1974; Geist 1974, 1977, 1978). Among bovids, for example, the sexes of small species (less than 20 kg) look alike whereas the sexes of medium to large species (80–300 kg) often show considerable dimorphism. In species with males weighing over 300 kg, both sexes tend to have horns and there is minimal sexual dimorphism, particularly among gregarious forms (Jarman 1983). Geist (1977) described numerous parallels in social and ecological specializations among gallinaceous birds and mammals. Gallinaceous birds are a useful analogy because, unlike other avian groups and similar to most bovids, they are characterized by minimal male parental care, high degrees of dimorphism, and complex varieties of intrasexual competition. As among ungulates, sexual homomorphism in gallinaceous birds appears in small, resource-defending forms and in highly gregarious species inhabiting open plains. In territorial species, males often resemble one another in appearance although they may be much larger than females.

Jarman (1983) predicted that sexual dimorphism frequently occurs under conditions where some males monopolize matings, where this monopolization is limited to certain age classes due to delayed growth of males, and where dimorphism is limited by the dispersion of females relative to males. Thus sexual dimorphism is greatest in species forming small groups, where large, adult males can restrict access of less mature males to females. Restriction in these cases is usually accomplished through a well-defined territory (under conditions of abundant, continuous resources) or a mobile harem (discontinuous resources).

Males adapt to female dispersion in various ways: monogamy and territoriality, polygamy and territoriality, mobile harems, and mixed-sex gregarious herds, to name a few possibilities. The evolution of territoriality may be directly related to predictability in social relationships. Among gregarious bovids, all-male bachelor herds often form in addition to mixed-sex aggregations, with male dominance hierarchies the norm in both instances. A principal reason for the evolution of rank indicators is that they permit the existence of open societies where conspecifics can join a group without having to engage in dangerous combat. Animals that defend resources are more likely to use damaging weapons, whereas damaging weapons will be reduced or absent among animals that depend on gregariousness as an antipredator strategy (Geist 1978). If productivity is reasonably continuous and abundant in an open plain, territoriality may arise; where resources cannot be defended and predicted, male rank hierarchies in mixed-sex herds are expected (Geist 1978). Dominance hierarchies can occur in the absence of overt rank indicators, but in these cases (e.g., many territorial bovids) they depend on individuals knowing each other's agonistic potential (Geist 1966).

Monospecific aggregations may be seasonal or perennial, and bisexual or monosexual (e.g., bachelor herds). Gregariousness is typical of large herbivores living in open terrain with patchy resources. With larger size comes increased conspicuousness and decreased advantage to systems of dispersal (Estes 1974). Herd structure provides cover, particularly for juveniles that can be positioned in the middle of the group. *Herd* is employed here in the sense of a large, socially cohesive unit organized under a system of dominance hierarchy. The formation of herds allows multiple pairs of eyes to watch for danger, thereby enabling individuals to spend more time feeding and less time scanning for predators (Jarman 1974). Jarman (1974) hypothesized that herding is often a response to predation pressure, while dispersion and availability of food items set the upper limits of herd size. When gregariousness evolves as a strategy against predation, weapons are selected that minimize intraspecific surface damage and retaliation (Geist 1977, 1978). Weapons and strategies that permit contests of strength, such as wrestling, are selected.

An analysis of horns in females by Kiltie (1985) suggests that, among bovid taxa, the percentage of females with well-developed horns increases with gregariousness, although female horns are generally more gracile and more variable. Differential development of horns is the most common source of sexual dimorphism, followed by color. Sexual dimorphism in body size and social organs decreases in highly gregarious ungulates, apparently through female mimicry of males. Reduced sexual dimorphism in gregarious forms may be advantageous to females, who have to compete directly with males for resources and fend off sexual advances from adolescent males (Jarman 1983; Kiltie 1985). Comparable size and weaponry might also be useful in avoiding predation by aiding combat against predators, allowing individuals to compete for a position at the center of the group, and/or confusing predators which might otherwise preferentially seek out females (Treisman 1975; Wilson 1975; Geist 1977; Jarman 1983). It is important to differentiate here between primary homomorphism (usually small, cryptic forms) and secondary homomorphism (large, gregarious, open-living forms) in which female social organs and body size closely approximate those of males.

In summary, neontological studies demonstrate that a number of socioecological factors are frequently correlated with bony morphology. For example, among large-bodied, gregarious herbivores, taxa characterized by territorial systems are often morphologically distinct from those with male rank hierarchy systems. Polygynous territorial systems are typically associated with extreme sexual dimorphism (often in body size as well as mating signals), minimal variation in adult male morphologies, and little or no retardation of growth. In contrast, large-bodied taxa forming nonterritorial, mixed-sex aggregations—from gallinaceous birds to macropodids to cervids and bovids—tend to be characterized by minimal sexual dimorphism (restricted largely to mating signals), delayed growth of mating signals in males, and social organizations structured upon rank hierarchies.

Ceratopsid Socioecology

Horns and Frills as Mating Signals

How do horned dinosaurs fit the patterns described above? All known ceratopsids are large-bodied herbivores (4–8 m long), and all possess outgrowths of the dermal skull roof, namely horns or bosses and adorned frills. For most of the twentieth century, paleontologists largely regarded these bizarre structures as weapons for predator defense (e.g., Colbert 1948). During the past few decades, a dramatic increase in our understanding of the behavior of living horned animals has resulted in a new perspective, with ceratopsid horns and frills now regarded as intraspecific social organs, or mating signals, for display and combat (Davitashvili 1961; Farlow and Dodson 1975; Molnar 1977; Spassov 1979; Sampson 1997). Support for this contention is derived from several lines of evidence, but by far the most convincing is that among extant animals with horns and hornlike organs—from ants and chameleons to cervids and bovids—in virtually all cases, these structures function first and foremost as mating signals (Geist 1966; Sampson 1997).

Gregarious Behavior

Mass death assemblages indicate that ceratopsids were gregarious, at least periodically. A paucispecific bone bed in Dinosaur Provincial Park, Alberta, dominated by remains of *Centrosaurus* (Currie 1981; Currie and Dodson 1984), covers approximately 1500 m² in area and, by conservative estimates, preserves at least 300 individuals (Ryan 1992). Other low-diversity bone beds preserve assemblages of *Chasmosaurus, Anchiceratops, Styracosaurus, Einiosaurus,* and *Pachyrhinosaurus,* some perhaps of equivalent scale to the *Centrosaurus* assemblage above (Lehman 1989; Sampson 1995; Sampson et al. 1997). The numerous occurrences of these low-diversity bone beds, including multiple taxa preserved in varying environments, argue against the rarity of these assemblages and strongly support the notion of gregarious behavior in ceratopsids. Although the argument is somewhat tautological, I would posit further that the occurrence of elaborate, species-specific mating signals subject to delayed growth provides additional evidence of gregarious behavior (see below).

Resource Exploitation and Habitat

As noted above, patterns of resource distribution appear to dictate, at least in part, dispersion of females, male mating strategies, and, ultimately, social organization. Unfortunately, our knowledge of the diet and habitat of ceratopsids (and dinosaurs generally) is still nascent. The highly derived dental battery of ceratopsids appears to be ideally suited to processing (slicing prior to ingestion) large quantities of poor-quality forage—that is, food with high fiber content. A distinct advantage of large body size, characteristic of all members of Ceratopsidae, is the ability to consume a lower-quality diet than small-bodied animals. Postulated food items of horned dinosaurs include palms and cycads (Ostrom 1966), as well as ferns (Coe et al. 1987). Like large mammalian herbivores today, ceratopsids and other large-bodied ornithischians likely possessed low mass-specific metabolic rates and may well have employed a gut microflora to aid in the fermentation and digestion of poor-quality fodder (Farlow 1987).

With regard to the distribution of such resources, the paleoflora associated with ceratopsids is perhaps best understood for the late Campanian Two Medicine and Judith River Formations (Lorenz 1981; Gavin 1986; Crabtree 1987; Horner 1989; Rogers 1990; Koppelhus 1997). Geological, palynological, and paleontological data suggest that the lowlands adjacent to the seaway, represented by the Judith River Formation, were temperate to subtropical, with large meandering rivers and abundant vegetation, particularly along the water courses. The western uplands, represented by the Two Medicine Formation, formed a well-drained, seasonal, semiarid environment with limited flora, subject to lengthy dry seasons and periodic drought (Horner 1989). In general, evidence from the western interior of North America during this period suggests a coastal plain environment with discontinuous, seasonally variable plant growth (but see Lehman 1997).

Ceratopsids inhabited a broad array of environments, ranging from as far north as present day Alaska and at least as far south as Texas. Lehman (1997; chap. 22, this volume) makes a strong argument for latitudinal zonation of terrestrial faunas in western North America. It may be that ceratopsids were only periodically gregarious, aggregating during the dry season and dispersing in the rainy season, as occurs today in many African herding animals (Western 1975; Rogers 1990). Several studies suggest that some ceratopsid taxa were relatively more abundant in coastal environments than in inland environments (Lehman 1987; Brinkman 1990; Brinkman et al. 1998; but see Lehman 1997). Interestingly, however, despite abundant micro- and macrovertebrate evidence in support of the coastal habitat hypothesis, Brinkman et al. (1998) found that paucispecific bone beds in the late Campanian sediments of Dinosaur Provincial Park, Alberta, are encountered more frequently in the lower part of the sequence, representative of more inland deposition. They interpret this anomalous finding as possible evidence of seasonal movements in response to climatic stress. That is, small groups of ceratopsids occupied coastal environments for part of the year, followed by the formation of large aggregations, which moved inland, perhaps in association with nesting.

Sexual Dimorphism

If the above described behavior-morphology correlations observed among extant vertebrates apply, ceratopsids should be characterized by minimal sexual dimorphism and the presence of horns in females, due to the correlation between large body size and gregariousness. Following the mammalian example, dimorphism, if present, is most likely to have been concentrated in the mating signals, namely horns and frills. Current evidence is consistent with these predictions, with ceratopsid males and females achieving similar body sizes and both sexes possessing hornlike organs (Sampson et al. 1997). There are no ceratopsid taxa known to have lacked these mating signals (horns or bosses) and, although sample sizes remain small, dimorphism in adult body size has yet to be been demonstrated. While occasional skulls do exhibit adult development of cranial characters along with somewhat smaller sizes than putative conspecific adults (e.g., Parks 1921), these differences may be the result of individual, temporal, and geographic variation. Bone beds are perhaps the best source for estimating dimorphism, given that they frequently appear to be dominated by single species, and perhaps even reflect local populational variation. Thus far, there is no convincing evidence among ceratopsids of sexual dimorphism in body size, and even that relating to mating signals is dubious (Sampson et al. 1997; Lehman 1990, 1998).

Kurzanov (1972) and Dodson (1976) employed morphometric analyses to skulls of the basal neoceratopsian *Protoceratops andrewsi,* concluding that sexual dimorphism was an important source of intraspecific variation. Significantly, the most reliable indicators of sex were the nasal prominence and the parietosquamosal frill. It would be interesting to determine if *Protoceratops* and other basal neoceratopsians

Figure 19.1. Reconstruction of *two* Einiosaurus *males battling for position in the dominance hierarchy (late Campanian, Montana). Original artwork by Michael Skrepnick.*

show greater sexual dimorphism than the larger, more derived ceratopsids. If so, this pattern would parallel that described above for other tetrapods, with dimorphism greatest in mid-size taxa and reduced in the largest forms. Of course, it must be remembered that dimorphism is commonly associated with nonosseous tissues preserved rarely if at all in the fossil record (e.g., keratinous horn sheaths, color, dewlaps, inflatable sacs). For example, keratinous sheaths covering the horns of bovids often possess complex textured surfaces and extra curves or twists not indicated by the horncore. Similarly, bright and/or contrasting colors, often present in gregarious open-country birds and mammals (Geist 1977), may have been used to highlight mating signals and increase their apparent size.

Retarded Growth of Mating Signals

With regard to growth patterns, Sampson et al. (1997) found that craniofacial mating signals (horns and frills), at least among centrosaurines, were subject to retarded growth, with adult morphologies developing close to the onset of adult size. Relative age was determined in this study using several indicators including relative size, degree of coossification of elements, presence of secondary ossifications, and ontogenetic variation in bone texture. Following the example observed in extant ungulates, delayed growth is interpreted here as evidence of heteromorphy, with the degree of mating signal development providing a clear indication of relative age. The retarded growth of mating signals is suggestive of an extended period of adolescence in males associated with a rank hierarchy structured on the basis of age-related differences (Sampson et al. 1997).

Social Organization: Herds or Infestations?

Given that direct preservation of behavior in extinct organisms is rare and relatively limited (e.g., trackways, nesting sites), it is difficult to envision what type of data would suffice as conclusive evidence of social organization in ceratopsids. One could envision multiple fossil localities preserving entire groups killed in an instantaneous event (e.g., volcanic eruption), with fully articulated and nontransported skeletons. Yet even in this best-case scenario, it would not be possible to determine with certainty that these animals did any more than die together. Thus we must rely on indirect indicators, preferably multiple lines of evidence that point to the same conclusion.

Lehman (1997) contrasted the dinosaur fauna from the late Campanian of the western interior of North America with extant megaherbivores in East Africa. He argued that ceratopsids exhibited much greater diversity and apparently higher population densities than occur today among extant large herbivores. On the basis of these putatively high population densities, Lehman postulated that ceratopsids likely did not possess elaborate social structures, or "herds," as some previous authors have posited. Rather, he argues, the evidence for gregariousness is better interpreted as periodic "infestations," with little or no complex social organization, as is said to occur today among crocodilians and tortoises.

Do the many paucispecific bone beds of ceratopsids preserve evidence of socially complex herds, or "infestations" akin to those observed in turtles (Lehman 1997)? In contrast to Lehman's argument, crocodilians, which form part of the archosaurian extant phylogenetic bracket for dinosaurs, do exhibit elaborate social structures. Although not referred to as herds, crocodilian aggregations are often organized around social hierarchies (e.g., references in Webb et al. 1987). Moreover, putting aside the pitfalls of calculating population densities for extinct taxa, Lehman's argument may well be flawed with regard to the extant taxa. Lehman derives his estimates of population densities among extant ungulates from the largest African forms (e.g., elephants, hippopotami). However, a number of other herd-forming ungulates (e.g., Aepycerotinae and members of Alcelaphinae) exhibit extremely high densities, at least periodically, that appear to meet or exceed those of ceratopsid and hadrosaurian dinosaurs.

I would argue further that the presence in ceratopsids of well-developed secondary sexual characters, combined with delayed maturation of these characters in at least some forms, provides strong evidence in favor of elaborate social organizations based on rank hierarchies. In my opinion, therefore, the current evidence is most consistent with herding in both ceratopsids and hadrosaurs. This is not to imply that all, or even most, ceratopsids shared a common social structure. Modes of social organization are clearly variable, both within and between species. Horned dinosaurs are a diverse group that inhabited a broad range of environments, and certainly would have varied in forms of social organization. My goal here has been to extrapolate possibilities and probabilities from the evidence at hand and discuss behavior-morphology correlations that might elucidate the nature of social structures. In short, despite superficial resemblances and frequent comparisons, rhinoceroses and ceratopsians likely represent distinct ecological strategies: the former relatively nongregarious, territorial, the latter gregarious, perhaps nonterritorial and with well-defined subadult stages and dominance hierarchies.

Predation Pressure

Antipredator strategies among ungulates are closely linked to body size, with larger-bodied species more likely to stand and face predators rather than attempt to flee (Eisenberg and Lockhart 1972; Geist 1974). Thus, as in ungulates, gregariousness may have evolved in horned dinosaurs as a direct response to predation pressure in an open environment. The large-bodied tyrannosaurs were the most likely predators, including such forms as *Daspletosaurus* (late Campanian) and *Tyrannosaurus* (late Maastrichtian). If indeed ceratopsid habitats were largely open rather than heavily vegetated, it is likely that concealment was typically not a viable strategy for avoiding predation. Moreover, functional locomotor studies indicate that tyrannosaurs are best regarded as relatively fleet-footed cursors whereas ceratopsians, like other large-bodied ornithischians, were graviportal (Coombs 1978; Carrano 1999). So, in strong contrast to many extant, open-living ungulates, it is equally unlikely that ceratopsids were able to rely on their locomotor abilities to escape predators. In addition to the sentry function afforded by the formation of mixed-sex groupings, gregariousness may have been an important predation deterrent, with larger adult animals protecting immature individuals.

Conclusion

Several lines of evidence—morphological, taphonomic, and paleoecological—suggest that the socioecology of (at least some) ceratopsid dinosaurs can be likened to that of large-bodied, open-living mammalian ungulates. Horns and frills are best regarded as mating signals used to attract and compete for mates. The mammalian analogue further indicates that ceratopsids engaged in relatively safe modes of intraspecific combat with the adorned skull used to catch or parry the attack of an opponent. As expected on the basis of extant analogues, sexual dimorphism appears to have been minimal in ceratopsids and, where present, concentrated in mating signals on the dermal skull roof. The presence of diverse cranial ornamentations, subject to delayed growth in some forms, is suggestive of a hierarchically structured form of social organization. These social organs, considered in conjunction with the abundance of low-diversity mass death assemblages, strongly support the notion that ceratopsids were gregarious, at least for a portion of the year, traveling in large, rank-structured, mixed-sex herds as a means of predator defense. Interestingly, Carrano et al. (1999) independently came to similar conclusions with regard to hadrosaurid dinosaurs. They hypothesized close ecomorphological parallels with ungulates, and postulated that at least the hadrosaurines were monomorphic, gregarious, and lived in open habitats.

Although one might criticize that several elements of the argument presented here are untestable, a number of predictions and expectations can be stated, and potentially falsified by additional fossil discoveries. For example: (1) Ceratopsids exhibit minimal sexual dimorphism in overall body size; dimorphism present will be concentrated in the horncore and frill morphologies, with females typically possessing more gracile mating signals. (2) Adult-size skulls with subadult features are males, due to the delayed development of horns and frills (Sampson et al. 1997); relative age determination should show these individuals to be older than some of the supposed female skulls showing fully developed horn and frill ornamentations. (3) Paleoecological evidence will show that ceratopsids inhabited an open terrain with discontinuous resource distribution. This prediction is consistent with gregariousness in large, mixed-sex herds.

Acknowledgments: I am very pleased to contribute to this volume in honor of Philip Currie, who has been a major inspiration for so many dinosaur enthusiasts, professionals and nonprofessionals alike. From a personal standpoint, I will be ever grateful to Phil; early in my graduate career, he applied an artful combination of reason, enthusiasm, and homemade libations to cement my resolve in pursuing dinosaur research, and he has been a stalwart supporter ever since. I sincerely thank Michael Skrepnick for providing the original artwork for figure 19.1.

References

Brinkman, D. B. 1990. Paleoecology of the Judith River Formation (Campanian) of Dinosaur Provincial Park, Alberta, Canada: Evidence from vertebrate microfossil localities. *Palaeogeography, Palaeoclimatology, Palaeoecology* 78: 37–54.

Brinkman, D. B., M. J. Ryan, and D. Eberth. 1998. The paleogeographic and stratigraphic distribution of ceratopsids (Ornithischia) in the Upper Judith River Group of western Canada. *Palaios* 13: 160–169.

Carothers, J. H. 1984. Sexual selection and sexual dimorphism in some herbivorous lizards. *American Naturalist* 124: 244–254.

Carrano, M. T. 1999. What, if anything, is a cursor? Categories versus continua for determining locomotor habit in mammals and dinosaurs. *Journal of Zoology* (London) 247: 29–42.

Carrano, M. T., C. M. Janis, and J. J. Sepkoski Jr. 1999. Hadrosaurs as ungulate parallels: Lost lifestyles and deficient data. *Acta palaeontologica polonica* 44 (3): 237–261.

Coe, M. J., D. Dilcher, J. O. Farlow, D. M. Jarzen, and D. A. Russell. 1987. Dinosaurs and land plants. In E. M. Friis, W. G. Chaloner, and P. R. Crane (eds.), *The Origins of Angiosperms and their Biological Consequences,* pp. 225–258. New York: Cambridge University Press.

Colbert, E. H. 1948. Evolution of the horned dinosaurs. *Evolution* 2: 145–163.

Coombs, W. P., Jr. 1978. Theoretical aspects of cursorial adaptations in dinosaurs. *Quarterly Review of Biology* 53: 393–418.

Crabtree, D. 1987. The Early Campanian flora of the Two Medicine

Formation, north central Montana. Ph.D. dissertation, University of Montana, Missoula.

Currie, P. J. 1981. Hunting dinosaurs in Alberta's huge bonebed. *Canadian Geographic Journal* 101 (4): 32–39.

Currie, P. J. 1989. Long-distance dinosaurs. *Natural History* 6 (89): 60–65.

Currie, P. J., and P. Dodson. 1984. Mass death of a herd of ceratopsian dinosaurs. In W.-E. Reif and F. Westphal (eds.), *Third Symposium on Mesozoic Terrestrial Ecosystems, Short Papers,* pp. 61–66. Tübingen: Attempto Verlag.

Davitashvili, L. S. 1961. *Teoriya polovogo otbora* [The theory of sexual selection]. Moscow: Izdatel'stvo Akademii Nauk [Academy of Sciences Press].

Demment, M. W., and P. J. Van Soest. 1985. A nutritional explanation for body-size patterns of ruminant and nonruminant herbivores. *American Naturalist* 125 (5): 641–672.

Dodson, P. 1976. Quantitative aspects of relative growth and sexual dimorphism in *Protoceratops. Journal of Paleontology* 50: 929–940.

Eisenberg, J. F., and M. Lockhart. 1972. An ecological reconnaissance of Wilpattu National Park, Ceylon. *Smithsonian Contributions in Zoology* 101: 1–118.

Emlen, S. T., and L. W. Oring. 1977. Ecology, sexual selection, and the evolution of mating systems. *Science* 197: 215–223.

Estes, R. D. 1974. Social organization of the African Bovidae. In V. Geist and F. Walther (eds.), *The Behaviour of Ungulates and Its Relation to Management. IUCN,* n.s., 24: 166–205.

Farlow, J. O. 1987. Speculations about the diet and digestive physiology of herbivorous dinosaurs. *Paleobiology* 13: 60–72.

Farlow, J. O., and P. Dodson. 1975. The behavioral significance of frill and horn morphology in ceratopsian dinosaurs. *Evolution* 29: 353–361.

Fitch, H. S., and R. W. Henderson. 1977a. Age and sex differences in the ctenosaur (*Ctenosaura similis*). *Milwaukee Public Museum Contributions to Biology and Geology* 11: 1–11.

Fitch, H. S., and R. W. Henderson 1977b. Age and sex differences, reproduction and conservation of *Iguana iguana. Milwaukee Public Museum Contributions to Biology and Geology* 13: 1–21.

Gavin, W. M. B. 1986. A paleoenvironmental reconstruction of the Cretaceous Willow Creek anticline dinosaur nesting locality, north central Montana. Master's thesis, Montana State University, Bozeman.

Geist, V. 1966. The evolution of horn-like organs. *Behaviour* 27: 173–214.

Geist, V. 1968. On delayed social and physical maturation in mountain sheep. *Canadian Journal of Zoology* 46: 899–904.

Geist, V. 1974. On the relationship of social evolution and ecology in ungulates. *American Zoologist* 14: 205–220.

Geist, V. 1977. A comparison of social adaptations in relation to ecology in gallinaceous bird and ungulate societies. *Annual Review of Ecology and Systematics* 8: 193–207.

Geist, V. 1978. On weapons, combat and ecology. In L. Krames, P. Pliner, and T. Alloway (eds.), *Aggression, Dominance, and Individual Spacing,* pp. 1–30. New York: Plenum Press.

Geist, V. 1987. On speciation in Ice Age mammals, with special reference to cervids and caprids. *Canadian Journal of Zoology* 65: 1067–1084.

Godsell, J. 1991. The relative influence of age and weight on the reproductive behaviour of male grey seals *Halichoerus grypus. Journal of Zoology* (London) 224: 537–551.

Horner, J. R. 1989. The Mesozoic terrestrial ecosystems of Montana. In J. D. McBane and P. B. Garrison (eds.), *Montana Geological Society 1989 Field Guidebook,* pp. 153–162. Billings: Montana Geological Society.

Iverson, J. B. 1979. Behavior and ecology of the rock lizard *Cyclura carinata. Bulletin of the Florida State Museum of Biological Sciences* 24: 175–358.

Janis, C. M. 1990. Correlation of reproductive and digestive strategies in the evolution of cranial appendages. In G. A. Bubenik and A. B. Bubenik (eds.), *Horns, Pronghorns, and Antlers,* pp. 114–133. New York: Springer Verlag.

Jarman, P. 1974. The social organisation of antelope in relation to their ecology. *Behaviour* 48 (3–4): 215–267.

Jarman, P. 1983. Mating systems and sexual dimorphism in large, terrestrial, mammalian herbivores. *Biological Reviews* 58: 485–520.

Kiltie, R. A. 1985. Evolution and function of horns and hornlike organs in female ungulates. *Biological Journal of the Linnean Society* 24: 299–320.

Koppelhus, E. B. 1997. Palynological data reveal clues to understand the paleoenvironments of the monogeneric *Centrosaurus* bonebeds in Dinosaur Park Formation, Alberta, Canada. *Journal of Vertebrate Paleontology* (abstract) 17 (suppl. to 3): 58A.

Kurzanov, S. M. 1972. Sexual dimorphism in protoceratopsians. *Palaeontologie zhurnal* 6: 91–97.

Lack, D. 1968. *Ecological Adaptations for Breeding in Birds.* London: Methuen Press.

Lehman, T. M. 1987. Late Maastrichtian paleoenvironments and dinosaur biogeography in the western interior of North America. *Palaeogeography, Palaeoclimatology, Palaeoecology* 60: 189–217.

Lehman, T. M. 1989. *Chasmosaurus mariscalensis,* sp. nov., a new ceratopsian dinosaur from Texas. *Journal of Vertebrate Paleontology* 9 (2): 137–162.

Lehman, T. M. 1990. The ceratopsian subfamily Chasmosaurinae: Sexual dimorphism and systematics. In K. Carpenter and P. J. Currie (eds.), *Dinosaur Systematics: Approaches and Perspectives,* pp. 211–229. New York: Cambridge University Press.

Lehman, T. M. 1997. Late Campanian dinosaur biogeography in the western interior of North America. In D. L. Wolberg, E. Stump, and G. D. Rosenberg (eds.), *Dinofest International,* pp. 223–240. Philadelphia: Academy of Natural Sciences.

Lehman, T. M. 1998. A gigantic skull and skeleton of the horned dinosaur *Pentaceratops sternbergii* from New Mexico. *Journal of Paleontology* 72 (5): 894–906.

Lorenz, J. C. 1981. Sedimentary and tectonic history of the Two Medicine Formation, Late Cretaceous (Campanian), northwestern Montana. Ph.D. dissertation, Princeton University.

Molnar, R. E. 1977. Analogies in the evolution of combat and display structures in ornithopods and ungulates. *Evolutionary Theory* 3: 165–190.

Orians, G. H. 1969. On the evolution of mating systems in birds and mammals. *American Naturalist* 103: 589–603.

Ostrom, J. H. 1966. Functional morphology and evolution of the ceratopsian dinosaurs. *Evolution* 20: 290–308.

Parks, W. A. 1921. The head and fore limb of a specimen of *Centrosaurus apertus. Transactions of the Royal Society of Canada,* section series, 3 (15): 53–63.

Rogers, R. R. 1990. Taphonomy of three dinosaur bone beds in the Upper Cretaceous Two Medicine Formation of northwestern Montana: Evidence for drought-related mortality. *Palaios* 5: 394–413.

Ryan, M., 1992. The taphonomy of a *Centrosaurus* (Reptilia: Ornithischia) bone bed (Campanian), Dinosaur Provincial Park, Alberta, Canada. Masters thesis, Calgary: University of Calgary.

Sampson, S. D. 1995. Two new horned dinosaurs from the Upper Cretaceous Two Medicine Formation of Montana; with a phylogenetic analysis of the Centrosaurinae (Ornithischia: Ceratopsidae). *Journal of Vertebrate Paleontology* 15 (4): 743–760.

Sampson, S. D. 1997. Bizarre structures and dinosaur evolution. In D. L. Wolberg, K. Gittis, S. Miller, L. Carey, and A. Raynor (eds.), *Dinofest International Proceedings,* pp. 39–45. Philadelphia: Academy of Natural Sciences.

Sampson, S. D. 1999. Sex and destiny: The role of mating signals in speciation and macroevolution. *Historical Biology* 13 (2–3): 173–197.

Sampson S. D., M. J. Ryan, and D. H. Tanke. 1997. Craniofacial ontogeny in centrosaurine dinosaurs (Ornithischia: Ceratopsidae): Taxonomic and behavioral implications. *Zoological Journal of the Linnean Society* 121: 293–337.

Schaffer, W., and C. A. Reed. 1972. The co-evolution of social behavior and cranial morphology in sheep and goats (Bovidae, Caprini). *Fieldiana Zoology* 61: 1–88.

Selander, R. K. 1965. On mating systems and sexual selection. *American Naturalist* 99: 129–141.

Selander, R. K. 1972. Sexual selection and dimorphism in birds. In B. G. Campbell (ed.), *Sexual Selection and the Descent of Man, 1871–1971,* pp. 180–230. Chicago: Aldine Press.

Spassov, N. B. 1979. Sexual selection and the evolution of horn-like structures of ceratopsian dinosaurs. *Paleontology, Stratigraphy, and Lithology* 11: 37–48.

Treisman, M. 1975. Predation and the evolution of gregariousness. Part 1, Models for concealment and evasion. *Animal Behavior* 23: 779–800.

Trivers, R. L. 1972. Parental investment and sexual selection. In B. G. Campbell (ed.), *Sexual Selection and the Descent of Man, 1871–1971,* pp. 136–179. Chicago: Aldine Press.

Walther, F. R. 1966. *Mit Horn und Huf.* Berlin: Paul Parey Verlag.

Webb, G. J. W., S. C. Manolis, and P. J. Whitehead. 1987. *Wildlife Management: Crocodiles and Alligators.* Chipping Norton, New South Wales: Surrey Beatty and Sons.

Western, D. 1975. Water availability and its influence on the structure and dynamics of a savannah large mammal community. *East African Wildlife Journal* 13: 265–286.

Wiley, R. H. 1974. Evolution of social organisation and life history patterns among grouse. *Quarterly Review of Biology* 49: 201–227.

Wilson, E. O. 1975. *Sociobiology: The New Synthesis.* Cambridge, Mass.: Harvard University Press.

Section IV.
Dinosaurian Faunas

20. Dinosaurs of Alberta (exclusive of Aves)

MICHAEL J. RYAN AND
ANTHONY P. RUSSELL

Abstract

Alberta has a rich and diverse collection of Late Cretaceous dinosaurs. Collection of specimens over the last 100 years has been done under an evolving understanding of the provincial stratigraphy leading to a number of different terms being used to describe the same rock units. This chapter attempts to list the dinosaur fauna known from each formation within Alberta using the most recent and accepted definition for each formation or higher group. The faunal list produced now gives us a clearer understanding of the stratigraphical distribution of the Albertan dinosaurs.

Introduction

Alberta has one of the richest and most diverse dinosaur assemblages from the Late Cretaceous in the world (see summary, table 20.1). Taxa are known from complete and partial skeletons and skulls, isolated bones, numerous bone beds, and microvertebrate assemblages. This material includes at least 38 type specimens from valid dinosaur taxa. Dinosaur taxa best known from multiple specimens come from the Oldman and Dinosaur Park formations of Dinosaur Provincial Park and area (upper Campanian), the Horseshoe Canyon Formation of the Drumheller Valley and Edmonton region (Maastrichtian), and the Scollard Formation (upper Maastrichtian), which outcrops intermittently in the central and southern portions of the province. Other formations, such as the Milk River (Deadhorse Coulee Member—

TABLE 20.1

Summary List of Alberta Dinosaurs

Milk River Formation

Hadrosauridae indet.
Pachycephalosauridae indet.
Ankylosauridae indet.
"protoceratopsid"
Ceratopsidae indet.
Tyrannosauridae indet.
Ornithomimidae indet.
cf. *Saurornitholestes langstoni*
Ricardoestesia gilmorei
Ricardoestesia n. sp.
Paronychodon lacustris

Foremost Formation

Hypsilophodontidae indet.
hadrosauridae indet.
Stegoceras sp.
Ankylosauridae indet.
Nodosauridae indet.
Ceratopsidae indet.
Tyrannosauridae indet.
Aublysodon sp.
Dromaeosaurus sp.
Saurornitholestes sp.
Ricardoestesia sp.
Paronychodon sp.

Oldman Formation

Hypsilophodontidae indet.
Brachylophosaurus canadensis
Hypacrosaurus stebingeri
Pachycephalosauridae indet.
Ankylosauridae indet.
Nodosauridae indet.
undescribed centrosaurine
Anchiceratops sp.
Tyrannosauridae indet.
Daspletosaurus n. sp.
Troodon sp.
Dromaeosaurus sp.
Saurornitholestes langstoni
Ricardoestesia sp.
Paronychodon sp.

Dinosaur Park Formation

Brachylophosaurus canadensis
Gryposaurus notabilis
"*Kritosaurus*" *incurvimanus*
Prosaurolophus maximus
Corythosaurus casuarius

Lambeosaurus lambei
L. magnicristatus
Lambeosaurus. n. sp.
Parasaurolophus walkeri
Thescelosaurus cf. *neglectus*
Stegoceras validum
Ornatotholus browni
Pachycephalosaurus
 wyomingensis
undescribed full-domed
 pachycepalosaurid?
Euoplocephalus tutus
Edmontonia rugosidens
Panoplosaurus mirus
cf. *Leptoceratops* sp.
Centrosaurus apertus
Styracosaurus albertensis
Chasmosaurus belli
C. russelli
undescribed new chasmosaurine
Dromiceiomimus samueli
Ornithomimus edmontonensis
Struthiomimus altus
Albertosaurus libratus
Aublysodon mirandus
Daspletosaurus torosus
indet. gracile tyrannosaurid
Chirostenotes pergracilis
Chirostenotes elegans
cf. *Erlikosaurus* sp.
Troodon formosus
Dromaeosaurus albertensis
Saurornitholestes langstoni
Ricardoestesia gilmorei
Ricardoestesia n. sp.
Paronychodon lacustris
Paronychodon-like

Bearpaw Formation

Hadrosauridae indet.
Brachylophosaurus sp.
Prosaurolophus sp.
Edmontonia sp.
Stegoceras sp.
Ceratopsidae indet.
Ornithomimidae indet.

Horseshoe Canyon Formation

Edmontosaurus regalis
Saurolophus osborni

TABLE 20.1 (cont.)

Hypacrosaurus altispinus
Parksosaurus warrenae
Thescelosaurus neglectus
Stegoceras edmontonense
Edmontonia longiceps
Pachyrhinosaurus canadensis
Anchiceratops ornatus
Arrhinoceratops brachyops
Dromiceiomimus brevetertius
Ornithomimus edmontonensis
Struthiomimus altus
Albertosaurus sarcophagus
Aublysodon mirandus
Daspletosaurus undescribed sp.
indet. gracile tyrannosaurid
Chirostenotes pergracilis
Troodon formosus
Dromaeosaurus albertensis
Saurornitholestes sp.
new velociraptorine
Ricardoestesia gilmorei
Ricardoestesia n. sp.
Paronychodon lacustris
Paronychodon-like

Scollard Formation

Hadrosauridae indet.
Thescelosaurus neglectus
Parksosaurus warrenae
Ankylosaurus magniventris
Leptoceratops gracilis
Triceratops horridus
indet. ceratopsian?
Ornithomimidae indet.
undescribed large ornithomimid
Tyrannosaurus rex

?Caenagnathidae n. sp.
cf. *Troodon formosus*
Dromaeosauridae indet.
Dromaeosaurus albertensis
cf. *Saurornitholestes langstoni*
cf. *Avimimus* sp.
Ricardoestesia gilmorei
Ricardoestesia n. sp.
Paronychodon-like

Belly River Group

Hadrosauridae indet.
Ornithomimidae indet.

St. Mary River Formation

Hadrosauridae indet.
Edmontosaurus sp.
Pachyrhinosaurus canadensis
Ornithomimidae indet.
Albertosaurus sp.
Saurornithoides-like
Troodon sp.

Willow Creek Formation

Hadrosauridae indet.
Montanoceratops sp.
Tyrannosaurus rex

Wapiti Formation

Hadrosauridae indet.
?Ankylosauridae
Pachyrhinosaurus n. sp.
Ornithomimidae indet.
Albertosaurus sp.
Saurornitholestes n. sp.
Troodon sp.

lowermost Campanian) and Foremost (upper Campanian), have fewer recorded taxa. This scarcity is primarily due to two compounding factors: these formations have fewer fossils relative to other formations, and little time is spent prospecting these formations due to their limited surface exposure or difficulty in reaching exposure. Similarly, few fossils have been collected from the Belly River Group (sensu Jerzykiewicz and Norris 1994) (Campanian) and the St. Mary River Formation (Maastrichtian) of southwestern Alberta, and the Wapiti Formation (Maastrichtian) of northwestern Alberta (with the exception of two extensive *Pachyrhinosaurus* bone beds in the latter formation).

The most intensely examined strata, and the source of the bulk of the dinosaur specimens collected in Alberta, are the Judith River Group of Dinosaur Provincial Park. In ascending order, the Judith River Group is composed of the Foremost, Oldman, and Dinosaur Park formations. Historically, the Judith River Group has a complex nomenclatural history (Eberth and Hamblin 1993) due, in part, to the artificial division of the Upper Cretaceous strata of the plains area of the North America by the Alberta-Montana border. In Alberta, the beds equivalent to the Judith River Formation (sensu Hayden 1871) in the type area near the Missouri River in Montana were divided into the Foremost and Pale Beds by Dowling (1915). The Pale Beds were later formally named the Oldman Formation by Russell and Landes (1940), but were referred to the Judith River Formation by McLean (1971). Subsequently, the subdivision between the Oldman and Foremost formations were accepted and carried further by Eberth and Hamblin (1993), who divided the Oldman Formation into the Oldman and Dinosaur Park formations.

With the recognition and description of these latter two formations came the realization that the long history of fossil collection from these two formations had left a confusing stratigraphic record for Albertan dinosaurs from the late Campanian. Recent work by Dr. Philip Currie, and others, has relocated many of the significant dinosaur quarries from Dinosaur Provincial Park in an attempt to, in part, determine which dinosaurs are found in which formation. We now recognize that not all dinosaurs in the Judith River Group are found in each of its formations, and that even within formations there are distributional differences—for example, the centrosaurine ceratopsids *Centrosaurus* and *Styracosaurus* appear to be stratigraphically isolated within the Dinosaur Park Formation. This work has progressed to the point where it is now possible to clarify the occurrences of dinosaurs from the Late Cretaceous of Alberta (app. 20.1).

Previous lists of Albertan dinosaur taxa include Béland and Russell 1978, Russell 1984, and various field guides produced by the Royal Tyrrell Museum of Palaeontology (e.g., Braman et al. 1995). Whenever possible, one reference specimen (the holotype where appropriate) has been listed for each taxon. Exact locality data can be obtained from the catalogue information on file with each reference specimen. Problematic specimens and taxa are listed separately. Occurrences of dinosaur eggshell and footprints are not listed, but will be dealt with at a later date. Taxonomic designations are made at the most inclusive level given in the literature or for the catalogued material.

Records for the dinosaur fauna of the Milk River (Baszio 1997a,b) and Foremost formations (Peng 1997) come primarily from microvertebrate fossils site studies, and, as such, consist primarily of teeth. The Bearpaw Formation represents a Late Cretaceous transgression of the Western Interior Seaway. Dinosaur specimens collected from this formation probably represent material washed to sea or reworked from older sediments. Records for these and other taxa are taken from archival collection data and field notes from the repository or collecting institutions.

Institutional Abbreviations: AMNH, American Museum of Natural History; BMNH, British Museum of Natural History; CMN, Canadian Museum of Nature (formerly the National Museum of Canada, NMC, and incorporating specimens from the Geological Survey of Canada, GSC); ROM, Royal Ontario Museum; TMP, Royal Tyrrell Museum of Palaeontology; UA, University of Alberta.

Other Abbreviations: Hg&s, holotype of genus and species; Hs, holotype of species; Lg&s, lectotype of genus and species.

References

Bakker, R. T., M. Williams, and P. J. Currie. 1988. *Nanotyrannus,* a new genus of pygmy tyrannosaur, from the latest Cretaceous of Montana. *Hunteria* 1: 1–30.

Baszio, S., 1997a. Palaeo-ecology of dinosaur assemblages throughout the Late Cretaceous of south Alberta, Canada. *Courier Forschungsinstitut Senckenberg* 196: 1–31.

Baszio, S., 1997b. Systematic palaeontology of isolated dinosaur teeth from the Latest Cretaceous of South Alberta, Canada. *Courier Forschungsinstitut Senckenberg* 196: 33–77.

Béland, B., and D. A. Russell. 1978. Paleoecology of Dinosaur Provincial Park (Cretaceous), Alberta, interpreted from the distribution of articulated vertebrate remains. *Canadian Journal of Earth Sciences* 15: 1012–1024.

Braman, D. R., P. A. Johnston, and W. M. Haglund. 1995. Upper Cretaceous paleontology, stratigraphy, and depositional environments at Dinosaur Provincial Park and Drumheller, Alberta. *Canadian Paleontology Conference Field Trip Guidebook No. 4, Fifth Canadian Paleontology Conference, Drumheller, Alberta, 29 September—2 October, 1995.*

Brown, B. 1908. The Ankylosauridae, a new family of armoured dinosaurs from the Upper Cretaceous. *Bulletin of the Museum of Natural History* 24: 187–201.

Brown, B. 1910. The Cretaceous Ojo Alamo beds of New Mexico with description of the new dinosaur *Kritosaurus. Bulletin of the American Museum of Natural History* 28: 267–274.

Brown, B. 1912. A crested dinosaur from the Edmonton Cretaceous. *Bulletin of the American Museum of Natural History* 31: 131–136.

Brown, B. 1913. A trachodont dinosaur, *Hypacrosaurus,* from the Edmonton Cretaceous of Alberta. *Bulletin of the American Museum of Natural History* 32: 395–406.

Brown, B. 1914a. *Anchiceratops,* a new genus of horned dinosaur from the Edmonton Cretaceous of Alberta. With discussion of the origin of the ceratopsian crest and the brain casts of *Anchiceratops* and *Trachodon. Bulletin of the American Museum of Natural History* 33: 539–548.

Brown, B. 1914b. *Corythosaurus casuarius,* a new crested dinosaur from the Belly River Cretaceous, with provisional classification of the family Trachodontidae. *Bulletin of the American Museum of Natural History* 33: 559–565.

Brown, B. 1914c. *Leptoceratops,* a new genus of Ceratopsia from the Edmonton Cretaceous of Alberta. *Bulletin of the American Museum of Natural History* 33: 567–580.

Brown, B. 1916. A new crested dinosaur, *Prosaurolophus maximus. Bulletin of the American Museum of Natural History* 35: 701–708.

Brown, B. 1933. A new longhorned Belly River ceratopsian. *American Museum Novitates* 669: 1–3.

Brown, B., and E. M. Schlaikjer. 1943. A study of the troödont dinosaurs with the description of a new genus and four new species. *Bulletin of the American Museum of Natural History* 82: 121–149.

Carr, T. D. 1999. Craniofacial ontogeny in Tyrannosauridae (Dinosauria, Coelosauria). *Journal of Vertebrate Paleontology* 19: 497–520.

Chapman, R. E., and M. K. Brett-Surman. 1990. Morphometric observations on hadrosaurid ornithopods. In K. Carpenter and P. J. Currie (eds.), *Dinosaur Systematics: Approaches and Perspectives,* pp. 163–201. Cambridge: Cambridge University Press.

Cope, E. D. 1876. Descriptions of some vertebrate remains from the Fort Union beds of Montana. *Proceedings of the Academy of Natural Sciences of Philadelphia* 1876: 248–261.

Cracraft, J. 1971. Caenagnathiformes: Cretaceous birds convergent on Dicynodont reptiles. *Journal of Paleontology* 45: 805–809.

Currie, P. J. 1989. The first records of *Elmisaurus* (Saurischia, Theropoda) from North America. *Canadian Journal of Earth Sciences* 26: 1319–1324.

Currie, P. J., and D. A. Russell. 1988. Osteology and relationships of *Chirostenotes pergracilis* (Saurischia, Theropoda) from the Judith River (Oldman) Formation of Alberta, Canada. *Canadian Journal of Earth Sciences* 25: 972–986.

Currie, P. J., J. K. Rigby Jr., and R. E. Sloan. 1990. Theropod teeth from the Judith River Formation of southern Alberta, Canada, pp. 107–125. In K. Carpenter and P. J. Currie (eds.), *Dinosaur Systematics: Approaches and Perspectives,* pp. 107–125. Cambridge: Cambridge University Press.

Dowling, D. B. 1915. Southern Alberta. *Geological Survey of Canada, Summary Report,* 1914, part L: 43–51.

Eberth, D. A., and A. P. Hamblin. 1993. Tectonic, stratigraphic, and sedimentological significance of a regional disconformity in the upper Judith River Formation (Belly River Wedge) of southern Alberta, Saskatchewan, and northern Montana. *Canadian Journal of Earth Sciences* 30: 174–200.

Galton, P. M., and H.-D. Sues. 1983. New data on pachycephalosaurid dinosaurs (Reptilia: Ornithischia) from North America. *Canadian Journal of Earth Sciences* 20: 462–473.

Gilmore, C. W. 1913. A new dinosaur from the Lance Formation of Wyoming. *Smithsonian Miscellaneous Collections* 61: 1–5.

Gilmore, C. W. 1924a. A new coelurid dinosaur from the Belly River Cretaceous of Alberta. *Bulletin of the Canadian Department of Mines, Geological Survey* 38: 1–12.

Gilmore, C. W. 1924b. On the skull and skeleton of *Hypacrosaurus,* a helmet-crested dinosaur from the Edmonton Cretaceous of Alberta. *Bulletin of the Canadian Department of Mines, Geological Survey* 38: 49–64.

Gilmore, C. W. 1930. On dinosaur reptiles from the Two Medicine Formation of Montana. *Proceedings of the United States National Museum* 77: 1–39.

Hayden, F. V. 1871. Geology of the Missouri valley. *Preliminary Report (4th Annual) of the U.S. Geological Survey of Wyoming and Portions of Contiguous Territories.*

Horner, J. R., and P. J. Currie. 1994. Embryonic and neonatal morphology of a new species of *Hypacrosaurus* (Ornithischia, Lambeosauridae) from Montana and Alberta. In K. Carpenter, K. F. Hirsch, and J. R.

Horner (eds.), *Dinosaur Eggs and Babies,* pp. 310–356. Cambridge: Cambridge University Press.

Horner, J. R., and R. Makela. 1979. Nest of juveniles provides evidence of family structure among dinosaurs. *Nature* 282: 296–298.

Horner, J. R., and D. Weishampel. 1988. A comparative embryological study of two ornithischian dinosaurs. *Nature* 332: 256–257.

Jerzykiewicz, T., and D. K. Norris. 1994. Stratigraphy, structure, and syntectonic sedimentation of the Campanian "Belly River" clastic wedge in the southern *Canadian Cordillera. Cretaceous Research* 15: 367–399.

Kurzanov, S. M. 1981. [On the unusual theropods from the Upper Cretaceous of Mongolia.] *Sovmestnaya sovetskogo-mongol'skaya paleontologicheskaya ekspeditsiya, trudy* 15: 39–50. (In Russian, with English summary.)

Lambe, L. M. 1902. New genera and species from the Belly River Series (mid-Cretaceous). *Contributions to Canadian Palaeontology, Geological Survey of Canada* 3: 25–81.

Lambe, L. M. 1904. On the squamoso-parietal crest of two species of horned dinosaurs from the Cretaceous of Alberta. *Ottawa Naturalist* 8: 81–84.

Lambe, L. M. 1910. Note on the parietal crest of *Centrosaurus apertus,* and a proposed new generic name for *Stereocephalus tutus. Ottawa Naturalist* 14: 149–151.

Lambe, L. M. 1913. A new genus and species from the Belly River Formation of Alberta. *Ottawa Naturalist* 27: 109–116.

Lambe, L. M. 1914a. On *Gyrposaurus notabilis,* a new genus and species of trachodont dinosaur from the Belly River Formation of Alberta, with a description of the skull of *Chasmosaurus belli. Ottawa Naturalist* 27: 145–155.

Lambe, L. M. 1914b. On the fore-limb of a carnivorous dinosaur from the Belly River Formation of Alberta, and a new genus of Ceratopsia from the same horizon, with remarks on the integument of some Cretaceous herbivorous dinosaurs. *Ottawa Naturalist* 27: 129–135.

Lambe, L. M. 1915. On *Eoceratops canadensis,* gen nov., with remarks on other genera of Cretaceous horned dinosaurs. *Geological Survey of Canada,* Geological ser., 24: 1–49.

Lambe, L. M. 1917. A new genus and species of crestless hadrosaur from the Edmonton Formation of Alberta. *Ottawa Naturalist* 31: 65–73.

Lambe, L. M. 1918. The Cretaceous genus *Stegoceras* typifying a new family referred provisionally to the Stegosauria. *Transactions of the Royal Society of Canada,* 3d ser., 12: 23–36.

Lambe, L. M. 1919. Description of a new genus and species (*Panoplosaurus mirus*) of armoured dinosaur from the Belly River beds of Alberta. *Transactions of the Royal Society of Canada,* 3d ser., 13: 39–50.

Lambe, L. M. 1920. The hadrosaur *Edmontosaurus* from the Upper Cretaceous of Alberta. *Memoirs of the Canadian Geological Survey* 120: 1–79.

Langston, W., Jr. 1976. A Late Cretaceous vertebrate fauna from the St. Mary River Formation in western Canada. In C. S. Churcher (ed.), *Athlon,* pp. 114–133. Toronto: Royal Ontario Museum.

Leidy, J. 1856. Notice of remains of extinct reptiles and fishes, discovered by Dr. F. V. Hayden in the badlands of the Judith River, Nebraska territory. *Proceedings of the Academy of Nature Sciences, Philadelphia* 8: 72–73.

Leidy, J. 1868. [Remarks on a jaw fragment of *Megalosaurus*.] *Proceedings of the Academy of Natural Sciences* (Philadelphia, 1870): 197–200.

Marsh, O. C. 1889. Notice of gigantic horned Dinosauria from the Cretaceous. *American Journal of Science*, 3d ser., 38: 173–175.

Marsh, O. C. 1890. Description of new dinosaurian reptiles. *American Journal of Science*, 3d ser., 39: 418–426.

Marsh, O. C. 1891. Notice of new vertebrate fossils. *American Journal of Science*, 3d ser., 42: 265–269.

Matthew, W. D., and B. Brown. 1922. The family Deinodontidae, with notice of a new genus from the Cretaceous of Alberta. *Bulletin of the American Museum of Natural History* 46: 367–385.

McLean, J. R. 1971. Stratigraphy of the Upper Cretaceous Judith River Formation in the Canadian Great Plains. *Saskatchewan Research Council, Geology Division Report* 11: 1–96.

Osborn, H. F. 1905. *Tyrannosaurus* and other Cretaceous carnivorous dinosaurs. *Bulletin of the American Museum of Natural History* 21: 259–265.

Osborn, H. F. 1917. Skeletal adaptations of *Ornitholestes, Struthiomimus, Tyrannosaurus. Bulletin of the American Museum of Natural History* 35: 733–771.

Parks, W. A. 1920. Preliminary description of a new species of trachodont dinosaur of the genus *Kritosaurus, Kritosaurus incurvimanus. Transactions of the Royal Society of Canada*, 3d ser., 13: 51–59.

Parks, W. A. 1922. *Parasaurolophus walkeri*, a new genus and species of crested trachodont dinosaur. *University of Toronto Studies*, Geological ser., 13: 1–32.

Parks, W. A. 1923. *Corythosaurus intermedius*, a new species of trachodont dinosaur. *University of Toronto Studies*, Geological ser., 15: 1–57.

Parks, W. A. 1925. *Arrhinoceratops brachyops*, a new genus and species of Ceratopsia from the Edmonton Formation of Alberta. *University of Toronto Studies*, Geological ser., 19: 5–15.

Parks, W. A. 1926. *Thescelosaurus warreni*, a species of ornithopodous dinosaur from the Edmonton Formation of Alberta. *University of Toronto Studies*, Geological ser., 21: 1–42.

Parks, W. A. 1928. *Struthiomimus samueli*, a new species of Ornithomimidae from the Belly River Formation of Alberta. *University of Toronto Studies*, Geological ser., 26: 1–24.

Parks, W. A. 1933. New species of dinosaurs and turtles from the Belly River formation of Alberta, with notes on other species. *University of Toronto Studies*, Geological ser., 34: 1–33.

Peng, J.-H. 1997. Palaeoecology of vertebrate assemblages from the Upper Cretaceous Judith River Group (Campanian) of southeastern Alberta, Canada. Ph.D. thesis, Calgary: University of Calgary.

Perle, A. 1981. [A new segnosaurid from the Upper Cretaceous of Mongolia.] *Sovmestnaya sovetskogo-mongol'skaya paleontologicheskaya ekspeditsiya trudy* 15: 50–59. (In Russian.)

Russell, D. A. 1970. Tyrannosaurs from the Late Cretaceous of western Canada. *National Museum of Natural Science Publications in Palaeontology* 1: 1–34.

Russell, D. A. 1972. Ostrich dinosaurs from the Late Cretaceous of western Canada. *Canadian Journal of Earth Sciences* 7: 181–184.

Russell, D. A. 1984. A checklist of the families and genera of North American dinosaurs. *National Museum of Natural Science, National Museums of Canada, Syllogeus*, 53: 1–35.

Russell, L. S., and Landes, R. W. 1940. Geology of the Southern Alberta Plains. *Geological Survey of Canada, Memoir* 221: 1–223.

Sternberg, C. M. 1928. A new armoured dinosaur from the Edmonton Formation of Alberta. *Canadian Field-Naturalist* 22: 93–106.

Sternberg, C. M. 1932. Two new theropods from the Belly River Formation of Alberta. *Canadian Field-Naturalist* 46: 99–105.

Sternberg, C. M. 1933. A new *Ornithomimus* with complete abdominal cuirass. *Canadian Field-Naturalist* 47: 79–83.

Sternberg, C. M. 1935. Hooded dinosaurs from the Belly River Series of the Upper Cretaceous. *Bulletin of the National Museum of Canada* 77: 1–37.

Sternberg, C. M. 1937. Classification of *Thescelosaurus:* A description of a new species. *Proceedings of the Geological Society of America* 1936: 375.

Sternberg, C. M. 1940a. Ceratopsidae from Alberta. *Journal of Paleontology* 14: 468–480.

Sternberg, C. M. 1940b. *Thescelosaurus edmontonensis,* n. sp., and classification of the Hypsilophodontidae. *Journal of Paleontology* 14: 481–494.

Sternberg, C. M. 1950. *Pachyrhinosaurus canadensis,* representing a new family of Ceratopsia. *Bulletin of the National Museum of Canada* 118: 109–120.

Sternberg, C. M. 1951. A complete skeleton of *Leptoceratops gracilis* Brown from the Upper Edmonton member on the Red Deer River, Alberta. *Bulletin of the National Museum of Canada* 123: 225–255.

Sternberg, C. M. 1953. A new hadrosaur from the Oldman Formation of Alberta: Discussion of nomenclature. *Bulletin of the Department of Natural Resources of Canada* 128: 1–12.

Sternberg, R. M. 1940. A toothless bird from the Cretaceous of Alberta. *Journal of Paleontology* 14: 81–85.

Sues, H.-D. 1978. A new small theropod dinosaur from the Judith River Formation (Campanian) of Alberta, Canada. *Zoological Journal of the Linnean Society* 62: 381–400.

Sues, H.-D. 1997. On Chirostenotes, a Late Cretaceous Oviraptorosaur (Dinosauria: Theropoda) from western North America. *Journal of Vertebrate Paleontology* 17: 698–716.

Sues, H.-D., and D. B. Norman. 1990. Hypsilophodontidae, *Tentontosaurus,* Dryosauridae. In D. B. Weishampel, P. Dodson, and H. Osmólska (eds.), *The Dinosauria,* pp. 498–509. Berkeley: University of California Press.

Wall, W. P., and P. M. Galton. 1979. Notes on pachycephalosaurid dinosaurs (Reptilia: Ornithischia) from North America, with comments on their status as ornithopods. *Canadian Journal of Earth Sciences* 16: 1176–1186.

Dinosaurs of Alberta by Formation, Exclusive of Aves

Formation	Reference Specimen	Other Specimens

——— *Alberta Plains* ———

MILK RIVER FORMATION *(Deadhorse Coulee member—lowermost Campanian)*

Hadrosauridae indet.	TMP 20001 (tooth)	teeth
Ankylosauria		
Ankylosauridae indet.[1]	MR-4:55[2] (tooth)	teeth
Pachycephalosauridae indet.	MR-4:63[2] (tooth)	teeth, skull fragments
Neoceratopsia		
"protoceratopsid"	MR-4:67[2] (tooth)	teeth
Ceratopsidae indet.	TMP 10032[12] (tooth)	teeth
Tyrannosauridae indet.	MR-4:74[2] (tooth)	teeth
Ornithomimidae indet.	uncatalogued TMP material	isolated phalanges
Dromaeosauridae		
Saurornitholestes Sues 1978		
cf. *Saurornitholestes langstoni*[3] Sues 1978	MR-4:1[2] (tooth)	teeth
Theropod incertae sedis		
Ricardoestesia[3] Currie et al. 1990		
Ricardoestesia gilmorei Currie et al. 1990	MR-4:18[2] (tooth)	teeth
Ricardoestesia n. sp. Baszio 1997b	MR-4:4[2] (tooth)	teeth
Paronychodon Cope 1876		
Paronychodon lacustris Cope 1876	MR-4:46[2] (tooth)	teeth

FOREMOST FORMATION *(upper Campanian)*

Hypsilophodontidae		
?Hysilophodontidae indet.	TMP 93.45.3 (tooth)	teeth, vertebra
Hadrosauridae		
Hadrosauridae indet.	TMP 96.81.1 (tooth)	teeth
Ankylosauria		
Ankylosauridae indet.	TMP 80.13.40 (tooth)	
Nodosauridae indet.	TMP 96.7.17 (tooth)	teeth
Pachycephalosauridae		
Stegoceras Lambe 1902		
Stegoceras sp.	TMP 86.146.2 (skull cap fragment)	
Neoceratopsia		
Ceratopsidae indet.	TMP 96.83.32 (tooth)	teeth, isolated elements
Tyrannosauridae		
Tyrannosauridae indet.	TMP 88.86.4 (tooth)	teeth
Aublysodon Leidy 1868		
Aublysodon sp.	uncatalogued TMP teeth	teeth
Dromaeosauridae		
Dromaeosaurus Matthew & Brown 1922		
Dromaeosaurus sp.	TMP 97.99.4 (phalanx)	
Saurornitholestes Sues 1978		
Saurornitholestes sp.	TMP 88.86.29 (tooth)	teeth
Theropoda incertae sedis		
Ricardoestesia Currie et al. 1990		
Ricardoestesia sp.	TMP 88.86.44 (tooth)	teeth
Ricardoestesia n. sp. Baszio 1997b	uncatalogued TMP teeth	
Paronychodon Cope 1876		
Paronychodon sp.	uncatalogued TMP teeth	teeth

Formation	Reference Specimen	Other Specimens
OLDMAN FORMATION *(upper Campanian)*		
Hypsilophodontidae indet.	TMP 87.62.21 (vertebra)	teeth, isolated elements
Hadrosauridae		
Hadrosaurinae		
Brachylophosaurus Sternberg 1953		
Brachylophosaurus canadensis Sternberg 1953	TMP 90.104.1 (skull & partial skeleton)	
Lambeosaurinae		
Hypacrosaurus Brown 1913		
Hypacrosaurus stebingeri 1994 Horner & Currie	TMP 87.79.227 (skull)	numerous isolated, embryonic & juvenile
Lambeosaurus n. sp. (Horner & Currie, in prep.)	TMP 78.16.1 (skeleton)	
Ankylosauridae indet.	TMP 88.82.8 (tooth)	teeth
Nodosauridae indet.	TMP 96.149.21 (tooth)	teeth
Pachycephalosauridae indet.	TMP 87.62.64 (tooth)	teeth, isolated skull fragments
Neoceratopsia		
Ceratopsidae		
Centrosaurinae		
undescribed centrosaurine	uncatalogued TMP material	bone bed material
Tyrannosauridae		
Tyrannosauridae indet.	TMP 92.30.219 (tooth)	teeth
Aublysodon Leidy 1868		
Aublysodon 1868 sp.	TMP 96.62.48	teeth
Daspletosaurus Russell 1970	CMN 8506	partial skeleton
Daspletosaurus torosus	CMN 8506[Hg&s] (skull & skeleton)	
Troodontidae		
Troodon Leidy 1856		
Troodon sp.	TMP 89.77.5 (tooth)	teeth
Dromaeosauridae		
Dromaeosaurus Matthew & Brown 1922		
Dromaeosaurus sp.	TMP 96.103.1 (tooth)	teeth
Saurornitholestes Sues 1978		
Saurornitholestes langstoni Sues 1978	TMP 94.144.105 (tooth)	teeth, partial skeleton
Theropod incertae sedis		
Ricardoestesia Currie et al. 1990		
Ricardoestesia sp.	TMP 89.79.62 (tooth)	teeth
Paronychodon Cope 1876		
Paronychodon sp.	TMP 92.77.6	teeth
DINOSAUR PARK FORMATION *(upper Campanian)*		
Hadrosauridae		
Hadrosaurinae		
Brachylophosaurus Sternberg 1953		
Brachylophosaurus canadensis Sternberg 1953	CMN 8893[Hg&s] (skeleton)	
Gryposaurus Lambe 1914a		
Gryposaurus notabilis Lambe 1914a	CMN 2278[Hg&s] (skull & postcrania)	skulls & associated postcrania

Formation	Reference Specimen	Other Specimens
"Kritosaurus" Brown 1910		
"Kritosaurus" incurvimanus Parks 1920		
(=unnamed gryposaur)	ROM 1918[Hs]	skull & associated postcrania
Prosaurolophus Brown 1916		
Prosaurolophus maximus Brown 1916	AMNH 5836[Hg&s] (skull)	skeletons
Lambeosaurinae		
Corythosaurus Brown 1914a		
Corythosaurus casuarius Brown 1914a	AMNH 5240[Hg&s] (skeleton)	skulls & associated postcrania
Lambeosaurus Parks 1923		
Lambeosaurus lambei Parks 1923	CMN 2869[Lg&s] (skull)	skulls & skeletons
Lambeosaurus magnicristatus Sternberg 1935	CMN 8705[Hs] (skull)	skulls & associated postcrania
Parasaurolophus Parks 1922		
Parasaurolophus walkeri Parks 1922	ROM 768[Hg&s] (skeleton)	
Hypsilophodontidae		
Thescelosaurus Gilmore 1913		
Thescelosaurus cf. *neglectus* Gilmore 1913	TMP 88.215.28 (tooth)	teeth, isolated elements
Ankylosauridae		
Euoplocephalus Lambe 1910		
Euoplocephalus tutus Lambe 1902 (=*Stereocephalus tutus* Lambe 1902)	CMN 210[Hg&s] (partial skull)	teeth, skulls, skeletons
Nodosauridae		
Edmontonia Sternberg 1928		
Edmontonia rugosidens Gilmore 1930 (=*Palaeoscincus rugosidens* Gilmore 1930)	AMNH 5665 (partial skeleton)	teeth, skeletons, isolated elements
Panoplosaurus Lambe 1919		
Panoplosaurus mirus Lambe 1919	CMN 2759[Hg&s] (skull & partial skeleton)	teeth, skeleton, isolated elements
Pachycephalosauridae		
Stegoceras Lambe 1918		
Stegoceras validum Lambe 1918	CMN 515[Hg&s] (skull cap)	teeth, skull caps, partial skeleton
Ornatotholus Galton & Sues 1983		
Ornatotholus browni Wall & Galton 1979 (=*Stegoceras browni* Wall & Galton 1979)	MNH 5450[Hg&s] (skull cap)	
Neoceratopsia		
Protoceratopsidae		
Leptoceratops Brown 1914c cf. *Leptoceratops* sp.	TMP 95.12.6 (dentary)	isolated elements
Ceratopsidae		
Centrosaurinae		
Centrosaurus Lambe 1904		
Centrosaurus apertus Lambe 1902 (immature specimens = *Monoclonius* Cope 1876)	CMN 971[Hg&s] (parietal)	skeletons, bone beds, isolated elements
Styracosaurus Lambe 1913		
Styracosaurus albertensis Lambe 1913	CMN 344[Hg&s] (skull & skeleton)	skeletons, bone beds, isolated elements

Formation	Reference Specimen	Other Specimens
Chasmosaurinae		
Anchiceratops Brown 1914a		
Anchiceratops sp.	CMN 9813	isolated elements
Chasmosaurus Lambe 1914a		
(=*Protorosaurus* Lambe 1914b;		
= *Eoceratops* Lambe 1915)		
Chasmosaurus belli Lambe 1914a		
(=*Monoclonius belli* Lambe 1902)	CMN 491[Hg&s] (skull)	skulls, skeletons, isolated elements
Chasmosaurus russelli Sternberg 1940a		
(includes *C. canadensis* Lambe 1902		
and *Monoclonius canadensis* Lambe 1902),		
Eoceratops canadensis Lambe 1915,		
(*C. kaiseni* Brown 1933)	CMN 8800[Hs] (skull)	skulls, skeletons, isolated elements
undescribed chasmosaurine	CMN 41357	skeleton, partial skull
Tyrannosauridae		
Albertosaurus Osborn 1905		
Albertosaurus libratus Lambe 1914b		
(=*Gorgosaurus libratus*[4] Lambe 1914b)	CMN 2120[Hs] (skull & skeleton)	skulls, skeletons, isolated elements
Aublysodon Leidy 1868		
Aublysodon mirandus Leidy 1868	TMP 82.19.367 (tooth)	teeth, isolated elements
Daspletosaurus Russell 1970		
Daspletosaurus n.sp. (Currie, pers. comm.)	TMP 85.62.1 (skull & skeleton)	4 skeletons
Troodontidae		
Troodon Leidy 1856		
Troodon formosus Leidy 1856	TMP 89.36.268 (tooth)	teeth, isolated element
Ornithomimidae		
Dromiceiomimus Russell 1972		
Dromiceiomimus samueli Parks 1928		
(=*Struthiomimus samueli* Parks 1928)	ROM 840[Hs] (partial skeleton)	
Ornithomimus Marsh 1890		
Ornithomimus edmontonensis Sternberg 1933	ROM 851 (skeleton)	
Struthiomimus Osborn 1917		
Struthiomimus altus Lambe 1902		
(=*Ornithomimus altus* Lambe 1902)	CMN 930[Hg&s] (skeleton)	
Avimimidae		
Avimimus Kurzanov 1981		
Avimimus sp.	TMP 98.8.28 (metatarsal)	isolated elements
Therizinosauroidea		
Erlikosauridae		
Erlikosaurus Perle 1981		
cf. *Erlikosaurus* sp.	CMN 12355 (frontal)	isolated elements
Oviraptorosauria		
Chirostenotes Gilmore 1924a		
Chirostenotes pergracilis Gilmore 1924a		
(=*Caenagnathus collinsi* R. M. Sternberg 1940)	CMN 2367[Hg&s] (lower jaw)	
Chirostenotes elegans Parks 1933		
(= *Caenagnathus sternbergi* Cracraft 1971;		

Formation	Reference Specimen	Other Specimens
Ornithomimus elegans Parks 1933; = *Elmisaurus elegans*[5] Currie 1989)	CMN 2690 (partial mandible)	metatarsals
Dromaeosauridae		
Dromaeosaurus Matthew & Brown 1922		
Dromaeosaurus albertensis Matthew & Brown 1922	AMNH 5356[Hg&s] (partial skull & postcranial)	isolated elements
Saurornitholestes Sues 1978		
Saurornitholestes langstoni Sues 1978	TMP 74.10.5[Hg&s] (partial skeleton)	partial skeletons, isolated elements
Theropod incertae sedis		
Ricardoestesia Currie et al. 1990		
Ricardoestesia gilmorei Currie et al. 1990	CMN 343[Hg&s] (paired dentaries)	teeth
Ricardoestesia n. sp. Baszio 1997b	TMP 96.142.19 (tooth)	teeth
Paronychodon Cope 1876		
Paronychodon lacustris Cope 1876	TMP 86.60.114 (tooth)	teeth

BEARPAW FORMATION *(upper Campanian to lowermost Maastrichtian)*
All occurrences are in marine shales

Hadrosauridae		
Hadrosauridae indet.	TMP 78.28.12 (femur)	isolated elements
Brachylophosaurus Sternberg 1953		
Brachylophosaurus sp.	TMP 98.50.1 (complete juvenile skeleton)	
Prosaurolophus Brown 1916		
Prosaurolophus sp.	TMP 83.64.3 (partial skeleton)	
Ankylosauria		
Nodosauridae		
Edmontonia Gilmore 1930		
Edmontonia sp.	TMP 83.64.91 (scute)	
?Pachycephalosauridae	TMP 90.108.1 (skull cap)	
Ceratopsidae		
Ceratopsidae indet.	TMP 91.36.416 (phalanx)	
Ornithomimidae indet.	TMP 78.28.16 (metatarsal)	isolated elements

HORSESHOE CANYON FORMATION *(Maastrichtian)*

Hadrosauridae		
Hadrosaurinae		
Edmontosaurus Lambe 1920		
Edmontosaurus regalis Lambe 1917	CMN 2288[Hg&s] (skull & skeleton)	skeletons, bone bed material
Saurolophus Brown 1912		
Saurolophus osborni Brown 1912	AMNH 5220[Hg&s] (skeleton)	skulls, skeleton, & isolated material
Lambeosaurinae		
Hypacrosaurus Brown 1913		
Hypacrosaurus altispinus Brown 1913	AMNH 5204[Hg&s] (postcrania)	skulls, associated post-crania, isolated elements
Ankylosauridae		
Euoplocephalus Lambe 1910		
Euoplocephalus tutus Lambe 1910	AMNH 5266 (partial skeleton)	teeth, skulls, skeletons, isolated elements

Formation	Reference Specimen	Other Specimens
Nodosauridae		
Edmontonia Sternberg 1928		
Edmontonia longiceps Sternberg 1928	CMN 8531[Hg&s] (skull & partial skeleton)	teeth, skulls, skeleton?, isolated elements
Pachycephalosauridae		
Stegoceras Lambe 1902		
Stegoceras edmontonense Brown & Schlaikjer 1943 (=*Troodon edmontonensis* Brown & Schlaikjer 1943)	CMN 8830[Hs] (frontoparietal)	isolated elements
Ceratopsidae indet.	TMP 1021	teeth
Centrosaurinae		
Pachyrhinosaurus Sternberg 1950		
Pachyrhinosaurus canadensis Sternberg 1950	CMN 8867[Hg&s] (skull)	isolated elements
Chasmosaurinae		
Anchiceratops Brown 1914b		
Anchiceratops ornatus Brown 1914b	AMNH 5251[Hg&s] (skull)	skulls, skeleton, isolated elements
Arrhinoceratops Parks 1925		
Arrhinoceratops brachyops Parks 1925	ROM 796[Hg&s] (skull)	
Tyrannosauridae		
Daspletosaurus Russell 1970		
Daspletosaurus sp.	CMN 11315 (partial skeleton)	
Albertosaurus Osborn 1905		
Albertosaurus sarcophagus Osborn 1905	CMN 5600[Hg&s] (skull)	skeletons, bone bed, isolated elements
Aublysodon Leidy 1868		
Aublysodon sp.	TMP 88.513.8 (tooth)	teeth
Troodontidae		
Troodon Leidy 1856		
cf. *Troodon formosus* Leidy 1856	TMP 83.12.11 (dentary)	teeth, isolated elements
Ornithomimidae		
Dromiceiomimus Russell 1972		
Dromiceiomimus brevetertius Parks 1926 (=*Struthiomimus brevetertius* Park 1926, including *S. ingens* Parks 1933)	ROM 797[Hg&s] (post crania)	
Ornithomimus Marsh 1890		
Ornithomimus edmontonensis Sternberg 1933	CMN 8632[Hs] (skeleton)	partial skeletons
Struthiomimus Osborn 1917		
Struthiomimus altus Lambe 1902 (= *Ornithomimus altus* Lambe 1902)	TMP 90.26.1 (partial skull & skeleton)	
Oviraptorosauria		
Chirostenotes Gilmore 1924		
Chirostenotes pergracilis Gilmore 1924a (=*Marcophalangia canadensis* C. M. Sternberg 1932,		

Formation	Reference Specimen	Other Specimens
Caenagnathus collinsi R. M. Sternberg 1940)	ROM 43250 (partial skeleton)	partial skeleton, isolated elements
Dromaeosauridae		
Dromaeosaurus Matthew & Brown 1922		
Dromaeosaurus albertensis Matthew & Brown 1922	TMP 1011[2] (tooth)	teeth
Saurornitholestes Sues 1978		
Saurornitholestes sp.[6]	TMP 1034[2] (tooth)	teeth
undescribed velociraptorine	uncatalogued	partial skull
Theropod incertae sedis		
Ricardoestesia Currie et al. 1990		
Ricardoestesia gilmorei Currie et al. 1990	UA 85[2] (tooth)	teeth
Ricardoestesia n. sp. Baszio 1997b	UA 112[2] (tooth)	teeth
Paronychodon Cope 1876		
Paronychodon lacustris Cope 1876	TMP 1041[2] (tooth)	teeth

SCOLLARD FORMATION *(upper Maastrichtian)*

Hypsilophodontidae		
Hypsilophodontidae indet.	TMP 98.8.22 (vertebra)	isolated elements
Thescelosaurus Gilmore 1913		
Thescelosaurus neglectus Gilmore 1913		
(=*Thescelosaurus edmontonensis* Sternberg 1940b)	CMN 8537[Hs] (skull& skeleton)	
Parksosaurus Sternberg 1937		
Parksosaurus warrenae Parks 1926		
(=*Thescelosaurus warrenae* Parks 1926)	ROM 804[Hg&s] (skeleton)	
Hadrosauridae[7] indet.	TMP 86.207.22 (vertebra)	teeth, isolated elements
Ankylosauridae		
Ankylosauridae indet.	TMP 86.207.26 (scutes)	scutes, teeth
Ankylosaurus Brown 1908		
Ankylosaurus magniventris Brown 1908	TMP 86.208.7 (scutes)	
Neoceratopsia		
Protoceratopsidae		
Leptoceratops Brown 1914c		
Leptoceratops gracilis Brown 1914c	AMNH 5205[Hg&s] (partial skeleton)	teeth, skeletons, isolated material
Ceratopsidae		
Ceratopsidae indet.	TMP 87.16.34 (partial dentary)	teeth, isolated elements
Chasmosaurinae		
Triceratops Marsh 1889		
Triceratops horridus Marsh 1889	TMP 98.102.1 (partial skull)	teeth, isolated elements
Tyrannosauridae		
Tyrannosauridae indet.	TMP 86.208.1 (tooth)	teeth, isolated elements
Tyrannosaurus Osborn 1905		
Tyrannosaurus rex[11] Osborn 1905	TMP 81.17.1 (partial skeleton)	skeletons, isolated elements
Troodontidae		
Troodon Leidy 1856		
cf. *Troodon formosus* Leidy 1856	TMP 94.106.1 (tooth)	teeth

Formation	Reference Specimen	Other Specimens
Ornithomimidae		
Ornithomimidae indet.	TMP 93.104.1 (partial skeleton)	isolated elements
Avimimidae		
Avimimus Kurzanov 1981		
cf. *Avimimus* sp.	TMP 98.8.28 (metatarsal)	
Dromaeosauridae		
Dromaeosauridae indet.	TMP 81.1.1 (ungual)	
Dromaeosaurus Matthew & Brown 1922		
Dromaeosaurus albertensis Matthew & Brown 1922	TMP 81.31.99 (tooth)	teeth
Saurornitholestes Sues 1978		
cf. *Saurornitholestes langstoni* Sues 1978 sp.	TMP 87.16.18 (tooth)	teeth
Theropod incertae sedis		
Ricardoestesia Currie et al. 1990		
Ricardoestesia gilmorei Currie et al. 1990	UA 85[2] (tooth)	teeth
Ricardoestesia n. sp. Baszio 1997b	UA 112[2] (tooth)	teeth
Paronychodon Cope 1876		
Paronychodon-like[8] sensu Currie et al. 1990	UA 121[2] (tooth)	teeth

——— *Southwestern Alberta* ———

BELLY RIVER GROUP *(Campanian)*

Hypsilophodontidae indet.		postcrania[9]
Hadrosauridae indet.		jaw
Neoceratopsia		
Montanoceratops Sternberg 1951		
Montanoceratops sp.	TMP 82.11.1 (partial skeleton)	

ST. MARY RIVER FORMATION (Maastrichtian)

Hadrosauridae indet.	TMP 97.66.1 (partial skeleton)	isolated elements
Edmontosaurus Lambe 1917		
Edmontosaurus sp.	CMN 10661 (squamosal)	isolated elements
Neoceratopsia		
Ceratopsidae		
Pachyrhinosaurus Sternberg 1950		
Pachyrhinosaurus canadensis Sternberg 1950	CMN 9485 (skull)	bone bed elements
Tyrannosauridae		
Tyrannosauridae indet.	CMN 9589 (tooth)	teeth
Albertosaurus Osborn 1905		
Albertosaurus sp.	TMP 87.85.10 (tooth)	
Ornithomimidae indet.	CMN 10653 (metatarsal)	
Troodontidae		
Troodon Leidy 1856		
Troodon sp.	CMN 10649 (tooth)	
Theropoda indeterminate		
Saurornithoides-like (sensu Langston 1976)	CMN 10674 (tooth)	

Formation	Reference Specimen	Other Specimens
WILLOW CREEK FORMATION *(upper Maastrichtian)*		
Hadrosauridae indet.	TMP 81.6.3 (ischium)	teeth
Tyrannosauridae		
Tyrannosaurus Osborn 1905		
Tyrannosaurus rex	TMP 81.6.1 (teeth, skull & partial skeleton)	isolated elements

——— *Northwestern Alberta* ———

WAPITI FORMATION *(Maastrichtian)*		
Hadrosauridae indet.	TMP 89.92.1 (partial skeleton)	isolated elements
Neoceratopsia		
Ceratopsidae		
Centrosaurinae		
Pachyrhinosaurus Sternberg 1950		
Pachyrhinosaurus n. sp.	TMP 86.55.258[Hs] (skull)[10]	elements from two bone beds
Tyrannosauridae indet.	TMP 89.62.4 (vertebra)	isolated teeth & elements
Albertosaurus Osborn 1905		
Albertosaurus sp.	TMP 89.55.1585 (tooth)	teeth
Ornithomimidae indet.	TMP 89.53.35 (vertebra)	isolated elements
Dromaeosauridae		
Saurornitholestes Sues 1978	TMP 89.55.1523 (tooth)	teeth, isolated elements
Saurornitholestes undescribed n. sp. (Currie, pers. comm.)	TMP 89.55.47 (frontal)	
Troodontidae		
Troodon Leidy 1856		
Troodon sp.	TMP 89.55.1008 (metatarsal)	teeth

——— *Problematic Specimens and Taxa* ———

Milk River Formation

1. Cf. *"Kritosaurus"* Brown 1910, TMP 83.18.1, partial skeleton. Identified in TMP collection notes by M. K. Brett-Surman; identification needs to be validated.
2. *Aublysodon* Leidy 1868, TMP 20021[12], a single tooth.

Oldman Formation

1. Cf. *Maiasaura peeblesorum* Horner & Makela 1979, phalanges—none catalogued—probably *Brachylophosaurus*. The distinctive ventral ridge on the phalanges of *Maiasaura* are also known from *Brachylophosaurus*.
2. *Orodromeus makelai* Horner & Weishampel 1988, anecdotal record only.
3. *Gravitholus albertae* Wall & Galton 1979 (TMP 72.27.1), surface collected from a region containing both Dinosaur Park and Oldman formation outcrop; the holotype and only specimen, a skull cap, can not be confidently referred to either formation but is clearly from one of them. Given that there are no occurrences of this taxon in the well-sampled Dinosaur Park Formation it is probable that *Gravitholus* is from the Oldman Formation.

Dinosaur Park Formation

1. *Pachycephalosaurus* Brown & Schlaikjer 1943, is known from a single skull cap (BMNH R8648) reported to have come from Dinosaur Provincial Park. There is some question as to the validity of this record.

Scollard Formation

1. *Anatotitan* Brett-Surman 1990 (in Chapman and Brett-Surman 1990), anecdotal record only.
2. Nodosauridae indet., TMP 94.86.18 (tooth). Specimen needs to be validated.
3. Pachycephalosauridae indet., TMP 93.87.2 (tooth), TMP 94.31.15 (tooth), TMP 94.125.449 (tooth). These specimens need to be validated.
4. Cf. *Torosaurus* Marsh 1891 sp., anecdotal record only
5. *Struthiomimus* Osborn 1917, TMP 86.47.4 (ungual) and isolated elements; generic designation needs to be validated.
6. *Aublysodon mirandus* Leidy 1868, TMP 91.87.54, known from a single catalogued tooth.
7. *Nanotyrannus*[9] Bakker et al. 1988, anecdotal record only.
8. Segnosauridae indet., TMP 86.207.17 (cervical vertebra). Known from this single specimen.

St. Mary River Formation

1. *Edmontonia* cf. *longiceps* Sternberg 1928, CMN 21864 (tooth). Only one record.
2. *Albertosaurus* Osborn 1905, TMP 87.85.10 (tooth). Only one record.
3. *Ricardoestesia* Currie et al. 1990, TMP 98.52.2 (tooth). Only one record.

NOTES

1. Baszio (1997b) referred his Ankylosauria teeth to Ankylosauridae not on diagnostic characters, but because members of this family are more common than those of the Nodosauridae.
2. These are temporary specimen numbers from the Royal Tyrrell Museum of Palaeontology (MR-#:#) and the University of Alberta (UA#), taken from Baszio (1997b).
3. *Ricardoestesia gilmorei* has proven difficult to distinguish from *Saurornitholestes* in the Milk River Formation (Baszio 1997a,b; Brinkman pers. comm.), and the designation of these two taxa in this formation is somewhat problematic.
4. Currie (in prep.) believes that *Albertosaurus libratus* of the Dinosaur Park Formation is distinguishable from *A. sarcophagus* of the Horseshoe Canyon Formation at the generic level, and is referable to *Gorgosaurus libratus* Lambe 1914b.
5. Sues (1997) refers *Elmisaurus elegans* Currie 1989 to be a junior subjective synonym of *Chirostenotes pergracilis,* and follows Currie and Russell 1988 in considering Elmisauridae a subjective junior synonym of Caenagnathidae. Currie (pers. comm.) asserts the validity of *Elmisaurus* as distinguishable from *Chirostenotes*.
6. The new veloceraptorine from the Horseshoe Canyon Formation may account for the *Saurornitholestes* sp. teeth catalogued from this formation.
7. Baszio (1997b) notes that the referred hadrosaur teeth from the Scollard Formation differ from those found throughout the rest of the Albertan Cretaceous in having relatively long, somewhat twisted crowns. He also concluded that their unique shape could be due to stream abrasion, and, additionally, he had difficulty separating these from ceratopsid teeth. Finally, the remaining few hadrosaur specimens from the Scollard need to be reexamined to establish whether this group is truly present in the Scollard Formation.
8. *Paronychodon lacustris* is a taxon of small theropod known only from teeth. The *Paronychodon*-like teeth found infrequently in the Cretaceous formations of Alberta appear to be caused by growth anomalies in teeth from other taxa (Currie et al. 1990; Baszio 1997b), but may also represent another distinct taxon in the Scollard Formation (Currie et al. 1990).
9. The hypsilophodontidae indet. from the Belly River Group are the postcranial elements described as ?*Laosaurus minimus* by Gilmore (1924b). This taxon was considered *nomen dubium* by Sues and Norman (1990), who also suggested that it might be referable to *Orodromeus* Horner and Weishampel 1988.
10. TMP 86.55.258 is to become the holotype of the new species of *Pachyrhinosaurus* from the Wapiti Formation currently in preparation (Tanke, pers. comm.).
11. Carr (1999) considers *Nanotyrannus lancensis* Bakker et al. 1988 to be a subjective junior synonym of *Tyrannosaurus rex* Osborn 1905.
12. Temporary specimen numbers as cited in Peng 1997.

21. Two Medicine Formation, Montana: Geology and Fauna

DAVID TREXLER

Abstract

The Two Medicine Formation of northwestern Montana contains evidence of unusual intraformational faunal turnover. Although exceptions exist, most dinosaur-bearing formations do not exhibit distinct, stratigraphically separate faunal groups. Rather, species tend to be present throughout a formation or be replaced one at a time; replacement of groups of species tends to occur between formations. However, at least three distinct hadrosaur groups with sharp occurrence boundaries have been identified within the Two Medicine Formation, and "transitional" species (species with no apomorphic characters and suspected to have been the result of anagenesis) have been reported.

Recent discoveries of additional identifiable specimens have allowed more detailed resolution of the transition areas in the strata and also more accurate geographic and stratigraphic ranges for each species. Possible mechanisms of hadrosaurian evolution and diversification in the Two Medicine Formation are examined in light of the new information.

Introduction

The Two Medicine Formation is a roughly 600-meter-thick wedge of sediments that extends from southern Alberta to central Montana along the eastern edge of the Rocky Mountains (fig. 21.1). This formation originated as a terrestrial deposit; preserved paleosols, fluvial deposits, and bentonitic layers are common. Vertebrate fossils found in the formation are most often preserved by CaCO$_3$ permineralization (pers. obs.). Although this process allows for excellent preservation of even microscopic detail, the specimens are easily and rapidly eroded or otherwise damaged once exposed. The predominantly fine-grained and loosely consolidated nature of the sediments that comprise the formation allow for relatively fast plant regrowth over badlands areas; only a few extensive badlands outcrops exist. Of these, two are located on the Blackfeet Indian Reservation (the area of the type section along the Two Medicine River and an area near Landslide Butte) and another is located west of Choteau. In other areas, small outcrops exist as isolated exposures surrounded by grasslands. Because of the limited, isolated exposures, most research has been primarily conducted in these three areas. Smaller outcrops have been, to a large degree, overlooked until fairly recently. Although precise stratigraphic correlations of isolated outcrops are virtually impossible, approximations were made by triangulation and interpolation from the logs of nearby oil wells for the stratigraphic correlations presented below.

The discovery of a nest containing the remains of baby dinosaurs in 1978 by Marion Brandvold and subsequent research led by John Horner (1984, 1992; Horner and Makela 1979; Horner and Weis-

Figure 21.1. Index map showing outcrop area of the Campanian formations near the Alberta-Montana border. Scale: 10 miles (16 km).

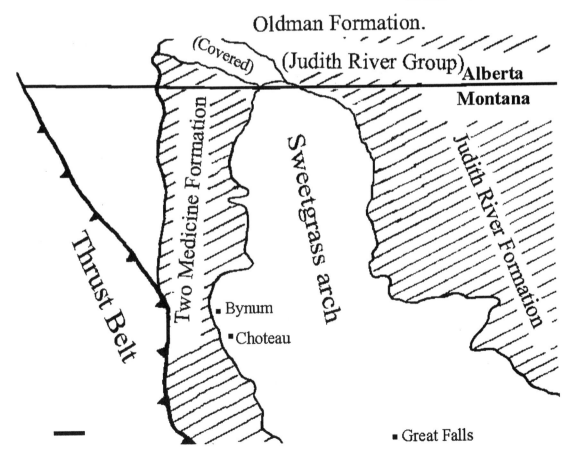

hampel 1988; Horner and Currie 1994) have fostered a renewed interest in the Two Medicine Formation and its fossil fauna.

Institutional Abbreviations: GM, Graves Museum; MOR, Museum of the Rockies; OTM, Old Trail Museum; ROM, Royal Ontario Museum; TA, Timescale Adventures.

Previous Work

Eugene Stebinger (1914) identified and described the Two Medicine Formation and also reported the first fossil remains from the strata. A U.S. Geological Survey crew headed by Stebinger and a U.S. National Museum crew under Charles Gilmore collaborated in recovering the first documented dinosaur remains from the formation in 1913. Gilmore continued his research in the area with fieldwork in 1928 and again in 1935 (Gilmore 1917, 1930, 1939). Although Gilmore's work in the region spanned more than twenty years, only three species were named, and only two of these new species, *Styracosaurus ovatus* and *Edmontonia rugosidens,* are still considered valid. In addition to Gilmore's research, Barnum Brown spent a portion of the summer of 1933 in the area. However, Brown did not report any significant discoveries, and research by both Brown and Gilmore was halted by World War II.

In 1977, I discovered hadrosaur remains in the upper portion of the formation west of Choteau, Montana. While working on this site the following year, Marion Brandvold discovered baby dinosaur bones a short distance from the quarry. These bones were later determined to be the remains of approximately 15 neonate hadrosaurs referred to *Maiasaura peeblesorum* by Horner and Makela (1979). The holotype of *Maiasaura peeblesorum* is an adult skull found in the same badlands in 1978 by Laurie Trexler. These discoveries became the basis for a renewed interest in the formation and its fauna that continues to the present. Several new species of dinosaurs have been recently described since *M. peeblesorum,* including *Orodromeus makelai* Horner and Weishampel 1988, *Prosaurolophus blackfeetensis* Horner 1992, *Gryposaurus latidens* Horner 1992, *Hypacrosaurus stebingeri* Horner and Currie 1994, *Einiosaurus procurvicornis* Sampson 1995, and *Achelousaurus horneri* Sampson 1995. All except *G. latidens* were discovered in the upper portion of the formation. These dinosaur species differ significantly from those in the time-equivalent and geographically adjacent portions of the Judith River and Oldman Formations (Horner 1984; Horner et al. 1992). Lorenz (1981) identified three lithofacies within the formation and Horner (1984) reported on separation of the dinosaur faunal communities over the facies boundaries.

Geological Setting

The Two Medicine Formation spans virtually the entire Campanian Stage. Rogers et al. (1993) obtained 40Ar/39Ar radiometric dates of 80 and 74 Ma (±0.1 Ma) from ash layers at approximately 105 meters above the base and approximately 10 meters below the top of the formation, respectively. The Two Medicine Formation overlies the

Virgelle Formation, and in turn is overlain by the marine shales of the Bearpaw Formation and the marine sandstones of the Horsethief Formation.

Toward the end of the Cretaceous period, the Colorado Sea began to recede, exposing additional land along its northern and western shores. The Virgelle Sandstone formed from the beach sands of this receding ocean. From the type section (along the Two Medicine River) northward, the sediments of the Two Medicine Formation directly above the Virgelle Sandstone are composed largely of blanket sandstones and interbedded lenticular mudstones (Lorenz and Gavin 1984). The depositional environment of the formation of these sediments has been interpreted as deltaic (Lorenz and Gavin 1984). Northward and eastward of the Two Medicine formation, the Virgelle Sandstone is overlain by other deltaic/shoreline sandstones and silty sediments. These other sediments, combined with those of the Virgelle, comprise the Eagle Formation.

The southern portion of the Two Medicine Formation exhibits basal sediments that more closely match the middle and upper sediments of the type section, which have been interpreted as upper coastal plain sediments (Lorenz 1981; Lorenz and Gavin 1984). The characteristics of the basal sediments in the southern portion of the formation suggest that a more upland environment existed in the south during deposition of the basal Two Medicine Formation. Several paleo stream channels have been identified within the formation that indicate a northeasterly flow (Gavin 1986, pers. obs.) from southwesterly located uplands.

A transgression of the Cretaceous Interior Seaway occurred shortly after deposition of the Two Medicine sediments began. The shallow marine sediments of the Claggett transgression are found directly underlying the Judith River Formation of Montana and the Oldman Formation of Alberta. These shallow marine sediments correlate with anomalous paralic sediments and isolated shale bodies approximately 100 meters above the base of the Two Medicine Formation (Lorenz and Gavin 1984). However, extensive marine facies within the formation have not been reported. Thus, if the area were inundated during that time, the period of submergence would have been extremely short.

The middle portion of the Two Medicine Formation begins approximately 100 meters above the Virgelle Sandstone and extends upward for approximately 225 meters. This portion of the formation was deposited during the regression of the Claggett Sea and the early portion of the transgression of the Bearpaw Sea (Varricchio 1993). This middle portion and a major section of the upper portion of the Two Medicine Formation are stratigraphically equivalent to the Judith River Formation of Montana and the Judith River Group (including the Oldman and Dinosaur Park Formations) of Alberta (Stebinger 1914; Eberth and Hamblin 1993). The sediments in this portion of the Two Medicine Formation consist primarily of bentonitic siltstones and mudstones, with occasional sandstone lenses. These sediments are interpreted to have been deposited on a broad upper coastal plain far removed from the Cretaceous Interior Seaway (Horner 1984).

The upper portion, roughly one-half of the formation, is distinguished from the middle portion by extensive red beds and caliche horizons within the otherwise similar sediments. The formation of red beds and caliche horizons indicate at least seasonally dry conditions (Lorenz 1981). Several dinosaurian bone beds discovered in this unit have been hypothesized to exhibit evidence for drought-related mortality (Rogers 1990). The uppermost 80 meters are thought to have been deposited after the Judith River–equivalent sediments had been inundated by the Bearpaw Sea, and are also thought to represent a depositional sequence of less than 0.5 million years' duration (Horner et al. 1992).

Bentonitic ash strata are common in the Two Medicine Formation. To the south, the emplacement of the Boulder Batholith and associated extrusive volcanic events, collectively referred to as the Elkhorn Volcanics, were coeval with Two Medicine deposition (Viele and Harris 1965). Explosive volcanic eruptions associated with the Elkhorn Volcanics were common throughout the time span of Two Medicine deposition (Viele and Harris 1965; Mudge 1972; Gill and Cobban 1973). Hooker (1987) also suggested close proximity of a volcanic source to a hadrosaur bone bed west of Choteau.

The top of the northern portion of the Two Medicine Formation is overlain by marine sediments of the Bearpaw Formation. These dark shales thicken northward. The southern portion of the Two Medicine Formation grades into a brackish water siltstone/sandstone series known as the Horsethief Formation. These sediments overlie the Two Medicine sediments west to the disturbed belt of the Rocky Mountains. Because the Horsethief Formation was laid down as very shallow marine sediments while the Bearpaw Shale was laid down in a deeper marine environment, the pinching out of the Bearpaw Shale to the south contributes additional evidence of a higher coastal plain in the southwest.

Two Medicine Fauna

Although Gilmore, Brown, and others worked in the Two Medicine Formation for many years, few dinosaurs were specifically identified. This was primarily due to the nature of fossil preservation in the formation. Very few articulated specimens have been found; most discoveries in the formation consist of isolated bones or of bone beds containing the disarticulated remains, often broken and poorly preserved, of several animals and commonly several species. Gilmore (1917, 1930, 1939) provided early lists of dinosaurs from the formation. However, most animals were classified only to the generic level, and the classifications assumed that animals identified from the Judith River Formation would also be present in the Two Medicine Formation.

More recent studies, beginning in 1978, have shown that most identified dinosaur specimens from the Two Medicine Formation belong to species unknown elsewhere (Horner and Makela 1979; Horner and Weishampel 1988; Horner 1992; Horner et al. 1992; Horner and Currie 1994; Sampson 1995, pers. obser.). Furthermore, recent discoveries of animals considered to be wide-ranging predators, such as

Daspletosaurus and *Albertosaurus* (also known as *Gorgosaurus*), show that these taxa exhibit species differentiation between the two formations (Bakker, pers. comm.; Currie [in prep.] has verified the validity of *Gorgosaurus,* but has not yet formalized this, so *Albertosaurus* is used). The species differences is unexpected considering the temporal and geographic proximity of the formation to the better-known fauna of the Oldman and Dinosaur Park Formations. No ecological barriers other than upland versus lowland habitat preferences (Horner 1984; Horner et al. 1992) have been postulated to exist between the formations.

Maiasaura unguals have been reported in the Dinosaur Park Formation of southern Alberta (Currie, pers. comm.) and the Claggett Formation of south-central Montana (Fiorillo 1990). However, the pedal unguals of *Brachylophosaurus* were, until very recently, unknown. A beautifully preserved, fully articulated specimen of *Brachylophosaurus,* discovered by Nate Murphy and currently being prepared by Museum of the Rockies, indicates that the plantar ridge once thought unique to *Maiasaura* unguals is also present in *Brachylophosaurus* (Murphy, pers. comm.; Horner, pers. comm.). There is thus no unequivocal evidence as yet reported of any faunal intermingling between the temporally equivalent and geographically adjacent areas represented by the aforementioned formations.

Dinosaurian remains are more common in the upper portion of the Two Medicine Formation than in the lower portions (Stebinger 1914; Gilmore 1917; Horner 1984, pers. obs.). Only one species of hadrosaurid dinosaur (*Gryposaurus latidens* Horner 1992) and no identified ceratopsian dinosaurs have been reported from the lower and middle portions of the formation. In contrast, the upper portion of the formation has yielded three hadrosaurian species *(Maiasaura peeblesorum* Horner and Makela 1979, *Prosaurolophus blackfeetensis* Horner 1992, *Hypacrosaurus stebingeri* Horner and Currie 1994), a nodosaur (*Edmontonia rugosidens* Gilmore 1930), and three currently valid ceratopsian species (*Styracosaurus ovatus* Gilmore 1930, *Einiosaurus procurvicornis* Sampson 1995, *Achelousaurus horneri* Sampson 1995).

During the past 20 years, additional dinosaur remains have been found within the Two Medicine Formation. These discoveries include:

• a single associated adult *Maiasaura peeblesorum* (OTM collections) collected approximately 100 meters lower in section and 31 kilometers north of the *M. peeblesorum* holotype locality (Trexler 1995).

• the second known remains of a single adult *Maiasaura peeblesorum* (ROM collections), consisting of a well-preserved and partially articulated skull and skeleton. The specimen was collected along the Two Medicine Formation type section, approximately 20 meters lower in section than the OTM specimen.

• the caudal half of an articulated *Hypacrosaurus sp.* (MOR collections). The specimen exhibits the diagnostic long dorsal neural spines on the proximal caudal vertebrae. The specimen was recovered 30 meters below and 0.5 kilometer south of the ROM *Maiasaura* site.

• isolated teeth of *Gryposaurus latidens,* discovered as a rare component of channel lag deposits throughout the middle portion of the formation along the Two Medicine River. Although hadrosaur teeth are generally not diagnostic at the specific level, this particular taxon exhibits extremely large teeth more closely resembling the iguanodont morphology than that of other hadrosaurs. This morphology is listed as one of the primary diagnostic characteristics of the species (Horner 1992).

• several specimens of two species of tyrannosaurid dinosaurs (cf. *Daspletosaurus*, MOR and TA collections; cf. *Albertosaurus* [= *Gorgosaurus*], GM and TA collections) from the middle and upper portions of the formation. Remains have been collected from along the Two Medicine River and from two locations west of Bynum.

• a new species of small theropod dinosaur (cf. *Saurornitholestes*, GM collections) from the lowest part of the upper portion of the formation. This specimen was collected west of Bynum.

• an unidentified lambeosaurine hadrosaur (in prep., TA collections) from a location west of Bynum, stratigraphically located in the lower part of the upper portion of the formation.

These discoveries extend the known stratigraphic range of *Hypacrosaurus, Maiasaura,* and *Gryposaurus* within the formation, and several new taxa are recognized in addition to those previously reported. It should be noted that it is possible, although unlikely, that the isolated *Gryposaurus* teeth represent a redeposit after initial fossilization. A list of dinosaur taxa by stratigraphic level and locality is presented in Table 21.1.

Faunal Turnover, Migration, and Evolution

Apparent faunal turnover in the fossil record can be produced by evolutionary change, migration, or selective preservation. Actual faunal turnover events, as opposed to inaccurate inference based on selective preservation or lack of data, can be the result of something as minor as a taxon migrating from one area to another, or as major as a catastrophe that wipes out the taxa in the region and allows repopulation from other areas. Unless the fossil and geologic records from the area are unusually intact, missing data can easily lead to a misinterpretation. Negative evidence alone (i.e., the absence of specimens of a taxon in a particular series of strata) does not provide a valid basis for inferring the absence of the species in the original ecosystem.

As can be seen from Table 21.1, specimens that allow full taxonomic diagnosis are rare in the lower and middle portions of the formation. However, fragmentary remains provide tantalizing evidence of a diverse fauna inhabiting the area at that time. Much of the limited data comes from the presence of teeth; sites where identifiable skeletons are preserved in the lowest 100 meters of the formation are virtually unknown, and in the middle portion are rare.

The deposition of the formation may be diachronous. It appears that *Maiasaura* remains are found higher in section in the Choteau area than along the Two Medicine River further north. The portion of the Two Medicine Formation where the faunal assemblage is best known is the upper portion from the area near Landslide Butte.

TABLE 21.1

Dinosaurs from the Two Medicine Formation by Stratigraphic Position and Locality

Locality	Taxa—well-preserved specimens, full diagnosis possible	Taxa—fragmentary remains and teeth, limited characters present
Landslide Butte		
Upper Portion (lower and middle portions not exposed)	*Hypacrosaurus stebingeri* *Prosaurolophus blackfeetensis* *Edmontonia rugosidens* cf. *Euoplocephalus* *Styracosaurus ovatus* *Einiosaurus procurvicornis* *Achelousaurus horneri* cf. *Leptoceratops*	*Troodon* sp. Tyrannosaurid incertae sedis Pachycephalosaurid indet. Dromaeosaurid indet. cf. *Saurornitholestes* Hypsilophodontid sp. cf. *Aublysodon*
Two Medicine River		
upper	*Hypacrosaurus stebingeri* *Prosaurolophus blackfeetensis* cf. *Euoplocephalus* *Troodon* sp. cf. *Daspletosaurus* *Maiasaura peeblesorum*	*Edmontonia rugosidens* Tyrannosaurid incertae sedis Pachycephalosaurid indet. Dromaeosaurid indet. cf. *Saurornitholestes* cf. *Aublysodon*
middle		*Gryposaurus latidens* *Troodon* sp. Tyrannosaurid incertae sedis Dromaeosaurid indet. cf. *Saurornitholestes* cf. *Aublysodon*
lower	*Gryposaurus latidens*	*Troodon* sp. Tyrannosaurid incertae sedis Dromaeosaurid indet. cf. *Saurornitholestes*
Choteau/Bynum		
upper	cf. *Hypacrosaurus* cf. *Daspletosaurus* *Maiasaura peeblesorum* *Orodromeus makelai* cf. *Saurornitholestes* cf. *Daspletosaurus* cf. *Albertosaurus* (*Gorgosaurus*)	*Troodon* sp. Pachycephalosaurid indet. Ankylosaurid indet. Dromaeosaurid indet. cf. *Aublysodon* cf. *Saurornitholestes* Ornithomimid sp.
middle	cf. *Albertosaurus* (*Gorgosaurus*)	*Troodon* sp. Ankylosaurid indet. Dromaeosaurid indet. Hadrosaurid indet. cf. *Aublysodon*
lower	Protoceratopsid incertae sedis	Tyrannosaurid incertae sedis Hadrosaurinae indet. cf. *Saurornitholestes* cf. *Aublysodon*

Due to poor preservation and the lack of specimens, we undoubtedly know little of the dinosaurian taxa that inhabited the region when the lower Two Medicine sediments were being deposited. Recent discoveries of *Hypacrosaurus* lower in the formation than previously known and well below the last occurrence of *Maiasaura* indicate that *Hypacrosaurus* coexisted with *M. peeblesorum* for some time. It is possible that other taxa known from the upper portions of the formation existed in the lower portion as well, and that there was no distinct separation of fauna between the middle and upper portions of the formation. The separation of fauna between the lower and middle portions of the formation is made more speculative by the lack of identifiable specimens from these sediments. However, the discovery of isolated teeth attributed to *G. latidens* well into the range of sediments where *M. peeblesorum* has been found suggests that this boundary may also be less distinct than once thought. From the list of taxa including fragmentary remains, the taxa separation between the lower, middle, and upper portions of the formation become indistinct at best. Teeth and fragmentary remains that could easily belong to the same taxa have been found throughout the formation.

An interesting aspect of the faunal turnover is the apparent diversification from the presence of *Maiasaura* as the only common large ornithischian in the early part of the upper portion, to the presence of a wide diversity of commonly identified large ornithischians, including at least two species each of hadrosaurs and ceratopsians, found in later sediments. The area along the Two Medicine River exhibits large exposures of the entire upper half of the formation. This area has been thoroughly examined, and preserved dinosaur remains are common throughout this area. Because of these factors, it seems likely that the increase in the number of large ornithischian taxa reflects a real condition rather than preservational bias.

A major problem with interpreting the relationship between the Two Medicine and Judith River Formations is the erosional loss of time-equivalent sediments between the two formations in Montana. In Alberta, it appears that the bentonitic mudstones that characterize the Two Medicine Formation occur in outcrops in the very south (i.e., the nesting locality near Warner), but the time-equivalent sediments a few kilometers further north (i.e., the Oldman exposures near Lethbridge) consist primarily of the large sandstone bodies consistent with the Judith River Formation (Judith River Group in Canada). Exposures between these two areas are frustratingly limited, and the postulated relationships between the formations is more inference than observation.

The current model of formational relationships identifies the Two Medicine paleoenvironment as an upper coastal plain with a dry, or at least seasonally dry, climate (Rogers 1990). The Judith River Group paleoenvironment is interpreted as a lower coastal plain/deltaic environment with a generally moist climate (Eberth and Hamblin 1993). No physical barrier to migration between the two areas has been identified, and biotic differences may have been caused by taxonomic preferences to particular habitats (Horner 1984).

Horner et al. (1992) suggested that "transitional" species are found in uppermost portions of the Two Medicine Formation, which represent the half-million-year span between the inundation of the area of the Judith River Formation and the later inundation of the Two Medicine area. They also suggested that these transitional species were due to forced evolution caused by the loss of primary habitat, the area represented by the Judith River Formation, during the later portions of the Bearpaw transgression. This hypothesis may prove valid for the cited examples of the ceratopsian and pachycephalosaurid taxa. However, the new data presented here extends the range of the cited examples of tyrannosaurid and hadrosaurid taxa to the earliest part of the upper portion of the formation. These taxa were extant well before the Bearpaw transgressive phase began, and thus could not have evolved in the manner postulated.

Conclusions

The Two Medicine Formation represents a reasonably long (~6 my) time span in which the region was subjected to a number of habitat-altering events. The earliest part represents a brief regression of the Cretaceous Interior Seaway, followed by the Claggett transgression. After the Seaway once again regressed, the area experienced a period of relative stability, although the proximity of the Elkhorn Volcanics likely had intermittent and potentially serious effects on the habitat. As the Bearpaw transgression inundated the areas to the east, available habitat was obviously affected. However, various hypotheses concerning the effects of these events on the dinosaur taxa present have often proven incorrect in the light of new discoveries.

Barring the discovery of a physical barrier to the intermingling of the dinosaur populations of the Two Medicine and Judith River regions, the most logical explanation to the species separations observed may be analogous to the ecological separation and speciation of the Galapagos iguana. Ecological separation during the Campanian may have allowed part of the dinosaur population to adapt to the higher, drier conditions of the Two Medicine region, where they speciated, while the other part adapted to the more moist lowland environment of the Judith River area. However, this explanation does not satisfy a number of observations, including:

• Taxon presence in both formations via migration or carcass transport should be possible, yet no well-known and adequately identified dinosaur taxon has been found in both formations.

• As the Bearpaw transgression occurred, taxa formerly inhabiting the lowland areas should have migrated to the newly formed lowlands that used to be highlands—the area represented by the uppermost Two Medicine Formation. Yet, even though literally hundreds of specimens have been collected from the upper Two Medicine sediments, no remains of any Judith River region taxon have been found.

• A generally northeasterly fluvial flow is postulated for the region during early to middle Two Medicine time, indicating that fluvial depositional facies in both the middle portion of the Two Medicine and lower portion of the Judith River Formations shared the same erosional source material. Yet there is no such similarity in the facies themselves.

• An arid region (the Two Medicine) and a moist lowands (the Judith River) exist within 100 kilometers of each other, under the same rain shadow effect of the high mountains to the west.

Even though the dinosaur fauna of the Two Medicine Formation is presently one of the most intensely studied, our understanding is still severely limited. Taxa range extensions show that what have seemed to be distinct faunal turnover boundaries may instead be artifacts of selective preservation. Resolution of these issues must await the discovery of diagnostic specimens, especially in the lower and middle portions of the formation.

Acknowledgments: I thank Phil Currie, Robert Bakker, Nate Murphy, and John Horner for their helpful comments concerning details of taxon identification, and Robert Gulbrandsen for his contributions to the geologic and stratigraphic work presented herein. Special thanks for the thoughtful and insightful comments from Kenneth Carpenter, Ray Rogers, and Darren Tanke in their reviews of this manuscript.

References

Eberth, D. A., and A. P. Hamblin. 1993. Tectonic, stratigraphic, and sedimentologic significance of a regional discontinuity in the upper Judith River Group (Belly River wedge) of southern Alberta, Saskatchewan, and northern Montana. *Canadian Journal of Earth Sciences* 30: 174–200.

Fiorillo, Anthony R. 1990. The first occurrence of hadrosaur (Dinosauria) remains from the marine Claggett Formation, Late Cretaceous of south-central Montana. *Journal of Vertebrate Paleontology* 10 (4): 515–517.

Gavin, W. M. 1986. A paleoenvironmental reconstruction of the Cretaceous Willow Creek Anticline dinosaur nesting locality, North Central Montana. Master's thesis, Montana State University.

Gill, J. R., and W. A. Cobban. 1973. Stratigraphy and geologic history of the Montana Group and equivalent rocks, Montana, Wyoming, and North and South Dakota. *Geological Survey Professional Paper* 776.

Gilmore, C. W. 1917. *Brachyceratops:* A ceratopsian dinosaur from the Two Medicine Formation of Montana, with notes on associated fossil reptiles. *United States Geological Survey Professional Paper* 103.

Gilmore, C. W. 1930. On dinosaurian reptiles from the Two Medicine Formation, upper Cretaceous of Montana. *Proceedings of the United States National Museum* 77. 39 pp.

Gilmore, C. W. 1939. Ceratopsian dinosaurs from the Two Medicine Formation, upper Cretaceous of Montana. *Proceedings of the United States National Museum* 87. 18 pp.

Hooker, J. S. 1987. Late Cretaceous ashfall and the demise of a hadrosaurian herd. *Geological Society of America, Rocky Mountain Section, Abstracts to Programs* 19: 284.

Horner, J. R. 1984. Three ecologically distinct vertebrate faunal communities from the late Cretaceous Two Medicine Formation of Montana, with discussion of evolutionary pressures induced by interior seaway fluctuations. *Montana Geological Society Field Conference Guidebook, Northwestern Montana:* 299–303.

Horner, J. R. 1992. Cranial morphology of *Prosaurolophus* (Ornithischia: Hadrosauridae) with descriptions of two new Hadrosaurid species and an evaluation of Hadrosaurid phylogenetic relationships. *Museum of the Rockies Occasional Paper* 2.

Horner, J. R., and P. J. Currie. 1994. Embryonic and neonatal morphology and ontogeny of a new species of *Hypacrosaurus* (Ornithischia: Lambeosauridae) from Montana and Alberta. In K. Carpenter, K. F. Hirsch, and J. R. Horner (eds.), *Dinosaur Eggs and Babies,* pp. 312–336. Cambridge: Cambridge University Press.

Horner, J. R., and R. Makela. 1979. Nest of juveniles provides evidence of family structure among dinosaurs. *Nature* 282: 296–298.

Horner, J. R., and D. B. Weishampel. 1988. A comparative embryological study of two ornithischian dinosaurs. *Nature* 332: 256–257.

Horner, J. R., D. J. Varricchio, and M. B. Goodwin. 1992. Marine transgressions and the evolution of Cretaceous dinosaurs. *Nature* 358: 59–61.

Lorenz, J. C. 1981. Sedimentary and tectonic history of the Two Medicine Formation, late Cretaceous (Campanian), northwestern Montana. Ph.D. dissertation, Department of Geology, Princeton University.

Lorenz, J. C., and W. M. Gavin. 1984. Geology of the Two Medicine Formation and the sedimentology of a dinosaur nesting ground. *Montana Geological Society Field Conference Guidebook, Northwestern Montana:* 175–185.

Mudge, M. R. 1972. Pre-Quaternary rocks in the Sun River Canyon area, northwestern Montana. *Geological Survey Professional Paper* 663–A.

Rogers, R. R. 1990. Taphonomy of three dinosaur bone beds in the upper Cretaceous Two Medicine Formation of northwestern Montana: Evidence for drought-related mortality. *Palaios* 5: 394–413.

Rogers, R. R., C. C. Swisher III, and J. R. Horner. 1993. 40Ar/39Ar age and correlation of the nonmarine Two Medicine Formation (Upper Cretaceous), northwestern Montana, U.S.A. *Canadian Journal of Earth Sciences* 30: 1066–1075.

Sampson, S. D. 1995. Two new horned dinosaurs from the Upper Cretaceous Two Medicine Formation of Montana, with a phylogenetic analysis of the Centrosaurinae (Ornithischia: Ceratopsidae). *Journal of Vertebrate Paleontology* 15: 743–760.

Stebinger, E. 1914. The Montana Group of northwestern Montana. *United States Geological Survey Professional Paper* 90–C: 60–68.

Trexler, D. 1995. A detailed description of newly discovered remains of Maiasaura peeblesorum (Reptilia: Ornithischia) and a revised diagnosis of the genus. M.Sc. thesis, Department of Biological Sciences, University of Calgary.

Varricchio, D. J., 1993. Montana climatic changes associated with the Cretaceous Claggett and Bearpaw transgressions. *Montana Geological Society Field Conference Guidebook, Energy and Mineral Resources of Central Montana:* 97–102.

Viele, G. W., and F. G. Harris. 1965. Montana Group stratigraphy, Lewis and Clark County, Montana. *Bulletin, American Association of Petroleum Geologists* 49: 379–417.

22. Late Cretaceous Dinosaur Provinciality

Thomas M. Lehman

Abstract

Late Cretaceous (Campanian-Maastrichtian) dinosaurs in the western interior of North America were remarkably provincial. Distinctive endemic associations of dinosaur herbivores exhibit a persistent latitudinal and altitudinal zonation. During late Campanian time, diverse and highly specialized hadrosaur-dominated faunas arose. However, in Maastrichtian time the Laramide Orogeny brought about environmental changes that led to the subordination of hadrosaurs, and the reduction in kinds and abundance of both hadrosaurs and ceratopsids, particularly among the more ornate and specialized lambeosaurines and centrosaurines. Sauropods, protoceratopsids, hypsilophodontids, and pterosaurs emerged from upland refugia as important "archaic" elements in the latest Cretaceous fauna. This Maastrichtian faunal turnover represents the most dramatic event that affected Late Cretaceous dinosaur communities in North America prior to their extinction.

Introduction

The pattern of dinosaur distribution and abundance during Late Cretaceous time in the western interior of North America provides evidence for altitudinal and latitudinal habitat zonation among dinosaurs at that time. The changing geographic distribution of dinosaurs plays a role in understanding their biology and phylogeny, and in particular the short interval between the late Campanian and the Maastrichtian, about 15 million years, has received extensive study.

Recent reviews have drawn attention to dinosaur diversity changes prior to their extinction (Russell and Dodson 1997; Dodson and Tatarinov 1990; Sloan et al. 1986; Bakker 1986). Although new discoveries are still being made, we may be approaching (if asymptotically) a limit in the documentation of Campanian-Maastrichtian faunas afforded by the stratigraphic record, at least in some areas. Of course there are many sources of error in an analysis of paleobiogeography (reviewed by Lehman 1997) and we may always lack sufficient knowledge of dinosaur distributions in most areas, including mountainous and other nondepositional regions, as well as for time increments between well sampled intervals.

Nevertheless, a broad review of late Campanian (Lehman 1997) and late Maastrichtian (Lehman 1987) dinosaur biogeography in western North America has been offered previously; and it is worthwhile to examine more closely the transition across this time interval. Observations made by those studying dinosaur assemblages in the various parts of the western interior region are reviewed, as are several hypotheses purporting to explain the biogeographic pattern that emerges. The provincial terms Judithian, Edmontonian, and Lancian are used in the following discussion to refer broadly to time increments of approximately 80 to 75 Ma, 75 to 70 Ma, and 70 to 65 Ma, respectively within the late Campanian–Maastrichtian interval (Russell 1975; Archibald 1996).

Endemism among Herbivorous Dinosaurs

In spite of their high mobility and typically large body size, most Late Cretaceous dinosaurs were decidedly not cosmopolitan in their distribution. Although most dinosaur families were widely distributed geographically and among environments, many dinosaur genera and species had remarkably small geographical ranges. Furthermore, in many cases it is the most conspicuous and abundant species that have the most restricted distributions. For example, *Corythosaurus* and *Centrosaurus* are unknown outside southern Alberta, where they are the most abundant Judithian dinosaurs; similarly *Pentaceratops* is unknown outside northern New Mexico, where it is the only known Judithian ceratopsian. Endemism is particularly evident among large herbivores such as hadrosaurs and ceratopsians, and stands in marked contrast to modern large-bodied mammalian herbivores, where typical geographic ranges span much of a continent.

For example, today there are 41 species of "large" mammals in North America (including bovids, cervids, antilocaprids, felids, canids, ursids, procyonids, a dasypodid, and some mustelids; range data from Burt and Grossenheider 1976). If our knowledge of the distribution of modern large North American mammals were limited to three hypothetical "fossil" assemblages, one in southern Alberta, one in northern New Mexico, and another in southwestern Texas, 34 of the 41 species could potentially be represented at these three sites; the remaining seven species have ranges outside these three areas (e.g., polar bear, musk ox).

About 20 species could potentially be represented at each of the three sites, and at least 11 of these species (perhaps as many as 16) would occur in all three areas. Moreover, these 11 would likely be the most abundant and conspicuous as fossils (e.g., bison, mule deer, white-tailed deer, coyote, black bear, raccoon). Only relatively rare taxa (some perhaps too rare for likely fossilization) would exhibit a restricted distribution (e.g., coati, peccary, armadillo, and ocelot might be diagnostic of a "southern" assemblage, while the mountain goat, moose, wolverine, and grizzly bear might be diagnostic of a "northern" assemblage). Among the most common taxa there would be little or no evidence for latitudinal faunal provinciality. This modern example illustrates that even with a strong latitudinal temperature gradient, many large mammals are widely distributed across a continent, in spite of their small size compared to dinosaurs.

In contrast, endemism is marked among the large herbivorous dinosaurs, suggesting that restrictions on their distribution may have been imposed by limited vegetational forage preferences, or narrow climatic or other environmental tolerances. They must have differed ecologically from large mammalian herbivores, whose widespread distribution may reflect in part the Neogene expansion of open savanna and plains habitats that were lacking in Late Cretaceous time. The limited range of dinosaur herbivores was likely not imposed by geographic barriers to dispersal. Carnivorous dinosaurs, particularly the smaller theropods, seem to have been more widely distributed; although they are known in most areas only by their shed teeth and so it is possible that when known from more complete material they also will prove to have more limited ranges. Hence, the present review focuses on the distribution of large herbivorous dinosaurs. Their restricted occurrence likely reflects a high degree of specialization, and this likely had an effect on their response to environmental change. Biotas characterized by a particular assemblage of dinosaurs are referred to here as ecological "associations," each identified by characteristic index species. The remarkable endemism among dinosaur herbivores appears to reflect both altitudinal and latitudinal habitat zonation.

Altitudinal and Transcontinental Life Zones

It has long been recognized that some dinosaurs appear to be restricted in their occurrence with respect to distance from the paleo-shoreline (e.g., contemporaneous faunas of the Two Medicine versus Judith River Formations; Horner 1984, 1989). Such observations have allowed for vague discrimination of "coastal" versus "inland" dinosaur assemblages (Lucas 1981; Lehman 1981). Furthermore, the entire Judithian through Lancian terrestrial sedimentary sequence in western North America is broadly regressive in character. Hence, the stratigraphic succession of fauna and flora preserved in each area records not only those evolutionary changes that occurred over time, but also the progression of altitudinal life zones that existed from coastal habitats at low elevations near sea level, to inland submontane habitats at higher elevation (e.g., Wheeler and Lehman 2000). Both phenomena allow for recognition of former altitudinal habitat zonation or "life zones."

Additionally, it is well known that the modern altitudinal biotic zonation mimics in many ways the broader latitudinal or transcontinental zonation. Life zones encountered where progressively ascending a mountain range resemble those observed moving poleward at low elevation, and species characteristic of high latitudes may extend their range to lower latitudes where a north-south trending mountain range creates an extension of comparable and suitable habitat. Hence, we may suppose that species collected from high Late Cretaceous paleolatitude might resemble those typical of piedmont environments at lower paleolatitude (e.g., occurrence of *Pachyrhinosaurus* in Alaska as well as Alberta; Brouwers et al. 1987).

Throughout the Judithian-Lancian time interval, the western interior region of North America was separated into distinct northern and southern biotic realms with a boundary in the vicinity of northern Colorado (Lehman 1987, 1997). Northern and southern faunal provinces correspond roughly with the *Aquillapollinites* and *Normapolles* palynofloral provinces, respectively. This persistent north-south provincialism in fauna and flora likely reflects the distinction between two latitudinal transcontinental life zones of Late Cretaceous time (Lehman 1997). Although the delineation of modern latitudinal "life zones" was once widely accepted, they are little used today because they are thought to oversimplify more complex biotic variation. Instead, large geographical areas with generally similar climate and characterized by a distinctive biotic community are recognized as "biomes." The two major western interior Late Cretaceous biomes appear to have been arrayed latitudinally as in transcontinental life zones, although given our present knowledge this is certainly an oversimplification. Horrell (1991) identified the northern interior (*Aquillapollinites*) life zone as part of the warm temperate everwet forest biome, and the southern interior (*Normapolles*) life zone as part of the warm dry winter wet shrubland biome. Very few dinosaurian herbivores cross over to occur in both northern and southern biomes.

In the northern biome, the late Campanian–Maastrichtian time interval generally records the transition from diverse, and in some ways extravagant Judithian dinosaur faunas to a Lancian fauna dominated by a single chasmosaurine ceratopsian (*Triceratops*) and hadrosaurine hadrosaur (*Edmontosaurus*), both of which were rather plain compared to their more ornamented predecessors. During this transition, the bizarrely specialized lambeosaurs and centrosaurs that so characterized Judithian time in the northern biome became extinct. At the same time, the southern biome witnessed the return of a "relict" fauna dominated by a titanosaurid sauropod (*Alamosaurus*). Lancian environments were visually, if not ecologically, dominated by a single large herbivore (*Triceratops* in the northern biome, *Alamosaurus* in the south). Although no conclusive judgment as to cause is offered here, it is useful to draw further attention to this faunal transition, and to pose several questions that arise from its consideration.

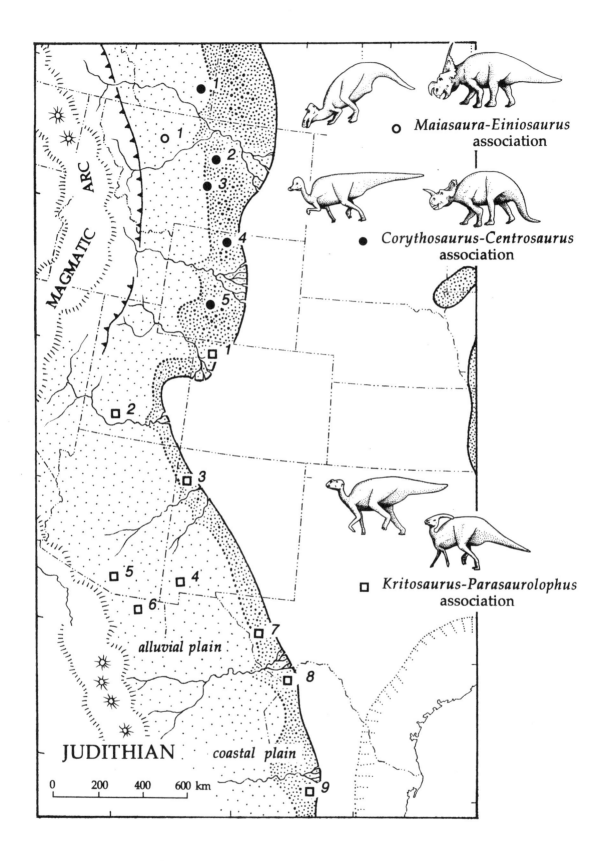

Maiasaura-Einiosaurus
association

Corythosaurus-Centrosaurus
association

Kritosaurus-Parasaurolophus
association

MAGMATIC ARC

JUDITHIAN

alluvial plain

coastal plain

0   200   400   600 km

Judithian Climax

The Judithian "age" (late Campanian, ca. 80 to 75 Ma) may have been the acme in dinosaur evolution in North America (fig. 22.1). The biogeography of this interval was reviewed in broad terms by Lehman (1997). In all areas the dinosaur fauna was dominated by hadrosaurs, which comprise over half of a typical assemblage; and in most environments varied hadrosaurs would certainly have been the most conspicuous inhabitants. The greatest generic diversity among large dinosaur herbivores was attained at this time, with as many as ten genera of hadrosaurs and ten genera of ceratopsians in Montana and Alberta alone.

In southern Alberta, the diverse dinosaur assemblage was dominated by lambeosaurine hadrosaurs and centrosaurine ceratopsians, with *Corythosaurus* and *Centrosaurus* perhaps most characteristic of this association (Dodson 1983). However, the dominance of lambeosaurine hadrosaurs is only evident in southern Alberta. Horner (1988, 1989) indicates that lambeosaurs are less abundant in nearby Judithian coastal deposits of Montana, only a few hundred kilometers away, and centrosaurines such as *Monoclonius* are found in lieu of *Centrosaurus* (Dodson 1986). The characteristic Albertan Judithian taxa are also not found in correlative inland facies (Two Medicine Formation; Horner 1983), where instead *Maiasaura* and early representatives of the pachyrhinosaur lineage (e.g., *Einiosaurus*; Sampson 1995) are dominant.

In the southern biome of Utah and northern New Mexico a fauna of lower diversity existed where lambeosaurines were subordinate and centrosaurines were absent (Lehman 1997). The hadrosaurine *Kritosaurus* (*Naashoibitosaurus* and *Anasazisaurus* are regarded as junior synonyms; Horner 1992; Hunt and Lucas 1993), the single lambeosaurine *Parasaurolophus*, and chasmosaurine *Pentaceratops* comprise the dominant taxa. In Texas and Mexico, a similar fauna is also dominated by *Kritosaurus* (Rowe et al. 1992). Both *Kritosaurus* and *Parasaurolophus* also occur infrequently in the northern biome, indicating that exchange was possible, but they are the dominant taxa in the southern biome. The giant eusuchian crocodile *Deinosuchus* is also a conspicuous member of the southern assemblage. Although also having some endemic forms, it seems likely that the dinosaur fauna of the Gulf and Atlantic coast of "Appalachia" (sensu Archibald 1996) was generally similar to that of Texas; however, lacking ceratopsians (Schwimmer 1997).

Edmontonian Transition

The Edmontonian (early Maastrichtian, ca. 75 to 70 Ma) may be recognizable as a distinct biostratigraphic interval only in the northern biome (fig. 22.2). In the southern realm (e.g., Texas and New Mexico), the Judithian-Edmontonian interval appears to be indivisible biostratigraphically. Here, the previous Judithian assemblage persists with few noticeable changes among dinosaurian taxa.

A critical time in the evolution of Late Cretaceous dinosaur com-

Figure 22.1. (opposite page) Paleogeographic reconstruction of the western interior region of North America during Judithian time (ca. 80–75 Ma, Late Campanian) adapted from Lehman (1997). Paleoshoreline configuration is approximately that of the Baculites scotti biozone, adapted from Dyman et al. (1994) and Cobban et al. (1994). Alluvial plain–coastal plain boundary is approximated by the limit of previous Early Campanian maximum transgression. Possible representatives of the Corythosaurus-Centrosaurus association (solid circles) are: (1) Dinosaur Provincial Park, (2) type area of Judith River Formation, (3) Wheatland County area, (4) Mesaverde Formation in Bighorn Basin, and (5) Wind River Basin. The Maiasaura-Einiosaurus association (open circle) is found in (1) the Two Medicine Formation. Possible representatives of the Kritosaurus-Parasaurolophus association (squares) are: (1) Williams Fork Formation, (2) Kaiparowits Formation (part), (3) Fruitland and lower Kirtland Formations, (4) Ringbone Formation, 5) Fort Crittenden Formation, (6) Cabullona Group, 7) San Carlos Formation, (8) lower Aguja Formation, and (9) Cerro del Pueblo Formation.

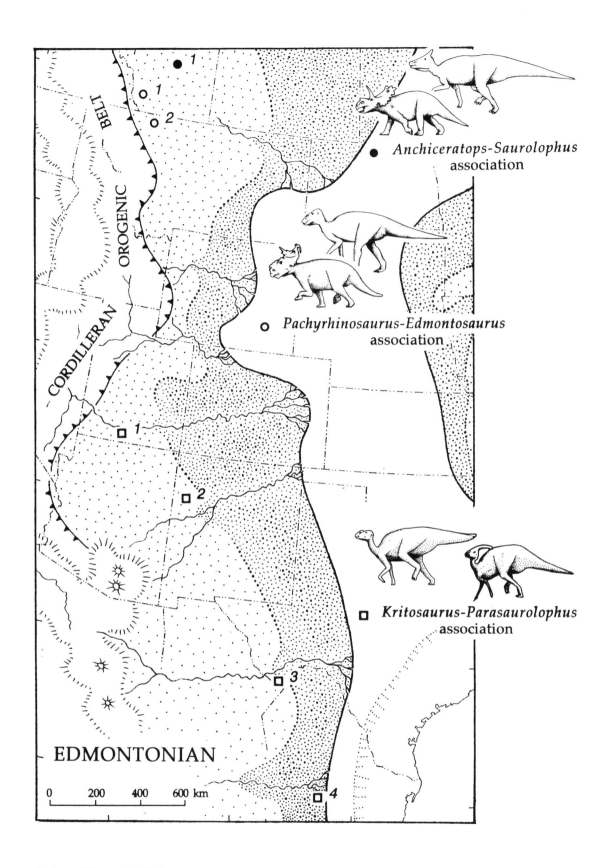

Anchiceratops-Saurolophus
association

Pachyrhinosaurus-Edmontosaurus
association

Kritosaurus-Parasaurolophus
association

CORDILLERAN OROGENIC BELT

EDMONTONIAN

0   200   400   600 km

munities occurred, however, during the Edmontonian transition. The northern Edmontonian assemblage commences a trend culminated in Lancian time, with a dramatic reduction of centrosaurines; only the single bizarre surviving Edmontonian genus *Pachyrhinosaurus;* and a similar reduction among lambeosaurines, represented by the sole survivor *Hypacrosaurus.* Neither group appears to have persisted later into Lancian time. An inland association characterized by *Saurolophus* and *Anchiceratops* coexisted alongside a coastal assemblage characterized by *Pachyrhinosaurus* (Langston 1975). The *Pachyrhinosaurus* association inhabited regions as far north as Alaska (Clemens and Nelms 1993; Brouwers et al. 1987). Its presence in Alberta suggests a southward deflection of Late Cretaceous transcontinental life zones along the cordillera, as observed in modern North America. The appearance of *Montanoceratops* with the inland Edmontonian fauna marks a late (re)appearance of basal neoceratopsians in North America, and along with hypsilophodontids (*Parksosaurus*) the presence of "archaic" elements in the upland fauna. *Arrhinoceratops* and *Edmontosaurus,* likely ancestors to Lancian dominants *Triceratops* and *Edmontosaurus* appear in the northern Edmontonian fauna.

Lancian Turnover

The culminating Lancian dinosaur faunas in North America (late Maastrichtian, ca. 70 to 65 Ma) differ markedly from their Judithian precursors (fig. 22.3). By this time, hadrosaurs are subordinant in all environments and are no longer the most characteristic species in any province. The northern *Triceratops-Edmontosaurus* association may represent the evolutionary culmination of surviving lineages of previous inhabitants of the northern coastal plains. Lancian environments had much lower diversity among large herbivores, with two surviving chasmosaurines (common *Triceratops,* rare *Torosaurus*) and one or two surviving hadrosaurines (common *Edmontosaurus,* rare "*Anatotitan*"); although *Triceratops* alone dominates this assemblage. Evidence for this decline in diversity or ecological "evenness" (sensu Bakker 1986) and its role in dinosaur extinction have been discussed by Bakker (1986), Sloan et al. (1986), Dodson and Tatarinov (1990), and Russell and Dodson (1997). Neither *Triceratops* or *Edmontosaurus* are as elaborate or bizarrely ornamented as their predecessors. In some ways this association is relict in aspect. In its domination by a single chasmosaurine and a single unornamented hadrosaurine, this association is superficially similar to the earlier Judithian-Edmontonian faunas of the southern biome. Protoceratopsids (*Leptoceratops*) maintain their presence in inland regions at this time, and hypsilophodontids (*Thescelosaurus*) remain a notable element in the coastal plain fauna. Both represent lineages prominent in earlier Cretaceous faunas.

In southern environments, the Lancian transition is even more dramatic, with the abrupt reemergence of a fauna having a superficially "Jurassic" aspect. This assemblage is overwhelmingly dominated by the titanosaurid sauropod *Alamosaurus,* and in Texas abundant ptero-

Figure 22.2. (opposite page) Paleogeographic reconstruction of the western interior region of North America during Edmontonian time (ca. 75–70 Ma, Early Maastrichtian). Paleoshoreline configuration is approximately that of the Baculites clinolobatus biozone, adapted from Lillegraven and Ostresh (1990) and Cobban et al. (1994). Alluvial plain–coastal plain boundary is approximated by the position of previous Judithian shoreline. The Anchiceratops-Saurolophus association (solid circle) is found in the (1) Horseshoe Canyon Formation. The Pachyrhinosaurus-Edmontosaurus association (open circles) is found in (1–2) St. Mary River Formation. Possible representatives of the Kritosaurus-Parasaurolophus association (squares) are: (1) Kaiparowits Formation (part), (2) upper Kirtland Formation, (3) upper Aguja Formation, and (4) Cerro del Pueblo Formation (part).

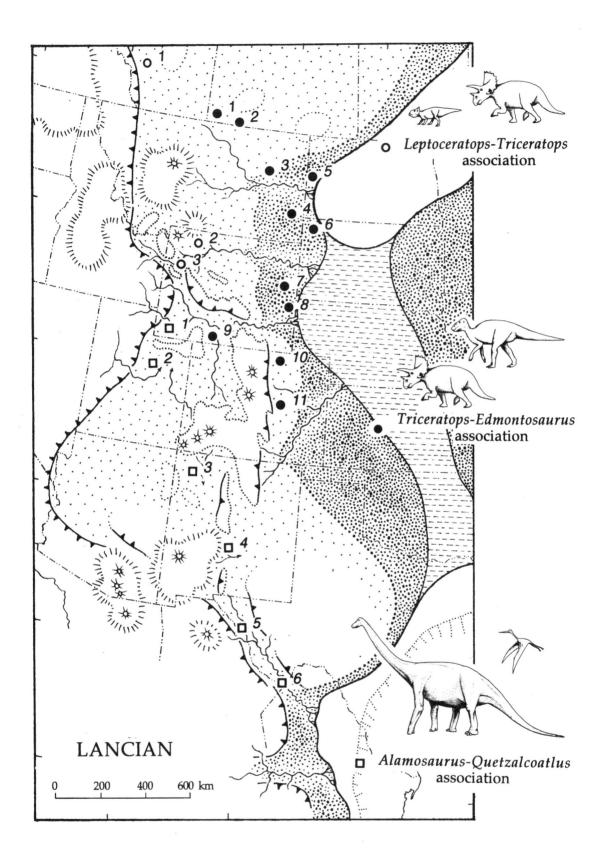

Leptoceratops-Triceratops
association

Triceratops-Edmontosaurus
association

Alamosaurus-Quetzalcoatlus
association

LANCIAN

0   200   400   600 km

saurs (*Quetzalcoatlus*). *Alamosaurus* was unquestionably the dominant animal in its environment; although ceratopsians and hadrosaurs are present, they are known from only a few specimens.

What Happened?

The subordination of hadrosaurs, reduction in diversity among both hadrosaurs and ceratopsids, particularly the extinction of more ornate and presumably specialized forms (centrosaurines and lambeosaurines), and the reemergence of sauropods, protoceratopsids, hypsilophodontids, and pterosaurs as important "archaic" elements in the terminal Cretaceous fauna evokes the expansion of a relict biota from earlier refugia. The most striking changes are evident among the large herbivorous dinosaurs, implying that a change in vegetation may have been the most immediate cause of this faunal turnover, though perhaps not the ultimate one.

This faunal turnover coincides in time with dramatic tectonism in the interior of North America—the onset of the Laramide Orogeny and uplift of the central Rocky Mountains (fig. 22.3). This event corresponds to a marked change in depositional systems, to a shift in paleocurrent orientation, sediment lithology, and facies in most areas, and to a marked relative fall in sea level. It seems hardly coincidental that a dramatic change in the North American dinosaur fauna occurred at this time, and it is convenient to relate the change in dinosaur distribution directly or indirectly to this event. It must have led at least to a shift in altitudinal life zones, and so the distribution of vegetation types that herbivorous dinosaurs utilized.

The Asian-American peninsula of Late Cretaceous time ("Laramidia" of Archibald 1996) was similar in physiographic aspect to the modern Mexican–Central American peninsula, and offered a much smaller area (about 10%) of terrestrial habitat than in North America today. At the end of Judithian time, North America had a land area of about 7.7 million km², and by the end of Lancian time about 17.9 million km², approaching the land area of North America today, 22.5 million km² (based on hypsographic data of Harrison et al. 1983; modern 610 m and 150 m elevation contours approximate Judithian and Lancian land areas). This kind of rapid expansion of terrestrial habitat (more than doubling from Judithian through Lancian time) is difficult to ignore; although the ecological consequences of simply increasing available terrestrial living space have not been explored in the same way that the effects of corresponding reduction in continental shelf area have on shallow marine ecosystems.

It has been postulated that uplift of the Rocky Mountains and regression of the interior epeiric sea through Campanian and Maastrichtian time may have affected dinosaur faunas in three major ways. It undoubtedly resulted in a rapid expansion of terrestrial habitat, and this likely had a subordinate effect on climate, leading to increasingly dry continental interior conditions and changes in vegetation ("loss of wetlands" hypothesis). At the same time, regression may have allowed

Figure 22.3. (*opposite page*) *Paleogeographic reconstruction of the western interior region of North America during Lancian time (ca. 70–65 Ma, Late Maastrichtian) adapted from Lehman (1987). Early paleoshoreline configuration is approximately that of the* Discoscaphites nebrascensis *biozone, later is approximately that at the Cretaceous-Tertiary boundary adapted from Lillegraven and Ostresh (1990) and Lehman (1987). Alluvial plain–coastal plain boundary is approximated by the position of previous Edmontonian shoreline. Possible representatives of the Triceratops-Edmontosaurus association (solid circles) are: (1–2) Frenchman Formation, (3–6) Hell Creek Formation, (7–9) Lance Formation, (10) Laramie Formation, and (11) Denver Formation. Possible representatives of the Leptoceratops-Triceratops association (open circles) are: (1) Scollard Formation, (2) Lance Formation in Bighorn Basin, and (3) Pinyon Canyon Formation. Possible representatives of the Alamosaurus-Quetzalcoatlus association (squares) are: (1) Evanston Formation, (2) North Horn Formation, (3) Naashoibito Member of Kirtland Shale, (4) McRae Formation, (5) El Picacho Formation, and (6) Javelina Formation.*

increased opportunity for immigration from other continents ("competition from invaders" hypothesis), and expansion of life zones to lower altitudes ("descent from the highlands" hypothesis).

Loss of Wetlands Hypothesis

Rapid expansion of terrestrial habitat and climate change attendant with a fall in sea level could provide a mechanism for change in flora and geographic and reproductive isolation of dinosaur populations. Archibald (1996) suggests that relative fall in sea level at the terminal Cretaceous may have lead to loss and fragmentation of coastal plain habitat, critical to many dinosaur species. Presumably this habitat fragmentation resulted in isolation of dinosaur populations below the level at which some species were viable. This seems particularly plausible when it is recognized that many dinosaur species had very limited geographic distributions in any case. If dinosaur species had small home ranges, a limited geographic distribution, perhaps with high nest-site fidelity, and were nonmigratory, this would have left them vulnerable to even localized environmental change. Species that are most adversely affected by such changes are typically those that have low dispersal power, limited geographical ranges, and narrow ecological tolerances.

Loss of wetlands could explain why hadrosaurs became subordinant among Lancian dinosaurs. However, even though the length of shoreline certainly decreased at this time, it is not clear that the actual area of coastal wetlands decreased, and in fact the area of low-lying coastal plain (elevations within about 150 m of sea level) must have expanded dramatically (fig. 22.3). Many dinosaurs were clearly not inhabitants of coastal wetland environments, and indeed these seem to have been among the species that prevailed in Lancian faunas. For example, there are numerous indications that the environments inhabited by the Lancian *Alamosaurus-Quetzalcoatlus* association were indeed semi-arid inland plains (Lehman 1989).

Horner et al. (1992) theorized that rising relative sea level between Judithian and Edmontonian time (the Bearpaw transgression) induced anagenetic change in several dinosaur lineages owing to adaptive pressure exerted by reduction of critical coastal plain habitat with rising sea level ("habitat bottlenecking," sensu Horner et al. 1992). However, Lillegraven and Ostresh (1990) demonstrated that Late Cretaceous transgressive-regressive cycles, including the Bearpaw transgression, were not in phase along the length of the western interior region. The Bearpaw transgression only affected the Montana and Alberta area. If the geographic ranges of dinosaur herbivores were truly as restricted as suggested here, and they were incapable of emigrating even a few hundred kilometers, then perhaps even a local transgressive event would force such adaptive pressure. If so, then alternatively falling sea level could have reduced the adaptive pressure for biological specialization, and led to rapid colonization by a few opportunistic dinosaur generalists, a pattern that seems evident in this case.

Competition from Invaders Hypothesis

Immigration from Asia could explain the late (re)appearance of basal neoceratopsians (*Montanoceratops, Leptoceratops*) in western North America, the introduction of dinosaurs resembling Asian forms (e.g., *Nodocephalosaurus;* Sullivan 1999), and the co-occurrence of such dinosaurs as *Saurolophus* in Asia as well as North America. Possible immigrants are particularly evident in upland habitats, rather than in coastal faunas of the time. All such occurrences could be evidence for the opening of an immigration corridor from Asia to North America in Edmontonian time. Bakker (1986) emphasized how immigrants might have brought competition, predation, and disease to drive the Maastrichtian faunal turnover among dinosaurs, and ultimately their extinction.

The appearance of *Alamosaurus* in the southern Lancian fauna is also generally thought to record an immigration event from South America (Kues et al. 1980). Such a southern connection could explain the possible occurrences of *Kritosaurus* and *Avisaurus* in South America as well (e.g., Rage 1986). However, if *Alamosaurus* were an immigrant from South America, at best it would have traveled by tropical "island hopping" (perhaps even requiring swimming). Yet all indications are that *Alamosaurus* was an inhabitant of relatively arid upland environments (Lehman 1989). Unlike coastal species, inhabitants of upland regions tend to be more endemic and less capable of dispersal over stretches of open water. Moreover, *Alamosaurus* was not simply a chance migrant or occasional exotic; this animal rapidly assumed dominance in its habitat; its introduction in the stratigraphic section records an abrupt event, not a gradual replacement. Some consider it possible alternatively that *Alamosaurus* was an immigrant from Asia (Wilson and Sereno 1998).

However, titanosaurid sauropods were already present in North America during Early Cretaceous time (Kirkland et al. 1997; Ostrom 1970), and fragmentary remains of a titanosaurid similar to, if not the same as *Alamosaurus*, are found in Santonian-Campanian deposits of southern Arizona (McCord 1997; Ratkevitch 1997). Hence, it is just as likely that the immediate ancestor of *Alamosaurus* was an indigenous inhabitant of North America. The same is true of the *Alamosaurus* associate *Quetzalcoatlus*. The occurrence of azhdarchid pterosaurs similar to *Quetzalcoatlus* in older Judithian deposits (Currie and Russell 1982; Padian 1984) suggests that the abundance of azhdarchids later in the Lancian of Texas may simply record the more widespread development of favorable semi-arid upland environmental conditions, and not a late immigration event. Furthermore, basal neoceratopsians are now known from Early Cretaceous deposits in North America, and it is not necessary to call upon a late immigration event to explain their reappearance here later (Chinnery et al. 1998). Hence, evidence for the importance of immigration in driving the turnover in dinosaur fauna is not particularly compelling. For the most part, possible dinosaur immigrants in western North America represent archaic lineages and it is difficult to envision them in direct competition with the typical North American hadrosaurs and ceratopsids in any case.

Descent from the Highlands Hypothesis

During Late Cretaceous time, rising relative sea level and progressive expansion of the angiosperm flora from coastal lowlands to inland environments may have resulted in isolation of highland refugia in the cordillera for conifer-dominated flora and archaic sauropod/protoceratopsid faunas. The "advanced" ceratopsid and hadrosaur species that dominated the Judithian landscape seem to have been adapted to coastal environments with angiosperm flora. The return of sauropods, protoceratopsids, hypsilophodontids, and azhdarchid pterosaurs in Lancian time could simply record a shifting of existing species to lower elevations as altitudinal life zones expanded with the regression of the interior epeiric sea. These supposed "relict" faunas may have inhabited highland regions, where they largely escaped the reach of the fossil record, persisting ultimately to descend into nearby inland areas and displace the more progressive hadrosaur-ceratopsian faunas which followed the retreating lowlands bordering the epeiric sea. As Cretaceous time drew to a close, the archaic highland faunas, perhaps also adapted to drier conditions, encroached to lower altitudes as the climate in the western interior became increasingly dry and continental in aspect. The typical form of a continental hypsographic curve (e.g., Harrison et al. 1983) indicates that a relatively small drop in sea level will result in a significant increase in the aerial extent of life zones, particularly at lower elevations.

Neoendemic, Paleoendemic, or Immigrant?

The possible effects of falling sea level, immigration, lowering of life zones, climate change, fragmentation of habitat, and loss of coastal wetlands are difficult to separate. Of course it is likely that some combination of these or other processes lead to the changing dinosaurian landscape. Proper evaluation of these and other likely processes also awaits the resolution of phylogenetic relationships for the dominant terminal Cretaceous dinosaurs. Oddly, all of the dominant terminal Cretaceous taxa (*Triceratops, Edmontosaurus, Leptoceratops, Alamosaurus*) are the subject of persistent phylogenetic uncertainty. Do these taxa represent (1) descendants of indigenous North American lineages, (2) immigrants introduced from other continents, or (3) survivors of relict endemic lineages whose ranges expanded to lower altitudes?

All seem to represent superficially plain or "simple" species, lacking the extravagant specializations found in their Judithian predecessors. But are these "advanced" species (representing the culmination of more derived lineages) only superficially convergent on basal morphology, or "archaic" holdovers (representing survivors of basal lineages)? Why were the survivors at the end of Lancian time relatively "plain" compared to their predecessors?

For example, *Triceratops* with its short, closed frill was for decades allied with the centrosaurine ceratopsians (e.g., Colbert 1948). However, today it is viewed alternatively as the culmination of the chasmosaurine lineage, with a secondarily shortened and closed frill (e.g., Dodson and Currie 1990; Lehman 1996), or as a "relict" basal ceratopsid (Penkalski and Dodson 1999) or basal chasmosaurine (Forster et al. 1993). If either of the latter hypotheses are correct, *Triceratops* represented the survivor of a long "ghost" lineage. If so, where were its ancestors? Likewise, was *Leptoceratops* the most primitive protoceratopsid survivor of a long ghost lineage (Dodson and Currie 1990), or a more derived immigrant of recent heritage (Chinnery and Weishampel 1998)? Similar debate surrounds the phylogenetic placement of *Edmontosaurus* (e.g., Hopson 1975) and *Alamosaurus* (Wilson and Sereno 1998). Determining whether these animals represent the successful competitors among diverse coastal plain lineages that remained in Lancian time, longtime inhabitants of cordilleran highlands where they mostly escaped the reach of the fossil record, descending at the end of Cretaceous time, or the products of multiple immigration events will certainly color our ultimate understanding of the Lancian faunal turnover.

A Recent Analog?

The North American Neogene large mammal fauna experienced a reduction in diversity, similar to that observed among terminal Cretaceous dinosaurs, particularly affecting the large herbivores, and occurring over a comparable time span (ca. 5 to 10 million years) from Late Miocene time to the present. For example, typical Late Miocene (Clarendonian-Hempillian) faunas of the Great Plains had a diverse assemblage of about 20 genera of sympatric large mammalian herbivores (e.g., proboscideans, rhinoceroses, camels, horses, peccaries, oreodonts, antilocaprids, and other horned ruminants; Schultz 1990). The fauna increased to include perhaps as many as 25 genera during Pleistocene time with the addition of immigrant bovids and cervids from Eurasia, as well as ground sloths and glyptodonts from South America. This diverse mammalian herbivore megafauna collapsed to as few as six sympatric genera during Holocene time, though typically only two or three genera dominate the modern Great Plains fauna (e.g., bison, deer, and pronghorn).

In several ways, the Clarendonian-Holocene turnover in North American mammalian herbivores is similar to that observed in dinosaurian herbivores during the Judithian-Lancian transition: (1) An earlier diverse assemblage of herbivores was replaced by a single dominant species (bison in the south, caribou in the north), (2) the most common, diverse, and extravagant of the earlier herbivore groups (elephants, rhinoceros, camels, horses) became extinct, (3) the turnover was preceded by an episode of immigration, and (4) associated with rapid expansion of terrestrial habitat (in this case resulting from deglaciation). The compression of terrestrial life zones during glaciation, and rapid expansion of terrestrial habitat during deglaciation perhaps produced effects similar in some ways to those resulting from the culminating Late Cretaceous transgressive-regressive cycle. However, in the case of the Clarendonian-Holocene transition, the surviving dominant herbivore taxa were clearly Eurasian immigrants (bovids and cervids), and early human inhabitants of North America played an uncertain role in this faunal transition (the "overkill" hypothesis; Martin 1973) . Nevertheless, the result was strikingly similar to that seen in the Judithian-Lancian turnover.

Conclusions

The Judithian dinosaur faunas of western North America reflect the progressive development of endemic biotas in northern and southern biomes following an earlier (ca. Cenomanian) dispersal event. Highly favorable environmental conditions led to increasingly complex and ornate dinosaur herbivores in each area, suggesting elaboration of biological interactions over physical adaptations. However, their coexistence may have been made possible by specialization of each to a limited range of available food resources. Migration may have been inhibited between such species-rich dinosaur communities by filling of all niches and efficient utilization of resources. The domination of hadrosaurs in all provinces may reflect a prevalence of coastal wetland habitats at that time. The persistent north-south latitudinal zonation apparent among dinosaurs appears comparable to modern latitudinal biotic provinciality, and probably reflects a similar circumstance. The northern and southern biomes likely reflected a unique set of environmental conditions to which species in other provinces were not adapted, though no physical barrier prevented their entering. An altitudinal zonation is also apparent, as in modern life zones. This suggests that more northerly latitudes may have been dominated by pachyrhinosaurs and protoceratopsians.

The diverse Judithian biota may record the "climax" in long-term development and "individuality" of dinosaur communities. It was brought to an end through a disturbance during Edmontonian time. The onset of the Laramide Orogeny led to a dramatic relative fall in sea level and rapid expansion of terrestrial habitat in western North America at that time. Opportunistic "weedy" and seemingly generalized herbivores rapidly colonized the disturbed habitat that emerged, resulting in domination of a single herbivore species in most environments over previously diverse and highly specialized species. During this transition, a diverse assemblage of ornate dinosaur forms was replaced by a few simple unornamented types. These seem to be survivors of indigenous lineages rather than immigrants. However, at the same time, dinosaur relicts—sauropods and protoceratopsids—descended from nearby cordilleran upland refugia, and appeared initially in areas of marginal habitat, or where no ecological equivalents were present, but spread over a broader area as altitudinal life zones expanded to lower elevations with falling sea level. By Lancian time, hadrosaurs were a subordinate part of the fauna, and both hadrosaurs and ceratopsians were displaced by sauropods in the southern biome. At the end of Cretaceous time, a single large herbivore species dominated the landscape in most areas, *Triceratops* in the north and *Alamosaurus* in the south.

Acknowledgments: Phil Currie, whose work has greatly expanded our knowledge of the northern Late Cretaceous biome, encouraged the author's studies of dinosaurs. The ideas presented here were also influenced by stimulating discussions with Jonathan Wagner, J. Jeff Anglen, and Alan Coulson, and the manuscript benefited from the comments of Dale Russell and Kenneth Carpenter. The illustrations are the work of the author.

References

Archibald, J. D. 1996. *Dinosaur Extinction and the End of an Era.* New York: Columbia University Press.

Bakker, R. T. 1986. *The Dinosaur Heresies.* New York: Kensington Publishing.

Brouwers, E. M., W. A. Clemens, R. A. Spicer, T. A. Ager, L. D. Carter, and W. V. Sliter. 1987. Dinosaurs on the North Slope, Alaska: High latitude, latest Cretaceous environments. *Science* 237: 1608–1610.

Burt, W. H., and R. P. Grossenheider. 1976. *A Field Guide to the Mammals.* Boston: Houghton Mifflin.

Chinnery, B. J., and D. B. Weishampel. 1998. *Montanoceratops cerorhynchus* (Dinosauria: Ceratopsia) and relationships among basal neoceratopsians. *Journal of Vertebrate Paleontology* 18: 569–585.

Chinnery, B. J., T. R. Lipka, J. I. Kirkland, J. M. Parrish, and M. K. Brett-Surman. 1998. Neoceratopsian teeth from the Lower to Middle Cretaceous of North America. *New Mexico Museum of Natural History and Science, Bulletin* 14: 297–302.

Clemens, W. A., and L. G. Nelms. 1993. Paleoecological implications of Alaskan terrestrial vertebrate fauna in latest Cretaceous time at high paleolatitudes. *Geology* 21: 503–506.

Cobban, W. A., E. A. Merewether, T. D. Fouch, and J. D. Obradovich. 1994. Some Cretaceous shorelines in the western interior of the United States. In M. Caputo, J. A. Peterson, and K. J. Franczyk (eds.), *Mesozoic Systems of the Rocky Mountain Region, USA,* pp. 393–413. Denver: Rocky Mountain Section, Society for Sedimentary Geology.

Colbert, E. H. 1948. Evolution of the horned dinosaurs. *Evolution* 2: 145–163.

Currie, P. J., and D. A. Russell. 1982. A giant pterosaur (Reptilia: Archosauria) from the Judith River (Oldman) Formation of Alberta. *Canadian Journal of Earth Sciences* 19: 894–897.

Dodson, P. 1983. A faunal review of the Judith River (Oldman) Formation, Dinosaur Provincial Park, Alberta. *Mosasaur* 1: 89–118.

Dodson, P. 1986. *Avaceratops lammersi:* A new ceratopsid from the Judith River Formation of Montana. *Academy of Natural Sciences of Philadelphia, Proceedings* 138: 305–317.

Dodson, P., and P. J. Currie. 1990. Neoceratopsia. In D. B. Weishampel, P. Dodson, and H. Osmólska (eds.), *The Dinosauria,* pp. 593–618. Berkeley: University of California Press.

Dodson, P., and L. P. Tatarinov. 1990. Dinosaur extinction. In D. B. Weishampel, P. Dodson, and H. Osmólska (eds.), *The Dinosauria,* pp. 55–62. Berkeley: University of California Press.

Dyman, T. S., E. A. Merewether, C. M. Molenaar, W. A. Cobban, J. D. Obradovich, R. J. Weimer, and W. A. Bryant. 1994. Stratigraphic transects for Cretaceous rocks, Rocky Mountains and Great Plains regions. In M. V. Caputo, J. A. Peterson, and K. J. Franczyk (eds.),

Mesozoic Systems of the Rocky Mountain Region, USA, pp. 365–392. Denver: Rocky Mountain Section, Society for Sedimentary Geology.

Forster, C. A., P. C. Sereno, T. W. Evans, and T. Rowe. 1993. A complete skull of *Chasmosaurus mariscalensis* (Dinosauria: Ceratopsidae) from the Aguja Formation (Late Campanian) of West Texas. *Journal of Vertebrate Paleontology* 13: 161–170.

Harrison, C., K. Miskell, G. Brass, E. Saltzman, and J. Sloan. 1983. Continental Hypsography. *Tectonics* 2: 357–377.

Hopson, J. A. 1975. The evolution of cranial display structures in hadrosaurian dinosaurs. *Paleobiology* 1: 21–43.

Horner, J. R. 1983. Cranial osteology and morphology of the type specimen of *Maiasaura peeblesorum* (Ornithischia: Hadrosauridae), with discussion of its phylogenetic position. *Journal of Vertebrate Paleontology* 3: 29–38.

Horner, J. R. 1984. Three ecologically distinct vertebrate faunal communities from the Late Cretaceous Two Medicine Formation of Montana, with discussion of evolutionary pressures induced by interior seaway fluctuations. In *Field Conference Guidebook,* pp. 299–303. Billings: Montana Geological Society.

Horner, J. R. 1988. A new hadrosaur (Reptilia, Ornithischia) from the Upper Cretaceous Judith River Formation of Montana. *Journal of Vertebrate Paleontology* 8: 314–321.

Horner, J. R. 1989. The Mesozoic terrestrial ecosystems of Montana. In *Field Conference Guidebook,* pp. 153–162. Billings: Montana Geological Society.

Horner, J. R. 1992. Cranial morphology of *Prosaurolophus* (Ornithischia: Hadrosauridae) with descriptions of two new hadrosaurid species and an evaluation of hadrosaurid phylogenetic relationships. *Museum of the Rockies, Occasional Paper* 2: 1–119.

Horner, J. R., D. J. Varricchio, and M. B. Goodwin. 1992. Marine transgressions and the evolution of Cretaceous dinosaurs. *Nature* 358: 59–61.

Horrell, M. A. 1991. Phytogeography and paleoclimatic interpretation of the Maastrichtian. *Palaeogeography, Palaeoclimatology, Palaeoecology* 86: 87–138.

Hunt, A. P., and S. G. Lucas. 1993. Cretaceous vertebrates of New Mexico. In S. G. Lucas and J. Zidek (eds.), *Vertebrate Paleontology in New Mexico,* pp. 77–91. *New Mexico Museum of Natural History Bulletin* 2.

Kirkland, J. I., B. Britt, D. L. Burge, K. Carpenter, R. Cifelli, F. DeCourten, J. Eaton, S. Hasiotis, and T. Lawton. 1997. Lower to Middle Cretaceous dinosaur faunas of the central Colorado Plateau: A key to understanding 35 million years of tectonics, sedimentology, evolution, and biogeography. *Brigham Young University Geology Studies* 42: 69–103.

Kues, B. S., T. M. Lehman, and J. K. Rigby. 1980. The teeth of *Alamosaurus sanjuanensis,* a Late Cretaceous sauropod. *Journal of Paleontology* 54: 864–869.

Langston, W., Jr. 1975. The Ceratopsian dinosaurs and associated lower vertebrates from the St. Mary River Formation (Maastrichtian) at Scabby Butte, southern Alberta. *Canadian Journal of Earth Science* 12: 1576–1608.

Lehman, T. M. 1981. The Alamo Wash local fauna: A new look at the old

Ojo Alamo fauna. In S. Lucas, K. Rigby, and B. Kues (eds.), *Advances in San Juan Basin Paleontology,* pp. 189–221. Albuquerque: University of New Mexico Press.

Lehman, T. M. 1987. Late Maastrichtian paleoenvironments and dinosaur biogeography in the western interior of North America. *Palaeogeography, Palaeoclimatology, Palaeoecology* 60: 189–217.

Lehman, T. M. 1989. Upper Cretaceous (Maastrichtian) paleosols in trans-Pecos Texas. *Geological Society of America Bulletin* 101: 188–203.

Lehman, T. M. 1996. A horned dinosaur from the El Picacho Formation of West Texas, and review of ceratopsian dinosaurs from the American Southwest. *Journal of Paleontology* 70: 494–508.

Lehman, T. M. 1997. Late Campanian dinosaur biogeography in the western interior of North America. In D. L. Wolberg, E. Stump, and G. D. Rosenberg (eds.), *Dinofest International,* pp. 223–240. Philadelphia: Academy of Natural Sciences.

Lillegraven, J. A., and L. M. Ostresh. 1990. Late Cretaceous (earliest Campanian/Maastrichtian) evolution of western shorelines of the North American Western Interior Seaway in relation to known mammalian faunas. In T. M. Bown and K. D. Rose (eds.), *Dawn of the Age of Mammals in the northern part of the Rocky Mountain Interior, North America.* Geological Society of America, Special Paper 243: 1–30.

Lucas, S. G. 1981. Dinosaur communities of the San Juan Basin: A case for lateral variations in the composition of Late Cretaceous dinosaur communities. In S. Lucas, K. Rigby, and B. Kues (eds.), *Advances in San Juan Basin Paleontology,* pp. 337–393. Albuquerque: University of New Mexico Press.

Martin, P. S. 1973. The discovery of America. *Science* 179: 969–974.

McCord, R. D. 1997. An Arizona titanosaurid sauropod and revision of the Late Cretaceous Adobe Canyon fauna. *Journal of Vertebrate Paleontology* 17: 620–622.

Ostrom, J. H. 1970. Stratigraphy and paleontology of the Cloverly Formation (Lower Cretaceous) of the Bighorn Basin Area, Wyoming and Montana. *Peabody Museum of Natural History Bulletin* 35: 1–234.

Padian, K. 1984. A large pterodactyloid pterosaur from the Two Medicine Formation (Campanian) of Montana. *Journal of Vertebrate Paleontology* 4: 516–524.

Penkalski, P., and P. Dodson. 1999. The morphology and systematics of *Avaceratops,* a primitive horned dinosaur from the Judith River Formation (Late Campanian) of Montana, with the description of a second skull. *Journal of Vertebrate Paleontology* 19: 692–711.

Rage, J. C. 1986. South American/North American terrestrial interchanges in the latest Cretaceous: Short comments on Brett-Surman and Paul (1985) with additional data. *Journal of Vertebrate Paleontology* 6: 382–383.

Ratkevich, R. P. 1997. Dinosaur remains of southern Arizona. In D. L. Wolberg, E. Stump, and G. D. Rosenberg (eds.), *Dinofest International,* pp. 213–221. Philadelphia: Academy of Natural Sciences.

Rowe, T., R. L. Cifelli, T. M. Lehman, and A. Weil. 1992. The Campanian Terlingua local fauna, with a summary of other vertebrates from the Aguja Formation, trans-Pecos Texas. *Journal of Vertebrate Paleontology* 12: 472–493.

Russell, D. A., and P. Dodson. 1997. The extinction of the dinosaurs: A dialogue between a catastrophist and a gradualist. In J. O. Farlow and M. K. Brett-Surman (eds.), *The Complete Dinosaur,* pp. 662–672. Bloomington: Indiana University Press.

Russell, L. S. 1975. Mammalian faunal succession in the Cretaceous System of Western North America. In W. G. E. Caldwell (ed.), *The Cretaceous System in the Western Interior of North America.* Geological Association of Canada, Special Paper 13: 137–161.

Sampson, S. D. 1995. Two new horned dinosaurs from the Upper Cretaceous Two Medicine Formation of Montana, with a phylogenetic analysis of the Centrosaurinae (Ornithischia: Ceratopsidae). *Journal of Vertebrate Paleontology* 15: 743–760.

Schultz, G. E. 1990. The Clarendonian faunas of the Texas and Oklahoma panhandles. In T. C. Gustavson (ed.), *Tertiary and Quaternary Stratigraphy and Vertebrate Paleontology of Northwestern Texas and Eastern New Mexico.* University of Texas, Bureau of Economic Geology, Guidebook 24: 83–94.

Schwimmer, D. R. 1997. Late Cretaceous dinosaurs in eastern U.S.A.: A taphonomic and biogeographic model of occurrences. In D. L. Wolberg, E. Stump, and G. D. Rosenberg (eds.), *Dinofest International,* pp. 203–211. Philadelphia: Academy of Natural Sciences.

Sloan, R. E., J. K. Rigby Jr., L. M. Van Valen, and D. Gabriel. 1986. Gradual dinosaur extinction and simultaneous ungulate radiation in the Hell Creek Formation. *Science* 232: 629–633.

Sullivan, R. M. 1999. *Nodocephalosaurus kirtlandensis,* gen. et sp. nov., a new ankylosaurid dinosaur (Ornithischia: Ankylosauria) from the Upper Cretaceous Kirtland Formation (Upper Campanian), San Juan Basin, New Mexico. *Journal of Vertebrate Paleontology* 19: 126–139.

Wheeler, E. A., and T. M. Lehman. 2000. Late Cretaceous woody dicots from the Aguja and Javelina Formations, Big Bend National Park, Texas, USA. *International Association of Wood Anatomists Journal* 21: 83–120.

Wilson, J. A., and P. C. Sereno. 1998. Early evolution and higher-level phylogeny of sauropod dinosaurs. *Society of Vertebrate Paleontology Memoir* 5: 1–68.

Section V.
Paleopathologies

23. Theropod Stress Fractures and Tendon Avulsions as a Clue to Activity

BRUCE ROTHSCHILD, DARREN H. TANKE, AND TRACY L. FORD

Abstract

A unique window into theropod behavior is afforded by the study of activity-related pathologies. Stress fractures and tendon avulsions provide evidence of activities overstressing regional mechanical strength of bone. Disjunction of pedal stress fracture localization with gait suggest alternative behavior-defined lesions. Identification of similar lesions in the manus and presence of upper extremity tendon avulsions support predation-related injuries. Such injuries are also the likely cause of the pedal stress fractures. The scavenger-predation debate thus has new evidence supporting a very active, probably predatory lifestyle for theropods.

Introduction

Theropod activity levels are a source of great interest, especially with respect to the question of predation or scavenging behavior (Currie 1997). If certain pathologies infer behavior, perhaps this question can be independently examined. One approach is to examine residual bony changes resulting from excessive forces to the bone-tendon continuum. In contrast to acute fractures, which are the result of acute trauma, stress or fatigue fractures occur from strenuous repetitive activities

(Daffner 1978; Morris and Blickenstaff 1967; Orova et al. 1978; Rothschild 1982; Rothschild and Martin 1993; van Hall 1982). Just as stress fractures in ceratopsians have been related to activity (Rothschild 1988; Rothschild and Tanke 1992), certain repetitive forceful activities are inferred for theropods with similar lesions. If the lesions were limited to the pes, the same running/migrating explanation could be explored. Occurrence in the manus indicates 'actively resisting' prey and therefore, predation. Avulsion injuries, in which the tendon actually rips out of the bone, are similarly indicative of predation, rather than scavenging behavior.

The appearance of a stress fracture is highly characteristic (Resnick and Niwayama 1988). Recognition of the classic "bump" on a theropod pedal phalanx in the collections at the University of Utah, stimulated a systematic survey for stress fractures in theropod pedal and manual elements and for evidence of forelimb tendon avulsions.

Methods

The phalanges and metapodials of theropods in various museum collections (listed in the Institutional Abbreviations) were examined macroscopically for surface abnormalities. Stress fractures are recognized on the basis of diaphyseal surface bulges, usually, but not invariably on the anterior surface. Radiologic examination reveals radiolucent clefts (areas of reduced X-ray beam attenuation) that appear as a clear zone angled through the bulge, but usually not visible on the surface (Resnick and Niwayama 1988; Rothschild 1988; Rothschild and Martin 1993).

Institutional Abbreviations: AMNH, American Museum of Natural History (New York); BHI, Black Hills Institute (South Dakota); BM, Blanding Museum (Utah); LACM, Los Angeles County Museum of Natural History (California); RTMP, Royal Tyrrell Museum of Palaeontology (Alberta).

Other Collections Examined: Brigham Young University Museum (Utah); Carnegie Museum of Natural History (Pennsylvania); College of Eastern Utah; Dinosaur National Monument (Utah); Mesa Southwest Museum (Arizona); Museum of the Rockies (Montana); National Museum of Natural History (Washington, D.C.); University of Kansas Museum of Natural History, Vertebrate Paleontology; University of Utah Museum of Natural History, Vertebrate Paleontology; Yale Peabody Museum (Connecticut).

Results

Diaphyseal bumps, characteristic of stress fractures, were noted in a wide spectrum of theropods (table 23.1). The frequency in *Allosaurus* manus and pes is significantly greater than that noted in *Albertosaurus, Archaeornithomimus,* and *Ornithomimus.* Stress fractures in *Allosaurus* pes were distributed to the proximal phalanges. They occur in 3% of medial, 10% of middle, and 10% of lateral digits. These frequencies are statistically indistinguishable (Fisher exact test, p = 0.196).

TABLE 23.1.
Distribution of Stress Fractures in Theropod Genera

Family	Genus	Pes (n)	Manus (n)	Specimen Affected
Herrerasauridae	*Herrerasaurus*	20	12	
Podokesauridae	*Coelophysis*	14		
Ceratosauridae	*Ceratosaurus*	1/1		
Halticosauridae	*Dilophosaurus*	60		
Carnosaur	gen. et sp. indet.	18	16	
Velocisauridae	*Velocisaurus*	12		
Alvarezsauridae	*Mononykus*	15		
Megalosauridae	*Megalosaurus*	16		
Allosauridae	*Allosaurus*	17/281	3/47	AMNH 324, 6128
Coeluridae	*Marshosaurus*	5		
	Ornitholestes	20		
Compsognathidae	*Compsognathus*	9		
Aublysodontidae	*Alectrosaurus*	23		
Tyrannosauridae	*Albertosaurus*	1/319	4	AMNH 5432
	Tyrannosaurus	1/81	10	LACM 23844
	Albertosaur	21	2	
	Gorgosaurus	54		
	gen. et sp. ident.	3/105	1/5	RTMP 81.16.328
				RTMP 79.14.694
				RTMP 89.36.343
	Tarbosaurus	18	1/10	Blanding II-2
Dromaeosauridae	*Utahraptor*	2		
	Deinonychus	52	43	
	Sauronitholestes	2/82	2/9	RTMP 89.172.32
				RTMP 94.172.32
				RTMP 81.19.97
	gen. et sp. indet.	4/17	4/12	RTMP 79.14.900
Therizinosauridae	gen. et sp. indet.	3		
Ornithomimidae	*Struthiomimus*	50		
	Ornithomimus	178		
	Archaeornithomimus	229		
	Dromiceiomimus	4		
	gen. et sp. indet.	1/15	8	
Caenagnathidae	*Chirostenotes*	1/17		RTMP 92.36.448 ? manus or pes
	Elmisaurus	23		
Troodontidae	*Troodon*	21		
Small theropod	gen. et sp. indet.	1/4		

Note: n = numbers of specimens examined. For fractional numbers, numerator is number affected, denominator is number examined.

Distal ungual pathologies were noted only in Dromaeosauridae, wherein it represented 50% of manual lesions. The low frequency in other affected families precludes meaningful distribution analysis.

Avulsion injuries were rare, noted only in *Allosaurus* and *Tyrannosaurus*. Strenuous muscle activity may at times overcome the strength of that muscle's bony attachment, allowing it to partially or fully tear free. The scars of such occurrences are localized to scapulae and humeri. The residual divot in the humerus of "Sue" FMNH PR 2081 (formerly BHI 2033) is characteristic of this pathology (see Carpenter and Smith, chap. 9 of this volume). Scapular distribution is to the distal lateral surface, where it appears as a 'divot' or surface discontinuity with raised edges and seems to represent the origin of the deltoid or teres major (Meers 1999; Surahya 1989).

Discussion

The appearance of theropod phalanges is pathognomonic (characteristic and unequivocal diagnostically) for stress fractures (Daffner 1978; Kroening and Shelton 1963; Morris and Blickenstaff 1967; Rothschild 1982; Rothschild and Martin 1993; Wilson and Katz 1969). These lesions are easily distinguished from osteomyelitis (bone infection) because of the lack of bone destruction (Daffner 1978; Rothschild 1982, 1988; Resnick and Niwayama 1988; Rothschild and Martin 1993). These lesions lack the sclerotic perimeter of benign bone tumors, such as osteoid osteoma. Osteoid osteoma can produce a bumplike structure, but it has a very thick margin with a central nidus. No perturbation of the internal bony architecture is identifiable to indicate a possible primary malignant bone tumor (Daffner 1978; Farlow et al. 1995; Huvos 1991; Resnick and Niwayama 1988; Rothschild and Martin 1993). Such a tumor is usually associated with spiculated or thin laminated periosteal reaction, but none were seen in the examined theropods elements. Metabolic disorders, such as hyperparathyroidism and hyperthyroidism, typically have subperiosteal reaction, but also were not seen in the specimens. Subperiosteal hematomas produce a thin shell, easily distinguished radiologically from the bone thickening and cleft formation seen in the specimens (table 23.1; Resnick and Niwayama 1988).

Currie (1997) noted that "as theropods became faster, they needed more control and better shock absorption in their feet," and that "the lower end of the third metatarsal would have contacted the ground first when a theropod was running." The lack of correlation of pes stress fractures in *Allosaurus* with the expected predisposition for such fractures in the third digit suggests an alternative explanation—that the stress fractures occurred during direct prey interactions. This damage is not related to direct trauma, such as occurs if the foot is stepped on, but rather, it occurs in the areas stressed as the foot is used to hold struggling prey. Such stress fracture distributions, which are not limited to primary weight-bearing forces, can be taken as evidence for predatory behavior.

Activities sufficient to overcome the strength of a muscle's bony attachment are suggested as etiologic of the scapular and humeral lesions. Tendon avulsions in *Tyrannosaurus* and *Allosaurus* imply vigorous resistance to the forelimbs. Such resistance would not be invoked in scavenging, but rather indicate predatory behavior with struggling prey (see Carpenter and Smith, chap. 9 of this volume). The myology for the scapular avulsion is intriguing. Birds have a relatively simplistic muscle design. The teres major covers three-quarters of the scapula, originating at the caudal lateral surface. Theropod scapular localization suggests a more complex, perhaps nonavian type of musculature. Further study is needed comparing the origin of teres major and deltoid muscles in crocodile and Komodo dragon (Meers 1999; Surahya 1989) with that in a theropod.

Acknowledgments: We wish to express our appreciation to Drs. Gordon Bell, David Berman, Christine Chandler, Dan Chure, Stephen Czerkas, Mary Dawson, Eugene Gaffney, John Heyning, Pat Holroyd, John (Jack) Horner, Nicholas Hotton, Neal Larson, Peter Larson, Rebecca Hanna, Larry Martin, Robert McCord, Samuel McLeod, Kevin Padian, Burkhard Pohl, Robert Purdy, Kenneth Stadtman, J. D. Stewart, Mary Ann Turner, and Richard Zakrzewski for access to and facilitation of examination of the collections they curate. Thanks to Lorrie McWhinney for her review comments. Finally, we thank Philip J. Currie for his continued support of our work.

References

Currie, P. J. 1997. Theropoda. In P. J. Currie and K. Padian (eds.), *Encyclopedia of Dinosaurs*, pp. 731–737. New York: Academic Press.

Daffner, R. H. 1978. Stress fractures: Current concepts. *Skeletal Radiology* 2: 221–229.

Farlow, J. O., M. B. Smith, and J. M. Robinson. 1995. Body mass, bone "strength indicator," and cursorial potential of *Tyrannosaurus rex*. *Journal of Vertebrate Paleontology* 15: 713–725.

Huvos, A. G. 1991. *Bone Tumors: Diagnosis, Treatment, and Prognosis*. 2d ed. Philadelphia: Saunders.

Kroening, P. M., and M. L. Shelton. 1963. Stress fractures. *American Journal of Roentgenology* 89: 1281–1286.

McIntosh, J. S., C. A. Miles, K. C. Cloward, and J. R. Parker. 1996. A new nearly complete skeleton of *Camarasaurus*. *Bulletin of the Gunma Museum of Natural History* 1: 1–87.

Meers, M. B. 1999. Evolution of the Crocodylian forelimb: Anatomy, biomechanics, and functional morphology. Ph.D. dissertation, Johns Hopkins University School of Medicine.

Morris, J. M., and L. D. Blickenstaff. 1967. *Fatigue fractures: A clinical study*. Springfield, Ill.: Charles C. Thomas.

Orova, S., J. Puranen, and L. Ala-Ketola. 1978. Stress fractures caused by physical exercise. *Acta orthopaedica scandinavia* 49: 19–27.

Resnick, D., and G. Niwayama. 1988. *Diagnosis of Bone and Joint Disorders*. Philadelphia: Saunders.

Rothschild, B. M. 1982. *Rheumatology: A Primary Care Approach*. New York: Yorke Medical Press.

Rothschild, B. M. 1988. Stress fracture in a ceratopsian phalanx. *Journal of Paleontology* 62: 302–303.

Rothschild, B. M., and L. D. Martin. 1993. *Paleopathology: Disease in the Fossil Record.* London: CRC Press.

Rothschild, B. M., and D. H. Tanke. 1992. Palaeopathology: Insights to lifestyle and health in prehistory. *Geosciences Canada* 19: 73–82.

Surahya, S. 1989. *Atlas Komodo: An Anatomical Study of Komodo Dragon and Its Position in Animal Systematics.* Yogyakarta: Gadjah Mada University Press.

van Hall, M. E. 1982. Stress fractures of the great toe sesamoids. *American Journal of Sports Medicine* 10: 122–128.

Wilson, E. S., and F. N. Katz. 1969. Stress fractures: An analysis of two hundred fifty consecutive cases. *Radiology* 92: 481–489.

24. Theropod Paleopathology: A Literature Survey

R. E. MOLNAR

Abstract

A survey of paleopathological features reported in the literature for theropod dinosaurs found that such features have occurred in at least 21 genera belonging to 10 families. Both large and small theropods exhibit pathologies, although pathologies are substantially less common (or less commonly preserved) among small theropods. Pathologies have been found in several parts of the skeleton, but appear absent (or very rare) from the major weight-bearing structures, such as the sacrum, femur, and tibia. They appear most commonly in the postcranial axial skeleton (especially the ribs and caudals), less commonly in the hind limbs and least (with about equal frequency) in skull and jaws and the forelimb.

Pathologies present include congenital malformations and evidence of infections, but mostly represent injuries. Features in the latter category include both fractures and pits or punctures possibly resulting from bites. Information on the frequency of fracture for various bones suggests that some elements (e.g., femora) were more strongly selected for resistance to breakage than others.

The study of congenital anomalies may be useful in illuminating the evolutionary history of developmental processes. Asymmetric and unusual fusions of cranial elements probably indicate gerontic individuals. There are two apparent instances of fluctuating asymmetry, mostly in the forelimb. Infections seem to have been quite localized.

Introduction

Although the first description of a large theropod dinosaur refers to pathologies being present on some of the elements (Eudes-Deslong-champs 1838), there actually have been few discussions of paleo-pathologies in theropods (Petersen et al. 1972; there is much information on theropods in Rothschild and Martin 1993; Rothschild 1997; Rothschild et al., chap. 23 of this volume). My work with theropods suggests that pathological features are not rare, thus the descriptive literature in English, French, German, Russian, Spanish was surveyed for such features. One Japanese and a few Chinese publications were also included. Pathologies have been reported in at least thirteen species belonging to thirteen genera. The survey covered the scientific literature, although instances from the popular literature are included when reported (or written) by trained personnel or supported by good illustrations. One electronic publication, the bibliography of dinosaurian paleopathology by Tanke and Rothschild (1999), is included.

A literature survey has several inherent limitations: detailed descriptions of the pathologies are often not presented, etiology is usually not discussed, and, in general, pathologies were probably often overlooked. Except for Osborn (1916), where pathologies were clearly illustrated but not mentioned in the text, it is difficult to document such omissions. Because the intent of description in paleontology for much of the past two centuries was to present information useful in recognizing and classifying the material under study, pathological features were ignored unless their consideration was necessary to assist comparison.

The practical difficulty of examining the large number of theropod specimens prohibits that kind of survey, and makes a literature survey the most practicable method of assessing the occurrence of theropod pathologies. However, this literature survey is supplemented by some personal observations.

Diffuse idiopathic skeletal hyperostosis (DISH) is here not considered pathological, following Rothschild (1987). Some apparently pathological features that show no evidence of healing, such as circular depressed fractures (presumably resulting from bites) and tooth scrapes, may have been made during scavenging. However, marks of this origin would be expected on bones associated with muscle tissue or viscera, such as femora or ribs, but not on phalanges or metapodials. Jaws and fore and hind limbs are used in combat, thus these structures are more likely to receive damage during life. Hence, circular fractures and scrapes found on elements associated with extensive soft tissues are not included, but those of jaws or teeth, and manual or pedal elements are.

An earlier manuscript version of this survey was cited by Rothschild and Martin (1993) as "Molnar (1992)" and by Rothschild (1997) as "Molnar, in press," however that version was never published. Many of the results were published by Rothschild and Martin (1993) and are extended and updated here.

Institutional Abbreviations: AMNH, American Museum of Natural History, New York; BHI, Black Hills Institute of Geological Research, Hill City, South Dakota; BMNH, Natural History Museum,

London; FMNH, Field Museum of Natural History, Chicago; IGM, Institute of Geology, Mongolia, Ulaanbaatar; IVPP, Institute of Vertebrate Paleontology and Paleoanthropology, Academia Sinica, Beijing; LACM, Los Angeles County Museum of Natural History; MIWG, Museum of Isle of Wight Geology, Sandown; MOR, Museum of the Rockies, Montana State University, Bozeman; NMC, National Museum of Natural History, Ottawa; OMNH, Oklahoma Museum of Natural History, University of Oklahoma, Norman; PVSJ, Museo de Ciencias Naturales, Universidad Nacional de San Juan; QM, Queensland Museum, Brisbane; ROM, Royal Ontario Museum, Toronto; SGM-Din, Ministère de l'Énergie et des Mines, Rabat; SMU, Shuler Museum of Paleontology, Southern Methodist University, Dallas; TMP, Royal Tyrrell Museum of Palaeontology, Drumheller, Alberta; UCM, University of Colorado Museum, Boulder; UCMP, University of California Museum of Paleontology, Berkeley; USNM, National Museum of Natural History, Washington; UUVP, Utah Museum of Natural History (Vertebrate Paleontology), University of Utah, Salt Lake City; YPM, Peabody Museum of Natural History, Yale University, New Haven; ZPAL—Zaklad Paleobiologii, Polish Academy of Sciences, Warsaw.

Survey Results
Skeletal

Pathologies have been reported in the following taxa in systematic order. Where a plausible minimum number of individuals with pathologies can be estimated, it is given at the end of the brief description. Specimen numbers are given where available.

Herrerasauridae

Herrerasaurus ischigualastensis: A pit, attributed by Sereno and Novas (1993) to a bite, is found in the dorsal margin of the supraoccipital ala of the left parietal of PVSJ 407. Two other pits occur in the ventral margin of left splenial. The features all show signs of healing, and the "porous swollen" bone around the pits was taken to indicate transient, hence presumably subacute, infection (Sereno and Novas 1993). Because of their size and the disparate directions of penetration, the pits are attributed to intraspecific fighting. Minimum number of individuals, 1.

Ceratosauridae

Ceratosaurus nasicornis: There has been disagreement over the status of the fused left metatarsals II to IV (the right were not preserved) of *Ceratosaurus nasicornis* (USMN 4735, the holotype) since Baur (1890). The material certainly looks pathological (cf. Gilmore 1920, pls. 24 and 25). However, following the cladistic study of the Saurischia, the fusion itself was regarded as not pathological (e.g., Rowe 1989) because it also occurred in related, more plesiomorphic taxa (e.g., *Syntarsus* spp.). The possibility that some pathology was present was not denied (Rowe and Gauthier 1990). Recent inspection by Tanke

(Tanke and Rothschild 1999) reveals that the fusion is indeed pathological: Baur (1890) attributed it to a healed fracture. Minimum number of individuals, 1.

Dilophosaurus wetherilli: A horizontal sulcus just below the pre-zygapophyseal facets on the anterior face of the neural arch of cervical 5 "is probably due to injury or crushing" (Welles 1984, 107). Two pits were found on the entotuberosity of the right humerus of UCMP 37302 "that seem to be abscesses, but might be artifacts" (Welles 1984, 128).

Syntarsus rhodesiensis: Healed fractures have been noted in the tibia (Raath, pers. comm. 1986) and metatarsus (Raath, pers. comm. 1999), but they are very rare (Raath, pers. comm. 1999).

Megalosauridae

Megalosaurus bucklandii: Tanke and Rothschild (1999) point out that a cervical rib of *Megalosaurus,* figured by Owen (1856; also in Owen 1884), shows an unusual swelling at the base of the capitular process that appears to be a healed fracture. Minimum number of individuals, 1.

Monolophosaurus jiangi: The neural spine of dorsal 10 (and possibly 11) of the IVPP 84019 had been broken, with spine 10 displaced and fused to 11 (Zhao and Currie 1993). Tanke and Rothschild (1999) report that one dentary of this specimen exhibits faint parallel ridges that may represent tooth marks. Minimum number of individuals, 1.

Poekilopleuron bucklandii: An anterior chevron is ankylosed to the succeeding caudal centrum with the development of an exostosis (Eudes-Deslongchamps 1838, pl. 2), and two phalanges also exhibit pathologies. One phalanx (probably pedal) seemingly has at least three low irregular exostoses or exostosislike projections (Eudes-Deslongchamps 1838, pl. 8, fig. 7), and a second phalanx (likely manal) exhibits a low rounded projection resembling a callus (Eudes-Deslongchamps 1838, pl. 8, fig. 8), perhaps indicating a healed fracture. Because the single specimen was destroyed by British bombing near the end of World War II, the causes of these conditions cannot be ascertained. Nonetheless it is noteworthy that a single individual had three anatomically independent pathologies. Minimum number of individuals, 1.

Allosauridae

Allosaurus fragilis: The left scapula (Gilmore 1915; 1920, pl. 5) and left fibula (Gilmore 1920, fig. 48) of USNM 4734 are pathological. Moodie (1923) attributed the scapular damage to a fracture (a conclusion also reached in Tanke and Rothschild 1999), and the damage to the fibula is consistent with such a cause. Examination of the gastralia of USNM 8367 by Tanke (reported in Tanke and Rothschild 1999) revealed several healed fractures near the middle of the elements, as well as poorly healed fractures that had formed pseudoarthroses.

Petersen et al. (1972) present a list of paleopathologies noted in the *Allosaurus* specimens from the Cleveland-Lloyd Quarry, Utah. These are: (1) willow breaks in two ribs (UUVP 1847 and 2252); (2) healed fractures of the humerus (UUVP 3435) and radius (UUVP 687); (3)

distortion of certain pedal joint surfaces, possibly osteoarthritic (UUVP 1848 and 4159; but see comments in the section on developmental anomalies); (4) similar distortions in the caudals (UUVP 1742 and 4895); (5) extensive "neoplastic" ankylosis of caudals, possibly traumatic (including fusion of chevrons to centra in UUVP 3773 and 5256), as well as more restricted "neoplastic" growth (UUVP 3811); (6) coossification of distal caudal centra, possibly regenerative (UUVP 177, 1849 and 1850); (7) amputation of a chevron (UUVP 837) and a pedal element (right IV-1; UUVP 1851), possibly resulting from bites; (8) extensive exostoses of a pedal phalanx III-1 (UUVP 1657); (9) lesions resembling those of osteomyelitis in two scapulae (UUVP 1528 and 5599; seemingly not from the same individual); and (10) various spurs recognized as pathologic only by comparison with homologous elements from other individuals (involving a premaxilla, UUVP 1852, unspecified ungual, UUVP 1853, and two metacarpals, UUVP 1854 and 1855). Some of these (1, 2—the radius only—and 5) are illustrated by Madsen (1976a), along with an extensive exostosis of a pedal phalanx that may be the one attributed to infectious disease by Rothschild and Martin (1993, 238). A metacarpal (YPM 4944) has a round depressed fracture.

Laws (1995, 1997) reported extensive paleopathologies of an apparently subadult male, MOR 693. These pathologies affected five dorsal ribs, cervical 6, dorsals 3, 8, and 13, caudal 2 and its chevron, the gastralia, right scapula, manual phalanx I-1, left ilium, metatarsals III and V, and pedal phalanges III-1 and II-3 (ungual; a second ungual was mentioned in 1995). Details were not given in the abstract, except to state that the conditions resulted from "trauma, infection, or aberrancy." Tanke and Rothschild (1999) report further details, given in a University of Wyoming website by R. Travsky. The injured ilium had sustained "a large hole . . . caused by a blow from above" (Tanke and Rothschild 1999, 106). A "prominent collar-like exostosis" (or involucrum) affected the proximal end of pedal phalanx III-1. A total of 14 injuries were recorded.

Recently Reid (1996) reported a fractured rib in a specimen from the Cleveland-Lloyd Quarry, and Rothschild (1997) mentions fused vertebrae in the distal tail and also observed fractured ribs in a different individual than that reported by Reid (Rothschild, pers. comm. 1999; reported in Hecht 1998). Minimum number of individuals, 6.

Neovenator salerii: Tanke and Rothschild (1999, 52) report that "the type specimen has numerous paleopathologies—midcaudal vertebrae fusion, healed fracture of mid-caudal vertebra transverse process; osteophytes affecting pedal phalanges; healed gastralia rib fractures, some forming false joints . . . [and] scapula fracture." The holotype is held in two museums as MIWG 6348 and BMNH R10001. Minimum number of individuals, 1.

Sinraptor dongi: The bones of one skull (IVPP 10600) show "a variety of gently curving tooth drags or gouges, shallow, circular punctures and one fully penetrating lesion" (Tanke and Currie 1995). An anterior left dorsal rib of this specimen was broken near the head and healed after some telescoping of the capitular shaft (Currie and Zhao 1993). Tanke and Rothschild (1999) report that other contiguous dorsal ribs also had healed fractures. Minimum number of individuals, 1.

Acrocanthosauridae fam. nov.

Acrocanthosaurus atokensis: The holotype skull (OMNH 8-0-59) has "exostotic material" (seemingly slight) on the articular surface of the squamosal for the quadrate (Stovall and Langston 1950). Tanke and Rothschild (1999) report that the neural spine of caudal 11 seemingly suffered a displaced fracture before healing and mention an unusual hooklike structure on the neural spine of caudal 3.

Newer material (SMU 74646), described by Harris (1998), shows more extensive pathologies. The neural spine of caudal 16 was broken and displaced, and bears a pit that may be the result of a bite. Harris points out that a thick bony mass at the flexure is probably due to infection. Nondisplaced, healed fractures of five ribs are interpreted by Harris as having been caused by a single incident. Another dorsal rib bears what seems to have been a pseudoarthrosis that ultimately rejoined. The position of this seeming pseudoarthrosis, at midlength of the rib, suggests that this pathology occurred in a separate incident from the other rib fractures, all of which occurred more distally on the ribs. The proximal end of dorsal rib 13 was also fractured, and a pit on the dorsal surface suggested to Harris that this might represent a healed bite. Other features are possibly pathological: a pair of apparently pseudoarthrotic gastralia and the neural spines of cervicals 3 and 4, that both deviate to the right. No other elements found in the immediate vicinity of these cervicals were deformed, so Harris suggested that these spines were curved during life (although there is at least one other curved neural spine shown in Harris 1998, fig. 20B).

In a popular publication, Larson (1998) reported several broken and healed right ribs and scapula in a third specimen, now at the North Carolina State Museum of Natural Sciences. This scapula also has "what appears to be either a puncture wound or a place of infection." Minimum number of individuals, 3.

Carcharodontosaurus saharicus: The skull (SGM-Din 1) shows a circular puncture wound in the nasal and an abnormal projection of bone on the antorbital rim (Sereno, pers. comm. 1999). Minimum number of individuals, 1.

Dromaeosauridae

Deinonychus antirrhopus: One pedal phalanx, II-2 (YPM 5205), has a healed fracture (Ostrom 1976). Minimum number of individuals, 1.

Velociraptor mongoliensis: Norell et al. (1995, 44, photo on 42) report a skull, IGM 100/976, bearing two parallel rows of small punctures. These, they note, match the spacing of the upper teeth in *Velociraptor,* thus they attribute the injuries to intraspecific combat. From the absence of evidence for healing, they suggest that this was the cause of death.

Undescribed dromaeosaurid: A bifurcated gastral rib (segment seven, of twelve, on the right) was found in an articulated, but incomplete skeleton of an immature dromaeosaurid from Tugrugeen Shireh (Mongolia) (Norell and Makovicky 1997). Minimum number of individuals, 1.

Oviraptoridae

"Oviraptorid most closely related to *Oviraptor*": Clarke et al. (1999), in describing the remains of a brooding oviraptorid (IGM 100/979), observe that the right ulna had been fractured and healed, leaving a callus and possibly a longitudinal groove, about two-thirds of the way distal along the shaft. The pieces of the shaft were not displaced, and the adjacent radius not broken. Minimum number of individuals, 1.

Pathological features are also present, but not yet described in the literature, in the pedal phalanges of oviraptorids (Currie, pers. comm. 1989).

Ornithomimidae

Unidentified ornithomimid: At least one ornithomimid shows a pathological pedal phalanx in which the distal end is expanded or "mushroomed," compared to normal phalanges (Tanke and Rothschild 1999; photo in Kawakami 1996, fig. 5). Minimum number of individuals, 1.

Deinocheiridae

Deinocheirus mirificus: The left manual phalanges III-1 and III-2 of ZPALNo.MgD-I/6 (the holotype) bear pits, presumably from an injury to the joint between them (Osmólska and Roniewicz 1970). The medial condyle of III-1 bears a deep pit and the corresponding sulcus of III-2 has a groove: both features have rounded margins. Minimum number of individuals, 1.

Troodontidae

Troodon formosus: One parietal (TMP 79.8.1) has a pathological aperture, apparently resulting from a cyst (Currie 1985) although Tanke and Rothschild (1999) suggest this feature may be an injury resulting from a bite. In addition, a possible congenital defect, "a peculiar dorsobuccal twist of the [mandibular] symphysis" (Carpenter 1982, 129), has been reported in a hatchling referred to *Troodon* (UCM 41666). Minimum number of individuals, 2.

Tyrannosauridae

Albertosaurus sarcophagus: An aperture (2.5 × 3.5 cm) penetrates the anteroventral process of the iliac blade (Parks 1928; not recognized as pathological by that author) of the holotype (ROM 807) of *A. arctunguis*. There is also a slight exostosis on the left metatarsal IV (Parks 1928; Russell 1970). Russell (1970) reports damage, of unspecified nature, to the humerus in two of the five specimens in which that element was known at the time. Minimum number of individuals, 2.

Albertosaurus sp.: Tooth marks have been seen on cranial elements (Tanke and Currie 1995).

Daspletosaurus torosus: The distal end of one humerus of NMC

8506 (the holotype) exhibits an unspecified pathology (Russell 1970). Minimum number of individuals, 1.

?*Daspletosaurus* sp.: Williamson and Carr (1999) report that in partial skeleton from the Kirtland Formation, New Mexico, the ectopterygoid seems to have sustained a puncture and became infected. In addition, one rib shows a healed fracture. Minimum number of individuals, 1.

Gorgosaurus libratus: In NMC 2120 (the holotype) the right dorsal rib 3, gastralia 13 and 14, and left fibula have healed fractures (Lambe 1917). The left metatarsal IV is "apparently diseased," with roughened exostoses on the lateral surface at midshaft and distally, the third phalanx of right pedal digit III is deformed (wider than the second phalanx, roughened and with a seemingly broadly convex distal articular surface) and the claw of that digit is "quite small and amorphous" (Lambe 1917, 80). Like the type of *Poekilopleuron bucklandii,* this individual seemingly had three anatomically independent pathologies (although the fractures may have been inflicted during a single encounter).

Currie, in a 1997 field report extensively summarized by Tanke and Rothschild (1999), reported that TMP94.12.602, a specimen over 8 m long, sustained several injuries. The midshaft of the right fibula had been broken longitudinally along at least 10 cm, and healed. Several dorsal ribs (number not given) from the middle of the sequence had well-healed fractures and a single gastralium had been broken and subsequently formed a pseudoarthrosis. Tanke and Rothschild report that "face-bite lesions . . . undergoing active healing" (1999, 27) were also present.

Another series of injuries in another specimen (TMP91.36.500) was described by Keiran (1999; Tanke and Rothschild 1999). This individual also sustained bites to the face, and a broken but well-healed right fibula. In addition it had a healed dentary fracture and "a mushroom-like hyperostosis of a right pedal phalanx" (Tanke and Rothschild 1999, 56), possibly like that mentioned above in an unidentified ornithomimid. McGowan (1991) also mentions a poorly healed fracture, resulting in a large callus, of the right fibula in an ROM specimen (given as *Albertosaurus,* but altered to *Gorgosaurus* in Tanke and Rothschild 1999). Minimum number of individuals, 4.

Tyrannosaurus rex: Dorsals 7 and 8 in AMNH 5027 exhibit pathological fusion of the centra (Newman 1970). This appears to be a congenital block vertebra (cf. Keats 1992). The centra of cervical 10 and dorsal 1 are also fused (Rothschild 1997). Moodie (1923) reported spondylitis deformans resulting in coalescence of the cervicals in *Tyrannosaurus,* presumably also of AMNH 5027. However, this may also be an instance of congenital block vertebra. There are pathological openings in the right surangulars of LACM 23844 and MOR 008 (Molnar 1991) discussed below. Also in MOR 008, the squamosal articular surface of the right quadrate is rugose and partly hemispherical in form (unlike that of LACM 23844) and thus it does not conform to the saddle-shaped (pseudospherical) articular surface on the corresponding squamosal. Rothschild et al. (1997) reported localized erosion in metacarpals I and II of FMNH PR2081, attributed to gout. Rothschild (1997) also reported one or more fractured fibulae. Glut (2000) notes that AMNH 5027 has fractured and healed ribs.

Larson (1991) very briefly reported healed injuries in a specimen, FMNH PR2081, in the skull, caudals, ribs, humerus, and fibula. Some of these were illustrated in the Sotheby's catalogue (Sotheby's 1997). In the popular literature generated by this specimen a number of pathological conditions were described or mentioned, unfortunately not always accurately (Tanke and Rothschild 1999). Six conditions have been either illustrated or clearly described: (1) a pathology on the right side of skull and right surangular (Webster 1999); (2) a pathology on the left side of skull (Tanke and Rothschild 1999, 48); (3) a twisted (and discolored) tooth (Webster 1999); (4) two adjacent pathological caudals (illustrated in Sotheby's 1997); (5) a right humerus with hooklike spur associated with larger rounded depression interpreted as an avulsion scar of the M. triceps humeralis (Carpenter and Smith, chap. 9 of this volume; Tanke and Rothschild 1999; illustrated in Sotheby's 1997); and (6) a left fibula broken and healed with extensive abnormal bone growth along the shaft (illustrated in Webster 1999; Sotheby's 1997). Even if some of the initial reports of pathologies are incorrect, this is still an impressive suite.

Another specimen collected by Larson, BHI-3033 ("Stan"), also exhibits pathological features. The popular literature (again reported by Tanke and Rothschild 1999) mentions broken ribs and ankylosed cervicals. Tanke and Rothschild note that a photograph of the skull, posted on the internet (also in *T. rex* 1995), has anomalous openings in the right jugal and surangular. Minimum number of individuals, 6.

Undescribed tyrannosaurid: This specimen in the Museum of the Rockies has three fractured and healed ribs, and one humerus also has a healed fracture. This humerus is shorter and more strongly curved than the normal one.

Tanke and Rothschild (1999) reported a fractured and healed gastralium in TMP97.12.229, in which the medial section suffered two breaks. This is also reported in a popular article by Grierson (1996).

Erosion, similar to that attributed to gout in FMNH PR2081, was seen in the pedal phalanx I-1 of an unidentified tyrannosaurid from Dinosaur Provincial Park, Alberta, in the Royal Tyrrell Museum (Rothschild et al. 1997). Minimum number of individuals, 2.

Family incertae sedis

Becklespinax altispinax: The three dorsals from Sussex referred incorrectly to *Altispinax dunkeri* by Huene (1923) exhibit marked irregular rugosities over the distal third of the neural spines (Owen 1884). Owen describes these as transversely and anteroposteriorly expanded, rugose, and thus unlike the laterally compressed, smooth basal portions of the spines. Owen notes that the cranial two spines were ankylosed. The cranial of the spines is about two-thirds as high as the others. Taken together these features suggest that the distal portions of the spines are likely pathological. Minimum number of individuals, 1.

Marshosaurus bicentesimus: One right ilium (UUVP 2742) is deformed, having an undescribed pathology associated with an anomalous ridge on the lateral face, thought to be the result of an injury (Madsen 1976b). A referred specimen includes a rib with an unspecified pathology (Glut 2000). Minimum number of individuals, 2.

Species undescribed: Heckert et al. (1999) refer to a (presumably) pathologically fused tibia-fibula-astragalus-calcaneum in a recently discovered large Triassic theropod. Minimum number of individuals, 1.

Species unknown: A pedal phalanx III-1 (QM F34621, cast) from the Early Cretaceous Strzelecki Group at Inverloch, Victoria (Australia), bears a depressed fracture on the plantar surface, just proximal to the distal articular trochlea. This phalanx is proportionately twice as long as that of adult *Allosaurus fragilis,* and may pertain to *Timimus hermani* (Rich and Vickers-Rich 1994) or a related taxon. Minimum number of individuals, 1.

Dental

Theropod dental abnormalities have been briefly summarized by Rothschild and Martin (1993). Abnormalities include damage from breaks and scratches as well as malformed (split) carinae and damage to cranial bones resulting from loss of teeth. Tanke and Rothschild (1997) report that large theropod teeth sometimes bear serration marks impressed by another tooth. The orientation of some of these marks suggests that they result from biting, presumably by a conspecific individual.

Ceratosauridae

Ceratosaurus sp.: A broken and then worn tooth is figured by Madsen (1976a, 17).

Allosauridae

Allosaurus fragilis: The holotype jaw of *Labrosaurus ferox* was distinguished from that of *Allosaurus* by being edentulous anteriorly—where there is a prominent concavity in the dorsal margin—and much deeper posteriorly (Marsh 1884). Unfortunately, the posterior part of the jaw is much restored with plaster (Glut 1997), so the reconstructed depth is probably unreliable. Rothschild (1997) attributes the anterior edentulous region to the loss of teeth, including replacement teeth so that bone was subsequently resorbed from the alveoli, resulting in the concavity. The jaw is probably from *A. fragilis,* in which the first four or five teeth were lost, presumably traumatically.

Split carinae occur in *Allosaurus* (Erickson 1995).

Tyrannosauridae

Albertosaurus sp.: Split carinae occur in this form (Erickson 1995), as well as transverse cuts, interpreted as tooth marks, and parallel sets of striae, interpreted as marks from serrations (Tanke and Currie 1995).

Daspletosaurus sp.: Split carinae occur in this form (Erickson 1995).

Tyrannosaurus rex: The fourth right premaxillary tooth of LACM 23844 was apparently broken and the broken face subsequently worn to an almost plane surface (Molnar 1991, pl. 1, fig. 3). That this was not a unique occurrence is shown by a similarly worn surface of an incomplete crown reported by Carpenter (1979). Such a tooth has been found in *Ceratosaurus* sp. (mentioned above) and Tanke and Rothschild (1997, 19) report that broken, then worn tyrannosaurid teeth (presumably not from *Tyrannosaurus*) are "not uncommon" in Alberta. The occurrence of this wear suggests that these broken teeth encountered resistant tissue sufficiently often to wear the surfaces. Because the broken surface of such a tooth was located well below the level of the tips of adjacent teeth, this suggests that either the resistant material was small enough to slip between the (unworn) teeth or the material contacted the tooth strongly enough and for long enough to result in wear during the period when one or more of the adjacent teeth had been shed but not yet replaced.

Rothschild (1997) figures a maxilla (MOR uncataloged) in which one crown is substantially inclined relative to the others. This could have resulted from biting something very resistant, such as bone, but an examination of the material to eliminate the possibility of postmortem damage is necessary before drawing any conclusions.

Some teeth show split carinae (Erickson 1995) that presumably resulted from anomalous development. Erickson discussed whether this might have arisen from injury to the dentigerous tissue, but concluded that it was more likely to have been genetic. Erickson also reported supernumerary cusps on teeth of *T. rex.*

Douglas and Young report that some teeth, presumably from *Tyrannosaurus,* "bear telltale marks made by the teeth of their fellows" (1998, 29).

Species unspecified: Fractured teeth from the Late Cretaceous deposits of Dinosaur Provincial Park, Alberta, were reported by Jacobsen (1996). Jacobsen reported that 29% of randomly collected teeth showed breakage with the subsequent development of wear, a proportion that Tanke and Rothschild, based on field observations, suggest is "artificially high" (1999, 52). Certainly the proportion of such teeth preserved in jaws is much lower. Jacobsen also reported teeth that had suffered marking by other, presumably tyrannosaurid, teeth.

Species unknown: Bohlin (1953) noted a split carina in a tooth from the Minhe Fm. of China. The crown is quite small, about 17 mm high.

Tracks

Pathologies may be discernible from tracks (Thulborn 1990). Tracks are not as easily interpreted in this respect as skeletal structures: it may be impossible to determine if anomalous features were due to physical abnormalities or unusual behaviors.

Anchisauripus: Some footprints attributed to this ichnotaxon in Norian sandstone at Glamorgan, southern Wales, indicate a malformed digit III in one individual (Tucker and Burchette 1977, fig. 3). The malformation consists of a consistent flexure of the distal end of the digit. These tracks could represent an individual with a physically deformed third digit or, alternatively, one in which the tip of that digit was rotated upon lifting the foot, thus giving the impression of a malformation (cf. Thulborn 1990, fig. 5.16).

Eubrontes: Abel (1935) documented a trackway of *Eubrontes* in which the impressions of the right foot are consistently didactyl, as if the animal had lost the second digit, or it was malformed (also discussed by Thulborn 1990).

Sauroidichnites abnormis: Hitchcock (1844) reported a trackway, in which one of the feet had an abnormally positioned toe. This may represent a physical injury or an unusual behavior in positioning or removing the foot.

"Coelurosaur": Jenny and Jossen (1982) reported and figured the trackway of a small theropod ("coelurosaur") from the Moroccan Jurassic that showed a limp, inferred from a trackway with alternating step length (also in Ishigaki 1986). Footprints show that the animal held pedal digits III and IV abnormally close together, presumably reflecting an injury also manifested in the limp. Dantas et al. (1995) discuss several such occurrences and conclude that without evidence of pathology from footprints, several nonpathological causes for alternating step length are possible.

Discussion

The discussion and analysis of pathologies are restricted to those of the skeleton (as opposed to those seen in the teeth or tracks).

Taxonomic distribution: This survey indicates that at least 21 species from at least 10 families of theropods show pathologies. Several individuals show pathologies of more than one element, and others (e.g., ROM 807 and the holotype of *P. bucklandii*) exhibit pathologies having two or more different etiologies. The discovery of such cases depends on having reasonably complete skeletons, as the occurrence of pathologies in several parts of the skeleton cannot be seen if one or more of these parts is not preserved. This restriction eliminates the majority of specimens from consideration, because most are incomplete. Thus, the survey is biased toward taxa, such as *Allosaurus* and the tyrannosaurids, known from relatively complete material. It also suffers from the "pull of the recent" (that the fossil record becomes generally more incomplete with increasing age): this is certainly the case for theropods (Molnar and Farlow 1990; Molnar 1997).

Of the families represented, the Tyrannosauridae has many specimens exhibiting paleopathologies (table 24.1). It is one of the few theropod families represented by more than one well-known genus. Other families, such as the Allosauridae, that are well represented in the fossil record also have many pathological specimens. Thus, it cannot be concluded that pathologies were more prevalent in tyrannosaurids than other theropod families, because most families are too poorly known for the extent of pathologies to be determined.

Some theropod families, so far, have few or no reported patholo-

TABLE 24.1
Synopsis of Reported Theropod Pathologies

Pathology	Element	Taxon	Minimum Apparent Number of Incidents	Reference
puncture(s)	cranial elements	*Carcharodontosaurus saharicus*	1	Sereno, pers. comm. 1999
		Sinraptor dongi	1	Tanke and Currie 1995
		Gorgosaurus libratus	2	Keiran 1999; Tanke and Rothschild 1999
		Albertosaurus sp.	?	Tanke and Currie 1995
	skull roof	*Velociraptor mongoliensis*	1	Norell et al. 1995
	parietal and splenial	*Herrerasaurus ischigualastensis*	3	Sereno and Novas 1993
(with secondary infection)	ectopterygoid	?*Daspletosaurus* sp.	1	Williamson and Carr 1999
	metacarpal	*Allosaurus fragilis*	1	original
	pedal phalanx	unknown species	1	original
possible puncture	ilium	*Allosaurus fragilis*	1	Laws 1995, 1997
fracture	dentary	*Gorgosaurus libratus*	1	Keiran 1999; Tanke and Rothschild 1999
	dorsal	*Monolophosaurus jiangi*	1	Zhao and Currie 1993
	caudals	*Neovenator salerii*	1	Tanke and Rothschild 1999
		Acrocanthosaurus atokensis	2	Harris 1998; Tanke and Rothschild 1999
	cervical rib	*Megalosaurus bucklandii*	1	Tanke and Rothschild 1999
		Allosaurus fragilis	1	Petersen et al. 1972
	dorsal ribs	*Allosaurus fragilis*	3	Petersen et al. 1972; Reid 1996; Hecht 1998
		Sinraptor dongi	1	Currie and Zhao 1993; Tanke and Rothschild 1999
		Acrocanthosaurus atokensis	4	Harris 1998; Larson 1998
		?*Daspletosaurus* sp.	1	Williamson and Carr 1999
		Gorgosaurus libratus	2	Lambe 1917; Tanke and Rothschild 1999
		Tyrannosaurus rex	2	Tanke and Rothschild 1999; Glut 2000
		undescribed tyrannosaurid	1	original
	gastralia	*Allosaurus fragilis*	1	Tanke and Rothschild 1999
		Neovenator salerii	1	Tanke and Rothschild 1999
		Gorgosaurus libratus	2	Lambe 1917; Tanke and Rothschild 1999
		unidentified tyrannosaurid	1	
	scapula	*Allosaurus fragilis*	1	Gilmore 1920
		Neovenator salerii	1	Tanke and Rothschild 1999
		Acrocanthosaurus atokensis	1	Larson 1998

TABLE 24.1 (cont.)

Pathology	Element	Taxon	Minimum Apparent Number of Incidents	Reference
	humerus	*Allosaurus fragilis*	1	Petersen et al. 1972
		undescribed tyrannosaurid	1	original
	ulna	oviraptorid	1	Clark et al. 1999
	radius	*Allosaurus fragilis*	1	Petersen et al. 1972
	tibia	*Syntarsus rhodesiensis*	?	Raath, pers. comm. 1986
	fibula	*Gorgosaurus libratus*	4	Lambe 1917; McGowan 1991; Keiran 1999; Tanke and Rothschild 1999
		Tyrannosaurus rex	1	Webster 1999
	metatarsal	*Syntarsus rhodesiensis*	?	Raath, pers. comm. 1999
		Ceratosaurus nasicornis	1	Gilmore 1920
	pedal phalanx	*Deinonychus antirrhopus*	1	Ostrom 1976
amputation	chevron	*Allosaurus fragilis*	1	Petersen et al. 1972
	pedal phalanx	*Allosaurus fragilis*	1	Petersen et al. 1972
avulsion scar	humerus	*Tyrannosaurus rex*	1	Carpenter and Smith, this volume; Tanke and Rothschild 1999
possible drag marks	dentary	*Monolophosaurus jiangi*	1	Tanke and Rothschild 1999
possible puncture	surangular	*Tyrannosaurus rex*	1	Molnar 1991
coossification	caudals	*Allosaurus fragilis*	3	Petersen et al. 1972
anomalous ridge	ilium	*Marshosaurus bicentesimus*	1	Madsen 1976b
erosive lesion	surangular	*Tyrannosaurus rex*	1	Molnar 1991
	scapula	*Allosaurus fragilis*	2	Petersen et al. 1972
	metacarpals	*Tyrannosaurus rex*	1	Rothschild et al. 1997
	pedal phalanx	unidentified tyrannosaurid	1	Rothschild et al. 1997
fusion	cervical 10 and dorsal 1	*Tyrannosaurus rex*	1	Rothschild 1997
	dorsals	*Tyrannosaurus rex*	1	Newman 1970
anomalous form	quadrate joint	*Tyrannosaurus rex*	1	original
	dentary	*Troodon formosus*	1	Carpenter 1982
	caudal joints	*Allosaurus fragilis*	2	Petersen et al. 1972
	pedal phalangeal joints	*Allosaurus fragilis*	2	Petersen et al. 1972

TABLE 24.1 (cont.)

Pathology	Element	Taxon	Minimum Apparent Number of Incidents	Reference
	pedal phalangeal and adjacent ungual joint	*Gorgosaurus libratus*	1	Lambe 1917
abnormal pits or apertures	cranial elements	*Tyrannosaurus rex*	2	Tanke and Rothschild 1999; Webster 1999
	jugal and surangular	*Tyrannosaurus rex*	1	Tanke and Rothschild 1999
	parietal	*Troodon formosus*	1	Currie 1985
	manual phalanges	*Deinocheirus mirificus*	1	Osmólska and Roniewicz 1970
	ilium	*Albertosaurus sarcophagus*		Parks 1928; Russell 1970
fusion	chevron with caudal	*Poekilopleuron bucklandii*	11	Eudes-Deslongchamps 1838
		Allosaurus fragilis	?	Petersen et al. 1972
	caudals	*Neovenator salerii*	1	Tanke and Rothschild 1999
		Tyrannosaurus rex	1	Anonymous, 1997a
	tibia, fibula, proximal tarsals	undescribed species	1	Heckert et al. 1999
anomalous bone growths	premaxilla	*Allosaurus fragilis*	1	Petersen et al. 1972
	squamosal	*Acrocanthosaurus atokensis*	1	Stovall and Langston 1950
	dorsals	*Becklespinax altispinax*	1	original
	caudal	*Allosaurus fragilis*	1	Petersen et al. 1972
		Acrocanthosaurus atokensis	1	Tanke and Rothschild 1999
	metacarpals	*Allosaurus fragilis*	?	Petersen et al. 1972
	manual(?) phalanx	*Poekilopleuron bucklandii*	1	Eudes-Deslongchamps 1838
	ungual	*Allosaurus fragilis*	1	Petersen et al. 1972
	fibula	*Allosaurus fragilis*	1	Gilmore 1920
	metatarsal	*Albertosaurus sarcophagus*	1	Parks 1928; Russell 1970
		Gorgosaurus libratus	1	Lambe 1917
	pedal(?) phalanx	*Poekilopleuron bucklandii*	1	Eudes-Deslongchamps 1838
	pedal phalanx	*Allosaurus fragilis*	2	Petersen et al. 1972; Laws 1995, 1997
		Neovenator salerii	1	Tanke and Rothschild 1999
		unidentified ornithomimid	1	Tanke and Rothschild 1999
		Gorgosaurus libratus	1	Keiran 1999; Tanke and Rothschild 1999
bifurcation	gastralium	undescribed dromaeosaurid	1	Norell and Makovicky 1997
unspecified	cervical	*Allosaurus fragilis*	1	Laws 1995, 1997
	dorsals	*Allosaurus fragilis*	1	Laws 1995, 1997
	caudal and chevron	*Allosaurus fragilis*	1	Laws 1995, 1997
	dorsal ribs	*Allosaurus fragilis*	1	Laws 1995, 1997

TABLE 24.1 (cont.)

Pathology	Element	Taxon	Minimum Apparent Number of Incidents	Reference
		Marshosaurus bicentesimus	1	Glut 2000
	gastralia	*Allosaurus fragilis*	1	Laws 1995, 1997
	scapula	*Allosaurus fragilis*	1	Laws 1995, 1997
	humerus	*Albertosaurus sarcophagus*	2	Russell 1970
		Daspletosaurus torosus	1	Russell 1970
	manual phalanx	*Allosaurus fragilis*	1	Laws 1995, 1997
	metatarsals	*Allosaurus fragilis*	1	Laws 1995, 1997
	pedal phalanx	*Allosaurus fragilis*	1	Laws 1995, 1997

Note: anomalous bone growths includes exostoses, spurs, and osteophytes. "?" was counted as 1.

gies (e.g., the abelisaurids). However, these tend to be poorly known families. Thus, it is premature to draw conclusions regarding the phyletic distribution of pathologies from this absence. If anything, the abundance of pathologies in the well-known taxa suggests that they were probably widespread among less well known theropods.

Pathologies have been found in both small and large theropods. They are substantially less common among small theropods (table 24.1). Large theropods are often represented by better preserved and more complete material than small theropods, and hence the opportunity for finding pathological features is greater in the larger specimens. Thus it is premature to conclude that large forms were more susceptible to pathology than small ones.

Causes: A total of 119 pathologies have been compiled (table 24.1). Those of Petersen et al. (1972) have been scored as one instance for each kind of pathology (almost certainly an underestimate), unless there was reason to believe that more than a single individual was involved. Pathologies interpreted to have resulted from separate events are counted as distinct. For example, the three tooth marks in *Herrerasaurus* (Sereno and Novas 1993) resulted from penetrations in three different directions, so are counted as three features. In the case of *Velociraptor* (Norell et al. 1995), where the marks probably result from a single event, they are counted as a single instance. *Dilophosaurus* is excluded because of uncertainty regarding the status of the features (Welles 1984), as are the developmental asymmetries discussed below. The attempt is made to count individual incidents (for fig. 24.1), rather than individual animals, because the incidents have the greater biological significance. Given the paradigm of evolution by natural selection, assumed here unless there is evidence to the contrary, the ability of an individual to survive and reproduce is affected by pathology. Damage to a bone that results in decreased effectiveness of finding food or mates or escaping predation is involved in selection for increased resistance to damage in that bone if those individuals with the more resistant bones are also more effective at finding food, mates, and so on, and hence

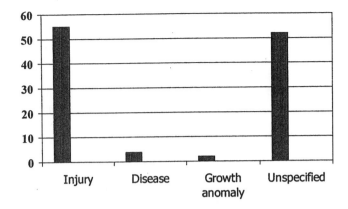

Figure 24.1. Histogram of reported etiologies of theropod pathologies.

reproduce more than their less equipped fellows. Similar considerations have been formulated in greater detail by Alexander (1981, 1984). Thus, the number of incidents during which bones are exposed to these risks potentially provides information on the relative resistance of bones in different individuals to incapacitating damage.

Of the cases, 58 are due to an undetermined (or at least unspecified) cause, 5 to disease, 2 to developmental conditions, and 55 to injury (Fig. 24.1; 2 cases involved both injury and subsequent secondary infection). The details of the injuries are not always given in the primary sources, but many are due to fractures of long bones or ribs (table 24.1). There are also depressed fractures (as in YPM 4944) and other apparent punctures (as in LACM 23844).

The latter conditions have not been detailed in the literature so a brief description is included here. The metacarpal of YPM 4944 (*Allosaurus*) bears a circular depressed fracture on the dorsal surface just proximal to midshaft. The size and configuration of the impression suggests that it was caused by a blow from a sharp object, such as a tooth. It is not entirely clear that this injury occurred during life, because there is no sign of callus formation indicative of healing. Hence, if it did occur during life, it was shortly before the animal's death. As mentioned above, this is unlikely to have been due to scavenging because of the absence from the manus of muscular or other soft tissues attractive to a scavenger. The depressed fracture of pedal phalanx QM F34621 from the Early Cretaceous of Victoria, Australia, is quite similar. It is a depressed fracture on the plantar surface, just proximal to the distal articular trochlea. This specimen has been crushed after burial, as evidenced by damage to both dorsal and ventral surfaces, but that damage is more extensive and lacks the concentric breakage of the depressed fracture. Like that of YPM 4944, it shows no indication of healing. The right surangular of LACM 23844 (*Tyrannosaurus rex*) has a slotlike aperture in the surangular buttress (shown in Molnar 1991, pl. 15). The buttress is dorsoventrally expanded to about twice its usual thickness, and rugose bone has formed below the aperture. This may well have resulted from an infected puncture, but further work is necessary to verify this. This surangular also has a large aperture near its anterior edge that appears pathological (as opposed to being the result of postmortem breakage) because there is indication of healing along its margins. These features may have been inflicted by biting.

In view of the great proportion (almost 50%) of pathologies of unspecified or doubtful etiology, the distribution of pathologies among the three causes is tentative. However, even if the unspecified pathologies include none due to injury, almost 50% of the pathologies were still due to that cause. Thus, we may regard injury as probably an important factor in the lives of theropods. However, not all instances have been thoroughly described and thus the attribution of injury as the cause in all these cases here accepted as injuries is also tentative: even with a thorough description the cause of a pathological feature may remain obscure.

Anatomical distribution: About 18% of the pathologies occurred in the head, 40% in the vertebral column or ribs, 17% in the forelimb, and 25% in the hind limb (fig. 24.2). In figures 24.2 through 24.6 the number of individual elements was scored, not the number of incidents, and the one element not identified as to limb was excluded. Vertebral pathologies tend to occur in the tail (fig. 24.3) and most of the pathologies of the postcranial axial skeleton involve the tail or ribs (fig. 24.4). Forelimb pathologies tend to be proximal, at the scapula or humerus, or distal, in the manus, but not in the antebrachium (fig. 24.5). In the hind limb, they are mostly distal, occurring in the pes (fig. 24.6). In the forelimb, pathologies have been reported in almost all elements (except the coracoid and carpals), while in the hind limb there are no reports of femoral pathologies and few of the tibia (fig. 24.6). This pattern suggests that injuries to these weight-bearing elements were, in general, not survivable in a bipedal animal (cf. Brandwood et al. 1986; Bulstrode et al. 1986). On the other hand, injuries to the fibula and foot seem not to have been fatal. The fibula was probably not a weight-bearing element, or did not bear a substantial proportion of the weight. Pedal injuries may have been more likely to occur because no single pedal element bears the entire body weight (as does the femur and, likely, the tibia) and pedal elements contact an often irregular substrate. Furthermore, the foot may have been used in hunting (Ostrom 1969) or feeding (Huene 1926) or even intraspecific combat, and hence have been more exposed to injury than more proximal hind limb elements (which are also deeply embedded in muscle). These considerations would also apply to the forelimb. Damage to the pes is consistent with footprint evidence for anomalous positions (or even loss) of toes.

A more speculative interpretation can be offered for depressed fractures, presumably tooth marks, in the pes. Kavanau (1987) reports that toe biting is a common interaction among parrots. In view of Kavanau's argument for the long-term stability of behavioral patterns (also found by Slikas 1998), these injuries of (nonavian) theropods suggest that toe biting was also practiced by them.

In view of the small sample size, all these results must be regarded as tentative.

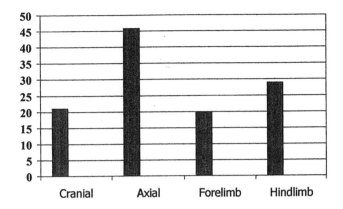

Figure. 24.2. Distribution of theropod pathologies among parts of the skeleton. Cranial *designates the skull and mandibles;* Axial, *the postcranial axial skeleton, including ribs and gastralia;* Forelimb *and* Hindlimb *include both girdles and free limbs.*

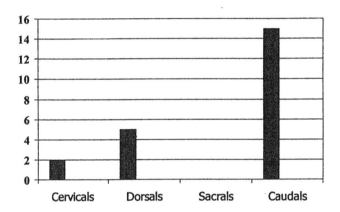

Figure 24.3. Distribution of theropod pathologies among parts of the vertebral column.

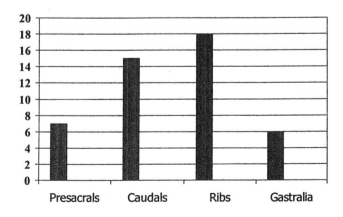

Figure 24.4. Distribution of theropod pathologies among presacral and caudal vertebrae, ribs, and gastralia.

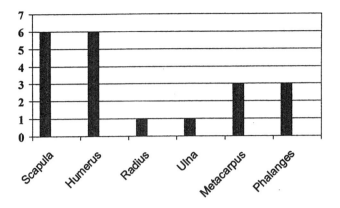

Figure 24.5. Distribution of theropod pathologies among elements of the forelimb. No reported pathologies involve the coracoid or carpus.

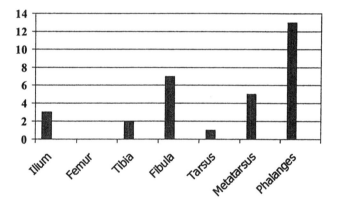

Figure 24.6. Distribution of theropod pathologies among elements of the hindlimb.

Developmental anomalies: The probable occurrence of a congenital block vertebra in *Tyrannosaurus rex,* a condition also found in humans (Keats 1992), has interesting implications. It suggests that dysfunctions of the developmental processes involved in vertebral development (in this case presumably a failure of the resegmentation of the sclerotomes) and hence the processes themselves (the resegmentation of the sclerotomes), have remained basically unaltered since the common ancestor of archosaurs and mammals in the Late Carboniferous. In this particular case, because the ontogenetic processes are similar in distantly related modern taxa, it is generally accepted that the processes have changed little over this period. For less well understood developmental processes, such paleopathologies can potentially be illuminating. The recognition of such anomalies can potentially permit the construction of a chronology of the sequential appearance of developmental processes in a phyletic lineage, much as mutations assist the construction of a chronology of developmental processes in an individual. Such a chronology will be useful in understanding the relationships of development to the changes in form and structure during evolution.

Much attention is being devoted to the recognition of juvenile and subadult individuals in the archosaur fossil record, but conversely little has been given to the recognition of old individuals. The anomalous fusion of cranial elements may permit recognition of gerontic individuals among saurischians. This is the case for MOR 008 (*Tyrannosaurus rex*), where fusions of the postorbital and jugal, angular and surangular, and prearticular and surangular are known (Molnar 1991). Each fusion occurs on only one side and none has previously been reported in a theropod. It has also been suggested (Hotton 1963) that arthritic conditions may have been more frequent in old individuals. However, there is now considerable doubt that any of the reported arthritic conditions have been properly diagnosed. Rothschild and Martin (1993) report that osteoarthritis is absent in *Allosaurus, Albertosaurus,* and *Tyrannosaurus.* Potentially, conditions that reflect individual age, even if only approximately, can provide further insight into the age structures of theropod populations, although there are practical difficulties of sampling. Ascertaining the age of individuals also permits some insight into the possible occurrence of selection, particularly reproductive selection: gerontic individuals probably had the opportunity for reproduction (whether or not they in fact did), but individuals that died before reaching sexual maturity did not reproduce.

It has been proposed that a developmental anomaly, fluctuating asymmetry, may provide insight into recognizing which populations were under stress, presumably by undergoing selection more intense than usual (e.g., Jones 1987; Leary and Allendorf 1988). Bilaterally symmetric animals can show asymmetries produced during the course of normal developmental processes, such as the development of the giant claw in fiddler crabs (Leary and Allendorf 1988). Fluctuating asymmetry is distinct from these kinds of asymmetries and is defined as asymmetry resulting from the disturbance of development (Van Valen 1962). It can be recognized as asymmetry that differs between comparable individuals (of the same sex, developmental state, etc.) of a population. Fluctuating asymmetry potentially can be seen in fossils, and so provides the opportunity of discerning unusually intense selective pressures. Unfortunately, it does not reveal the nature of these pressures.

There are several possible instances of fluctuating asymmetry among theropods. In these cases more than a single individual is known, and the others do not show the asymmetry. In one specimen of *Dilophosaurus wetherilli* "the left humerus is smaller and more delicate than the right, [but] the reverse is true of the forearm, where the left propodials are larger and stouter" (Welles 1984, 128). The right radius and left ulna of the holotype of *Struthiomimus currelli* (now *Ornithomimus edmontonensis*) are about 80% as long as their counterparts (Parks 1933). Raath (1969) reported asymmetric development of the supporting buttresses of the second sacral rib in *Syntarsus rhodesiensis*. Further study is desirable, because these conditions might also have resulted from traumatic events early in development. Such traumatic events presumably would have been randomly distributed in time, whereas fluctuating asymmetry is expected to have preferentially occurred during times of environmental stress. So far, examples of fluctuating asymmetry in fossil theropods are too few to reveal any general patterns of unusual selection, but examination of skeletons from periods of extinction might prove interesting. If the causal factors of the extinction acted over a prolonged period prior to the extinction, indications of fluctuating asymmetry would be expected in the skeletons of individuals living (and dying) at this time: if the extinction was abrupt and catastrophic, fluctuating asymmetry would not be expected.

Disease: There is no indication of anything other than very localized infective processes among theropods, although evidence of more extensive infections has been noted in other dinosaurs (e.g., Swinton 1934). An example of such localized infections is provided by MOR 008 (*Tyrannosaurus rex*). The right surangular has a small bowl-like depression of smoothly surfaced bone just below the anterior end of the surangular buttress. A small opening into the depression penetrates the element. This is basically similar to modern deformations resulting from abscesses, although how the infective organism was introduced in this case is unclear.

Injury: Injuries directly reflect behavior, and hence can illuminate the behavior of extinct animals. Depressed fractures and other puncture wounds may have resulted from combat. Cott (1961) showed that intraspecific combat is a significant source of injury in *Crocodylus niloticus,* and Webb and Manolis (1989) have presented similar evidence for *Crocodylus porosus*. Injuries of similar magnitude to those that have been reported among these crocodilians, such as limb or jaw amputations, have not been seen in theropods. This may suggest that theropods were not as aggressive in intraspecific conflict, or that individuals suffering such damage did not survive, and died under circumstances not conducive to their preservation as fossils. An analogous example of trauma-induced death in crocodiles was reported by Webb and Manolis (1989).

Fractures are very rare in *Syntarsus rhodesiensis* (Raath, pers. comm. 1999). Injuries reported in the literature occur less than one-fifth as frequently in small as in large theropods, although large theropod specimens seem no more abundant than small. This suggests that large size may have been a factor in the fractures, although, in view of the small sample size, caution is appropriate in interpreting this result.

The consistent absence of fractures on certain elements (e.g., the femur) suggests that these elements were under selection regarding their architecture and strength (i.e., resistance to fracture). If, as has been asserted (Bulstrode et al. 1986), modern animals in the wild do not usually survive major fractures of the limb bones, then similar selection would be expected for extinct creatures. It seems likely that such elements did fracture, and that individuals suffering these fractures are not represented in the fossil record because they died shortly thereafter in situations not amenable to preservation. The absence of fractures in certain bones permits ranking anatomical elements with respect to the importance of selection in their design, as opposed to developmental or structural influences. In theropods, it would seem that the sacrum, femur, and tibia experienced the greatest selective pressure in this regard.

Acknowledgments: Robert Allen, Leslie Drew, Jack Horner, John Ostrom, Michael Raath, Paul C. Sereno, and Masahiro Tanimoto all assisted this study by bringing to my attention specimens not reported in the literature, and in other ways; I am very grateful for their thoughtfulness. Jerry Harris, Robert Paterson, Mark Norell, Bruce Rothschild, Tony Thulborn, and Lorrie McWhinney provided appreciated assistance. I also much benefited from discussions with Philip J. Currie and Kenneth Carpenter.

References

Abel, O. 1935. *Vorzeitliche Lebensspuren.* Jena: Gustav Fischer Verlag.

Alexander, R. McN. 1981. Factors of safety in the structure of animals. *Science Progress* 67: 109–130.

Alexander, R. McN. 1984. Optimum strength for bones liable to fatigue and accidental damage. *Journal of Theoretical Biology* 109: 621–636.

Baur, G. 1890. A review of the charges against the Paleontological Department of the U. S. Geological Survey, and of the defence made by Prof. O. C. Marsh. *American Naturalist* 24: 298–304.

Bohlin, B. 1953. Fossil reptiles from Mongolia and Kansu. In *Reports from the Scientific Expedition to the North-Western Provinces of China under Leadership of Dr. Sven Hedin,* 37. Vertebrate Palaeontology 6: 1–113; Stockholm: Statens Etnografiska Museum.

Brandwood, A., A. S. Jayes, and R. McN. Alexander. 1986. Incidence of healed fracture in the skeleton of birds, molluscs, and primates. *Journal of Zoology* 208: 55–62.

Bulstrode, C., J. King, and B. Roper. 1986. What happens to wild animals with broken bones? *Lancet,* January 4, pp. 29–31.

Carpenter, K. 1979. Vertebrate fauna of the Laramie Formation (Maastrichtian), Weld County, Colorado. *Contributions to Geology* (University of Wyoming) 17: 37–49.

Carpenter, K. 1982. Baby dinosaurs from the Late Cretaceous Lance and Hell Creek formations and a description of a new species of theropod. *Contributions to Geology* (University of Wyoming) 20: 123–134.

Clark, J. M., M. A. Norell, and L. M. Chiappe. 1999. An oviraptorid skeleton from the Late Cretaceous of Ukhaa Tolgod, Mongolia, preserved in an avianlike brooding position over an oviraptorid nest. *American Museum Novitates* 3265: 1–36.

Cott, H. B. 1961. Scientific results of an inquiry into the ecology and economic status of the Nile Crocodile (*Crocodilus niloticus*) in Uganda and Northern Rhodesia. *Transactions of the Zoological Society of London* 29: 211–356.

Currie, P. J. 1985. Cranial anatomy of *Stenonychosaurus inequalis* (Saurischia, Theropoda) and its bearing on the origin of birds. *Canadian Journal of Earth Sciences* 22: 1632–1658.

Currie, P. J., and Zhao X.-J. 1993. A new carnosaur (Dinosauria, Theropoda) from the Jurassic of Xinjiang, People's Republic of China. *Canadian Journal of Earth Sciences* 30: 2037–2081.

Dantas, P., V. F. dos Santos, M. G. Lockley, and C. A. Meyer. 1995. Footprint evidence for limping dinosaurs from the Upper Jurassic of Portugal. *Gaia* 10: 43–48.

Douglas, K., and S. Young. 1998. The dinosaur detectives. *New Scientist* 158 (2130): 24–29.

Erickson, G. M. 1995. Split carinae on tyrannosaurid teeth and implications of their development. *Journal of Vertebrate Paleontology* 15: 268–274.

Eudes-Deslongchamps, J.-A. 1838. *Memoir sur le* Poekilopleuron bucklandii, *grand saurien fossile, intermédiare entre les crocodiles et les lézards.* Caen: A. Hardel, Imprimeur de l'Académie.

Gilmore, C. W. 1915. On the fore-limb of *Allosaurus fragilis*. *Proceedings of the United States National Museum* 49: 501–513.

Gilmore, C. W. 1920. Osteology of the carnivorous Dinosauria in the United States National Museum, with special reference to the genera *Antrodemus* (*Allosaurus*) and *Ceratosaurus*. *Bulletin of the United States National Museum* 110: 1–159.

Glut, D. F. 1997. *Dinosaurs: The Encyclopedia.* Jefferson, N.C.: McFarland.

Glut, D. F. 2000. *Dinosaurs: The Encyclopedia, Supplement 1.* Jefferson, N.C.: McFarland.

Grierson, B. 1996. Writing down the bones. *Western Living* 23 (8): 107–108, 110, 114, 177.

Harris, J. D. 1998. A reanalysis of *Acrocanthosaurus atokensis*, its phylogenetic status, and paleobiologic implications, based on a new specimen from Texas. *Bulletin, New Mexico Museum of Natural History and Science* 13: 1–75.

Hecht, J. 1998. The deadly dinos that took a dive. *New Scientist* 158 (2130): 13.

Heckert, A. B., L. F. Rinehart, S. G. Lucas, A. Downs, J. W. Estep, P. K. Reser, and M. Snyder. 1999. A diverse new Triassic fossil assemblage from the Petrified Forest Formation (Revueltian: Early-Mid Norian) near Abiquiu, New Mexico. *New Mexico Geology* 21: 42. (Abstract.)

Hitchcock, E. 1844. Report on ichnolithology, or fossil footmarks, with a description of several new species, and the coprolites of birds, from the valley of the Connecticut River, and of a supposed footmark from the valley of the Hudson River. *American Journal of Science* 47 (2): 292–322.

Hotton, N., III. 1963. *Dinosaurs.* New York: Pyramid Publications.

Huene, F. 1923. Carnivorous Saurischia in Europe since the Triassic. *Bulletin of the Geological Society of America* 34: 449–458.

Huene, F. 1926. The carnivorous Saurischia in the Jura and Cretaceous formations principally in Europe. *Revista de Museo de la Plata* 29: 1–167.

Ishigaki S. 1986. *Morocco no Kyouryu.* Tokyo: Tsukiji Shokan.

Jacobsen, A. R. 1996. Wear patterns on tyrannosaurid teeth: An indication of biting strategy. *Journal of Vertebrate Paleontology* 16 (suppl. to no. 3): 43A. (Abstract.)

Jenny, J., and J. A. Jossen. 1982. Découverte d'empreintes de pas de dinosauriens dans le Jurassique inferieur (Pliensbachien) du Haut Atlas central (Maroc). *Comptes rendus hebdomadaire des séances de l'Académie des Sciences* (Paris) 294: 223–226.

Jones, J. S. 1987, An asymmetrical view of fitness. *Nature* 325: 298–299.

Kavanau, J. L. 1987. *Lovebirds, Cockatiels, Budgerigars: Behavior and Evolution.* Los Angeles: Science Software Systems.

Kawakami, G. 1996. Report on the study at the Royal Tyrrell Museum of Palaeontology, Alberta, Canada. *Bulletin of the Hobetsu Museum* 12: 9–15.

Keats, T. E. 1992. *Atlas of Normal Roentgen Variants That May Simulate Disease*. St. Louis: Mosby Yearbook.

Keiran, M. 1999. *Discoveries in Palaeontology—Albertosaurus—Death of a Predator*. Vancouver: Raincoast Books. (Not seen.)

Lambe, L. M. 1917. The Cretaceous theropodous dinosaur *Gorgosaurus*. *Geological Survey of Canada, Memoirs* 100: 1–84.

Larson, P. L. 1991. The Black Hills Institute *Tyrannosaurus:* A preliminary report. *Journal of Vertebrate Paleontology* 11 (suppl. to no. 3): 41A-42A. (Abstract.)

Larson, P. L. 1998. A stitch in time. *Lapidary Journal* January, pp. 42–46.

Laws, R. R. 1995. Description and analysis of the multiple pathological bones of a sub-adult *Allosaurus fragilis* (MOR 693). Rocky Mountain Section, Geological Society of America 47th annual meeting, p. 43. (Abstract with program.)

Laws, R. R. 1997. Allosaur trauma and infection: Paleopathological analysis as a tool for lifestyle reconstruction. *Journal of Vertebrate Paleontology* 17 (suppl. to no. 3): 59A-60A. (Abstract.)

Leary, R. F., and F. W. Allendorf. 1988. Fluctuating asymmetry as an indicator of stress: Implications for conservation biology. *Trends in Ecology and Evolution* 4: 214–217.

Madsen, J. H., Jr. 1976a. *Allosaurus fragilis:* A revised osteology. *Utah Geological and Mineral Survey, Bulletin* 109: 1–163.

Madsen, J. H., Jr. 1976b. A second new theropod dinosaur from the Late Jurassic of east central Utah. *Utah Geology* 3: 51–60.

Marsh, O. C. 1884. Principal characters of American Jurassic dinosaurs. Part 8, The order Theropoda. *American Journal of Science* 27 (3): 329–341.

McGowan, C. 1991. *Dinosaurs, Spitfires, and Sea Dragons*. Cambridge, Mass.: Harvard University Press.

Molnar, R. E. 1991. The cranial morphology of *Tyrannosaurus rex*. *Palaeontographica* A, 217: 137–176.

Molnar, R. E. 1997. Biogeography for dinosaurs. In J. O. Farlow and M. K. Brett-Surman (eds.), *The Complete Dinosaur*, pp. 581–606. Bloomington: Indiana University Press.

Molnar, R. E., and J. O. Farlow. 1990. Carnosaur paleobiology. In D. B. Weishampel, P. Dodson, and H. Osmólska (eds.), *The Dinosauria*, pp. 210–224. Berkeley: University of California Press.

Moodie, R. L. 1923. *Paleopathology*. Urbana: University of Illinois Press.

Newman, B. 1970. Stance and gait in the flesh-eating dinosaur *Tyrannosaurus*. *Biological Journal of the Linnean Society* 2: 119–123.

Norell, M. A., and P. J. Makovicky. 1997. Important features of the dromaeosaur skeleton: information from a new specimen. *American Museum Novitates* 3215: 1–28.

Norell, M. A., E. S. Gaffney, and L. Dingus. 1995. *Discovering Dinosaurs in the American Museum of Natural History*. New York: Knopf.

Osborn, H. F. 1916. Skeletal adaptation of *Ornitholestes, Struthiomimus, Tyrannosaurus*. *Bulletin of the American Museum of Natural History* 35: 733–771.

Osmólska, H., and E. Roniewicz. 1970. Deinocheiridae, a new family of theropod dinosaurs. *Palaeontologia polonica* 21: 5–19.

Ostrom, J. H. 1969. Osteology of *Deinonychus antirrhopus,* an unusual theropod from the Lower Cretaceous of Montana. *Bulletin of the Peabody Museum of Natural History* 30: 1–165.

Ostrom, J. H. 1976. On a new specimen of the Lower Cretaceous theropod dinosaur *Deinonychus antirrhopus*. *Breviora* 439: 1–21.

Owen, R. 1856. *Monograph on the Fossil Reptilia of the Wealden and Purbeck Formations*. London: Palaeontographical Society.

Owen, R. 1884. *A History of British Fossil Reptiles*. London: Cassell.

Parks, W. A. 1928. *Albertosaurus arctunguis*, a new species of theropodous dinosaur from the Edmonton Formation of Alberta. *University of Toronto Studies*, Geological ser., 25: 1–42.

Parks, W. A. 1933. New species of dinosaurs and turtles from the Upper Cretaceous formations of Alberta. *University of Toronto Studies*, Geological ser., 34: 1–33.

Petersen, K., J. I. Isakson, and J. H. Madsen Jr. 1972. Preliminary study of paleopathologies in the Cleveland-Lloyd dinosaur collection. *Utah Academy Proceedings* 49: 44–47.

Raath, M. 1969. A new coelurosaurian dinosaur from the Forest Sandstone of Rhodesia. *Arnoldia* 28: 1–25.

Reid, R. E. H. 1996. Bone histology of the Cleveland-Lloyd dinosaurs and of dinosaurs in general. Part 1: Introduction to bone tissues. *Brigham Young University Geology Studies* 41: 25–71.

Rich, T. H., and P. Vickers-Rich. 1994. Neoceratopsians and ornithomimosaurs: Dinosaurs of Gondwana origin? *National Geographic Research* 10: 129–131.

Rothschild, B. M. 1987. Diffuse idiopathic skeletal hyperostosis as reflected in the paleontologic record: Dinosaurs and early mammals. *Seminars in Arthritis and Rheumatism* 17: 119–125.

Rothschild, B. M. 1997. Dinosaurian paleopathology. In J. O. Farlow and M. K. Brett-Surman (eds.), *The Complete Dinosaur*, pp. 426–448. Bloomington: Indiana University Press.

Rothschild, B. M., and L. D. Martin. 1993. *Paleopathology*. Boca Raton: CRC Press.

Rothschild, B. M., D. H. Tanke, and K. Carpenter. 1997. Tyrannosaurs suffered from gout. *Nature* 387: 357–358.

Rowe, T. 1989. A new species of the theropod dinosaur *Syntarsus* from the early Jurassic Kayenta Formation of Arizona. *Journal of Vertebrate Paleontology* 9: 125–136.

Rowe, T., and J. Gauthier. 1990. Ceratosauria. In D. B. Weishampel, P. Dodson, and H. Osmólska (eds.), *The Dinosauria*, pp. 151–168. Berkeley: University of California Press.

Russell, D. A. 1970. Tyrannosaurs from the Late Cretaceous of western Canada. *National Museum of Natural History, Publications in Palaeontology* 1: 1–34.

Sereno, P., and F. E. Novas. 1993. The skull and neck of the basal theropod *Herrerasaurus ischigualastensis*. *Journal of Vertebrate Paleontology* 13: 451–476.

Slikas, B. 1998. Recognizing and testing homology of courtship displays in storks (Aves: Ciconiiformes: Ciconiidae). *Evolution* 52: 884–893.

Sotheby's. 1997. *Tyrannosaurus rex*, a highly important and virtually complete fossil skeleton. (Sotheby's Catalog, sale 7045) New York: Sotheby's.

Stovall, J. W., and W. Langston Jr. 1950. *Acrocanthosaurus atokensis*, a new genus and species of Lower Cretaceous Theropoda from Oklahoma. *American Midland Naturalist* 43: 696–728.

Swinton, W. E. 1934. *The Dinosaurs*. London: Thomas Murby.

Tanke, D. H., and P. J. Currie. 1995. Intraspecific fighting behavior inferred from toothmark trauma of skulls and teeth of large carnosaurs (Dinosauria). *Journal of Vertebrate Paleontology* 15 (suppl. to no. 3): 55A. (Abstract.)

Tanke, D. H., and B. M. Rothschild. 1999. DINOSORES: An annotated bibliography of dinosaur paleopathology and related topics, 1838–1999. (On disk, distributed by Tanke: dtanke@dns.magtech.ab.ca.)

Thulborn, R. A. 1990. *Dinosaur Tracks*. London: Chapman and Hall.

[*The T. rex World Exposition.*] 1995. Tokyo: TBS. (In Japanese.)

Tucker, M. E., and T. P. Burchette. 1977. Triassic dinosaur footprints from south Wales: Their context and preservation. *Palaeogeography, Palaeoclimatology, Palaeoecology* 22: 195–208.

Van Valen, L. 1962. A study of fluctuating asymmetry. *Evolution* 16: 125–142.

Webb, G., and C. Manolis. 1989. *Crocodiles of Australia*. French's Forest, N.S.W.: Reed Books.

Webster, D. 1999. A dinosaur named Sue. *National Geographic Magazine* 195 (6): 46–59.

Welles, S. P. 1984. *Dilophosaurus wetherilli* (Dinosauria, Theropoda) osteology and comparisons. *Palaeontographica* A, 185: 85–180.

Williamson, T. E., and T. D. Carr. 1999. A new tyrannosaurid (Dinosauria: Theropoda) partial skeleton from the Upper Cretaceous Kirtland Formation, San Juan Basin, New Mexico. *New Mexico Geology* 21: 42–43. (Abstract.)

Zhao, X.-J., and P. J. Currie. 1993. A large crested theropod from the Jurassic of Xinjiang, People's Republic of China. *Canadian Journal of Earth Sciences* 30: 2027–2036.

25. Dinosaurian Humeral Periostitis: A Case of a Juxtacortical Lesion in the Fossil Record

LORRIE MCWHINNEY, KENNETH CARPENTER, AND BRUCE ROTHSCHILD

Abstract

Juxtacortical (surface) lesions originate from a variety of etiologies (causes) that can be either tumor or tumorlike in appearance. The resulting periostitis pattern is from the disease process and not the periosteum. A juxtacortical lesion is noted on an adult *Camarasaurus grandis* right humerus located along the distal anterior diaphysis and terminating at the metadiaphyseal junction. A resulting long-term or chronic periostitis would cause secondary inflammation of the muscles (myositis) and fascia (fascitis). The increased pressure in the affected area would then have the potential to compress the neurovascular system. The inflammatory process ultimately would cause decreased range of motion of the flexor and extensor muscles. Based on the location of the pathology, the primary muscles affected by the periostitis would be M. brachialis and M. brachioradialis, with secondary involvement of the M. biceps brachii. The distally sloping of the juxtacortical lesion and exostosis on the humerus follows the direction of the muscle bundles. The occurrence of this type of pathology on the humerus is caused by a stress injury or repetitive overexertion of the muscles resulting in an avulsion with fracture. The periosteal mass and spur represents the reparative process after a posttraumatic event.

Introduction

A nearly complete right humerus of an adult *Camarasaurus grandis* (DMNH 2908) was found in 1992 along with an associated dorsal vertebra and several caudal centra (Virginia Tidwell, pers. comm.). The specimens were collected in the Bryan Small *Stegosaurus* quarry in the Morrison Formation, near Cañon City, Colorado. There is some preparation damage that has exposed some of the trabecular bone, most notably at the pectoral crest, and along the proximal and distal articular surfaces. The humerus is 95 cm long. The greatest proximal breadth (GP) is 36 cm and the greatest distal breadth (GD) is 30.8 cm. The least circumference (LC) was 42.5 cm, as measured near a break ~35 cm from the distal end of the humerus.

A juxtacortical lesion is noted on the humerus along the distal anterior diaphysis and terminating at the metadiaphyseal junction (fig. 25.1). Juxtacortical lesions originate from a variety of etiologies that can be either tumor or tumorlike in appearance. Juxtacortical is a broad-based term to describe a variety of surface lesions that are of extracortical origin (Kenan et al. 1993), independent of their exact relative location to the periosteum. This term can be used when the "point of origin" of the lesion cannot be determined. The point of origin can be either cortical, subperioteal, periosteal, or parosteal (from the outer fibrous layer of periosteum). The cortical processes are included in the differential diagnosis because these lesions break through the cortex into the subperiosteal space. The resulting periostitis pattern is from the disease process and not the periosteum.

The periosteum is a thin membranous layer of connective tissue that covers the entire cortical bone except for the articular surfaces in the body. In adults, the periosteum is separated into two indistinct layers, the external and the deep layer. The external layer contains a dense network of connective tissue and the blood supply. There are four sets of vessels that supply the periosteum with blood: the (1) intrinsic periosteal system, (2) musculoperiosteal system, (3) fascioperiosteal system, and (4) cortical capillary anastomoses (Simpson 1985). The deep layer is composed of loosely arranged collagenous bundles and thin elastic fibers. The periosteum is "a multipotential membrane" (Kenan et al. 1993) whose cellular composition is most likely to provide the genesis to neoplasms and other tumorlike conditions.

Periostitis is considered a chronic inflammatory process of the periosteum at the origin of Sharpey's fibers, occurring secondary to some predisposing event (Meese and Sevastianelli 1996). It is considered a nonspecific response to various etiologies. Many of the recognized bony characteristics of periostitis can overlap in the various etiologies. In primary periostitis, the etiology is by trauma or infection. This process does not typically affect the entire bone but rather is more localized or unevenly distributed. It is usually limited to the metaphysis and diaphysis of bones. There is an appearance of either thickened bone or additional layers in the affected area. In the active stage, the cortical bone has the appearance of reactive, unhealed bone, which is very porous and lamellar. In the inactive stage, the characteristics most likely to be seen are a "denser, less porous and more sclerotic" lesion (Schwartz 1995). In secondary periostitis, also called periostitis ossificans or osseous periostosis, the etiology is a direct result of or occurs in conjunction with other diseases, such as neoplasm (tumor), metabolic disease (e.g., rickets [children], osteomalacia [adults], Paget's disease, and arthritis), systemic disease (e.g., hypertrophic pulmonary osteoarthropathy [HOA], syphilis [caused by *Treponema pallidum*], and tuberculosis) (Edeiken 1981; Greenfield 1986).

Primary periostitis can be either with or without infection and represents an inflammatory process, by which the thin membranous layer of the periosteum is lifted away from the cortex. In infectious periostitis, the bone infection can result from a direct extension from the adjacent soft tissues. Since the periosteum is the first structure encountered along the bone surface, the infection will produce pyogenic material (pus), which elevates the periosteal tissue and stimulates the periosteum to produce sclerosis (new bone). This is in contrast to the hematogenous (blood-related) spread of osteomyelitis, which would originate in the cortical and medullary bone.

Recognition of periostitis on bone is "on the basis of expansion of bone contour related to alteration of bone surface texture" (Rothschild and Rothschild 1998). Two different types of textures can be seen on the bone surface. "Surface" periosteal reaction is macroscopically inseparable from the bony cortex, with margins that blend with the cortical surface (Rothschild and Rothschild 1998). The second type of periosteal reaction is referred to as an "appliqué" form. It can be recognized by new bone having the appearance of being pasted on the parent bone. The margins are sharply defined between the periosteal reaction and the cortical bone with a noticeable line of demarcation at the process margins. Radiographic examination concurs with the macroscopic evaluation of both types of periosteal textures. While the surface periosteal reaction cannot be differentiated from the cortical bone, the appliqué form reveals a radiolucent line, which separates the periosteal reaction from the cortical bone.

Evidence of periostitis has been found in a study of *Leidyosuchus* (= *Borealosuchus*) *formidabilis* by Sawyer and Erickson (1998). Of 7,154 total elements reviewed in this study, 236 were noted to have pathologies. Periostitis was the most common pathology, with 134 elements. Forelimb periostitis was limited to 6 elements, including 1 humerus. A right humerus was noted to have a smooth surface periostitis on the dorsal surface. The mass, measuring 4.5 cm long and 2.5 cm wide, was located slightly distal to the midshaft of the humerus. In a separate example, a ceratopsian scapula exhibited bony characteristics identified as infectious periostitis (Rothschild and Martin 1993).

Institutional abbreviation: DMNH, Denver Museum of Natural History.

Materials and Methods

A nearly complete right humerus of an adult *Camarasaurus* (DMNH 2908) humerus showed evidence of periostitis. The specimen was cleaned using 5% acetic acid, followed later by baking soda airbrade. Dissections were made of a caiman (*Caiman*) and a turkey (*Meleagris*) wing in order to map soft tissue and anatomy used in figures 25.3 and 25.4. CT scans of the pathological region using both sagittal and axial views were performed in 3 mm increments. Measurements made on the *Camarasaurus* humerus (DMNH 2908) were based on measurements used by Wilhite (1999) on the same element.

Description

The *Camarasaurus* humerus (DMNH 2908) has an juxtacortical mass distal to the pectoral crest along the distal anterior diaphysis and terminating at the metadiaphyseal junction (fig. 25.1). The pathology is an unusual case of upper-arm periostitis. The sclerotic (hard) and dense mass has a fibrous-woven appearance, indicating an area of active periostitis (fig. 25.1B). In humans, the occurrence of this type of woven bone matrix is a pathologic condition in adults, noted in disease or disorders that form reactive new bone. In dinosaurs, the woven bone normally occurs in accessory dental bone (Rothschild and Martin 1993).

Fusion of the juxtacortical lesion and the lack of porosity at the proximal and distal ends, indicates areas of inactive or healed periosteal reaction. The mass appears to extend from the surface of the cortex and is solid and localized. The solid periosteal appearance is a "hallmark of a benign process" (Edeiken 1981). The dense mass is elliptical. The relative thickness or density of the mass most likely represents the progress and age of the periosteal response (Edeiken et al. 1966) and correlates to a chronic condition. Near the middle of the process located along the lateral margin, is an area of gradational "blending" of remodeled bone growth into normal cortex (fig. 25.2B). At the apex of the periosteal mass, in cross-section, a thin layer of matrix is seen separating the mass from the parent bone (fig. 25.2C, D). This may represent an area of periosteal stripping from an episode of violent injury to the humerus. There are undulating fibrous bundles that are oriented in the direction of the M. brachialis (figs. 25.1B, 25.2A).

The juxtacortical lesion measures ~25 cm long by 18 cm wide. It is ~ 42 cm from the distal end of the humerus along the anterior diaphysis and terminates at the metadiaphyseal junction. In cross-section, along a transverse break in the bone, the mass extends laterally and distally beyond the parent bone to form an exostosis. The exostosis also projects distally, which gives it an appearance of a hook or spur (fig. 25.2A,C). This process is not considered an osteochondroma, which grows at right angles on the metaphysis. The exostosis was formed at the origin of M. brachioradialis. The axis of the spur (exostosis) is 25° relative to the vertical axis of the humerus, and represents the direction of pull for M. brachioradialis (fig. 25.1C) This extension of bone from the periosteal mass is most likely due to an osteoblastic response caused by the flexor motion of the M. brachioradialis. The orientation of some of the muscle fibers has been preserved on the periosteal mass near the origin of M. brachialis (fig. 25.1B).

Figure 25.1. (next page) Pathologic right humerus of an adult Camarasaurus grandis *(DMNH 2908). (A) anterior view of the right humerus; (B) close-up image of the periosteal reaction, showing woven bone and orientation of the bundles of fibrous tissue; (C) lateral image of the right humerus, showing that the axis of the spur (exostosis) is 25° relative to the axis of the humerus, and represents the direction of the origin of M. brachioradialis muscle pull. Scale: 10 cm.*

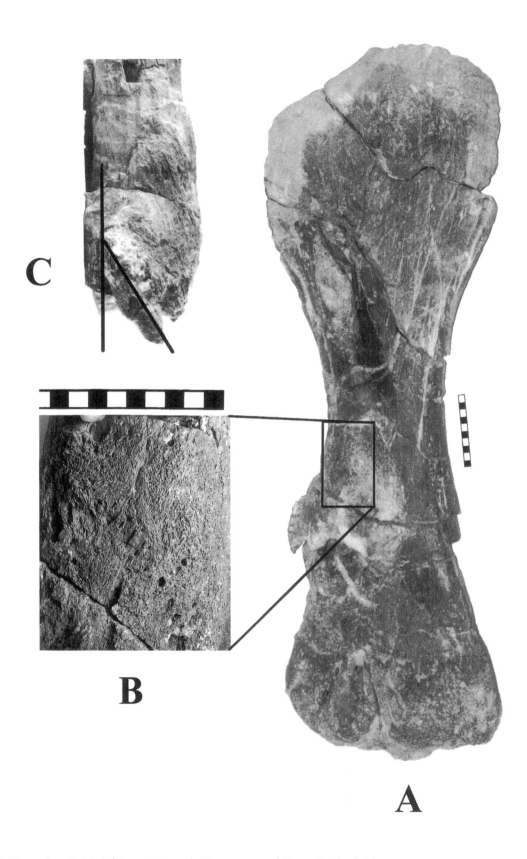

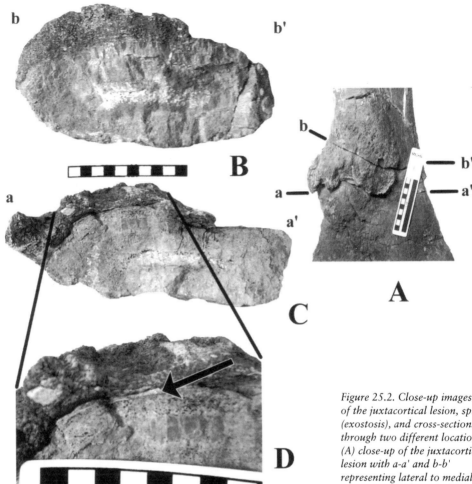

Figure 25.2. Close-up images of the juxtacortical lesion, spur (exostosis), and cross-sections through two different locations. (A) close-up of the juxtacortical lesion with a-a' and b-b' representing lateral to medial cross-sections through the pathology; (B) cross-section of an area where the lesion has fused with the cortical (parent) bone. The lateral (left) portion of the pathology has a gradational "blending" appearance through the cortical bone, whereas the medial (right) portion of the pathology has an abrupt change between the cortical bone and the periostitis; (C) cross-section through the juxtacortical lesion, showing an area where the pathology is unfused. There is a distinctive line of demarcation, which was originally filled with matrix; (D) Close-up image of cross-section a-a'. Arrow indicates the line of demarcation separating the cortical (parent) bone with the periostitis. Scale in cm.

Muscle reconstruction in the area of the pathology shows that the M. brachialis and M. brachioradialis were the primary muscle groups most likely involved in the pathology, with possible secondary involvement of the M. biceps brachii (figs. 25.3, 25.4). The M. brachialis and M. brachioradialis muscle groups are distally placed, as in birds, not more proximally placed, as in crocodiles. The nerves associated with this area include the radial collateral branch of N. radialis (radial nerve), the lateral antebrachial cutaneous branch of N. musculocutaneous, and N. medianus (medial nerve). Only the radial collateral nerve is directly impacted; the other nerves have secondary involvement in the pathologic process. The vascular supply for the affected area would include the A. brachialis (brachial artery) and radial collateral branch of A. radialis (radial artery) (fig. 25.4). The focal area of periostitis noted on the humerus confirms that both the osteology and myology should be considered when rendering a diagnosis in a dry bone. Carpenter and Smith (chap. 9 of this volume) comment on the importance of viewing the intimate connection of bone and muscle in their discus-

*Figure 25.3. Schematic drawing
of proposed locations of M.
biceps, M. brachialis, M.
brachioradialis superimposed on
bone and pathology in context to
the reconstructed forelimb.
(A) anterior presentation of the
pathology on the right humerus;
(B) lateral presentation of the
pathology on the right humerus;
(C) anterior placement of the M.
biceps, M. brachialis, and M.
brachilradialis overlying the
pathology; (D) lateral
presentation of same muscle
groups; (E) anterior placement of
the deeper muscle groups directly
involved with the pathology;
(F) lateral view of the deeper
muscle groups directly involved
with the pathology. Abbreviations: a, tendon insertion; b, M.
brachialis; bi, M. biceps; br, M.
brachioradialis; pa, pathology.*

sion on the forelimb of *Tyrannosaurus*. Although only the right humerus of *Camarasaurus* (DMNH 2908) was available to review, it can
be surmised that the inflammation of both the periosteum and muscles
produced additional or secondary complications involving the lower
forelimb as well.

CT scans were performed to rule out any other pathology (e.g.,
stress fracture, infection). Both sagittal and axial views were performed
in 3 mm sections through the pathology. There was no evidence of a
fracture or infectious process (osteomyelitis or infectious periostitis).
The porosity noted in the mass does not extend subperiosteally and
appears to be not of an infectious origin. The internal bony structure of
the parent bone appeared normal in all except one CT image. In that
image, close to the proximal aspect of the mass, a radiolucent area is
seen that measures less than 3 mm. The radiolucency appears to be intracortical. There is a lack of sclerotic material surrounding the radiolucent area. The significance of this process is indeterminate at this time
although the radiolucency probably represents an artifact and does not
represent a nidus. The CT scans confirm that the area of periostitis
represents a hyperdense mass that has a radiolucent line between the
mass and the cortex of the humerus. The mass has a uniform radio-

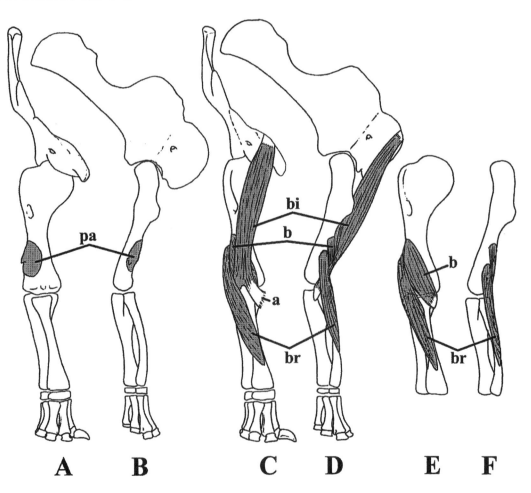

A   **B**   **C**   **D**   **E**   **F**

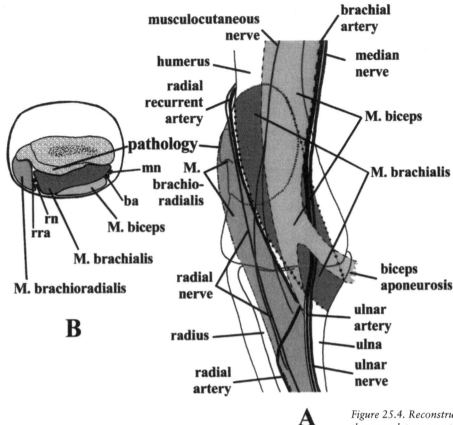

density. The radiolucent line parallels the contour of the diaphysis. At present, no histological study of the humerus has been undertaken.

Discussion

In a review of the literature on periosteal reactions many etiologies need to be considered. Etiologies can be influenced by the approximate age (juvenile, subadult, or adult) of the specimen, the lesion's location on the bone (e.g., diaphysis), and presence of other pathologic processes (McCarthy and Frassica 1998). There is a preference of some bony pathology to manifest in certain age groups. The location of a lesion is also an important diagnostic tool. Some processes are only found in certain areas of the bone. Also, the relationship of the lesion in conjunction to the anatomical structures (e.g., medullary cavity) of the bone should be considered when making a diagnosis of the pathology.

The periostitis noted on the adult *Camarasaurus* (DMNH 2908) humerus has some defining bony characteristics that limit the differential diagnoses. These differential diagnoses include: (1) hypertrophic osteoarthropathy (HOA), (2) osteiod osteoma, (3) an arm equivalent of shinsplints (Tibial Stress Syndrome, or TSS), (4) myositis ossificans traumatica (circumscripta), and (5) avulsion injury. The differential di-

Figure 25.4. Reconstruction of the musculature near the elbow. The reconstruction is only an approximation in order to show the soft tissue impacted by the pathology. (A) anterior view showing major muscle groups, based on muscle scars, and also major nerves and arteries, based on their relative positions to the muscles groups in extant vertebrates. Based on the muscle scars, the M. brachialis and M. brachioradialis are distally placed, as in birds, and not more proximally placed, as in crocodiles; (B) cross-section of humerus across the middle of the pathology showing the relative positions of muscles, nerves, and arteries. Note that the pathology would press against the M. brachialis and M. brachioradialis, and these in turn would constrict between them the radial nerve and radial recurrent artery. Abbreviations in B: ba, brachial artery; mn, median nerve; rn, radial nerve; rra, radial recurrent artery.

agnoses share some common overlapping bony characteristics (table 25.1). Except for the possibility of an osteoid osteoma or HOA, no other pathological processes associated with secondary periostitis are considered at this time.

Hypertrophic osteoarthropathy (HOA) or pulmonary hypertrophic osteoarthropathy is a secondary periosteal response that causes enlargement of the extremities due to chronic lung diseases. This disease process can be found "equally divided between the distal diaphysis and diffuse distribution" on the bone (Rothschild and Rothschild 1998). Proximal involvement was rarely noted. In a study done by Rothschild and Rothschild (1998) on HOA, the rare occurrence of both the appliqué and surface forms of periosteal reaction was noted on the same bone. The humerus was noted to have a higher percentage of appliqué periosteal reaction versus the surface form. A layer of thin, smooth bone is the first periosteal reaction recognized in the HOA. With the increasing density, the deeper portion of the periosteal reaction goes through lamellar reconstruction and can merge (fuse) with the cortex (Edeiken 1981; Greenfield 1986). A radiolucent line separating the appliqué periosteal reaction and the cortical bone can be seen by radiographic correlation. While 90% of HOA was attributed to intrathoracic pathology (Rothschild and Rothschild 1998), the location of the reaction can indicate either chronic pulmonary infection (tubercular and nontubercular), cancer, or endocarditis. This study found that the involvement of the distal portion of the humerus had the highest percentage of cases (88%) relating to a tubercular form of chronic pulmonary infection. Of note, with the resolution of the pulmonary infection by surgery, the hypertrophic ostearthropathy symp-

TABLE 25.1
Shared Bony Characteristics Found on the *Camarasaurus* Humerus
and the Differential Diagnoses of the Pathology

Camarasaurus humerus (DMNH 2908)	Hypertrophic Osteoarthropathy	Osteoid Osteoma	Arm Equivalent of Shinsplints	Myositis Ossificans Traumatica (circumscripta)	Avulsion Injuries
Solid/fused		Solid	Solid	Solid	
Localized	Localized or diffused	Localized with nidus		Localized	Localized
Elliptical and undulating		Elliptical			
Dense/sclerotic		Dense or thin	Sclerotic	Sclerotic	
Radiolucent line	Radiolucent line	Radiolucent line	Radiolucent line	Radiolucent line	
Diaphysis/ metadiaphyseal junction	Only proximal involvement; rarely seen	Site preference depends on bone type	Diaphysis	All sites	All sites

toms disappear within 24 hours unless there is extensive accumulation of periosteal reaction. At that point the mass becomes incorporated into the cortex.

Some of the characteristics of HOA, such as a radiolucent line, fusion, and location of the pathology are similar to those observed on the *Camarasaurus* humerus. But due to the extensive periosteal reaction (spur) seen on the lateral aspect of the humerus, the occurrence of HOA as an etiology for the pathologic process on the *Camarasaurus* humerus is not highly probable.

Periosteal reactions are typical of bone lesions. These reactions radiologically are described as either a solid (benign) or interrupted process (malignant). Solid reactions produce a single layer of new bone that is uniform in shape and is in excess of 1 mm thick. There is a correlation between the density and thickness of the periosteal reaction to the aggressive behavior of the irritant. Solid reactions are highly suggestive of a benign process. Examples of lesions that have a solid periosteal reaction are "eosinophilic granuloma, fracture, osteomyelitis, hemorrhage, hypertrophic pulmonary osteoarthropathy, osteoid osteoma, and vascular diseases" (Edeiken et al. 1966). It should be noted that eosinophilic granuloma and osteomyelitis represent the few benign processes that can also look malignant.

In malignant processes, the lesions are characterized by interrupted periosteal reactions. The new bone formation is not uniform, and is pleomorphic (many shapes). They are typically lamellated or perpendicular (e.g., sunburst) patterns. This type of periosteal reactive is indicative of a rapid and progressive process. Malignant tumors such as Ewing's sarcoma and osteosarcoma, as well as infection and repeated hemorrhage, are typical.

Osteoid osteoma is a benign, surface lesion of bone characterized by either a thin or dense uniform periosteal reaction. The dense reaction is elliptical. Elliptical reactions can vary in size from 2 mm to 1 cm. They are thickest near the apex of the lesion, tapering toward the smooth edges. The periosteal reaction surrounds a nidus (nucleus), which may or may not be seen radiographically, depending on the density of the lesion. The nidus is a "small focus of benign tumor, usually replaced by osteoid and fibrous tissue" (Jacobson 1985), and can measure 0.5 to 1.0 cm in size. The nidus can be located in the sub-periosteal, intracortical, or subcortical areas of the bone (McCarthy and Frassica 1998). This type of tumor is a benign process of osteoblastic material. There is radiographic evidence of a radiolucent line noted in this type of lesion, which disappears with time.

The presence of an osteoid osteoma commonly occurs between the second or third decades of life and is twice as common in males as females (Jacobson 1985). These lesions may be found in any bone, but are predominately seen in the femur and tibia (all areas of the bone) in 50% of cases (McCarthy and Frassica 1998). Other common sites include the neural arch of the spine, the distal humerus, hands, and feet. The vertebral body of the spine, ribs, and innominate bone are less likely to be involved in this type of process. Osteoid osteomas can be highly irritating to the adjacent tissues, causing edema. Although this differential diagnosis represents many of the bony characteristics seen on the *Camarasaurus* humerus, it does not account for the osteoblastic formation of the spur noted on the lateral aspect.

The occurrence of periostitis in the humerus caused by stress or repetitive overexertion of the muscles can be considered an equivalent of shinsplints normally found in the lower leg (Greyson 1995). Periostitis seen in shinsplints is a secondary reaction to the periosteum being pulled away from the bone by the overuse of the muscle, causing myositis. To adapt to the increased stress, the injured area will produce an osteoblastic response in an attempt to remodel and strengthen the bone. Without this response, the increased stress on the bone may result in a stress fracture. In a study done by Greyson (1995), the brachialis muscle was the source of the "arm-splints" phenomenon noted in the humerus. The initial laminated periosteal response looks like a thin layer resembling an onion skin. As the process progresses to a chronic stage, additional "onion layers" fill in the space between the elevated periosteum and the cortical bone.

The periosteal reaction runs parallel to the long axis of the bone involved and can occur either anteriorly or posteriorly. Initially, the plain radiographs will show no bony changes. Radiographic change is evident at 10 days to three weeks following the initial injury (Ragsdale et al. 1981). Greyson (1995) and Meese and Sevastianelli (1996) concluded that periosteal reaction is apparent radiographically approximately four to six weeks after the initial injury. The diagnosis can be confirmed with a radionuclide three-phase bone scan, correlating with the physical findings in the absence of other radiographic evidence of periosteal reaction. The bone scan will show the periostitis as an area of increased radionuclide uptake resembling linear streaking on the images. On plain films, the periostitis seen in shin- and arm-splints appears as a uniform linear density and has a sclerotic and solid appearance, producing a radiolucent line.

Either discontinuation of the physical activity or rest will result in the resolution of this disorder. Although basically unheard of in current medical literature, in untreated occurrences of "arm-splints" the ongoing periosteal reaction would create a "cause-and-effect" response to include the muscles and neurovascular supply of the forelimb (Thomas Barsch, M.D., pers. comm.). The constant irritation caused by the forelimb in motion would affect the bony architecture, which in turn would cause inflammation of the muscles in the region and constrict the neurovascular supply. This would then produce an altered range of motion for the entire forelimb as it compensated for this process. The altered range of motion for the extensor and/or flexor muscles of the humerus would have been dependent on the extent of involvement of either the N. radialis or N. medialis. Considering the very active lifestyle of the *Camarasaurus*, injuries producing a arm-splint phenomenon is a possible etiology for the pathology observe on the humerus. With the exception of the spur, the pathologic characteristics of this phenomenon and those observed on the humerus are similar.

Myositis ossificans traumatica (circumscripta) is a heterotopic (abnormal location of normal tissue) formation of bony material in soft tissue after trauma in 60 to 75% of the cases (McCarthy and Frassica 1998). The trauma may be from either a traction/tug injury or a direct blow (Isaacs, pers. comm.) to the area. The etiology of myositis ossificans traumatica (circumscripta) is usually from an avulsion at the site of either tendinous and/or muscle attachments (Aufderheide and Rodriguez-Martin 1998). Muscle crushed against bone can occasionally produce this type of tumorlike lesion. A juxtacortical lesion can appear just days after the trauma and grow to approximately 4 to 10 cm in diameter. This type of juxtacortical lesion develops next to any bone or joint.

Ossification may occur at the M. adductor longus (rider's bone), M. brachialis (fencer's bone), or M. soleus (dancer's bone) (Edeiken 1981), as well as other sites. Radiographic features change as the lesion matures. A radiolucent line separates the lesion from the underlying bone, but can eventually disappear as the residual lesion decreases in size. Some lesions have been known to attach to the adjacent bone structure, eventually blending with the cortex (McCarthy and Frassica 1998) or simply disappear. Soft-tissue edema (swelling) and hemorrhage (bleeding) can occur due to the trauma adjacent to the lesion, and may result in a decreased range of motion due to pain or mass effect (Isaacs, pers. comm.). Even with partial healing, the range of motion in this area may still be limited. If the hypertrophic (increased size) changes in the muscles were significant, compression of the nerves would eventually lead to muscle atrophy. Based on the bony characteristics and possible muscle involvement observed on the *Camarasaurus* humerus, myositis ossificans traumatica is included as a differential etiology and may be associated with an avulsion injury.

Avulsion injuries are a direct result of a traumatic event that "pulls off a portion of periosteum" (McCarthy and Fressia 1998). The resultant periosteal reaction that can be misinterpreted as a neoplasm (tumor) or osteomyelitis (El-Khoury et al. 1997). These injuries occur with or without a fracture. The fracture can either partially or completely remove a segment of bone and is associated with significant damage to the surrounding soft tissue. The resulting fragment could resemble an exostosis (El-Khoury et al. 1997). When there is an involvement of a fracture, it is not uncommon for the avulsed bone to fail to unite, especially if there is a poor vascular supply or the fragment displaced.

An avulsion injury can be caused by an episode of either overstretching of the musculotendinous or ligamentous attachment, direct trauma, or sudden deceleration. This type of injury would produce a focal area of edema and hemorrhage. Repeated microtrauma to an injured area is another mechanism that can produce an avulsion. This allows the tissue repair to be "outpaced by the recurring injury" (El-Khoury et al. 1997). Usually, an avulsed area with full or complete tear of the tendinous, ligamentous, and/or muscle attachments leaves a cuplike depression where the bone has been pulled away (avulsion fracture) by following a violent or explosive traction force similar to that seen on the right humerus of the *Tyrannosaurus* (Carpenter and Smith, chap. 9 of this volume). Avulsion injuries may or may not produce a secondary-pathology myositis ossificans traumatica (circum-scripta).

An avulsion injury at the origin of M. brachialis and a partial tear at the origin of M. brachioradialis is considered the primary etiology for the pathology seen on the *Camarasaurus* humerus. The direction of the tear to the muscles involved in this injury produced a downward-sloping elliptical mass. The partial tear of the M. brachioradialis produced a bone spur, which most likely represents a fragment of bone that was not completely avulsed from the diaphysis. The origin of the M. brachialis and M. brachioradialis does not attach to the bone by way of a tendon but directly to the bone by way of Sharpey's fibers to the periosteum. This allowed a broader-based area to respond to the avulsion injury.

Conclusions

Making a final diagnosis on fossil bone material can be quite problematic, since we have no additional information (e.g., physical history or additional associated bony material) to correlate with the physical examination of the specimen. When available and appropriate, it is quite helpful to use all resources (e.g., X-rays, CT scans, histological thin-sections, etc.). In the case of trying to determine the etiology of the periostitis noted on the adult *Camarasaurus* (DMNH 2908) right humerus, there is a further complication arising from the fact that periosteal reaction has a similar response to many etiologies. Perhaps the best we can hope for is to define the process seen in this case as either a benign or malignant juxtacortical lesion.

Based on the physical findings and CT scan of the humerus, the pathological process is most likely due to primary periostitis, with a high probability that the etiology of the humeral periostitis seen on this specimen is due to an avulsion injury with partial fracture near the site of origin of M. brachioradialis. This process would represent a benign tumorlike juxtacortical lesion. Due to the overlapping of bony characteristics, the differential diagnoses would include an avulsion injury in conjunction with myositis ossificans traumatica (circumscripta) or, less likely, an arm equivalent of shinsplints, or osteoid osteoma. Although hypertrophic osteoarthropathy (HOA) is also considered for this pathology, it is unlikely. Perhaps in the future, a histological study of the *Camarasaurus* humerus would further define the bony characteristics seen by visual examination and CT scans.

Because of the long-term nature of the periostitis and the likelihood of complications arising from muscle inflammation and the possible compression of the neurovascular supply, the mobility of the *Camarasaurus* right forelimb was moderately to severely restricted. This in turn caused limping and made everyday activities such as foraging for food and escaping predators harder to accomplish.

Acknowledgments: McWhinney owes a great debt to the people who advised her on the pathology noted in this case study. She would like to thank the radiologists at Kaiser Permanente, in particular Thomas Barsch M.D., John Bair M.D., Darwin Kuhlmann M.D., and

Margaret Montana M.D. To the CT technologist, Kim Hardy, for volunteering her spare time after work, "thanks" is never enough. A special thank you goes to Kim Adcock M.D. and Dave Bellamy for supporting McWhinney's paleopathology work. McWhinney is very grateful for the assistance of her friend Pamela Isaacs D.O., radiologist, University Hospital, Denver.

References

Aufderheide, A. C., C. Rodriguez-Martin, and O. Langsjoen. 1998. *The Cambridge Encyclopedia of Human Paleopathology*. Cambridge: Cambridge University Press.

Edeiken, J. 1981. *Roentgen Diagnosis of Diseases of Bone*. Baltimore: Williams and Wilkins.

Edeiken, J., P. J. Hodes, and L. H. Caplan. 1966. New bone production and periosteal reaction. *American Journal of Roentgenology Radium Therapy and Nuclear Medicine* 97 (3): 708–718.

El-Khoury, G. Y., W. W. Daniel, and M. H. Kathol. 1997. Acute and chronic avulsion injuries. *Radiologic Clinics of North America* 35 (3): 747–766.

Greenfield, G. B. 1986. *Radiology of Bone Disease*. Philadelphia: Lippincott.

Greyson, N. D. 1995. Humeral stress periostitis: The arm equivalent of "shin splints." *Clinical Nuclear Medicine* 20: 286–287.

Jacobson, H. G. 1985. Dense bone—too much bone: Radiological considerations and differential diagnosis. Part 2. *Skeletal Radiology* 13: 97–113.

Kenan S., I. F. Abdelwahab, M. J. Klein, G. Herman, and M. M. Lewis. 1993. Lesions of juxtacortical origin (surface lesions of bone). *Skeletal Radiology* 19 (5): 337–357.

McCarthy, E. F., and F. J. Frassica. 1998. *Pathology of Bone and Joint Disorders, with Clinical and Radiographic Correlation*. Philadelphia: Saunders.

Meese, M. A., and W. J. Sevastianelli. 1996. Periostitis of the upper extremity. *Clinical Orthopaedics and Related Research* 324: 222–226.

Ragsdale, B. D., J. E. Madewell, and D. E. Sweet. 1981. Radiologic and pathologic analysis of solitary bone lesions. Part 2, Periosteal reactions. *Radiologic Clinics of North America* 19 (4): 749–778.

Rothschild, B. M., and L. D. Martin. 1993. *Paleopathology: Disease in the Fossil Record*. Boca Raton: CRC Press.

Rothschild, B. M., and C. Rothschild. 1998. Recognition of hypertrophic osteoarthropathy in skeletal remains. *Journal of Rheumatology* 25 (11): 2221–2227.

Sawyer, G. T., and B. R. Erickson. 1998. Paleopathology of the paleocene crocodile *Leidyosuchus* (=*Borealosuchus*) *formidabilis*. *Science Museum of Minnesota Paleontology Monograph* 4: 1–38.

Schwartz, J. H. 1995. *Skeleton Keys: An Introduction to Human Skeletal Morphology, Development, and Analysis*. New York: Oxford University Press.

Simpson, A. H. 1985. The blood supply of the periosteum. *Journal of Anatomy* 140 (4): 697–704.

Wilhite, R. 1999. Ontogenetic variation in the appendicular skeleton of the genus Camarasaurus. M.S. thesis, Brigham Young University.

26. Pathological Amniote Eggshell—Fossil and Modern

KARL F. HIRSCH[*]

Abstract

Pathologic eggshell results from abnormal biological or environmental conditions. Biological conditions, such as oviduct retention of amniote eggs beyond the time of normal oviposition (egg laying), may lead to multilayered eggshells in reptiles and *ovum in ovo* in birds. Fossil multilayered eggshells were first reported from the Upper Cretaceous of France and, later, from the Upper Cretaceous of India and Argentina. More recently, such eggshells have been recognized in the Upper Cretaceous of Montana and the Upper Jurassic of Utah. Ovum in ovo, on the other hand, has a low preservation potential and has not yet been observed in the fossil record. Abnormal environmental conditions, such as the presence of DDT, causes thinning of eggshells and changes in the microstructure of condor eggs today. Environmental conditions may have caused thinning in fossil dinosaur eggshells; however, its role the extinction of dinosaurs is disputed. Recognition of fossil pathological eggshell, except for the obvious multilayered specimens, is difficult due to variations in preservation and diagenesis.

*Deceased.

Introduction

Pathologic eggshells are an indication of abnormal biological or environmental conditions. Most of our knowledge about such shells in modern eggs is derived from the poultry industry, especially for chicken eggs (Nathusius 1868; Asmundson 1931, 1933; Romanoff and Hutt 1945; Romanoff and Romanoff 1949; Schmidt 1943, 1964; Woodward and Mather 1964; Erben 1970, 1972). Ringleben (1966) and Schmidt (1967) described an "ovum in ovo" from a wild blackbird and Sauer et al. (1975) reported abnormal shell in struthious eggs. Pathological eggshell has also been reported from reptiles (mostly zoo animals), as in chelonian eggs (Schmidt 1943; Cagel and Tihen 1948; Erben 1970, 1972; Erben et al. 1979; Ewert et al. 1984; Schleich and Kaestle 1988; and Hirsch 1989). Pathologic squamate and crocodilian eggs have been mentioned only briefly by Erben (1972) and Schleich and Kaestle (1988). Only a few cases of pathological eggshells have been reported from the wild.

Pathological fossil eggshell has only recently been recognized, and can be understood only with adequate knowledge of the processes that lead to pathological shell in modern eggs. Even so, the recognition of fossil pathological eggshell is made more difficult by geological processes that may alter the fossil eggshell, such as diagenesis, erosion, and crushing. These difficulties have resulted in a considerable amount of the misinformation discussed below. The terminology for eggshell structure follows Hirsch and Packard (1987).

Institutional Abbreviations: MOR, Museum of the Rockies, Bozeman, Montana; MWC, Museum of Western Colorado, Grand Junction; TMP, Royal Tyrrell Museum of Palaeontology, Drumheller, Alberta; UCM, University of Colorado Museum, Boulder; UUVP, University of Utah, Vertebrate Paleontology, Salt Lake City.

Other Abbreviations: CL, cathodoluminescence; HEC, Hirsch Eggshell Catalogue; PLM, polarizing light microscope; SEM, scanning electron microscope.

External Abnormalities

Irregularities of the eggshell surface and unusual egg shapes have been described in detail by Nathusius (1868), Romanoff and Romanoff (1949), and Schmidt (1964). Most of these phenomena are probably caused by convulsive uterus contractions, or by temporary or permanent uterus deformation (Romanoff and Romanoff 1949). The outer surface of the eggshell may be wrinkled (fig. 26.1) or show ridges (fig. 26.2). Aggregations of grains, calcitic nodules, or bulges are occasionally found on the shell surface. The shape of the egg may be constricted (fig. 26.3), bound (fig. 26.4), or truncated (fig. 26.5), or the egg may have shell-like or membranelike attachments, or other unusual forms. The abnormally curved parts of these eggs, as well as grains and bulges, often have abnormal shell microstructure and the shell units are not tightly interlocked, as in normal avian eggshell (figs 26.6, 26.7; Schmidt 1957). Although these pathological phenomena are easy to recognize, they have not yet been reported in fossil eggshell.

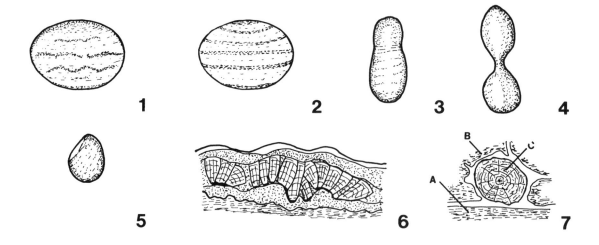

Abnormal Shell Thickness

Abnormal changes in eggshell thickness probably encompass the largest and best-known type of pathology in modern specimens. Such pathologies are generally easy to recognize because a broad database exists for normal eggshell. Shell pathologies are more difficult to recognize in fossil eggshell because the broad database for normal eggshell is not as well known. A shell that is thinner or thicker than normal disturbs the delicate equilibrium of the various functions and morphological features needed for the successful development of the embryo, and for the hatchling to escape from the egg (Ar et al. 1979).

Thin Eggshell

Abnormally thin eggshell leads to excessive evaporation of the egg content, causing either dehydration of the embryo, or dehydration of the eggshell membrane and subsequent loss of gas permeability. The shell may also become so thin that it collapses. Thinning of eggshell in captive animals is usually caused by improper diet, leading to calcium deficiency in the mother. Most studies on the thinning of avian eggshell focus on environmental pollutants, especially insecticides (e.g., McFarland et al. 1971; Cooke 1973, 1975; Kiff et al. 1979). Cooke (1973, 1975) noted that "in a laying bird, organochlorine residues affect many biochemical mechanisms known to be essential for proper shell formation." Kiff et al. (1979) documented the presence of DDE, the principle metabolite of DDT, in thin-shelled condor eggs (*Gymnogyps californianus*).

A study of California condor eggs demonstrates the changes in eggshell thickness before DDT was used, during DDT use, and after DDT use (table 26.1). Radial thin sections from five different condor eggs (fig. 26.8) show thinning of the shell layers, as well as changes in the microorganic (proteinaceous) network within the shell as represented by horizontal growth lines. SEM micrographs of thin eggshell are illustrated in figure 26.9. Micrographs of the outer shell surface of these

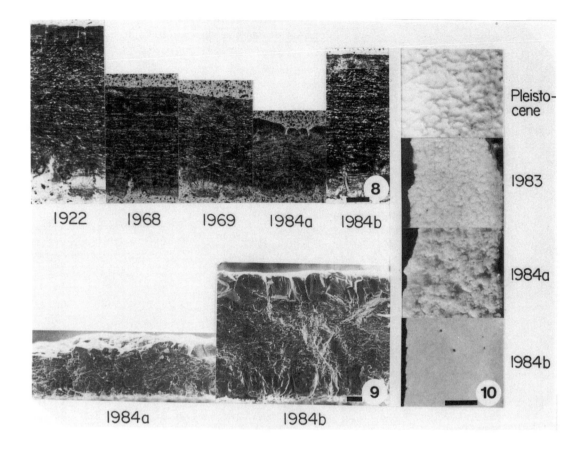

1922 1968 1969 1984a 1984b

8

Pleisto-
cene

1983

1984a

1984b

9

1984a 1984b

10

two eggs (fig. 26.10) show the contrast between normal eggshells (1984b) and one where DDT has caused surface deterioration (1984a).

Abnormally thin eggshell attributed to the Late Cretaceous sauropod *Hypselosaurus priscus* has been reported by Erben et al. (1979). They ascribed the extinction of this dinosaur to thinning of the eggshell by climatic and vegetation changes, and to physical stress due to overcrowding. However, there are several other factors not considered by Erben and his colleagues. Penner (1983) noted resorption craters in the basal caps at the base of the columns, indicating that the eggs had hatched. He also suggested that the thinner eggshell represented different taxa, which was confirmed by Vianey-Liaud et al. (1994). Further, among modern eggs of a single species, egg size and shell thickness is variable (Schmidt 1943; Romanoff and Romanoff 1949), with, for example, older animals laying larger and thicker-shelled eggs than younger animals.

Figures 26.8–10. Eggshell of condor eggs (Gymnogyps californianus) (HEC 390 = UCM OS1132). 26.8–26.9, eggshells show thinning and deterioration caused by DDT. 26.8, radial thin sections of condor eggs from five different females. Scale: 100μ; 26.9, micrographs of eggs 1984a and 1984b of fig. 26.8 showing different aspects of shell structure. Scale: 100μ. 26.10, outer shell surface of normal condor egg 1984b, of condor egg 1984a affected by DDT, and of Pleistocene and 1983 weathered condor eggs. Scale: 1 mm.

Thick Eggshell

Unintended retention of eggs in the oviduct will also lead to pathology. Due to different processes of shell development and the ovarian function, this condition differs between reptiles and birds. In turtles, the eggs are shelled in one section of the uterus, whereas in birds and

TABLE 26.1
Effects of DDT on Eggshell Thickness of California Condor,
Gymnogyps californianus

Year	Thickness
1922 (pre-DDT)	0.7mm
1968 (DDT in use)	0.46mm
	0.43mm
1984 (post-DDT)	0.30mm
	0.64mm

crocodiles the eggshell is formed sequentially in several specialized regions of the uterus. (Taylor 1970; Aitken and Solomon 1976; Hirsch et al. 1989; Palmer and Guillette 1992). However, birds lay only one egg at a time, whereas crocodiles and the other reptiles ovulate an entire clutch at one time. Eggs may be retained in reptiles in order to enhance the survivorship of the embryo. The most common reason for short-term egg retention is poor weather. However, abnormally extended retention may be caused by stress, illness, lack of nesting sites, or other environmental factors. This type of retention may lead to the deposition of additional shell layers over the eggs. The embryos in such multilayered eggs would suffocate due to a lack of oxygen because of the misalignment of pore canals in the different shell layers. Multilayered eggshells have been described for modern turtles, crocodiles, and geckos (Schmidt 1943; Erben 1970, 1972; Erben et al. 1979; Ewert et al. 1984; Schleich and Kaestle 1988; Hirsch 1989). An example of a modern multilayered egg is shown in figures 26.11 and 26.12. The specimen is from a Galápagos tortoise (*Geochelone elephantopus*) from the San Diego Zoo. One egg (UCM HES1144) has seven to nine shell layers, having a total thickness of 4 to 5 mm. The pathological calcareous and shell membrane layers are not all complete. Magnified, they show an extensive variation in their microstructure (fig. 26.13). The cause for the pathology is unknown but may have been stress related, due perhaps to lack of suitable laying ground or to overcrowding.

Eggshell with a secondary layer has been reported in modern birds (Solomon et al. 1987; Jackson et al. 1998). It is not clear, however, whether the pathological layer is a structured layer like true multilayered eggshell or if it is an amorphous secondary layer similar to that previously reported by Romanoff and Romanoff (1949). Additionally, this pathology was artificially induced in chickens by adrenaline injections (Solomon et al. 1987) and has not been reported in eggshells as a result of naturally stimulated adrenaline due to stress.

Multilayered fossil eggshells were first reported from the Cretaceous of France and Spain (fig. 26.14) (e.g., Thaler 1965; Erben 1970, 1972; Erben et al. 1979), and later from Mongolia (Sochava 1971), India (Mohabey 1984; 1998), Argentina (Powell 1987; Ribeiro 1999), Canada (Zelenitsky and Hills 1997), Montana (fig. 26.15), and the

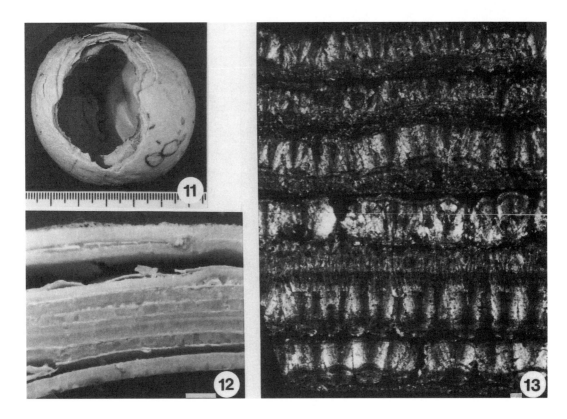

Figures 26.11–26.13.
Pathological egg of Galapagos
turtle (Geochelone elephantopus)
(HEC 326 = UCM OS1144) with
up to nine aragonitic shell layers
and egg membranes. 26.11,
whole egg. Scale in mm. 26.12,
photograph of eight shell layers.
Scale: 1 mm. 26.13, radial thin
section of six shell layers. Note
differences in structure and size
of layers and shell units. Scale:
100μ.

Jurassic of Utah (Hirsch et al. 1989). Most of these occurrences are megaloolithid eggshells having a discretisspherulitic morphotype that superficially resemble pliable and rigid turtle eggshells. Other pathological eggshells occur in prismatic, filispherulitic, dendrospherulitic, and prolatospherulitic morphotypes (e.g., figs. 26.15, 26.16). As in modern shells, the fossil pathological shell layers are not always complete, vary more or less in their structure, and the pores (air canals) do not line up with those of the primary layer. Their shell membrane has been dissolved or replaced by secondary calcite. In the specimens from Spain and Montana, the pathological layer is as thick as the primary shell layer, whereas in the specimen from Alberta it is about three-quarters the thickness. The pathological layer is only half as thick as the primary layer in a specimen from Utah. The specimen from Utah is from a complete egg that is split open but still connected along one side. Peculiarities of this egg suggested that the egg was still in the body of the mother when buried (Hirsch et al. 1989).

Schleich and Kaestle (1988) described 27 double-layered gecko eggs and one multilayered, nongecko egg from the Oligocene of West Germany. They stated that most of the eggs had hatched, "as there are typical openings or breaks" (1988, 102). However, that seems very unlikely. Even if the pore system was functional allowing the embryo to breathe, the double shell wall would be too thick to allow the hatchling to break out.

Birds generally do not retain their eggs, even for a short time. When

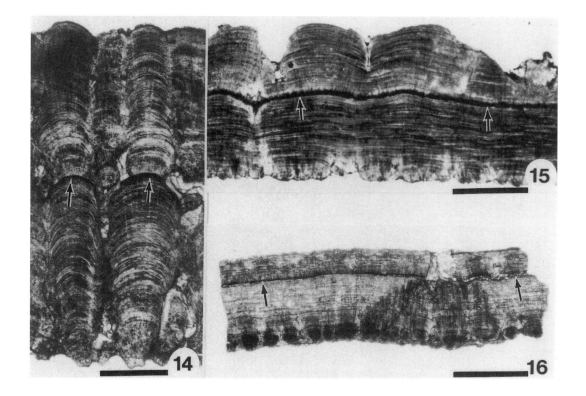

Figures 26.14–26.16.
Multilayered dinosaur eggshell;
radial thin sections. Note
replaced egg membrane layer
between the two shell layers
(arrows). 26.14, Megaloolithus
sp., Upper Cretaceous, Spain
(HEC 68 = UCM 54472). Scale:
1 mm. 26.15, Spheroolithus
?maiasauroides, Upper
Cretaceous, near Egg Mountain,
Montana (HEC 490 UCM
59321). Scale: 1 mm. 26.16,
Preprismatoolithus coloradensis,
Upper Jurassic, Cleveland-Lloyd
Dinosaur Quarry, Utah (HEC
464 = UUVP 11584). Note
pathological layer only half as
thick as primary layer; diagenetic
change of structure in primary
shell below pore opening in outer
layer. Scale: 1 mm.

retention does occur, it usually leads to the death of the female (Riddle 1923). One exception is the cowbird, which retains eggs for a short time until a suitable host nest is found. When the egg is retained unintentionally in birds due to stress or other causes, this generally does not lead to multilayered shell as in reptiles, but to ovum in ovo. This "egg-within-an-egg" occurs because reverse peristalsis sends the egg back to the anterior uterus, where it meets the next descending egg. Together, these two eggs may descend back into the shell-secreting region, where a shell is secreted around both (fig. 26.17; Asmundson 1931, 1933; Romanoff and Hutt 1945; Romanoff and Romanoff 1949). The eggshells stick together only where the shell of the inner egg touches the shell of the outer egg. An ovum in ovo has never been found as a fossil, although the term has been incorrectly used for multilayered dinosaur eggshell (e.g., Erben at al. 1979).

Recognition and Identification of Pathological Eggshell

In many cases, it is difficult to recognize pathological eggshell in the fossil record. Most pathological eggshell has been reported from Cretaceous and Jurassic strata, almost exclusively in dinosaur eggshell. Recognition of pathological eggshell can be difficult due to crushing, taphonomic phenomena, and diagenetic alterations. However, use of modern analytical methods discussed below can help distinguish these features.

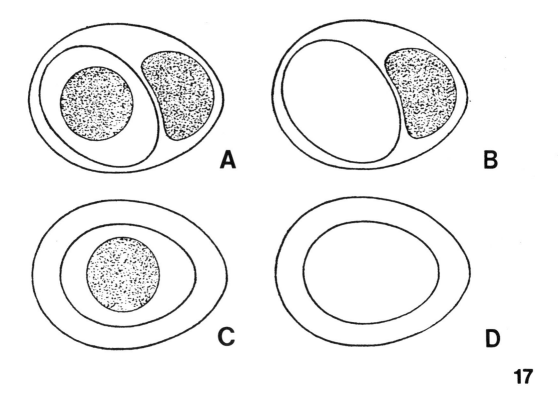

17

Outer Shell Surface

Four specimens of Condor eggshells can be used to demonstrate the difficulties in discriminating between pathological and weathered shell surfaces in modern eggshell. As can be seen in figure 26.10, eggshell 1984b has a normal outer surface, whereas the Pleistocene and 1983 fragments show different stages of weathering; fragment 1984a is a pathological shell affected by DDT (fig. 26.10). The unexposed internal structure of an eggshell is not as easily affected by weathering and diagenesis as is the outer surface. Radial thin sections of the specimens show that the first three fragments have a normal shell structure, verifying that in two cases weathering alone affected the outer surface. The inner structure of the fourth eggshell (1984a) is as abnormal as the outer surface, demonstrating a pathological condition. DDT residues were found within this shell fragment.

Extraspherulitic Growth Units

Occasionally, abnormal shell units and abnormal extraspherulitic units are found in modern eggshell (figs. 26.7, 26.18; Schmidt 1964), but also in fossil eggshell (figs. 26.19, 26.20; Vianey-Liaud et al. 1987, 1994; Zhao et al. 1991; Zhao 1994; Zelenitsky et al. in press). The factors causing these phenomena are not known. In otherwise normal eggshell from the Jurassic of Colorado (fig. 26.19), abnormal spherulites occur rarely, whereas in the shell fragment from Montana they occupy almost the entire shell layer (fig. 26.20). The eggshell from the

Figure 26.17. Ovum in ovo, schematic drawing of most common forms (after Romanoff and Romanoff 1949).

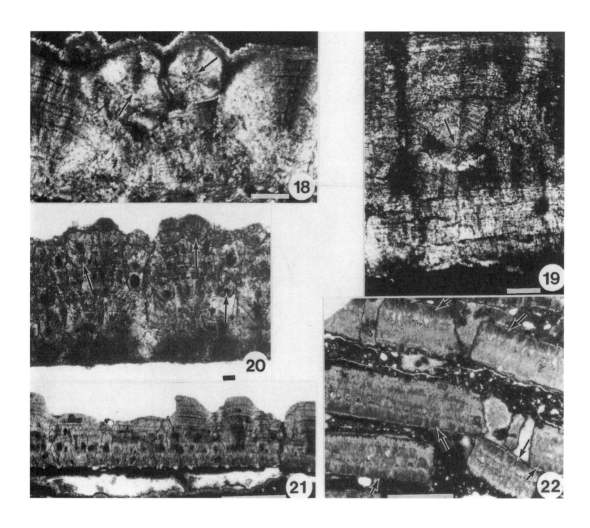

Milk River area of Alberta (fig. 26.21) was probably partially diagenetically altered by dissolution, then had secondary calcite (diagenetic) units deposited on it.

Multilayered and Stacked Eggshell

Multilayered fossil eggshell are the best known and most described pathological phenomenon (e.g., Kerourio 1981). However, if the pathological layers do not stay attached to the primary shell, or to each other, it is hard to recognize them as pathological shell. The outer and inner surface of these shells may also show alterations in the shell structure due to inconsistencies in the shape and size of shell units in each layer, such as seen in the *Geochelone* egg (figs. 26.13, 26.18). Such multiple eggshell might be misinterpreted as a new eggshell type (Hirsch 1989). Stacked eggshells (fig. 26.22), most often found in nesting areas, look superficially like multilayered eggshells (e.g., Tazaki et al. 1994). However, in thin section, the shell surfaces are not positioned in the same way relative to each other. Opposing mammillae in contact with each other indicate a collapsed and compressed egg, whereas outer shell surfaces in contact indicate crushing of multilayered clutches. With many of these "multilayered" specimens, careful examination of thin sections may show that the thin layer between the shell layers is actually sediment. In true multilayered eggshell, the various layers are arranged in the same direction (i.e., outer surface in contact with the mammillae of the overlying layer), and any separation between the shell layers from the replaced egg membrane will be similar to that of the primary shell, not the surrounding sediment.

Pathologically and Diagenetically Altered Eggshell

The recognition of pathological eggshell is very difficult when the eggshell has been altered by diagenesis. Various combinations of recrystallized eggshell or eggshell partially dissolved and coated with diagenetic deposits may occur. Cathodoluminescence (CL) has become a valuable a tool to solve this problem (Zinkernagel 1978; Marshall 1988). All rigid eggshell, with the exception of that of turtles, is composed of calcite. In fossil eggshell, the calcareous eggshell structure is remarkably well preserved (Hirsch and Packard 1987; Mikhailov 1997). Calcitic or aragonitic biogenetic structures with trace magnesium do not luminesce. However, they do luminesce when they are replaced by a different mineral or if the shell structure is altered (Richter and Zinkernagel 1981). CL shows clearly the replacement of magnesium-calcite by manganese-calcite (figs. 26.23–26). Traces of manganese have been found with microprobe analysis of areas that luminesce bright red-orange or yellow-orange. In well-preserved fossil eggshell, the horizontal layering (organic or proteinaceous network) is replaced by brightly orange luminescing manganese-calcite, and the mammillae, rich in organic matter, are also replaced (fig. 26.23). In one shell fragment (fig. 26.24), partial dissolution and diagenetic alteration can be observed within the eggshell. Recrystallization of the shell structure and secondary diagenetic calcite deposit on the outer surfaces are visible in figure 26.25. In multilayered eggshell, the different shell layers have the same color, whereas the egg membrane between these layers is replaced by manganese-calcite that luminesces bright orange (fig. 26.26). The blue color of CL in initially nonluminescing calcium carbonate (figs. 26.23, 26.24) may be due to bombardment by electrons causing a slight, irreversible change of the carbonate lattice structure (Richter and Zinkernagel 1981).

Luminescence of calcite is caused by or inhibited by several trace elements, with manganese as the main activator and iron as the main inhibitor. However, no previous CL work on eggshell has been reported, and it is not known how often these trace elements are present in eggshell. Thus, future work must demonstrate the reliability of this tool for separating pathological and diagenetic changes in eggshells.

Figures 26.18–26.22. (opposite page) Eggshell with pathological extra growth units; viewed in radial thin sections. Arrows point to nucleation centers of units. 26.18, shell fragment from Galápagos tortoise in figs. 26.11–26.13. Note disturbed shell structure below extra units. Scale: 100μ. 26.19, shell fragment of Preprismatoolithus coloradensis, Upper Jurassic, Colorado (HEC 418 = MWC 122.2.3), otherwise normal structure in most of shell fragment. Scale: 100μ. 26.20, shell fragment from an unidentified dinosaur egg, Upper Cretaceous, Landslide Butte, Montana (HEC 422 = UCM 59319). Note shell consists almost completely of abnormal, irregularly placed extra growth units. Scale: 100μ. 26.21, shell fragment from same area as in fig. 26.20 (HEC 426 = UCM 59320). Fragment was probably partially dissolved and then built up by diagenetic growth units. Scale: 1 mm. 26.22, stacked eggshells (Spheroolithus albertensis) from hadrosaur nest, Upper Cretaceous, Alberta, Canada (HEC 471-1-2 = TMP 87.79.161). These stacked eggshells give the impression of multilayered eggshell, however note that the arrows point to different cone layers facing different directions. Note also the sediment that separates the shell pieces. Scale: 1 mm.

Discussion

Pathologies are found in both modern and fossil eggshell. However, most pathological phenomena known in modern eggshell are not

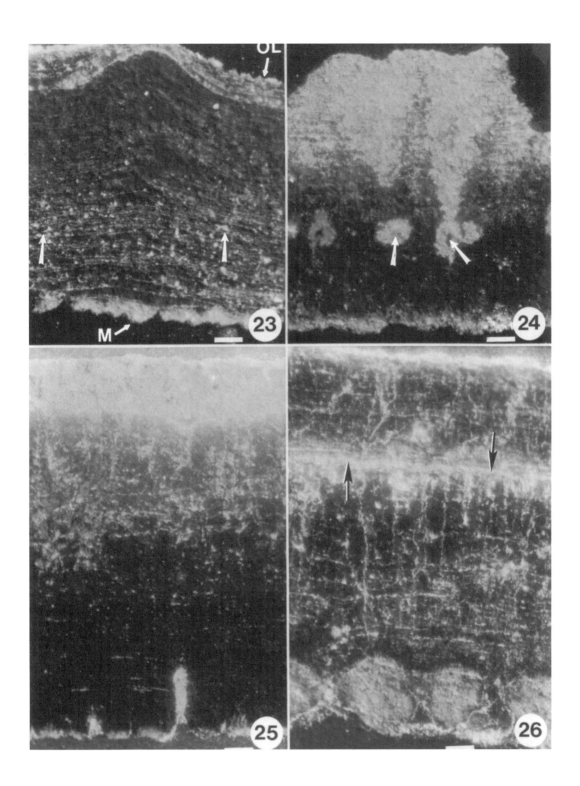

represented in the fossil record. The two recognized pathological forms, multilayered eggs and eggshells with abnormal structure, have been described almost exclusively from dinosaur eggs or eggshells.

The rarity of pathological fossil eggshells, the difficulties in recognizing and identifying them, and the limited state of our understanding of their causes restricts their usefulness in paleoenvironmental and paleobiological studies. Zhao et al. (1991) has attempted to document an increase in eggshell pathologies with paleoenvironmental change based on trace mineral analysis, but the correlation is weak (Carpenter 1999).

The fossil record does show that multilayered eggshell is not restricted to only one shell morphotype or to one taxonomic group of egg-laying animals. The multilayered eggshell, however, does appear to be more abundant in discretispherulitic morphotype and less common in other morphotypes. On the surface, this might imply that discretispherulitic eggs (probably sauropod, Chiappe et al. 1998) were more prone to pathologies, but is actually probable due to the greater sample size for discretispherulitic eggs. The pathological egg from Utah, which was probably still in the mother when she died, is a unique occurrence (Hirsch et al. 1989).

Acknowledgments: A special thank you to the keepers of the Reptile House of the San Diego Zoo for the *Geochelone* egg. I am also indebted to Dave Budd and John Drechsler with the Geology Department of the University of Colorado, Boulder, for introducing me to cathodoluminescence and for their help with the use of the equipment; to Lou Taylor of Texaco, Denver, and Bob McGrew of the Colorado School of Mines for elemental and mineralogical analysis. Judith Harris (University of Colorado Museum), Richard Stucky (Denver Museum of Natural History), Hans-Peter Schultze (Humbolt Museum of Natural History, Berlin), and Art Binkley reviewed the manuscript, helped with advice, and kindly revised the English. Kenneth Carpenter and Darla Zelenitsky revised the manuscript. This study was supported by my personal funds and help from the University of Colorado, Denver Museum of Natural History, Texaco, and the Colorado School of Mines.

Figures 26.23–26.26. (opposite page) Eggshell viewed in radial thin sections under cathodoluminescence. Calcite that contains manganese will luminesce from bright red-orange to yellow-orange. Blue hue in eggshell could be due to changing carbonate levels. Scale: 100μ. 26.23–26.24, Spheroolithus sp., Upper Cretaceous, Milk River area, Alberta, Canada (HEC 472 = TMP 87.77). 26.23, normal eggshell, mammillae (cone layer) (M) and outer layer (OL) completely or partially replaced by manganese-calcite, fine horizontal, orange-luminescing lines (small arrows) represent replaced network of organic (proteinaceous) matter. Luminescing spots (large arrows) are possible nucleation centers for diagenesis. 26.24, eggshell from same nest with nucleation centers (arrows) of diagenetically changed portions which luminesce bright orange. 26.25, Prismatoolithus levis, Upper Cretaceous, Egg Mountain, Montana (HEC 371 = MOR 247). Bright yellow is secondary diagenetically deposited outer layer (OL); below is partially replaced (recrystallization) shell layer grading into normal shell layer in lower half with replaced growth lines and mammillae tips (m), partial pore canal. 26.26, Preprismatoolithus coloradensis, Upper Jurassic, Cleveland-Lloyd Quarry, Utah (HEC 464 = UUVP 11584). Double-layered eggshell with replaced egg membrane of pathological eggshell (arrows), with replaced organic lines, mammillae, and vertical fractures.

References

Aitken, R. N. C., and S. E. Solomon. 1976. Observations on the ultrastructure of the oviduct of the Costa Rican green turtle (*Chelonia mydas* L.). *Journal of Experimental Marine Biology and Ecology* 21: 75–90.

Ar, A., H. Rahn, and C. V. Paganelli. 1979. The avian egg: Mass and strength. *Condor* 81: 331–337.

Asmundson, V. S. 1931. The formation of the hen's egg. *Scientific Agriculture* 11: 1–50.

Asmundson, V. S. 1933. The formation of eggs within eggs. *Zoologischer Anzeiger* 104: 209–217.

Cagle, F. R., and J. Tihen. 1948. Retention of eggs by the turtle *Deirochelys reticularia*. *Copeia* 1: 66.

Carpenter, K. 1999. *Eggs, Nests, and Baby Dinosaurs*. Bloomington: Indiana University Press.

Chiappe, L. M., R. A. Corio, L. Dingus, F. Jackson, A. Chinsamy, and M. Fox. 1998. Sauropod embryos from the Late Cretaceous of Patagonia. *Nature* 396: 258–261.

Cooke, A. S. 1973. Shell thinning in avian eggs by environmental pollutants. *Environmental Pollution* 4: 85–157.

Cooke, A. S. 1975. Pesticides and eggshell formation. *Symposium Zoological Society London* 35: 339–361.

Erben, H. K. 1970. Ultrastrukturen und Mineralisation rezenter und fossiler Eischalen bei Vogeln und Reptilien. *Biomineralisation* 1: 2–66.

Erben, H. K. 1972. Ultrastrukturen und Dicke der Wand pathologischer Eischalen. *Akademie der Wissenschaften und Literatur, Abhandlungen mathematische und naturwissenschaftliche Klasse* 6: 193–216.

Erben H. K., J. Hoefs, and K. H. Wedepohl. 1979. Paleobiological and isotopic studies of eggshells from a declining dinosaur species. *Paleobiology* 5: 380–414.

Ewert, M. A., S. J. Firth, and C. E. Nelson. 1984. Normal and multiple eggshells in batagurine turtles and their implications for dinosaurs and other reptiles. *Canadian Journal of Zoology* 62: 1834–1841.

Hirsch, K. F. 1989. Interpretations of Cretaceous and pre-Cretaceous eggs and shell fragments. In D. D. Gillette and M. G. Lockley (eds.), *Dinosaur Tracks and Traces,* pp. 89–97. Cambridge: Cambridge University Press.

Hirsch, K. F., and M. J. Packard. 1987. Review of fossil eggs and their shell structure. *Scanning Microscopy* 1: 383–400.

Hirsch, K. F., K. L. Stadtman, W. E. Miller, and J. H. Madsen Jr. 1989. Upper Jurassic dinosaur egg from Utah. *Science* 243: 1711–1713.

Jackson, F. D., S. E. Solomon, and D. Varricchio. 1998. Pathological eggshell: New implications for dinosaur reproduction. *Journal of Vertebrate Paleontology, Abstracts with Program* 18: 53A.

Kerourio, P. 1981. La distribution des "Coquilles d'oeufs de Dinosauriens multistratifiées" dans le Maestrichtien continental du Sud de la France. *Geobios* 14: 533–536.

Kiff, L. F., D. B. Peakall, and S. R. Wilbur. 1979. Recent changes in California condor eggshells. *Condor* 81: 166–172.

Marshall, D. J. 1988. *Cathodoluminescence of Geological Materials.* Boston: Unwin Hyman.

McFarland, L. Z., R. L. Garrett, and J. A. Nowell. 1971. Normal eggshells and thin eggshells caused by organochlorine insecticides viewed by the scanning electron microscope. *Scanning Electron Microscopy* 1: 377–384.

Mikhailov, K. E. 1997. Fossil and recent eggshell in amniotic vertebrates: Fine structure, comparative morphology, and classification. *Special Papers in Palaeontology* 56: 1–80.

Mohabey, D. M. 1984. Pathologic dinosaurian eggshell from Kheda district, Gujarat. *Current Science* 53: 701–702.

Mohabey, D. M. 1998. Systematics of Indian Upper Cretaceous dinosaur and chelonian eggshells. *Journal of Vertebrate Paleontology* 18: 348–362.

Nathusius, W. von. 1868. Über die Bildung der Schale des Vogeleies. *Zeitschrift für die gesammten Natürwissenschaften,* 31: 1921.

Palmer, B. D., and L. J. Guillette Jr. 1992. Alligators provide evidence for the evolution of an archosaurian mode of oviparity. *Biology of Reproduction* 46: 39–47.

Penner, M. M. 1983. Contribution à l'étude de la microstructure des coquilles d'oeufs de dinosaures du Crétacé superieur dans la bassin d'Aix-en-Provence: Application biostratigraphique. Ph.D. thesis, Pierre and Marie Curie University, Paris.

Powell, J. E. 1987. The Late Cretaceous fauna of Los Alamitos, Patagonia, Argentina. Part 6, The Titanosaurids. *Revista del Museo Argentino de Ciencias Naturales, Paleontologia* 3: 147–153.

Ribeiro, C.1999. Occurrence of pathological eggshells in the Allen Formation, Late Cretaceous, Argentina. *Abstracts, Seventh International Symposium on Mesozoic Terrestrial Ecosystems* 31.

Richter, D. K., and U. Zinkernagel. 1981. Zur Anwendung der Kathodolumineszenz in der Karbonatpetrographie. *Geologische Rundschau* 70: 1276–1302.

Riddle, O. 1923. Asphyxial death of embryos in eggs abnormally retained in the oviduct. *American Journal of Physiology* 66: 309–321.

Ringleben, H. 1966. Ein Ei im Ei der Schwarzdrossel. *Der Falke* 13: 167.

Romanoff, A. L., and F. B. Hutt. 1945. New data on the origin of double avian eggs. *Anatomical Record* 91: 143–154.

Romanoff, A. L., and A. J. Romanoff. 1949. *The Avian Egg.* New York: Wiley.

Sauer, E. G. F., E. M. Sauer, and M. Gebhardt. 1975. Normal and abnormal patterns of struthious eggshells from South West Africa. *Biomineralization* 8: 32–54.

Schleich, H.-H., and W. Kästle. 1988. *Reptile Egg-Shells: SEM Atlas.* Stuttgart: Gustav Fischer Verlag.

Schmidt, W. J. 1943. Über den Aufbau der Kalkschale bei den Schildkroteneiern. *Zeitschrift für Morphologie und Ökologie der Tiere* 40: 1–16.

Schmidt, W. J. 1957. Über den Aufbau der Schale des Vogeleies nebst Bemerkungen über kalkige Eischalen anderer Tiere. *Bericht der Oberhessische Gesellschaft für Natür- und Heilkunde, Giessen, Natürwissenschaftliche Abteilung* 28: 82–108.

Schmidt, W. J. 1964. Über die Struktur einiger abnormer Vogel-Eischalen nebst Bemerkungen zu neueren Auffassungen betreffend Bau und Bildung der Kalkschale. *Zeitschrift für Morphologie und Ökologie der Tiere* 53: 311361.

Schmidt, W. J. 1967. Die Eischalenstruktur eines "Ovum in ovo" von *Turdus merula*. *Zoologischer Anzeiger* 181: 185–190.

Sochava, A. V. 1971. Two types of eggshell in Senonian dinosaurs. *Paleontological Journal* 3: 353–361.

Solomon, S. E., B. O. Hughes, and A. B. Gilbert. 1987. Effects of a single injection of adrenaline on shell ultrastructure in a series of eggs from domestic hens. *British Poultry Science* 28: 585–588.

Taylor, T. G. 1970. How an eggshell is made. *Scientific American* 22: 89–95.

Tazaki, K., M. Aratani, S. Noda, P. J. Currie, and W. S. Fyfe. 1994. Microscopic and chemical composition of duckbilled dinosaur eggshell. *Science Report, Kanazawa University* 39: 17–37.

Thaler, L. 1965. Les oeufs des dinosaures du Midi de la France livrent le secret de leur extinction. *La Nature,* February, pp. 41–48.

Vianey-Liaud, M., S. L. Jain, and A. Sahni. 1987. Dinosaur eggshells (Saurischia) from the Late Cretaceous Intertrappean and Lamenta Formations (Deccan, India). *Journal of Vertebrate Paleontology* 7: 408–424.

Vianey-Liaud, M., P. Mallan, O. Buscail, and C. Montgelard. 1994. Review of French dinosaur eggshells: Morphology, structure, mineral, and organic composition. In K. Carpenter, K. Hirsch, and J. R. Horner (eds.), *Dinosaur Eggs and Babies,* pp. 151–183. New York: Cambridge University Press.

Woodward, A. E., and F. B. Mather. 1964. The timing of ovulation, movement of the ovum through the oviduct, pigmentation, and shell deposition in Japanese Quail (*Coturnix coturnix japonica*). *Poultry Science* 43: 1427–1432.

Zelenitsky, D. K., and L. V. Hills. 1997. Normal and pathological eggshells of *Spheroolithus albertensis,* sp. nov., from the Oldman Formation (Judith River Group, Late Campanian), southern Alberta. *Journal of Vertebrate Paleontology* 17: 167–171.

Zelenitsky, D. K., K. Carpenter, and P. J. Currie. In press. First record of elongatoolithid (Dinosauria: Theropoda) eggshell from North America. *Journal of Vertebrate Paleontology.*

Zhao Z. 1994. Dinosaur eggs in China: On the structure and evolution of eggshells. In K. Carpenter, K. Hirsch, and J. R. Horner (eds.), *Dinosaur Eggs and Babies*, pp. 184–203. New York: Cambridge University Press.

Zhao Z., Ye J., Li H., Zhao Z., and Yan Z. 1991. Extinction of the dinosaurs across the Cretaceous-Tertiary boundary in Nanxiong Basin, Guandong Province. *Vertebrata PalAsiatica* 29: 1–20.

Zinkernagel, U. 1978. Cathodoluminescence of quartz and its application to sandstone petrology. *Contributions to Sedimentology* 8: 1–69.

Section VI.
Ichnology

27. The Impact of Sedimentology on Vertebrate Track Studies

G. C. NADON

Abstract

Vertebrate tracks are important sources of information on both animal locomotion and sedimentary conditions at the time of impact by the feet. Recognition of the most favorable sites for the formation and preservation of tracks can help researchers exploit this resource more fully. Models of track formation show clearly that the layer upon which the foot descends retains the most information of the impactor. Stresses are distributed radially away from the impact site and decrease exponentially with distance. Application of the models to natural track sites shows that the vast majority of tracks are not underprints or transmitted prints but are instead true tracks. The absence of fine details in tracks is a result of a combination of either unsuitable substrate or infill, or simply covering of the foot by mud from a previous step.

Introduction

Vertebrate tracks are the most common product of vertebrates found in the rock record (Gillette and Lockley 1989; Lockley 1991). Tracks are valuable in part because they are the only fossil record of an animal's actions. They therefore provide a glimpse into locomotion and behavior (Padian and Olsen 1989; Thulborn 1989; Gatesy et al. 1999), as well as a unique record of soft-tissue morphologies that are otherwise absent in the fossil record. Tracks pass through a different taphonomic filter than bone material both during and after formation and burial. Tracks may be the only vertebrate record in some formations (Lockley and Hunt 1995).

Vertebrate tracks are also a valuable sedimentological tool. The impact of feet and paws on unlithified sediments can be viewed as paleo-engineering tests of the substrate. These fossilized tests provide subtle but important details on the spatial and temporal variations in the geotechnical properties of floodplain and coastal environments (Allen 1989, 1997; Nadon and Issler 1997; Nadon 1998).

Despite their utility, tracks and traces are not often exploited to their full potential because of problems of recognition and interpretation. Sedimentologists are not accustomed to looking for a biological explanation for mesoscale sedimentary features; paleontologists are not accustomed to considering the sediment surrounding a track as critical to track interpretation. Each group may underestimate the number and quality of tracks in sedimentary deposits. Full use of the vertebrate track record requires an understanding of the mechanisms of sediment deposition and deformation as well as vertebrate biology. Criteria are presented for track recognition from a sedimentological perspective with the goal of increasing the information gathered from this valuable resource.

Track Formation and Recognition

Problems in track recognition occur on several levels. First, workers who do not expect to find tracks do not apply the proper search image when studying sediments. Second, sediments are more commonly exposed in cross-section than in plan view. Researchers encountering tracks in cross-section have either ignored them or interpreted the structures as the product of physical sedimentary processes (Nadon 1993). Third, the lack of detail within most tracks has resulted in misunderstandings with respect to what constitutes tracks relative to deformed sediments attributable to underprints, transmitted prints, ghost prints, and overprints (Lockley 1989; Thulborn 1990; Lockley and Hunt 1995). Consideration of what constitutes the common track-forming environments, the depositional processes involved in the formation and preservation of tracks, and the geotechnical properties of sediment under loads can resolve these problems.

Optimal Track Environments

The best setting for track formation and preservation is one in which the sediments are strongly heterolithic and rapidly aggrading. The heterolithic nature of the strata is important for two reasons. First, variations in grain size and color between the substrate and the infill provide a visual contrast that is important in track recognition. Second, the coarser grain sizes within the heterolithic strata are commonly better cemented than the finer grain sizes. This allows the tracks, whether molds or casts, to be preserved during exhumation. Rapid aggradation is required to ensure preservation potential. Exposed tracks degrade rapidly after formation (Laporte and Behrensmeyer 1980), especially if the substrate is sand that becomes fully saturated with water prior to burial.

Two depositional environments that best meet the criteria above are tidal and anastomosed fluvial systems. The intertidal region is subjected to high-frequency flooding events with enough time between inundations to partially dewater the sediments. The deposits of the intertidal zone are also fine grained and commonly finely laminated (Tucker and Burchette 1977; Allen 1997). However, intertidal deposits are not regionally extensive.

The other depositional environment that meets the conditions described above is the anastomosed fluvial system (Currie et al. 1991; Nadon 1993, 1994). Anastomosed fluvial deposits, which are the products of suspended-load rivers, record the presence of numerous stable channels separating lowlands comprising extensive, seasonally dry, vegetated floodplains. The floodplains of such river systems are prone to the formation of thick heterolithic crevasse splay deposits. The combination of vegetation and water to attract animals, the mud, and seasonal inundation allows for the optimum formation and preservation of tracks.

The relatively common anastomosed fluvial deposits occur in sediments from the Devonian to Recent (Nadon 1994). Published descriptions of these deposits commonly refer to the presence of deformed sediment at the base of crevasse splay deposits generally termed convolute lamination, contorted bedding, or load structures. However, most of these structures do not contain the requisite deformed laminations. In such cases the horizontal layers of the infilling sediments shows that these deformation structures commonly are tracks (Nadon 1993).

Track Formation

Van der Lingen and Andrews (1969) and Lewis and Titheridge (1978) illustrated many of the key points of identification of tracks in cross-section. These criteria include compression of the strata directly under the foot, folding of strata adjacent to the foot, a rapid decrease in disturbance with depth, and the horizontal layering of the track infill. Loope (1986) provided a more comprehensive analysis of track cross sections and provided examples that illustrate the variations in track morphology in three dimensions. The modeling experiments by Allen (1989, 1997) show clearly how cross sections of tracks in sediments of moderate cohesion are demonstrably different from naturally occurring soft-sediment deformation (fig. 27.1). These experiments also document the presence of small-scale faults surrounding tracks, which may be a useful criterion for track discrimination, and illustrate the style of deformation of strata above and below the layer in contact with the bottom of the foot.

Tracks versus Underprints, Transmitted Prints, Ghost Prints, and Overprints

Sedimentology can also refine interpretations of plan-view exposures of tracks. In addition to the various biological factors that can

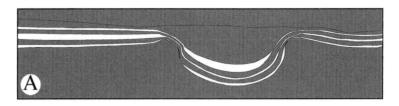

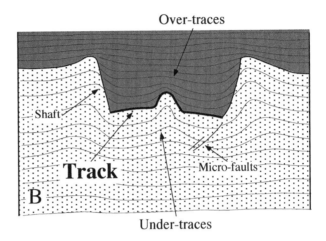

Figure 27.1. Comparison of the cross-sectional geometry between soft-sediment deformation and a footprint. (A) The laminations follow the lower, down-warping surface. The deformed layers were deposited prior to deformation. (B) The more or less horizontal infill of the track shows that the sediment is filling a preexisting void (after Allen 1989).

affect track shape in plan view (Lockley 1989), physical sedimentary conditions, such as grain size and water content, cause variations in both track morphology and preservation (Tucker and Burchette 1977; Scrivner and Bottjer 1986; Allen 1997; Gatesy et al. 1999). The identification of which deformed lamination or bed was in contact with the animal is less straightforward.

Hitchcock (1858) first defined this problem in a classic study of vertebrate tracks of the Newark Basin. Through a series of diagrams Hitchcock illustrated how he thought tracks could be impressed on strata below the surface and retain the same outline as the surface track but lose the finest details, such as skin impressions (see esp. Hitchcock 1858, figs. 1, 2, pl. 6). His text and figures show that he was considering sediment layers 0.2 to 0.25 m thick.

The lack of fine detail, such as skin impression, within most plan-view tracks continues to be cited as evidence that tracks were formed in one or more layers below the actual surface in contact with the foot. These traces are termed underprints (Lockley 1989, fig. 50.3; Lockley 1991, fig. 3.1; Lockley and Hunt 1995, fig. 1.11) and as transmitted or ghost prints (Thulborn 1990, fig. 2.6). They are interpreted to be formed in beds below the foot-sediment interface by the transmission of stresses through the intervening sediment. Thulborn (1990, fig. 2.5) used the term *underprint* in a different sense, to refer to incomplete tracks resulting from erosion of track casts. Removal of the upper parts of the (in situ) cast leave only pad impressions at the base. None of the figures cited above contain scales but all follow Hitchcock (1858) by

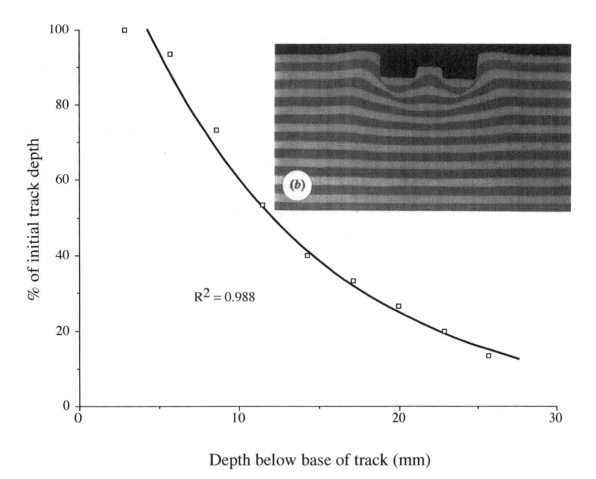

$R^2 = 0.988$

Depth below base of track (mm)

implying that underprints (a) occur directly below the foot impact site, (b) faithfully depict gross anatomical details, such as toes, and (c) can occur at significant (but unspecified) depths below the exposure surface. However, sediments subjacent to tracks do not deform in the way depicted and do not separate or erode in the manner suggested. None of the scenarios presented by Hitchcock, Lockley, or Thulborn is sedimentologically reasonable and some of the interpretations of track formation suggested by Hitchcock (1858) are physically impossible.

Experimental data that are supported by numerous engineering studies show that the process of making tracks in cohesive sediments transmits stresses throughout the sediment in a radial manner (Allen 1989, 1997, fig. 1). The result is a rapid nonlinear decrease in deformation and information with depth. Even a few centimeters below a track, there is no preservation of the details of foot morphology (fig. 27.2). Allen's work shows that the underprint/transmitted print scenarios of sediment deformation, even allowing for artistic license, are incorrect.

The important consideration is the rate of information loss with depth. The rapid and irreversible attenuation of the "signal" means that,

Figure 27.2. The loss of track information with depth. The data are from an experiment by Allen (1997, fig. 9b—see inset). The loss of information on foot morphology, which is shown the decrease in layer deflection with depth relative to the original track impression, is well fit by an exponential function. This loss of detail is irreversible. The undeformed layers within the inset photo are 2.85 mm thick. Photo reproduced by permission of the Royal Society and J. R. L. Allen.

Figure 27.3. Examples of the variations in track preservation as a function of sediment strength and grain size within the St. Mary River Formation (Upper Cretaceous), southwestern Alberta. (A) Mold of a hadrosaur foot on the top of a crevasse splay sandstone. The shallowness of the track is solely a function of the sediment rheology. The substrate was just moist enough to preserve a shallow print. This is not an underprint. Ice ax is 0.8 m long. (B) Cast of a hadrosaur track. The lack of preserved detail is a function of the coarse grain size of the sediment infill and the length of time between track formation and burial. Hammer is 0.3 m long. (C) Cast of a large theropod foot. Arrows indicate termination of toes. The shallowness of the track indicates the substrate was largely dewatered prior to the footfall. Ice ax is 0.8 m long.

in a suitable substrate, positive tracks that preserve any details, such as the clear outline of toes (fig. 27.3A), are not underprints or transmitted prints. The shallowness of the tracks in figure 27.3A is a function of the strength of the substrate, not erosion (cf. Allen 1997, fig. 16f).

The loss of information is also irreversible. This means that impressions below the impact layer cannot have more detail preserved than is present at the foot-sediment interface. The "book of tracks" of Hitchcock (1858, fig. 6, pl. 53) is impossible. Even Hitchcock (p. 33) had difficulty rationalizing the variation in track location between the "leaves." Unconsolidated sand will transmit a footprint less efficiently than the material used by Allen (1997), and therefore the signal will diminish at an even higher rate with depth than illustrated in figure 27.2. Thulborn (1990) cites Van Dijk (1978) as an illustration that the layers below the track can preserve more detail than the track itself. Van Dijk (1978) describes the surface track as obscured by runzel marks, a wrinkling of the surface formed by wind stress on saturated mud (Reineck and Singh 1975). The runzel marks cannot have been made on the track itself because the impact of the foot would compact the sediment beyond the capacity of the wind to shear. Sediments are not elastic media and do not expand after compression. Therefore, a more reasonable interpretation is that the runzel marks were made on laminae that draped the track (the overprints of Allen 1989). The "transmitted" track in this case must be the true track.

The absence of the finest details on most tracks is the result of a more prosaic process. The data of Allen (1997) show that fine-grained sediments will faithfully document the details of a footprint if the foot was clean when the track was formed. Most tracks, even in a suitable substrate, lack fine-scale details simply because the animal's foot was already covered by mud from a previous step. Only when mud has dropped from the foot as a step was taken will the details of the base of the foot be formed and possibly preserved (fig. 27.4; Lockley 1989; Currie et al. 1991). These conditions are unlikely and therefore skin impressions within tracks are not common.

Similar arguments refute the possibility that footprint casts represent either underprints or transmitted prints. Casts are formed by the influx of sediment into the mold made by the foot. The infill that makes the cast is commonly a single event, although exceptions can occur in the case of tidal deposits (Allen 1997). If the track was formed in mud, then the casting medium must be of a coarser grain size in order for the trace to be preserved and exposed. The lack of detail preserved in many tracks is mainly a function of the coarse grain size of the infilling material (compare figs. 27.3, 27.4; Tucker and Burchette 1977).

Optimizing Track Information

Optimizing track information involves determining which strata contain tracks, examining the types and distribution of tracks from a paleontological perspective, and viewing the tracks as paleopenetrometers to refine sedimentological interpretations of floodplain environ-

Figure 27.4. Sandstone cast of a dinosaur track also in the St. Mary River Formation, southwestern Alberta. (A) Skin impressions are present in three locations. The preservation of these details of the bottom of the foot are a result of a cohesive substrate combined with a fine-grained infill. The absence of fine details on the rest of the foot is most likely a result of mud clinging to the base of the foot prior to impact. Lens cap is 5.2 cm in diameter. (B) Detail of the larger patch of skin impression. Note that the striations (arrow) along the edge of the footprint. These striations, as well as the skin impression, indicate the sediment was stiff; perfect for the preservation of fine details.

Figure 27.5. Tracks from Brushy Basin Member of the Morrison Formation, above the Cleveland-Lloyd Dinosaur Quarry, Utah. (A) The facies pattern and the abundance of mudstones suggested this was an anastomosed fluvial deposit prompted an examination of the strata by the author for footprints. (B,C) Deformed sediment lobes (white arrows) that represent tracks in cross-section. These structures have different geometries than expected from soft-sediment deformation (see fig. 27.1).

Figure 27.6. (A) Well-exposed tridactyl track from the Brushy Basin Member found after the track cross sections confirmed the presence of tracks in the area. Lens cap is 5.2 m in diameter. (B) Well-preserved manus and pes of a juvenile sauropod in the Brushy Basin Member. Person in background for scale. Photo courtesy of John Bird.

ments. Each step in the process has potential feedback to paleonto-logists and sedimentologists. Ideally, a simultaneous exploration for tracks by both paleontologists and sedimentologists would maximize information recovery because their training provides each with a slight-ly different search image and bias.

Understanding the relationship between tracks and depositional environment can help to locate likely track-bearing horizons with a minimum of effort. The initial discovery of tracks in the Brushy Basin Member of the Morrison Formation above the Cleveland-Lloyd Dinosaur Quarry was a result of first recognizing the depositional environment as anastomosed fluvial (Kantor et al. 1995), and then looking for track cross sections (fig. 27.5). Once tracks were known to be present, plan-view tracks were located (fig. 27.6). There is now a rich ichnological as well as paleontological record at the quarry site.

Tracks were first used as paleopenetrometers to assess the spatial variations in sediment consistency by Lockley (1987). The variations in track depth provided Lockley with a means to interpret small-scale lateral changes in sedimentary environment that would have otherwise been impossible. This concept was extended by Allen (1997), who demonstrated how an understanding of the variations in track formation and preservation can assist in reconstructing paleocommunities. Nadon (1998) used the presence of tracks on the top of coal seams as paleopenetrometers to draw inferences on the magnitude and timing of peat compaction.

Variations in track morphology on a smaller scale can also yield useful information. Gatesy et al. (1999) have used the variation in track preservation due to sediment consistency to model variations in foot movement during the stride of a small theropod. The same can be applied to figure 27.4. The striations formed by the tubercles on the side of the foot (fig. 27.4B) show that the foot did not have any lateral movement as it entered the substrate. The presence of skin impression (fig. 27.4A) means that there was no lateral motion after initial contact with the substrate or when the animal extracted the foot. Any motion after initial contact or upon extraction would have smeared the fine detail. The skin impression in three locations means that this bipedal animal did not pivot the foot during a stride.

Conclusions

Understanding both the gross depositional environment and the geotechnical properties of sediment deformation obtained from the strata adjacent to vertebrate tracks can help constrain interpretations of this important paleontological resource. The rapid and irreversible attenuation of stresses with depth means that tracks which show evidence of anatomical features, such as well-defined digit outlines, are not underprints or transmitted tracks but the actual product of a foot impacting a subaerially exposed surface. Similarly, researchers should consider tracks as a viable alternative to convolute lamination in fluvial sediments that are not conducive to the formation of such structures. While it is possible to form convolute lamination in floodplain deposits, deformed bedding in floodplain sediments in all Late Paleozoic and younger fluvial deposits should be examined as potential vertebrate tracks.

Acknowledgments: I thank Phil Currie for the interest and encouragement he has shown in both the sedimentology and paleontology of the St. Mary River Formation. The patience and insights of paleontologists, such as Martin Lockley, John Bird, James MacEachern, and George Pemberton, as well as Phil, have always aided me in my attempts to understand sediment deposition and deformation. My thanks to J. R. L. Allen and the Royal Society of London for permission to reproduce figure 27.9b from Allen 1997.

References

Allen, J. R. L. 1989. Fossil vertebrate tracks and indenter mechanics. *Journal of the Geological Society of London,* 146: 600–602.

Allen, J. R. L. 1997. Subfossil mammalian tracks (Flandrian) in the Severn Estuary, S.W. Britain: Mechanics of formation, preservation, and distribution. *Philosophical Transactions of the Royal Society of London,* ser. B, 352: 481–518.

Currie, P. J., Nadon, G. C., and M. G. Lockley. 1991. Dinosaur footprints with skin impressions from the Cretaceous of Alberta and Colorado. *Canadian Journal of Earth Sciences,* 28: 102–115.

Gatesy, S. M., K. M. Middleton, F. A. Jenkins Jr., and N. H. Shubin. 1999. The three-dimensional preservation of foot movements in Triassic theropod dinosaurs. *Nature* 399: 141–144.

Gillette, D. D., and M. G. Lockley, eds. 1989. *Dinosaur Tracks and Traces.* New York: Cambridge University Press.

Hitchcock, E. 1858. *Ichnology of New England: A Report on the Sandstone of the Connecticut Valley, Especially Its Fossil Footmarks.* Boston: W. White.

Kantor, D. C., C. W. Byers, and G. C. Nadon. 1995. The Upper Brushy Basin Member and Buckhorn Conglomerate at the Cleveland-Lloyd Dinosaur Quarry: Transition from an anastomosed to a braided fluvial deposit. *Geological Society of America, Abstracts with Program* 27 (6): 277.

Laporte, L., and A. K. Behrensmeyer. 1980. Tracks and substrate reworking by terrestrial vertebrates in quaternary sediments of Kenya. *Journal of Sedimentary Petrology* 50: 1337–1346.

Lewis, D. W., and D. G. Titheridge. 1978. Small scale sedimentary structures resulting from foot impressions in dune sands. *Journal of Sedimentary Petrology* 48: 835–838.

Lockley, M. G. 1987. Dinosaur trackways and their importance in paleoenvironmental reconstruction. In S. Czerkas and E. C. Olson (eds.), *Dinosaurs Past and Present,* pp. 81–95. Los Angeles: Los Angeles County Museum.

Lockley, M. G. 1989. Summary and prospectus. In D. D. Gillette and M. G. Lockley (eds.), *Dinosaur Tracks and Traces,* pp. 441–447. New York: Cambridge University Press.

Lockley, M. G. 1991. *Tracking Dinosaurs: A New Look at an Ancient World.* New York: Cambridge University Press.

Lockley, M. G., and A. P. Hunt. 1995. Ceratopsid tracks and associated ichnofauna from the Laramie Formation (Upper Cretaceous: Maastrichtian) of Colorado. *Journal of Vertebrate Paleontology* 15: 592–614.

Loope, D. B. 1986. Recognizing and utilizing vertebrate tracks in cross section: Cenozoic hoofprints from Nebraska. *Palaios* 1: 141–151.

Nadon, G. C. 1993. The association of anastomosed fluvial deposits and dinosaur tracks, eggs, and nests: Implications for the interpretation of floodplain environments and a possible survival strategy for ornithopods. *Palaios* 8: 31–44.

Nadon, G. C. 1994. The genesis and recognition of anastomosed fluvial deposits: Data from the St. Mary River Formation, southwestern Alberta, Canada. *Journal of Sedimentary Research* B64: 451–463.

Nadon, G. C. 1998. Geometrical constraints on the timing and magnitude of peat-to-coal compaction. *Geology* 26: 727–730.

Nadon, G. C., and D. R. Issler. 1997. The compaction floodplain sediments: What's wrong with this picture? *Geoscience Canada* 10: 38–42.

Padian, K., and P. E. Olsen. 1989. Ratite footprints and the stance and gait of Mesozoic theropods. In D. D. Gillette and M. G. Lockley (eds.), *Dinosaur Tracks and Traces,* pp. 231–241. New York: Cambridge University Press.

Reineck, H.-E., and I. B. Singh. 1975. *Depositional Sedimentary Environments.* New York, Springer Verlag.

Scrivner, P. J., and D. J. Bottjer. 1986. Neogene avian and mammalian tracks from Death Valley National Monument, California: Their context, classification, and preservation. *Palaeogeography, Palaeoclimatology, Palaeoecology* 57: 285–331.

Thulborn, R. A. 1989. The gaits of dinosaurs. In D. D. Gillette and M. G. Lockley (eds.), *Dinosaur Tracks and Traces,* pp. 39–50. New York: Cambridge University Press.

Thulborn, R. A. 1990. *Dinosaur Tracks.* London: Chapman and Hall.

Tucker, M. E., and T. P. Burchette. 1977. Triassic dinosaur footprints from South Wales: Their context and preservation. *Palaeogeography, Palaeoclimatology, Paleoecology* 22: 195–208.

Van der Lingen, G. J., and P. B. Andrews. 1969. Hoof-print structures in beach sand. *Journal of Sedimentary Petrology* 39: 350–357.

Van Dijk, D. E. 1978. Trackways in the Stormberg. *Palaeontologia africana* 21: 113–120.

28. *Acrocanthosaurus* and the Maker of Comanchean Large-Theropod Footprints

James O. Farlow

Abstract

Variations in pedal phalangeal proportions do not closely track phylogenetic relationships of theropod dinosaurs, particularly large theropods. Species in a given large-theropod clade often have phalangeal proportions as much like those of species in a different clade as like those of other species of their own clade. Consequently we cannot assign large-theropod footprints to any particular large-theropod zoological taxon on the basis of footprint morphology alone. However, a given footprint morphotype can be interpreted as having been made by a specific large-theropod taxon if the footprint type and the zoological taxon occur in rocks of comparable age in the same geographic area, and if the size and shape of the footprint and the putative trackmaker's foot skeleton are consistent with each other. On this basis, large, tridactyl Comanchean (Early Cretaceous, Texas) footprints could well have been made by the large theropod *Acrocanthosaurus atokensis*. Given the relationship between body size and home range area and daily movement distance in extant vertebrate predators, *Acrocanthosaurus* is expected to have been a mobile animal, ranging over large areas and different habitat types. If *Acrocanthosaurus* was indeed the Comanchean large-theropod trackmaker, the occurrence of its footprints in rocks formed in carbonate coastal mudflats, and of its skeletal remains in clastic sedimentary rocks, are consistent with predictions based on the peregrinatory nature of extant large predators.

Introduction

Tridactyl dinosaur footprints were discovered in Comanchean (Early Cretaceous, Albian, ca. 111 Ma) rocks near Glen Rose, Texas, early in the 20th century (Shuler 1917, 1935, 1937) and have since been discovered at sites all over the central part of the state (Langston 1974; Farlow 1981, 1987; Pittman 1989; Hawthorne 1990). Most of these footprints have the appearance of theropod prints, and Langston (1974) proposed that the likely maker of large-theropod footmarks in Comanchean rocks was *Acrocanthosaurus atokensis,* a big theropod known at that time from rather scrappy material (Stovall and Langston 1950). Langston's interpretation has been followed by subsequent workers (Farlow 1987; Pittman 1989).

In recent years, new skeletal material of *Acrocanthosaurus* has been discovered (Harris 1998b; Currie and Carpenter in press). These fossils permit a critical comparison of the foot of *Acrocanthosaurus* with large tridactyl Comanchean footprints. Particular emphasis will be placed on NCSM 14345, one of the most complete skeletons of *Acrocanthosaurus* presently known (fig. 28.1).

Dinosaur taxa are named on the basis of osteological remains, so identifying the taxon responsible for a footprint is an exercise in correlating between a footprint shape and the shape of a foot skeleton. The ultimate limit in our ability to assign a dinosaur footprint to a trackmaker candidate therefore is the degree to which the dinosaur's foot skeleton recognizably differs from that of alternative trackmakers. Ideally the comparison should be made on the basis of features that potentially could be recorded in footprints. The obvious features to consider in this regard are the lengths and widths of digits, and of the individual phalanges in those digits. Digital and phalangeal lengths could conceivably be estimated from phalangeal pads in very well preserved footprints (Farlow and Lockley 1993).

I first consider the extent to which theropod foot shapes track the phylogenetic relationships of theropod taxa, to see (a) if pedal proportions are characteristic of particular taxa within a theropod clade, and distinctly different from pedal proportions of other members of the same clade, and (b) if pedal proportions of the members of a theropod clade are more like those of each other than like those of theropods in other clades. I then examine whether the size and shape of Comanchean large-theropod footprints are consistent with the foot skeleton of *Acrocanthosaurus*. Finally, I speculate on the paleoecological implications of the conclusion that Comanchean large-theropod prints were indeed made by *Acrocanthosaurus*.

Institutional Abbreviations: AMNH, American Museum of Natural History; BHI, Black Hills Institute of Geological Research; CEU, College of Eastern Utah; DPP, Dinosaur Provincial Park field station of the Royal Tyrrell Museum of Palaeontology; GIN, Geological Institute, Mongolian Academy of Sciences; Ghost Ranch, Ruth Hall Museum, Ghost Ranch; GSC, Geological Survey of Canada; LACM, Natural History Museum of Los Angeles County; MNA, Museum of Northern Arizona; MOR, Museum of the Rockies; MUC, Museo de Ciencias

Figure 28.1. Cast of the skeleton of NCSM 14345, Acrocanthosaurus atokensis. *Length of reconstructed skeleton = 11 m. Glenoacetabular length = 2100 mm. Femur length ≈ 1165 mm (midshaft circumference = 425 mm); tibia length ≈ 850 mm; metatarsal III length ≈ 473 mm. Measurements and photograph by Dale Russell.*

Naturales de la Universidad Nacional del Comahue, Argentina; NCSM, North Carolina State Museum of Natural Sciences, North Carolina State University; PVSJ, Museo de Ciencias Naturales, Universidad Nacional de San Juan, Argentina; ROM, Royal Ontario Museum; SMP, State Museum of Pennsylvania; TMP, Royal Tyrrell Museum of Palaeontology; SMA, Sauriermuseum Aathal; UCMP, University of California Museum of Paleontology; USNM, U.S. National Museum (Smithsonian Institution); Western Paleo, Western Paleontological Laboratories, Orem, Utah; YPM, Yale Peabody Museum.

Methods
Analysis of Pedal Proportions

Pedal phalangeal measurements were made by me, or by others following my directions. Phalangeal lengths were measured along both the medial and lateral sides of bones (where possible), from the dorsoventral midpoint (or near it) of the concave proximal edge of the bone to the dorsoventral midpoint of the bone's convex distal end. Medial and lateral values were then averaged. Unguals were measured in a

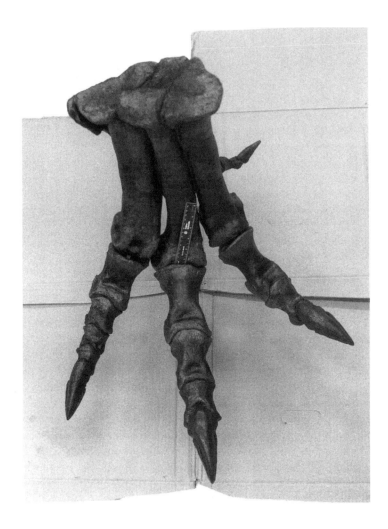

28.2. Cast of the right foot of NCSM 14345 (Acrocanthosaurus); photograph taken from nearly directly above the toes and obliquely to the metatarsus. Phalanx III1 length = 143 mm. Phalanges II1, II2, the proximal half of II3, III1, III2, IV1–3, and the proximal half of IV4 are real in the actual specimen; the other phalanges are artistic reconstructions.

straight-line fashion (i.e., NOT following the curvature of the bone) from the dorsoventral midpoint of the concave proximal edge to the ungual tip, again with medial and lateral values averaged. Widths of nonungual phalanges were measured as the maximum transverse dimension across the distal articular end of the bone; ungual widths were not used. All phalangeal measurements were made to the nearest millimeter. The data on which my analyses were based are available on request.

Missing elements are the bane of any morphometric study of dinosaur bones, and all my analyses represented some compromise between the equally desirable but unfortunately mutually exclusive goals of obtaining complete sets of measurements for a given specimen, but at the same time having a large sample size of specimens. For some analyses I used only phalangeal lengths, but in others I also employed phalangeal widths. In some analyses I restricted the number of phalanges used, and in others I used all the phalanges of the skeleton.

Although size is clearly an important consideration in matching dinosaur foot skeletons with footprints, it is also true that individuals of the same theropod species could vary considerably in size due to growth, individual variation, and perhaps sexual dimorphism (Russell 1970; Carpenter 1990; Colbert 1990; Raath 1990; Smith 1998), and different species of the same clade also differ in body size. It is therefore necessary to consider foot shape apart from size in ascertaining whether feet of theropods of the same clade are more like each other than like feet of taxa of other theropod clades.

Consequently I analyzed phalangeal lengths in two ways. First, I performed a principal components analysis (PCA) of log-transformed phalangeal lengths, using a covariance matrix. Data used were lengths of only those phalanges that were preserved with NCSM 14345: phalanges II1, II2, III1, III2, and IV1–IV3 (fig. 28.2).

In my second analysis, I first scaled each phalangeal length to keep it in proper proportion when phalanx III1 length was assigned an arbitrary value of 100 mm:

observed phalanx length × (100 mm/observed phalanx III1 length)

Thus all phalangeal lengths of all feet were scaled to the same III1 length—the numerical equivalent of manipulating photographs of different-size specimens to make them look the same size.

Hierarchical cluster analysis of individual theropod feet was then done using the scaled phalangeal lengths. Clusters were created by the unweighted pair-group method using arithmetic averages (UPGMA), based on the squared Euclidean distance (Norušis 1988). Data used were for those phalanges (other than III1, which now had identical values for all feet) known for NCSM 14345.

I then did additional cluster analyses, adding scaled phalangeal widths and additional phalanges. Obviously the sample size on which each analysis was based became smaller as more measurements were tossed into the analysis.

Digital lengths were estimated by adding the lengths of their component phalanges. These will necessarily be underestimates because they do not take into account the horny claws (see below) or the soft tissues that separate pedal phalanges in the joints. However, in my observations of dinosaur toe skeletons preserved in articulation, adjacent pedal phalanges are separated from their partners in joints by only a few millimeters. For some analyses lengths of entire digits were estimated, but in others I used the aggregate lengths of only those phalanges known for NCSM 14345.

A brief comment on tyrannosaurid systematics must be inserted here to forestall confusion. Russell (1970) considered *Gorgosaurus* to be so similar to *Albertosaurus* that he relegated the former to junior synonymy with the latter. This usage has been followed by many subsequent workers (e.g., Carr 1999), but recently some theropod specialists have again employed the name *Gorgosaurus* (Bakker et al. 1988; Carpenter 1997). Although a complete defense of this proposal has not yet appeared, I will employ *Gorgosaurus* for those tyrannosaurid specimens (*G. libratus* from the Dinosaur Park Formation of

Alberta) that Currie (pers. comm.) considers most appropriately assigned to this genus. I retain *Albertosaurus* for other tyrannosaurid specimens that have been identified as that genus, but recognize that this may not be the name finally applied to those specimens once the dust of tyrannosaurid systematics has settled.

Comanchean Theropod Footprints

Measurements of tridactyl dinosaur footprints were taken at numerous tracksites in the Glen Rose Limestone and Fort Terrett Formation of Texas (Farlow 1987; Pittman 1989; Hawthorne 1990). I measured footprint lengths for trackways comprising at least two prints in succession. Where possible, I averaged lengths for more than one footprint in a trackway. Footprint lengths were taken from the "heel" to the tip of digit III. For some of the prints from the bed of the Paluxy River (Dinosaur Valley State Park, Glen Rose), footprint lengths are underestimates: At the time these footprints were made, the sediment behaved in a very cohesive, plastic fashion. When a dinosaur's foot was withdrawn, sediment collapsed around the toemarks, reducing their length in surface expression. The toemarks thus extend some distance forward as tunnels beneath the surface of the rock containing the footprints.

Casts were made of the better large tridactyl theropod prints, and some of these were used to compare footprint size and shape with large-theropod foot skeletons. One important consideration is whether *Acrocanthosaurus* would have made a footprint comparable in size to Comanchean large-theropod footmarks. Because the foot of NCSM 14345 is incomplete, I first estimated the lengths of phalanges II3, III3, and III4 by comparison with those of other large theropods (table 28.1).

Following Baird (1957), I assume that the impression of digit II in a theropod footprint can be approximated by adding half the length of phalanx II1 to the combined lengths of phalanges II2 and II3. In like manner, I assume that the toemark of digit III would consist of half the length of phalanx III1 plus the combined lengths of III2–III4. These estimates do not include the contribution of the horny claw to the digit impression. Olsen et al. (1998) assumed that the ungual tips of theropod toes would terminate about midway along the lengths of the clawmarks of footprints made by those feet. This is not an unreasonable assumption, but in my observations of the feet of extant ground birds and alligators, the ungual often extends nearly to the tip of the claw. I therefore made no allowance for toemark enlargement due to horny sheaths around unguals in comparing the foot of NCSM 14345 to Comanchean large-theropod prints. Consequently there is a possibility that I have somewhat underestimated the lengths of the marks that digits II and III of the living NCSM 14345 would have made.

To translate digit II and III impression lengths of NCSM 14345 into an estimate of overall footprint length, toemark lengths (measured from the proximal to the distal end of each toe impression; "digit length" of Leonardi 1987, pl. 5G) were compared with overall footprint lengths in five well-preserved Comanchean theropod footprints

TABLE 28.1
Estimates of Footprint Length for *Acrocanthosaurus*

Phalanx	Phalanx Length (mm)				
	Acrocanthosaurus NCSM 14345	*Allosaurus* MOR 693	*Gorgosaurus* ROM 1247	*Albertosaurus* MOR 657	*Daspletosaurus* MOR 590
II1	141	109	139	143	130
II2	85	74	94	90	77
II3		71	95	89	77
III1	145	110	138	137	117
III2	95	79	93	86	75
III3		57	76	67	61
III4		89	76	103	91

In all cases II2 length $\cong$ II3 length. For *Acrocanthosaurus*, assume II3 length = 85mm

Estimating III3 length of *Acrocanthosaurus* from III3/III1 length ratio in other theropods:

From *Allosaurus:* 57/110 = III3/145; III3 length = 75 mm

From *Gorgosaurus:* 76/138 = III3/145; III3 = 80 mm

From *Albertosaurus:* 67/137 = III3/145; III3 = 71 mm

From *Daspletosaurus:* 61/117 = III3/145; III3 = 76 mm

For *Acrocanthosaurus*, assume III3 length = 75 mm (a "consensus" of the four preceding estimates)

Estimating III4 length of *Acrocanthosaurus* from III4/III1 length ratio in other theropods:

From *Allosaurus:* 89/110 = III4/145; III4 length = 117 mm

From *Gorgosaurus:* 76/138 = III4/145; III4 length = 80 mm

From *Albertosaurus:* 103/137 = III4/145; III4 length = 109 mm

From *Daspletosaurus:* 91/117 = III4/145; III4 length = 113 mm

For *Acrocanthosaurus*, assume III4 length = 110 mm (a rough "consensus" of the four preceding estimates)

Estimated length of digit II impression in an *Acrocanthosaurus* footprint:

(141/2) + 85 + 85 mm = 241 mm

Estimated length of digit III impression in an *Acrocanthosaurus* footprint:

(145/2) + 95 + 75 + 110 mm = 353 mm

Comparison with Comanchean large theropod footprints:

Footprint	Digit II Impression Length	Digit III Impression Length (mm)	Overall Footprint Length (mm)	Estimated Overall Footprint Length (mm) for an *Acrocanthosaurus* Print with an Overall Print Length–Toe Length Ratio Comparable to That Seen in the Footprint	
				Based on Digit II	Based on Digit III
Hamilton County	230	310	465	487	529
South San Gabriel River print 1	~230	330	535	561	572
Paluxy River CT1	240	310	490	492	558
Davenport Ranch print 7	275	~345	~530	464	542
F[6] Ranch G46	260	360	525	487	515

TABLE 28.2
Principal Components Analysis (Using a Covariance Matrix) of Log-
Transformed Lengths of Those Phalanges Known from NCSM 14345
across Theropod Taxa. Number of specimens = 32.

Phalanx	Component 1 Loading	Component 2 Loading
II1	0.990	0.118
II2	0.992	-0.062
III1	0.996	0.085
III2	0.985	0.150
IV1	0.993	0.014
IV2	0.993	-0.093
IV3	0.984	-0.173
Cumulative variance explained	98.079	99.285

Kaiser-Meyer-Olkin measure of sampling adequacy = 0.844;
Bartlett's test of sphericity: chi-square = 718.963, $p < 0.001$

(table 28.1). On the assumption that a print of NCSM would have had similar proportions, I estimated the length of footprints made by this animal as 46–57 cm. Thus 50–55 cm seems a reasonable "consensus" estimate for the overall length of a footprint made by NCSM 14345, and because this individual was only a bit larger than previously described specimens of *Acrocanthosaurus* (Harris 1998b; Currie and Carpenter in press), these individuals would have made footprints only slightly smaller.

Results

Comparisons of Foot Shapes within and across Theropod Taxa

About 98% of the variance in log-transformed pedal phalangeal lengths is accounted for by the first component, which shows high loadings with all phalangeal lengths (table 28.2). Thus absolute size is far more important than shape in accounting for differences in phalangeal lengths among theropod taxa. This result is reminiscent of Smith's (1999) observations on intraspecific variability within elements in *Allosaurus fragilis*, in which pedal phalanges showed a dominant influence of size, as opposed to shape, in affecting element dimensions.

The second component in the PCA does show a contrast (fig. 28.3) between taxa with relatively long phalanges II1, III1, and III2 (ornithomimids) and those with relatively long phalanges II2, IV2, and IV3 (*Mononykus, Deinonychus*). However, these are all small to medium-size theropods. Large forms (*Dilophosaurus*, allosaurids, tyrannosaurids) are less variable in the contrast embodied by component 2.

Rather similar results are obtained with a cluster analysis of scaled phalangeal lengths (fig. 28.4). Large theropods, regardless of clade, are very similar and tend to cluster together, albeit along with some smaller

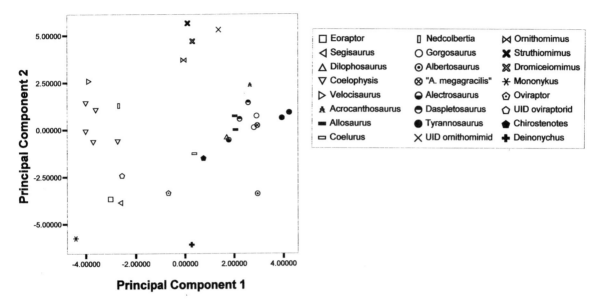

Figure 28.3. Principal components analysis of log-transformed lengths of those pedal phalanges known for NCSM 14345 (Acrocanthosaurus) and other theropods. The first component relates mainly to overall foot size. Component 2 contrasts specimens that have relatively long phalanges II1, III1, and III2 (positive values) with specimens that have relatively long phalanges IV2, IV3, and II2 (negative values).

taxa. Although the two specimens of *Tyrannosaurus* are very similar, *Gorgosaurus* and *Allosaurus* specimens do not cluster with their respective conspecifics apart from other large theropods. Members of some small-bodied clades (e.g., ornithomimids, ceratosaurs), tend to be more like each other than like other clades, but there are some odd pairings at higher levels in the dendrogram (e.g., *Nedcolbertia* and ornithomimids with ceratosaurs).

If phalangeal widths are added to lengths (fig. 28.5), much of the pattern remains. Most large theropods group together (and *Acrocanthosaurus* is more like a tyrannosaurid than like *Allosaurus*), although *Gorgosaurus* and *Dilophosaurus* now cluster together with a heterogeneous assemblage of small theropods. Adding more phalangeal measurements (figs. 28.6, 28.7) produces some changes in dendrogram topology (due at least in part to loss of specimens because of incomplete preservation of foot skeletons), but several groupings persist: *Nedcolbertia* always clusters close to ornithomimids, *Dilophosaurus* stays close to *Gorgosaurus,* some tyrannosaurids cluster with *Allosaurus* apart from other tyrannosaurids, and in all comparisons *Deinonychus* is very different from all other theropods.

Although some small theropods differ markedly in the relative lengths of the three digits (fig. 28.8), large theropods appear more uniform, with no obvious differences among clades. Among large theropods, *Gorgosaurus* is a rather slim-toed form (fig. 28.9), but otherwise there is little difference among large-theropod clades in the relationship between digit length and width.

These results indicate a relative homogeneity of pedal proportions of large theropods, particularly in comparison with smaller-bodied theropods. Overall foot shape does not seem to correlate closely with the presumed phylogenetic affinities of big carnivorous dinosaurs.

Of course, it may be unfair to expect otherwise. Theropod clades are recognized on the basis of synapomorphies throughout the skeleton (Holtz 1994; Padian et al. 1999; Sereno 1999), while my analyses

416 • James O. Farlow

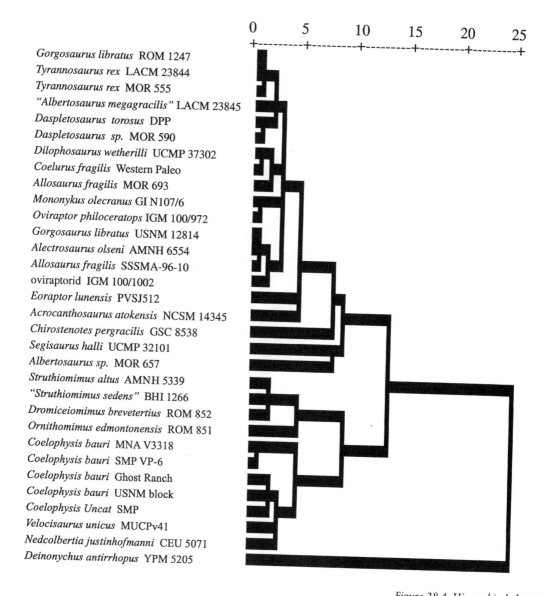

Figure 28.4. Hierarchical cluster analysis of pedal phalangeal length proportions in theropods. All phalangeal lengths were scaled relative to a common phalanx III1 length prior to analysis. Distances between clusters are expressed in terms of values between 0 and 25. Many of the theropod taxa (particularly large-bodied forms) have phalangeal proportions so similar that the dendrogram cannot show the topology of the actual clustering schedule (particularly in its first few steps). Analysis based on phalanges known for NCSM 14345 (II1–II2, III2, IV1–IV3).

probably hang on a mixture of both primitive and derived pedal phalangeal proportions. My analyses were done this way because comparisons between footprint shapes and skeletal shapes traditionally are based on overall proportions. Although it might be possible to recognize synapomorphies in phalangeal proportions (Olsen et al. 1998), I suspect that these will be useful mainly at higher taxonomic levels (e.g., in distinguishing between feet and footprints of theropods and ornithopods), and will seldom be helpful in distinguishing among clades within major dinosaur groups. I am especially doubtful that synapomorphies in pedal proportions can be recognized in clades of large theropods; note, for example, how similar pedal phalangeal proportions of NCSM 14345 are to those of tyrannosaurids (table 28.1).

My data suggest that pedal phalangeal skeletons of large cerato-

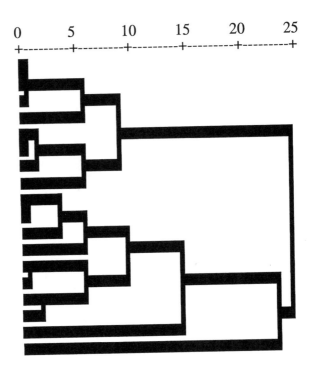

Tyrannosaurus rex MOR 555
Tyrannosaurus rex LACM 23844
Daspletosaurus sp. MOR 590
Acrocanthosaurus atokensis NCSM 14345
Allosaurus fragilis SSSMA-96-10
Alectrosaurus olseni AMNH 6554
Allosaurus fragilis MOR 693
Mononykus olecranus GI N107/6
Dilophosaurus wetherilli UCMP 37302
Gorgosaurus libratus ROM 1247
Coelurus fragilis Western Paleo
Eoraptor lunensis PVSJ512
Dromiceiomimus brevetertius ROM 852
"Struthiomimus sedens" BHI 1266
Coelophysis bauri Ghost Ranch
Nedcolbertia justinhofmanni CEU 5071
Chirostenotes pergracilis GSC 8538
Deinonychus antirrhopus YPM 5205

Figure 28.5. Hierarchical cluster
analysis based on the phalanges
known for NCSM 14345, using
scaled phalangeal lengths and
widths.

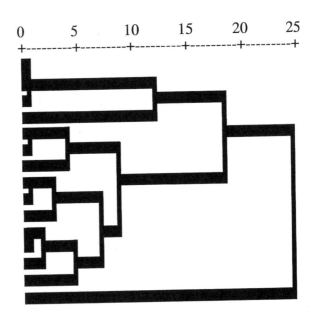

Tyrannosaurus rex MOR 555
Tyrannosaurus rex LACM 23844
Daspletosaurus sp. MOR 590
Mononykus olecranus GI N107/6
Dromiceiomimus brevetertius ROM 852
"Struthiomimus sedens" BHI 1266
Nedcolbertia justinhofmanni CEU 5071
Dilophosaurus wetherilli UCMP 37302
Gorgosaurus libratus ROM 1247
Coelurus fragilis Western Paleo
Allosaurus fragilis SSSMA-96-10
Alectrosaurus olseni AMNH 6554
Allosaurus fragilis MOR 693
Eoraptor lunensis PVSJ512
Deinonychus antirrhopus YPM 5205

Figure 28.6. Hierarchical cluster
analysis based on scaled lengths
and widths of all nonungual
phalanges.

Dilophosaurus wetherilli UCMP 37302
Gorgosaurus libratus ROM 1247
Eoraptor lunensis PVSJ512
Nedcolbertia justinhofmanni CEU 5071
"Struthiomimus sedens" BHI 1266
Allosaurus fragilis MOR 693
Allosaurus fragilis SSSMA-96-10
Daspletosaurus sp. MOR 590
Deinonychus antirrhopus YPM 5205

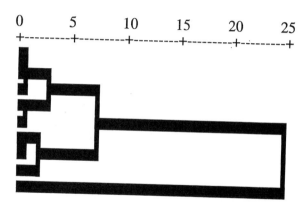

Figure 28.7. Hierarchical cluster analysis based on scaled lengths and widths of all nonungual phalanges, and lengths of unguals.

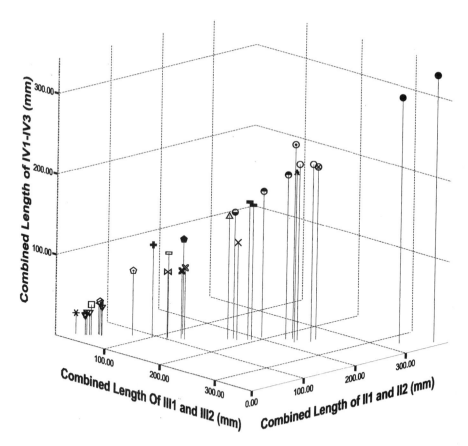

Figure 28.8. Aggregate lengths of phalanges of digits II-IV, using the phalanges known for NCSM 14345. Taxon symbols as in figures 28.3 and 28.9.

Acrocanthosaurus and the Maker of Comanchean Large-Theropod Footprints • 419

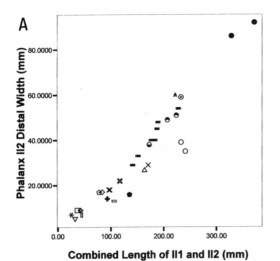

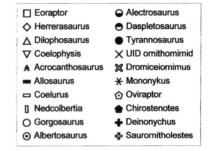

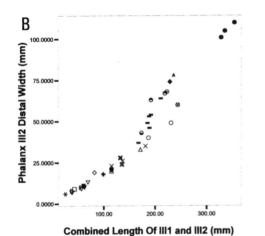

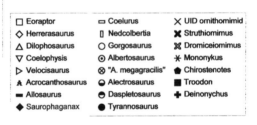

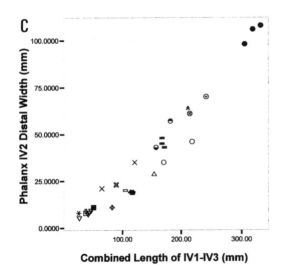

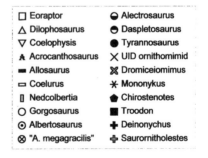

saurs, allosaurs, and tyrannosaurs are indistinguishable. That being the case, it is probably impossible to correlate large-theropod footprints with the clades of their makers on the basis of print shape alone. Identifying large-theropod trackmakers will instead be a matter of determining whether a footprint shape is consistent with the pedal skeleton of a zoological taxon known from the same or a correlative stratigraphic unit from the same region. A footprint identical in size and shape found in rocks from a different time or place will not necessarily have been made by a member of the same large-theropod clade.

If this conclusion is valid, then using large-theropod ichnotaxa to make intercontinental correlations (e.g., Lockley et al. 1996; Lockley 1998) is a procedure that should be done with considerable caution. Footprints that on morphological grounds can be placed in the same ichnotaxon might have been made by large theropods that were not closely related. An analogous situation occurs with early Tertiary tridactyl footprints attributed to ungulates (Lockley et al. 1999); northern hemisphere specimens were likely made by perissodactyls, but very similar South American footprints were probably made by notoungulates or litopterns.

Figure 28.9. (opposite page) Digit width (as indicated by the distal width of the second phalanx) as a function of digit length in theropods, using the phalanges known for NCSM 14345. (A) Digit II, (B) Digit III, (C) Digit IV.

Comparison of the Foot of Acrocanthosaurus *with* Comanchean Large-Theropod Footprints

A size-frequency histogram of footprint lengths of Comanchean tridactyl trackmakers (fig. 28.10) is at least bimodal. A single trail records an enormous creature that would have dwarfed NCSM 14345 (an isolated footprint of a second, equally large animal occurred at the same site, but as a "singleton" was not included in my size-frequency analysis), with a footprint size rivaling that of a footmark attributed to *Tyrannosaurus* (Lockley and Hunt 1994). However, Farlow and Hawthorne (1989) thought the Texas trail to have been made by a huge ornithopod.

The strongest size class is centered on footprint lengths of 50 cm, a nice match with my estimate of the length of footprints that *Acrocanthosaurus* would have made. There is a suggestion of another mode at 25–30 cm, and perhaps yet another at 35–40 cm. Whether these smaller prints were made by young individuals of the same species responsible for the larger size classes, or whether they represent different species, is beyond the scope of this chapter. The most important point for now is that the most commonly observed size class is consistent with expectations for the size of *Acrocanthosaurus* footprints.

Pittman (1989) concluded that the pedal phalanges of either *Allosaurus* or *Acrocanthosaurus* could be arranged to fit the toemarks of Comanchean large-theropod footprints. I concur with his interpretation. There is nothing in the shape of these footprints (fig. 28.11) that would preclude *Acrocanthosaurus* as their maker. However, even though the size and shape of Comanchean large-theropod prints are consistent with the hypothesis that *Acrocanthosaurus* was their maker, this does not eliminate the possibility that some or all of them were in fact made by some other large theropod.

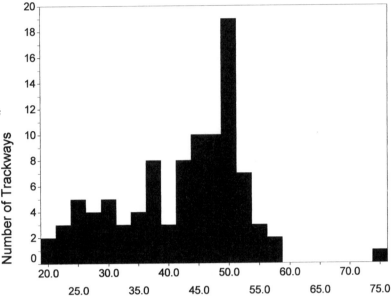

Figure 28.10. Size-frequency
distribution of footprint lengths
of Comanchean tridactyl
dinosaurs, based on trackways
comprising at least two footprints
in sequence. Number of trails =
97.

Mean Print Length in Trail (cm)

Figure 28.11. (opposite page)
Examples of well-preserved
Comanchean tridactyl footprints
attributed to large theropods; see
Pittman (1989) and Hawthorne
(1990) for locality information.
All specimens are casts (negative
copies) of concave epirelief
footprints; as photographed, the
prints come out of the plane of
the page toward the viewer, and
left-right symmetry is reversed
from the actual footprints.
Footprints from the Glen Rose
Limestone: (A) Left footprint
from an unspecified locality,
Hamilton County, Texas; (B) left
footprint from a long trackway,
South San Gabriel River,
Williamson County, Texas;
(C) right footprint CT1, Paluxy
River, Dinosaur Valley State
Park, Somervell County, Texas;
(D) left footprint from a long
trackway, West Verde Creek,
Davenport Ranch, Medina
County, Texas. The distal end of
digit III is damaged in this cast.
Fort Terrett Formation: (E) Left
footprint G46, Middle Copperas
Creek, F[6] Ranch, Kimble County,
Texas.

It is not unusual to find more than one large-theropod species in the same formation, and even in the same fossil quarry. *Gorgosaurus* (or *Albertosaurus*) *libratus* and *Daspletosaurus torosus* co-occur in the Dinosaur Park Formation (Russell 1970), and species of *Allosaurus, Ceratosaurus, Torvosaurus,* and *Saurophaganax* are found together in the Morrison Formation (Foster and Chure 1998, in press; Bilbey 1999; Turner and Peterson 1999). Even so, one species is usually numerically dominant over the others *(Allosaurus fragilis* in the Morrison Formation, and *Gorgosaurus libratus* in the Dinosaur Park Formation).

Acrocanthosaurus atokensis was not the only large theropod in the Early Cretaceous of North America (Harris 1998a), and so it is possible that some or all Comanchean large-theropod footprints were actually made by some other large-theropod species. However, *Acrocanthosaurus* is the only Early Cretaceous large theropod presently known from the region of Texas and Oklahoma, and the fact that it is known from four skeletons suggests that if it was not the only large theropod living in that area at that time, it was probably the most common. Thus *Acrocanthosaurus* is the most likely maker of Comanchean large-theropod footprints.

Discussion

Acrocanthosaurus was one of the largest theropods (Currie and Carpenter in press). From the femoral midshaft circumference and the equation of Anderson et al. (1985), the live mass of NCSM 14345 can

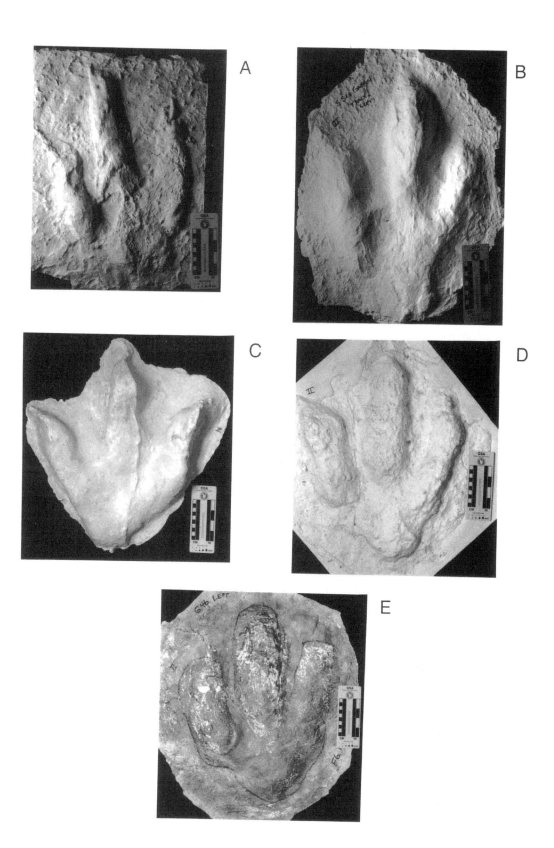

TABLE 28.3
Empirical Relationships between (Y) Home Range Area (km²) and (M) Body Mass (kg) in Extant Predatory Mammals, Birds, and Lizards, with Predictions for a 2400-kg Animal

These predictions should be regarded with considerable caution, because 2400 kg is considerably beyond the range of data used to generate the regression equations. Equations from Peters (1983).

Group	Regression Equation	Predicted Home Range Area for a 2400-kg Animal
Mammals	$Y = 1.39 \times M^{1.37}$	59,400
Birds	$Y = 8.3 \times M^{1.37}$	355,000
Lizards	$Y = 0.12 \times M^{0.95}$	195

be estimated as 2400 kg, and SMU 74646 would have massed about 1900 kg (Currie and Carpenter in press; Harris 1998b). However, this equation may underestimate the body masses of large theropods (Farlow et al. 1995), in which case *Acrocanthosaurus* would have been even heavier.

Big animals require a lot of space, endotherms need more living space than ectotherms, and carnivores need bigger home ranges than herbivores (Peters 1983). Empirical relationships between home range area and body mass in extant predatory mammals, birds, and lizards can be used to speculate about the magnitude of the home range size required by a 2400-kg predator (table 28.3). A minimum estimate, assuming *Acrocanthosaurus* had the space needs of a gargantuan lizard, would put its home range area in the tens or hundreds of square kilometers. If instead the dinosaur had the habitat area requirements expected for a predatory endotherm, a single acrocanthosaur might have had a home range encompassing tens of thousands, or hundreds of thousands, of square kilometers.

From an equation published by Garland (1983), we would expect a hypothetical 2400-kg predatory mammal to move about 21 km/day in foraging and other activities. Adult Komodo dragons (*Varanus komodoensis*), with body masses of about 50 kg, generally move a couple kilometers in a single day, but have been observed to travel as much as 10 km/day (Auffenberg 1981).

It therefore seems reasonable to suppose that individual acrocanthosaurs, being huge carnivores, would have been wide-ranging animals, and that acrocanthosaur populations would have been spread over considerable areas on a landscape basis, and thus across many kinds of habitat. Although Comanchean large-theropod footprints dominate dinosaur footprint assemblages in what Lockley et al. (1994) termed the *Brontopodus* ichnofacies, it would be astonishing if the large-theropod trackmakers had been restricted to such carbonate mudflat situations. If *Acrocanthosaurus* indeed was the Comanchean large-theropod trackmaker, the occurrence of skeletal fossils of this dinosaur in clastic sedimentary rocks provides prima facie support for this conclusion. As argued by Meyer and Pittman (1994) for sauropod footprints, the association of dinosaur footprint types with particular sedimentary facies may often reflect preservational bias as much as real habitat preferences on the part of trackmakers.

Acknowledgments: Given his contributions to the study of theropods and dinosaur footprints, it is a particular pleasure to have this paper included in a volume honoring Phil Currie. Dale Russell provided stimulating discussion about acrocanthosaur paleobiology. Numerous other colleagues made specimens, information, and measurements available to me over the course of this study; among them, Kenneth Carpenter, Dan Chure, Phil Currie, Ronnie Hastings, Mike Hawthorne, Jack Horner, Glen Kuban, Wann Langston, Peter Larson, Martin Lockley, Peggy Maceo, and Jeff Pittman have been particularly helpful. Dave and Margaret Akers and Billy and Pam Baker have frequently afforded hospitality during fieldwork. This research was supported by grants from the National Science Foundation, American Philosophical Society, and Indiana University–Purdue University, Fort Wayne.

References

Anderson, J. F., A. Hall-Martin, and D. A. Russell. 1985. Long-bone circumference and weight in mammals, birds, and dinosaurs. *Journal of Zoology* (London) 207: 53–61.

Auffenberg, W. 1981. *The Behavioral Ecology of the Komodo Monitor.* Gainesville: University Presses of Florida.

Baird, D. 1957. Triassic reptile footprint faunules from Milford, New Jersey. *Bulletin of the Museum of Comparative Zoology, Harvard University* 117: 449–520.

Bakker, R. T., M. Williams, and P. Currie. 1988. *Nanotyrannus,* a new genus of pygmy tyrannosaur, from the latest Cretaceous of Montana. *Hunteria* 1 (5): 1–30.

Bilbey, S. A. 1999. Taphonomy of the Cleveland-Lloyd Dinosaur Quarry in the Morrison Formation, central Utah: A lethal spring-fed pond. In D. D. Gillette (ed.), Vertebrate Paleontology in Utah, pp. 121–133. *Utah Geological Survey Miscellaneous Publication 99–1.*

Carpenter, K. 1990. Variation in *Tyrannosaurus rex.* In K. Carpenter and P. J. Currie (eds.), *Dinosaur Systematics: Perspectives and Approaches,* pp. 141–145. Cambridge: Cambridge University Press.

Carpenter, K. 1997. Tyrannosauridae. In P. J. Currie and K. Padian (eds.), *Encyclopedia of Dinosaurs,* pp. 766–768. San Diego: Academic Press.

Carr, T. D. 1999. Craniofacial ontogeny in Tyrannosauridae (Dinosauria, Coelurosauria). *Journal of Vertebrate Paleontology* 19: 497–520.

Colbert, E. H. 1990. Variation in *Coelophysis bauri.* In K. Carpenter and P. J. Currie (eds.), *Dinosaur Systematics: Perspectives and Approaches,* pp. 81–90. Cambridge: Cambridge University Press.

Currie, P. J., and K. Carpenter, in press. A new specimen of *Acrocanthosaurus atokensis* from the Lower Cretaceous Antlers Formation (Lower Cretaceous, Aptian) of Oklahoma, USA. *Bulletin du Muséum National d'Histoire Naturelle* (Paris).

Farlow, J. O. 1981. Estimates of dinosaur speeds from a new trackway site in Texas. *Nature* 294: 747–748.

Farlow, J. O. 1987. *Lower Cretaceous Dinosaur Tracks, Paluxy River Valley, Texas.* Waco: South-Central Section, Geological Society of America, Baylor University.

Farlow, J. O., and J. M. Hawthorne. 1989. Comanchean dinosaur footprints. In D. A. Winkler, P. A. Murry, and L. L. Jacobs (eds.), *Field Guide to the Vertebrate Paleontology of the Trinity Group, Lower Cretaceous of Texas,* pp. 23–30. Dallas: Institute for the Study of Earth and Man, Southern Methodist University.

Farlow, J. O., and M. G. Lockley. 1993. An osteometric approach to the identification of the makers of early Mesozoic tridactyl dinosaur footprints. In S. G. Lucas and M. Morales (eds.), *The Nonmarine Triassic,* pp. 123–131. Albuquerque: New Mexico Museum of Natural History and Science Bulletin 3.

Farlow, J. O., M. G. Smith, and J. M. Robinson. 1995. Body mass, bone "strength indicator," and cursorial potential of *Tyrannosaurus rex. Journal of Vertebrate Paleontology* 15: 713–725.

Foster, J. R., and D. J. Chure. 1998. Patterns of theropod diversity and distribution in the Late Jurassic Morrison Formation, western U.S.A. *Abstracts and Program, Fifth International Symposium on the Jurassic System, International Union of Geological Sciences Subcommission on Jurassic Stratigraphy,* Vancouver, British Columbia, pp. 30–31.

Foster, J. R., and D. J. Chure, in press. An ilium of a juvenile *Stokesosaurus* (Dinosauria, Theropoda) from the Morrison Formation (Upper Jurassic: Kimmeridgian), Mead County, South Dakota. *Brigham Young University Geology Studies.*

Garland, T., Jr. 1983. Scaling the ecological cost of transport to body mass in terrestrial mammals. *American Naturalist* 121: 571–587.

Harris, J. D. 1998a. Large, Early Cretaceous theropods in North America. In S. G. Lucas, J. I. Kirkland, and J. W. Estep (eds.), *Lower and Middle Cretaceous Terrestrial Ecosystems,* pp. 225–228. Albuquerque: New Mexico Museum of Natural History and Science Bulletin 14.

Harris, J. D. 1998b. *A Reanalysis of* Acrocanthosaurus atokensis, *Its Phylogenetic Status, and Paleobiogeographic Implications, Based on a New Specimen from Texas.* Albuquerque: New Mexico Museum of Natural History and Science Bulletin 13.

Hawthorne, J. M. 1990. *Dinosaur Track-Bearing Strata of the Lampasas Cut Plain and Edwards Plateau, Texas.* Waco: Baylor Geological Studies Bulletin 49.

Holtz, T. R., Jr. 1994. The phylogenetic position of the Tyrannosauridae: Implications for theropod systematics. *Journal of Paleontology* 68: 1100–1117.

Langston, W., Jr. 1974. Nonmammalian Comanchean tetrapods. *Geoscience and Man* 8: 77–102.

Leonardi, G. (ed.). 1987. *Glossary and Manual of Tetrapod Footprint Palaeoichnology.* Brasília: Republica Federativa do Brasil, Ministério das Minas e Energia, Departamento Nacional da Produção Mineral.

Lockley, M. G. 1998. The vertebrate track record. *Nature* 396: 429–432.

Lockley, M. G., and A. P. Hunt. 1994. A track of the giant theropod dinosaur *Tyrannosaurus* from close to the Cretaceous/Tertiary boundary, northern New Mexico. *Ichnos* 3: 213–218.

Lockley, M. G., A. P. Hunt, and C. A. Meyer. 1994. Vertebrate tracks and the ichnofacies concept: implications for palaeoecology and palichnostratigraphy. In S. Donovan (ed.), *Paleobiology of Trace Fossils,* pp. 241–268. New York: Wiley.

Lockley, M. G., C. A. Meyer, and V. F. dos Santos. 1996. *Megalosauripus, Megalosauropus,* and the concept of megalosaur footprints. In M.

Morales (ed.), *The Continental Jurassic,* pp. 113–118. Flagstaff: Museum of Northern Arizona Bulletin 60.

Lockley, M. G., B. D. Ritts, and G. Leonardi. 1999. Mammal track assemblages from the Early Tertiary of China, Peru, Europe, and North America. *Palaios* 14: 398–404.

Meyer, C. A., and J. G. Pittman. 1994. A comparison between the *Brontopodus* ichnofacies of Portugal, Switzerland, and Texas. *Gaia* 10: 125–133.

Norušis, M. J. 1988. *SPSS-X Advanced Statistics Guide.* Chicago: SPSS.

Olsen, P. E., J. B. Smith, and N. G. McDonald. 1998. Type material of the type species of the classic theropod footprint genera *Eubrontes, Anchisauripus,* and *Grallator* (Early Jurassic, Hartford and Deerfield Basins, Connecticut and Massachusetts, U.S.A.). *Journal of Vertebrate Paleontology* 18: 586–601.

Padian, K., J. R. Hutchinson, and T. R. Holtz Jr. 1999. Phylogenetic definitions and nomenclature of the major taxonomic categories of the carnivorous Dinosauria (Theropoda). *Journal of Vertebrate Paleontology* 19: 69–80.

Peters, R. H. 1983. *The Ecological Implications of Body Size.* Cambridge: Cambridge University Press.

Pittman, J. G. 1989. Stratigraphy, lithology, depositional environment, and track type of dinosaur track-bearing beds of the Gulf Coastal Plain. In D. D. Gillette and M. G. Lockley (eds.), *Dinosaur Tracks and Traces,* pp. 135–153. Cambridge: Cambridge University Press.

Raath, M. A. 1990. Morphological variation in small theropods and its meaning in systematics: Evidence from *Syntarsus rhodesiensis.* In K. Carpenter and P. J. Currie (eds.), *Dinosaur Systematics: Perspectives and Approaches,* pp. 91–105. Cambridge: Cambridge University Press.

Russell, D. A. 1970. *Tyrannosaurs from the Late Cretaceous of Western Canada.* Ottawa: National Museum of Natural Sciences, National Museums of Canada, Publications in Palaeontology 1.

Sereno, P. C. 1999. The evolution of dinosaurs. *Science* 284: 2137–2147.

Shuler, E. W. 1917. Dinosaur tracks in the Glen Rose Limestone near Glen Rose, Texas. *American Journal of Science* 44: 294–298.

Shuler, E. W. 1935. Dinosaur track mounted in the bandstand at Glen Rose, Texas. *Field and Laboratory* 4: 9–13.

Shuler, E. W. 1937. Dinosaur tracks at the fourth crossing of the Paluxy River near Glen Rose, Texas. *Field and Laboratory* 5: 33–36.

Smith, D. K. 1998. A morphometric analysis of *Allosaurus. Journal of Vertebrate Paleontology* 18: 126–142.

Smith, D. K. 1999. Patterns of size-related variation within *Allosaurus. Journal of Vertebrate Paleontology* 19: 402–403.

Stovall, J. W., and W. Langston Jr. 1950. *Acrocanthosaurus atokensis,* a new genus and species of Lower Cretaceous Theropoda from Oklahoma. *American Midland Naturalist* 43: 696–728.

Turner, C. E., and F. Peterson. 1999. Biostratigraphy of dinosaurs in the Upper Jurassic Morrison Formation of the western interior, U.S.A. In D. D. Gillette (ed.), Vertebrate Paleontology in Utah, pp. 77–114. *Utah Geological Survey Miscellaneous Publication* 99–1.

29. Trackways of Large Quadrupedal Ornithopods from the Cretaceous: A Review

MARTIN G. LOCKLEY AND
JOANNA L. WRIGHT

Abstract

Trackways of large quadrupedal ornithopods attributed to iguanodontids and hadrosaurs have been reported from at least a dozen localities in the Lower and Upper Cretaceous of Europe, North and South America, and eastern Asia. Although trackways attributed to ornithopods are known from the Jurassic, they represent small, gracile animals that were mainly bipedal. In contrast, Cretaceous ornithopod tracks are often large and frequently indicate quadrupedal animals with well-padded hind feet and small forefeet. The maximum size of these tracks tends to be larger through time, and the heel pad of later, larger forms tends to be broader. Manus impression shape varies from subtriangular or crescentic to subrounded. although these different morphologies do not seem to have any direct correlation to pes morphology. Manus emplacement occurs along an arc from lateral to anterior of the pes impressions, although they are generally placed anterolaterally. *Amblydactylus, Caririchnium,* and *Iguanodontipus* are valid ornithopodan ichnotaxa and *Camptosaurichnus, Hadrosaurichnoides, Hadrosaurichnus,* and *Iguanodonichnus* are not considered to represent the tracks of ornithopods.

Introduction

Trackways of quadrupedal ornithopods were first reported by Norman (1980) from England, although Newman (1990) and others inferred a theropod trackmaker for these tracks (see Wright 1999 and Lockley 1987 for discussion). Partly as a result of the track evidence, Norman (1980) inferred that large iguanodontids, such as *Iguanodon bernissartensis,* may have been at least facultatively quadrupedal. This view, which is now accepted as correct, was at the time somewhat radical, given that the large, historically famous ornithopods *Iguanodon* and *Hadrosaurus* had, for more than a century, been interpreted as bipeds.

In the 1980s trackways of large quadrupedal ornithopods were reported from the Cretaceous of British Columbia (Currie 1983, 1995), Brazil (Leonardi 1984), Colorado (Lockley 1987), and Texas (Pittman 1989). The Brazilian trackways were at first identified as stegosaurian. The Texas tracks have since been shown to be sauropodan (Lockley et al. 1994). Further reports in the 1990s indicated the presence of trackways of quadrupedal ornithopods in Alberta (Currie et al. 1991), Spain (Moratalla et al. 1992, 1993; Pérez-Lorente et al. 1997), New Mexico (Lockley and Hunt 1995), and China (You and Azuma 1995). Subsequently, trackways of quadrupedal ornithopods have been identified at sites in Germany and South Dakota (described below). The new trackways range in age from Berriasian to Maastrichtian. This survey of quadrupedal ornithopod trackways summarizes some recent discoveries as well as correcting previously published errors and omissions. All trackway specimens except those from Brazil and China have been examined by one or both authors.

Trackways from England and Germany

A pair of trackways from the Early Cretaceous (early Berriasian) part of the Purbeck Limestone Group of England (Wright 1999) and four specimens from the Wealden Group (mid-late Berriasian) at the famous Münchehagen dinosaur tracksite, near Hannover (Fischer 1998) are among the oldest and best-dated Early Cretaceous trackways of quadrupedal ornithopods. The English specimen has had a checkered career, having been incorrectly interpreted as a single trackway or as two theropod trackways, and as a theropod and an iguanodontid trackway side by side (see Lockley 1991 and Wright 1999 and references therein). We follow Wright's (1999) interpretation of the trackways as those of two quadrupedal ornithopods. The trackway on the left in figure 29.1A is poorly preserved, with respect to the outline of the pes, but shows manus tracks associated with many of the 26 pes tracks. (Pes tracks are connected by linear marks that are subparallel to the trackway axis and may be interpreted as toe or tail drag marks or even as a later water-erosion feature). By contrast, the trackway on the right (also in fig 29.1A) has clearer tridactyl pes outlines but only one manus

Figure 29.1. (A) two parallel iguanodontid trackways from the Purbeck Limestone Formation, England. All manus impressions belong to the left-hand trackway apart from the one marked m. (Redrawn from Wright 1999.) (B) three trackway segments from the Bückeburg Formation. Both samples are Berriasian in age. Scale: 1 m.

track. The British trackways have the distinction of being the longest sequences with associated manus tracks. These trackways also show that iguanodontids sometimes placed their mani well outside their pedes when walking, and that there was slight inward rotation of the axis of pes digit III.

Sarjeant et al. (1998) recently assigned British ornithopod (*Iguanodon*?) tracks from the same general geographic and stratigraphic location to the ichnospecies *Iguanodontipus burreyi*, and noted that "should manual impressions be discovered" in association with this ichnospecies "the combined ichnogeneric and ichnospecific diagnosis will need to be augmented" (Sarjeant et al. 1998, 199). Because the specimen with manus tracks (fig. 29.1A) was not included on the synonymy list for *Iguanodontipus burreyi*, the initial implication is that it does not represent the same ichnospecies. However, given that it is considered of iguanodontid affinity it might be referable to *Iguanodontipus* sp. Certainly we can see no compelling reason to suggest that it should be assigned to a new or existing ichnogenus.

The German manus tracks were discovered by us in May 1999 while studying the tracksite at Münchehagen, Germany. The site is an old quarry where a single large surface in the Bückeburg Formation (Wealden subdivision 3, sensu Fischer 1998; mid-late Berriasian) is exposed. Many of the in situ trackways at this site are attributed to sauropods, but a number of track-bearing slabs, recovered during excavation of overlying beds, reveal isolated ornithopod track casts and occasional short trackway segments. Among some 20 slabs examined, we recorded four with manus-pes sets (three are shown in Fig. 29.1B). Two specimens have consecutive pes tracks and thus the pace length could be measured. Both trackways show strong inward rotation of the axis of pes digit III. In one specimen the manus was placed in front of pes digit IV, but the other three are similar to the British specimens, where the manus fell outside pes digit IV. The shape of the manus in both the German and British specimens is subtriangular (cf. Wright 1999, fig. 5).

Trackways from Spain

Trackways of quadrupedal ornithopods are known from three sites in Spain (Moratalla et al. 1992, 1993; Pérez-Lorente et al. 1997; Lockley and Meyer 1999). The oldest, from the Cerradicas locality, is preserved in the Villar del Arzobispo Formation and is assigned a lower-middle Berriasian age (Pérez-Lorente et al. 1997). Based on measurements provided by these authors, these tracks are relatively diminutive (pes length and width 23 cm) in comparison to the Berriasian tracks from England and Germany, which are approximately 27–35 cm in length. No information is given on the shape of the manus other than line drawings that show that they were oval and are consistently placed close to the anterior extremity of pes digit III. The trackway shows inward rotation of the axis of pes digit III. Pérez-Lorente et al. (1997) suggest that the tracks could have been made by ornithopods such as *Camptosaurus dispar* or *Iguanodon atherfieldensis*.

Trackways from Regumiel de la Sierra (fig. 29.2B) and Cabezón de Cameros (fig. 29.2C) are considerably larger than those from Cerradicas, with mean pes lengths of 45 and 58.4 cm respectively (Moratalla et al. 1993, 1992). The age of these tracksites is not well constrained. The age of the Regumiel de la Sierra site is given as Neocomian but described as "very debatable" (Moratalla et al. 1993, 1). The age of the Cabezón de Cameros site is somewhat more confidently placed as "probably Hauterivian" (Moratalla et al. 1992, 150). Such data lead to the tentative inference that the maximum size of ornithopod tracks

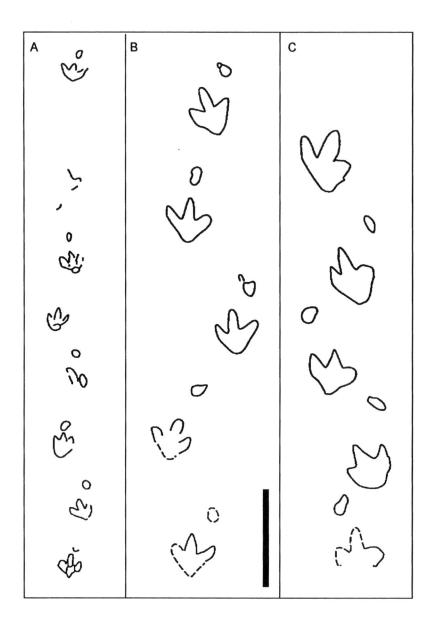

Figure 29.2. Three trackways from the Lower Cretaceous of Spain. (A) Berriasian trackway from Cerradicas, (B) trackway from Regumiel de la Sierra, (C) trackway from Cabezón de Cameros. (B) and (C) have been attributed to iguanodontids (cf. Iguanodontipus). Scale: 1 m. (After Lockley and Meyer 1999.)

increases through time during the Lower Cretaceous, as also noted in the British succession (Wright 1996). The usual placement of the manus in these trackways (fig. 29.2B,C) is anterior to the apex of digit IV. There is, however, a slight asymmetry to the Regumiel de la Sierra trackway, where the left manus is placed anterior to the apex of digit III (fig. 29.2 B). The reason for this is not known; the arrangement of the pes impressions shows no asymmetry. Both trackways show strong inward rotation of the axis of pes digit III.

Sarjeant et al. (1998) referred the Regumiel de la Sierra and the Cabezón de Cameros trackways to *Iguanodontipus burreyi* although we are uncertain whether enough morphological information can be extracted from the British and Spanish material to substantiate this ichnospecies comparison. Further discussion of the ichnotaxonomy of Spanish ornithopod tracks proposed by Moratalla et al. (1993) is summarized by Lockley and Meyer (1999).

Trackways from Brazil

Leonardi (1984) reported the trackway of a quadrupedal dinosaur from the Lower Cretaceous Antenor Navarro Formation, Paraiba state, Brazil, which he named *Caririchnium magnificum* (fig. 29.3A). Al-

Figure 29.3. (A) Caririchnium magnificum *from Brazil (after Leonardi 1984); (B) unnamed trackway from China (after You and Azuma 1995). Scale: 1 m.*

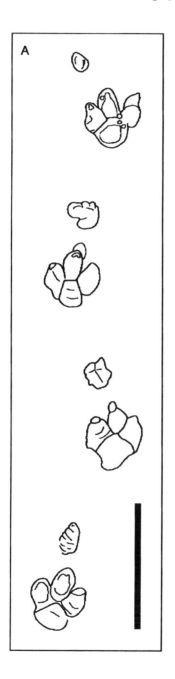

though he initially attributed it to a stegosaur (Leonardi 1984; Lockley 1987), he subsequently reinterpreted it as the trackway of an ornithopod (Leonardi 1994). The pes is large (about 50 cm long), with the suboval manus situated more or less anterior to the apex of pes digit III. The pes had a quadripartite configuration with a rounded heel pad, and there is inward rotation of the axis of pes digit III. This is the only report of a trackway of a quadrupedal ornithopod from South America.

Trackways from China

You and Azuma (1995) described two trackways from the Early Cretaceous Xiguayuan Formation of Hebei province, and suggested that among potential trackmakers "iguanodonts [would] be the most reasonable candidate" (You and Azuma 1995, 155). We agree with this interpretation. The trackways represent large animals (foot length 47–50 cm; foot width 45 cm). One trackway (fig. 29.3B) shows three consecutive manus-pes sets with manus impressions anterior or lateral to the apex of pes digit IV. The second trackway only shows one manus-pes set with the manus in a similar position. The manus appears to be subcrescentic and anterolaterally convex and posteromedially concave. Both trackways show inward rotation of the axis of pes digit III.

Trackways from the Lower Cretaceous of North America

The first report of trackways of quadrupedal ornithopods from North America was published by Currie (1983), who described several trackways of Aptian age from the Gething Formation of British Columbia (see also Lockley 1989; Pérez-Lorente et al. 1997). These trackways have been assigned to the ichnogenus *Amblydactylus*. Illustrations of representative trackways with clear manus and pes impressions (Currie 1983; Pérez-Lorente et al. 1997; see fig. 29.4A) indicate large pes tracks (typically >50 cm long) and a manus that is subcrescentic and situated anterior or lateral to the apex of pes digit III. Trackways show inward rotation of the axis of pes digit III.

Trackways of quadrupedal ornithopods are now known from the Lakota Formation of probable Barremian age, near Rapid City, South Dakota (Lockley et al., chap. 30 of this volume). At least five trackway segments with manus impressions are known from this locality. The most complete (fig. 29.4B) reveals four consecutive manus-pes sets with a quadripartite pes about 35 cm long and suboval to subtriangular manus that is situated anterior to the apex of pes digits III and IV. Other trackways reveal the manus situated more laterally. Trackways show inward rotation of the axis of pes digit III.

Trackways from the Upper Cretaceous of North America

The first report of trackways of quadrupedal ornithopods from the Late Cretaceous of North America was published by Lockley (1985,

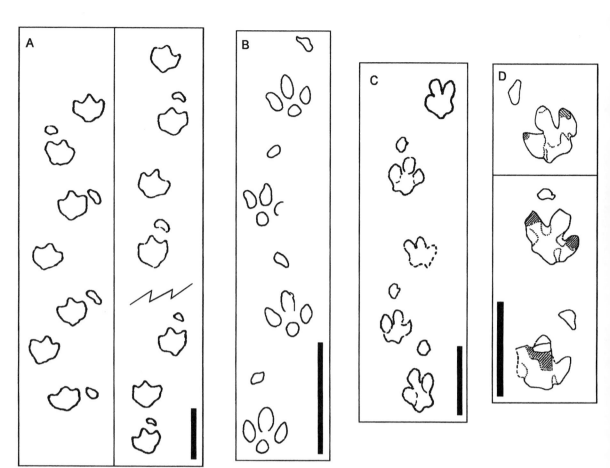

Figure 29.4. (A), (B) Lower Cretaceous trackways from British Columbia and South Dakota, respectively; (C) Caririchnium leonardii *from the Cenomanian of Colorado; (D) hadrosaur tracks from the Maastrichtian of Canada. Scale: 1 m. (29.4A after Currie 1983; 29.4B after Lockley et al., this volume; 29.4C, D after Currie et al. 1991).*

1987), who described several trackways of early Cenomanian age from the Dakota Group of Colorado (fig. 29.4C). The trackways were first reported from the Alameda Parkway locality (now known as Dinosaur Ridge, Lockley and Hunt 1995), and a locality near Lamar (Lockley 1987). Subsequent studies of this stratigraphic unit have revealed trackways of quadrupedal ornithopods at several other sites: Eldorado Springs, Turkey Creek, Roxborough State Park, and Apishapa (all in Colorado), and Clayton Lake State Park and Mosquero Creek in New Mexico (Lockley et al. 1992; Lockley and Hunt 1995). Trackways of quadrupedal ornithopods are now known from at least eight different geographic localities within the Dakota Group stratigraphic complex, which has been designated a "megatracksite" (Lockley and Hunt 1995). Some sites reveal dozens of trackways of quadrupedal ornithopods, so collectively the Dakota Group reveals by far the largest sample currently known.

Almost all the trackways of quadrupedal ornithopods from the Dakota reveal a suboval manus placed relatively close to the midline (i.e., close to the apex of digits III and IV). Many manus impressions show a medial protuberance that may have been made by manus digit II. Because of similarities with the Brazilian material, most of the Dakota Group tracks have been assigned to the ichnospecies *Caririchnium leonardii* (Lockley 1987), although ornithopod tracks from one site were assigned to the ichnogenus *Amblydactylus* (Lucas et al. 1989). Despite the large sample and considerable size range (pes 20–50 cm approx.), no manus tracks have been found situated outside the margins of pes digit IV. The only exception is the trackway of a "limping" individual (Lockley and Hunt 1995, fig. 5.22) in which the left manus is situated slightly lateral to pes digit IV. This asymmetric configuration can be compared to the Regumiel de la Sierra trackway (fig. 29.2B). All trackways show inward rotation of the axis of pes digit III.

The final example of trackways of quadrupedal ornithopods from the Upper Cretaceous was reported by Currie et al. (1991) from the St. Mary River Formation (Maastrichtian) of Alberta (fig. 29.4D). These tracks are preserved as natural casts with skin impressions (skin impressions have also been noted in natural impressions from the Turkey Creek site). The pes tracks are large (55 cm long by 60 cm wide) and the manus tracks are described as "semilunate" (Currie et al. 1991, 110); they are very similar in outline to the Berriasian tracks from England (fig. 29.1A) and the Barremian tracks from South Dakota (fig. 29.4B). The manus tracks are located anterior to the apex of digits III and IV, and trackways show inward rotation of the axis of pes digit III. Pes tracks show a very broad and flattened posterior margin to the heel pad which is slightly anteriorly concave giving the posterior margin of the heel a somewhat bilobed appearance.

Discussion

The only known tracks of quadrupedal ornithopods not discussed above are those assigned to the Lower Jurassic ichnogenus *Anomoepus* (Hitchcock 1848; Lull 1953) from North America and the related ichnogenus *Moyenosauripus* from southern Africa (Ellenberger 1972). Both these ichnogenera occur in other regions at this time, but have not been positively identified from younger deposits. Both ichnogenera are also relatively small in relation to the Cretaceous trackways described, and have pes digits that are segmented, and often display a clear hallux (digit I) impression. In addition, they often display manus tracks that are clearly pentadactyl, or at least tetradactyl when not fully impressed. They are therefore fundamentally different from the trackways of large Cretaceous quadrupedal ornithopods.

Few other tracks from the Jurassic have been assigned with any confidence to ornithopods. One possible exception is the ichnogenus *Dinehichnus* from the Upper Jurassic of Utah (Lockley et al. 1998), which has a distinctive quadripartite pes morphology reminiscent of certain larger Cretaceous ornithopod tracks (see fig 29.4B). They are generally small (less than 20 cm long) except for one trackway in which the pes measures 28 cm in length. *Dinehichnus* also lacks well-defined digital pads, and has been tentatively attributed to a dryosaurid (Lockley et al. 1998), whereas similar quadripartite tracks from the Lower

Cretaceous of Europe have been attributed to hypsilophodontids such as *Hypsilophodon* (Aguirrezabala et al. 1985). None of these trackways show evidence of quadrupedal progression and so only the pes morphology can be compared with large Cretaceous ornithopod trackways. *Dinehichnus* and purported hypsilophodontid tracks represent mainly bipedal species far more gracile than any inferred from the fleshy, robust quadruped footprints described above.

Thus the Jurassic ichnogenera *Anomoepus, Moyenosauripus,* and *Dinehichnus* were produced by species (presumably ornithopods) significantly different from the trackmakers of *Iguanodontipus, Caririchnium,* and *Amblydactylus*. These differences include the fleshy nature of the single, broad pads on the heel and on each digit of the pes, and the equally fleshy nature of the manus. Subtle differences in pes track morphology include the degree to which heel and digit pads appear to be coalesced, and the extent to which the posterior margin of the heel is rounded or bilobed. There is some indication that the latter morphology only appears in the Late Cretaceous (Cenomanian-Maastrichtian) ornithopods, and so could be a feature characteristic of hadrosaurs rather than iguanodontids. There is some indication, however, that Late Cretaceous ornithopod (presumably hadrosaurid) tracks may have both rounded and bilobed heels (cf. Langston 1960).

Manus tracks are subtriangular, semilunate, or oval. All manus tracks tend to have their long axis oriented anteromedially to posterolaterally. Subtriangular tracks, such as those reported from England, Germany (fig. 29.1), and South Dakota, are frequently concave along the posteromedial margins and have a pronounced bump or protuberance on the anterolateral margin. This concave posteromedial margin is also evident in the semilunate manus tracks from the Lower Cretaceous of China (fig. 29.3B) and from the Lower and Upper Cretaceous of Canada (fig. 29.4A,D). In contrast, tracks from two of the Spanish localities (fig. 29.2B, C), Brazil (fig. 29.3A) and Colorado–New Mexico, are oval and without a concave posterolateral margin. In manus impressions from all three of these areas, however, there are signs of some degree of development of an anteromedial protuberance.

The position of the manus relative to the pes varies, occupying a moderately wide arc from a position lateral to digit IV to one in front of digit III (fig. 29.5). We produced these diagrams by superimposing tracings and outlines of manus-pes sets (all corrected to the left side), using as our reference point the midline of the more or less symmetrical pes. Our results are presented in approximate stratigraphic order so as to show all examples at the same scale.

Comparison of Berriasian trackways from England and Germany shows that the manus in the English trackways is situated far from the midline. When the German tracks are compared with those of comparable size from the Barremian of South Dakota, the manus tracks of the latter group are situated in a more anterior position. In most other examples cited we have dealt with single trackways or small samples (e.g., two trackways), so we have plotted the relative position of manus and pes tracks in the same trackway. This had the unexpected benefit of highlighting asymmetric trackways. For example, the small trackway

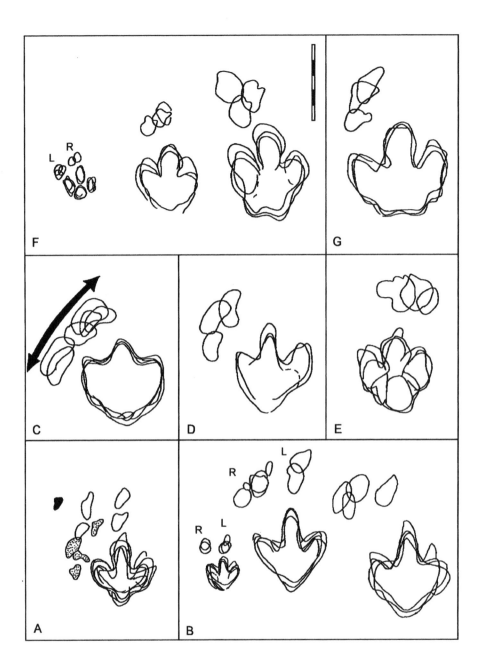

from Cerradicas and the larger one from Regumiel de la Sierra (both in Spain) and the small trackway from Colorado (figs. 29.5B left and center, 29.5F left) all show a clear separation of left and right footprints.

These comparisons also give some indication as to the variation in size and morphology of quadrupedal ornithopod trackways, although, as is evident from the Colorado sample, there is considerable intrasample variation in large samples. It is outside the scope of this chapter to do more than point to certain features that need further consideration:

• the possibility of systematic increases in the maximum size of tracks during the Lower and Upper Cretaceous, possibly reflecting reiterations of evolutionary size increase in both iguanodontids and hadrosaurs;

• the possibility of the evolution of a broad bilobed heel in Late Cretaceous ornithopods (possibly accentuated in larger individuals in any given population; see, for example, fig. 29.5F);

• variation between tracks that have a quadripartite configuration and those in which the digit and heel pads are coalesced (possibly a function of both size and preservation);

• variation in manus shape from subtriangular to semilunate or oval, and the relationship of this variation to pes morphology, age of sample, size of individuals, preservational context, etc.

There are several other features that need to be considered. One is the consistency of the breadth (foot length usually as wide or wider than long), the relatively short pace length, and the inward rotation of the pes in most if not all trackways. Is this a feature of the trackways of large quadrupedal ornithopods, of all large ornithopods, or of ornithopods in general? Are such features useful in distinguishing the trackways of ornithopods from those of theropods, which generally have more elongate footprints and longer pace lengths, with little or no inward rotation of the pes impressions? As indicated below, such distinctive characteristics are important in identifying ornithopod trackways correctly, and ultimately in our choice of ichnotaxonomic names.

Systematic Note

Caution should be exercised in the naming of tracks, especially when good material is not available. As summarized by Sarjeant (1989), one of the obvious recommendations is that tracks should not be named without adequate reference to existing literature. We stress this point because several purported ornithopod tracks have been named without adequate study and attention. For example, Casamiquela and Fasola (1968) named *Iguanodonichnus* for a purported Cretaceous ornithopod trackway from Chile, which turned out to be that of a late Jurassic sauropod (Sarjeant et al. 1998). They also named *Camptosaurichnus*, even though *Camptosaurus* is not known from South America. Similarly, Alonso (1980) named *Hadrosaurichnus* for tracks from Argentina that in all probability are theropod tracks, as indicated by their elongate shape and long step. Casanovas Cladellas et al. (1993) introduced the name *Hadrosaurichnoides* for Lower Cretaceous tracks that are purported to have webbed feet. These tracks may also be of theropod origin because they are longer than wide and the trackway is narrow.

Given the dubious status of the ichnogenera *Iguanodonichnus*, *Camptosaurichnus*, *Hadrosaurichnus*, and *Hadrosaurichnoides*, we can exclude these ichnogenera from this discussion of quadrupedal ornithopods. The ichnogenera *Iguanodontipus*, *Amblydactylus*, and *Caririchnium* have the distinction of representing tracks that really are attributable to large ornithopods. As knowledge increases, we hope that any ichnotaxonomic revisions will be based on well-preserved

Figure 29.5. (opposite page) Position of manus tracks in relation to pes, all corrected to left side, in approximate stratigraphic order. (A)–(E) Lower Cretaceous; (F), (G) Upper Cretaceous. (A) shows English sample (black) and German sample (stippled), both from Berriasian with South Dakota Barremian sample (white): all tracks represent separate trackways; (B)–(F): each composite is based on an individual trackway (G= two trackways). (B): from left to right, three Lower Cretaceous trackways from Spain (corresponding to trackways A–C in fig. 29.2); note that left and right manus are clearly separated, in two cases indicating asymmetric trackways. (C) Gething Fm., Canada, with black arrow showing typical arc of movement; (D) China and (E) Brazil; (F) three sizes of (Caririchnium) trackway from Colorado (left is a "limper" with left and right manus clearly separated); (G) hadrosaur tracks, St. Mary River Fm., based on two trackways. All scales 50 cm.

samples that are adequately described with respect to both pes and (where appropriate) manus tracks.

References

Aguirrezabala, L. M., J. A. Torres, and L. I. Viera. 1985. El Weald de Igea (Cameros–La Rioja): Sedimentología, bioestrategrafía, y paleoichnología de grandes reptiles (dinosaurios). *Munibe* (Sociedad de Ciencias Naturales Aranzadi, San Sebastián) 32: 257–79.

Alonso, R. 1980. Icnitas de dinosaurios (Ornithopoda, Hadrosauridae) en el Cretácico superior del norte Argentina. *Acta geologica lilloana* 15: 55–63.

Casamiquela, R. M., and A. Fasola. 1968. Sobre pisadas de dinosaurios del Cretácico Inferior de Colchagua (Chile). *Publicaciones Departamento de Geología, Chile Universidad* 30: 1–24.

Casanovas Cladellas, M. L., R. Ezquerra Miguel, A. Fernández Ortega, F. Pérez-Lorente, J. V. Santafé Llopis, and F. Torcida Fernández. 1993. Tracks of a herd of webbed ornithopods and other footprints found in the same site (Igea, la Rioja, Spain). *Révue de paléobiologie spécial* 7: 29–36

Currie, P. J. 1983. Hadrosaur trackways from the Lower Cretaceous of Canada. *Acta paleontologica polonica* 28: 63–73

Currie, P. J. 1995. Ornithopod trackways from the Lower Cretaceous of Canada. In W. A. S. Sarjeant (ed.), *Vertebrate Fossils and the Evolution of Scientific Concepts*, pp. 431–443. Reading, England: Gordon and Breach.

Currie, P. J., G. Nadon, and M. G. Lockley. 1991. Dinosaur Footprints with Skin Impressions from the Cretaceous of Alberta and Colorado. *Canadian Journal of Earth Sciences* 28: 102–115.

Ellenberger, P. 1972. Contribution a la classification des piste de vertebres du Trias: Les types du Stormberg d'Afrique du Sud (1). *Palaeovertebrata, Mémoire extraordinaire,* 117: 1–30.

Fischer, R. 1998. Die Saurierfährten im Naturdenkmal Münchehagen. *Mitteilungen aus dem Institut für Geologie und Paläontologie der Universität Hannover* 37: 3–59.

Hitchcock, E. 1848. An attempt to discriminate and describe the animals that made the fossil footmarks of the United States, and especially of New England. *Transactions of the American Academy of Arts and Sciences,* n.s., 3: 129–256.

Langston, W., Jr. 1960. A hadrosaur ichnite. *National Museum Canada Natural History Papers* 4: 1–9.

Leonardi, G. 1984. Le impronte di dinosauri. In J. F. Bonaparte et al. (eds.), *Sulle orme dei dinosauri,* pp. 333. Venice: Erizzo.

Leonardi, G. 1994. *Annotated Atlas of South America Tetrapod Footprints (Devonian to Holocene).* Brasília: Companhia de Pesquisa de Recursos Minerais.

Lockley, M. G. 1985. Vanishing tracks along Alameda Parkway: Implications for Cretaceous dinosaurian paleobiology from the Dakota Group, Colorado. In C. D. Chamberlain, E. G. Kauffman, L. M. W. Kiteley, and M. G. Lockley (eds.), *A Field Guide to Environments of Deposition (and Trace Fossils) of Cretaceous Sandstones of the Western Interior,* pp. 3.131–142. Denver: Midyear Meeting Field Guides.

Lockley, M. G. 1987. Dinosaur Footprints from the Dakota Group of Eastern Colorado. *Mountain Geologist* 24: 107–122.

Lockley, M. G. 1989. Tracks and traces: New perspectives on dinosaurian behavior, ecology, and biogeography. In K. Padian and D. J. Chure (eds.), *The Age of Dinosaurs. Paleontological Society, Short courses in Paleontology* 2: 134–145.

Lockley, M. G. 1991. *Tracking Dinosaurs: A New Look at an Ancient World*. Cambridge: Cambridge University Press.

Lockley, M. G., and A. P. Hunt. 1995. *Dinosaur Tracks and Other Fossil Footprints of the Western United States*. New York: Columbia University Press.

Lockley, M. G., and C. A. Meyer. 1999. *Dinosaur Tracks and Other Fossil Footprints of Europe*. New York: Columbia University Press.

Lockley, M. G., J. Holbrook, A. P. Hunt, M. Matsukawa, and C. Meyer. 1992. The dinosaur freeway: A preliminary report on the Cretaceous megatracksite, Dakota Group, Rocky Mountain Front Range and High Plains; Colorado, Oklahoma, and New Mexico. In R. Flores (ed.), *Mesozoic of the Western Interior*, pp. 39–54. SEPM Midyear Meeting Fieldtrip Guidebook.

Lockley, M. G., J. G. Pittman, C. A. Meyer, and V. F. Santos. 1994. On the common occurrence of manus-dominated sauropod trackways in Mesozoic carbonates. *Gaia: Revista de Geociencias, Museu Nacional de Historia Natural* 10: 119–124.

Lockley, M. G., V. F. Santos, C. A. Meyer, and A. P. Hunt. 1998. A new dinosaur tracksite in the Morrison Formation, Boundary Butte, Southeastern Utah. In K. Carpenter, D. Chure, and J. Kirkland (eds.), The Upper Jurassic Morrison Formation: An interdisciplinary study. *Modern Geology* 23: 317–330.

Lucas, S. G., A. P. Hunt, and K. K. Kietze. 1989. Stratigraphy and age of Cretaceous dinosaur footprints in northeastern New Mexico and northwestern Oklahoma. In D. D. Gillette and M. G. Lockley (eds.), *Dinosaur Tracks and Traces*, pp. 217–221. Cambridge: Cambridge University Press.

Lull, R. S. 1953. Triassic life of the Connecticut Valley. *Bulletin of the Connecticut State Geological and Natural History Survey* 181: 1–331.

Moratalla, J. J., J. L. Sanz, and S. Jiménez. 1993. Dinosaur Tracks from the Lower Cretaceous of Regumiel de la Sierra (province of Burgos, Spain): Inferences on a new quadrupedal ornithopod trackway. *Ichnos* 2: 1–9.

Moratalla, J. J., J. L. Sanz, S. Jiménez, and M. G. Lockley. 1992. A quadrupedal ornithopod trackway from the Early Cretaceous of La Rioja (Spain): Inferences on gait and hand structure. *Journal of Vertebrate Paleontology* 12: 150–157.

Newman, B. H. 1990. A dinosaur trackway from the Purbeck Beds of Swanage, England. *Palaeontolografica africana* 27: 97–100.

Norman, D. B. 1980. On the ornithischian dinosaur *Iguanodon bernissartensis* of Bernissart (Belgium). *Mémoirs de l'Institut Royal des Sciences Naturelle de Belgique* 178: 1–105.

Pérez-Lorente, F., C. Cuenca-Bescos, M. Aurell, J. I. Canudo, A. I. Soria, and J. I. Ruíz-Omenaca. 1997. Las Cerradicas tracksite (Berriassian, Galve, Spain): Growing evidence for quadrupedal ornithopods. *Ichnos* 5: 109–120.

Pittman, J. 1989. Stratigraphy, lithology depositional environment, and track type of dinosaur track-bearing beds of the Gulf Coastal Plain. In D. D. Gillette and M. G. Lockley (eds.), *Dinosaur Tracks and Traces*, pp. 135–153. Cambridge: Cambridge University Press.

Sarjeant, W. A. S. 1989. Ten palichnological commandments: A standardized procedure for the description of fossil vertebrate footprints. In D. D. Gillette and M. G. Lockley (eds.), *Dinosaur Tracks and Traces*, pp. 269–370. Cambridge: Cambridge University Press.

Sarjeant, W. A. S., J. B. Delair, and M. G. Lockley. 1998. The footprints of *Iguanodon*: A history and taxonomic study. *Ichnos* 6: 183–202.

Wright, J. L. 1996. Fossil terrestrial trackways: Function, taphonomy, and paleoecological significance. Ph.D. thesis, University of Bristol.

Wright, J. L. 1999. Ichnological evidence for the use of the forelimb in iguanodontoid locomotion. *Special Papers in Palaeontology* 60: 209–219.

You H., and Y. Azuma. 1995. Early Cretaceous dinosaur footprints from Luanping, Hebei province, China. In Sun A. and Wang Y. (eds.), *Sixth Symposium of Mesozoic Terrestrial Ecosystems and Biota*, pp. 151–156. Beijing: China Ocean Press.

30. First Reports of Bird and Ornithopod Tracks from the Lakota Formation (Early Cretaceous), Black Hills, South Dakota

Martin G. Lockley, Paul Janke, and Leon Theisen

Abstract

Bird and ornithopod trackways are reported for the first time in the Lakota Formation and added to previous reports of theropod tracks. The bird tracks are the oldest known in North America (Barremian). The ornithopod trackways provide the oldest evidence of quadrupedal progression by members of this group in the Cretaceous of North America, and suggest that this mode of locomotion was common. Given the rarity of Neocomian ichnites, especially in North America, this relatively diverse and distinctive Lakota track assemblage adds significantly to our knowledge of ichnofaunas at this time.

Introduction

Dinosaur tracks were first reported from the Lakota Formation by O'Harra (1917) and Anderson (1939). These tracks from the Burton Quarry site near Rapid City and the Grace Coolidge Creek site, near Hermosa, 25 km to the south, appear to have been those of theropods.

Anderson (1939) made casual reference to tracks of herbivorous dinosaurs at the latter site, but did not indicate what type they might be. A large slab with well-preserved theropod tracks (fig. 30.1) is on display at the Rapid City Regional Airport with a label indicating that it originated from a locality near Hermosa approximately 25 km south of Rapid City. This is evidently incorrect; the specimen probably came from the Burton Quarry site, which is "about one and a half miles northwest of the business section of Rapid City" (Anderson 1939, 361). This conclusion is based on the lithology of the Burton Quarry track-bearing surface, which shows conspicuous desiccation cracks of a type not seen at the Hermosa site. As noted below this slab also reveals at least one faint track that we regard as being of ornithopod affinity.

Little else is currently known about the Lakota tracks or how they compare with ichnofaunas from other regions. Given that nothing has been published since 1939, and that the study of Anderson (1939) does little to indicate the abundance or diversity of tracks at the Grace Coolidge Creek site, preliminary results of a study is presented. Based on studies of invertebrate remains (Sohn 1979) and remains of *Iguanodon* (Weishampel and Bjork 1989) a Barremian age is inferred for the Lakota Formation (Lucas 1993).

In the last fifteen years, interest in fossil footprints from the western United States (Lockley and Hunt 1995) has highlighted the need to reevaluate the Lakota dinosaur tracks and place them in the broader context of Cretaceous ichnology. For example Aptian, Albian, and Cenomanian ichnofaunas are now quite well known from western

Figure 30.1. Tracks on display at Rapid City Regional Airport are mainly attributable to theropods, though one (lower left) is of ornithopod affinity. Tracks are from the Burton Quarry site in northwest Rapid City.

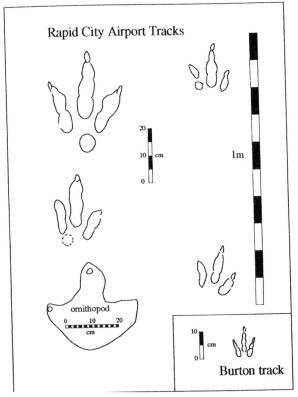

Rapid City Airport Tracks

ornithopod

Burton track

North America, as discussed below, but Barremian ichofaunas are still largely unknown. Further impetus for this study came from the discovery of bird tracks that are the oldest known from North America, and from the study of abundant ornithopod tracks, hitherto unreported from the Lakota Formation.

Description of Material

We confine ourselves to the preliminary documentation of a large track assemblage from a locality, on private land, near Hermosa (the Grace Coolidge Creek site). The tracks at this locality are preserved on a number of large and small sandstone blocks that have fallen from a cliff exposure. Most of the best-preserved tracks are preserved as impressions (positive relief), and the horizon from which the tracks originated has been established—a task which Anderson (1939, 363) stated that he failed to accomplish. The Lakota Formation in this area is relatively well sorted and clean washed with well-preserved ripple marks, some desiccation cracks, and invertebrate traces at various horizons. However, as noted by Anderson (1939) the track-bearing surface at the Grace Coolidge Creek site lacks the conspicuous "mud cracks" (desiccation cracks in sandstone) seen on the track-bearing surface at the Burton Quarry site. In fact, many of the dinosaur tracks at the Grace Coolidge Creek site are quite deep, which suggests that the substrate had a high water content at the time most of the tracks were made. Only one track-bearing layer is known at this site. Some bird tracks also show relief indicative of a soft substrate; others, however, are preserved as traces without relief and are distinguished primarily by color contrast with the surrounding matrix.

We documented the majority of tracks by tracing them on acetate film, though some small slabs with bird and dinosaur tracks were collected for further study in the lab. Ongoing efforts to remove lichens from the track surface reveal many of the smaller tracks—especially bird footprints. We have also made replicas of representative footprints for several repositories (University of Colorado at Denver, Black Hills Museum of Natural History: see acknowledgments). In addition to the Burton Quarry site specimen, located at the Rapid City Regional Airport, which belongs to the South Dakota School of Mines collection, we have located a number of isolated specimens, some from private collections, from which we have been able to obtain replicas.

Bird Tracks

Bird tracks are quite abundant at the Hermosa site, and often occur in trackways. For example, one slab already collected shows two clear trackway segments (fig. 30.2). Tracks on this slab have no relief and are distinguished from the surrounding surface only by color differences. The tracks on this slab are white and the intervening surface is brown. Tracks on other blocks nearby, but representing the same track-bearing surface, show clear relief and are not differentiated from the surrounding surface by color contrasts.

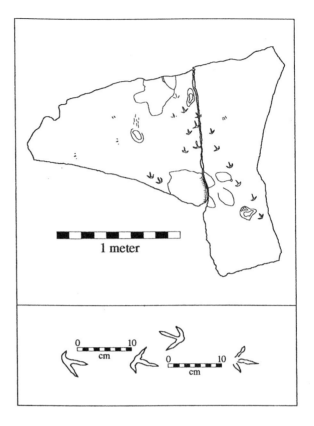

Figure 30.2. Slab with bird trackways, Grace Coolidge Creek site, near Hermosa site, with detail of tracks from an adjacent slab representing the same surface.

The tracks are tridactyl, without hallux impressions. Most tracks are about 5.0 cm long and wide with variable step lengths (13.0 cm and 19.0 cm in the example illustrated, fig. 30.2). Tracks show wide digit divarication angles; up to 150–160° (for digits II–IV) in some cases. Claw impressions are very sharp and pointed in most cases. The digits are slender but wider proximally than at the midpoint or distally. Discrete digital pads are seen in some examples.

Because many new Cretaceous bird track sites have been reported in recent years (Lockley, Yang, et al. 1992), it is possible to compare the Lakota tracks with those from other sites. Until recently, the only well-documented Lower Cretaceous bird tracksite from North America was an Aptian site from the Gething Formation of British Columbia, from which the type material of *Aquatilavipes* derives (Currie 1981). Tracks from the Lakota are similar to *Aquatilavipes* in size and general morphology. Bird tracks assigned to the ichnogenus *Ignotornis* are also known from the early Cenomanian of Colorado (Mehl 1931; Lockley, Yang, et al. 1992) and large tracks assigned to the ichnogenus *Magnoaivpes* (Lee 1997) have been reported from the Cenomanian of Texas. Both these reports are from the basal part of the Upper Cretaceous. The Lakota tracks are considerably older and cannot be assigned to either of these ichnogenera on the basis of morphological similarity. Indeed, we doubt that *Magnoavipes* can confidently be assigned to a bird (Lockley et al., 1999).

Recent studies of ichnofaunas from the Gates Formation of Western Canada reveal at least two distinct avian ichnites that can be assigned a Lower or Middle Albian age (McCrea and Sarjeant, chap. 31 of this volume). At least one of these is similar to *Aquatilavipes*. Previous reports of this ichnogenus from the Gates Formation at a different locality (Lockley, Yang, et al., 1992) now place these tracks in ?middle Albian Gladstone Formation, after stratigraphic reevaluation (Richard McCrea, pers. comm. 1999). Thus, we conclude that the Lakota bird tracks, which are probably Barremian in age, are considerably older than the three reports of *Aquatilavipes* and *Aquatilavipes*-like ichnites from western Canada, which are all Aptian or Albian in age.

There are only two reports of bird tracks as old or older than Barremian from localities outside North America. The first, from the Valanginian of Japan (Lockley, Yang, et al. 1992) are small tridactyl footprints that again resemble *Aquatilavipes*. The second, from the Berriasian of Spain, are large bird tracks assigned to the ichnogenus *Archaeornithipus* (Fuentes Vidarte 1996).

Theropod Tracks

Theropod tracks known from the Hermosa site range in size from approximately 10 cm to 35 cm (foot length). The largest tracks occur in trackways with a step of 100–110 cm. Intermediate-size tracks (foot length about 20 cm) occur in trackways with step lengths of about 90 cm. Small tracks with foot length of 10–12 cm include forms that show tapering digits and wide divarication angles and forms with less tapered and less divergent digits (fig. 30.3). Comparison of these tracks with those described by Anderson (1939) and illustrated in figure 30.1 suggest an abundance of small and medium-size theropod tracks, most of which were smaller than the ornithopod tracks described below.

The Lakota tracks might fruitfully be compared with theropod tracks from the Aptian Gething Formation (Sternberg 1932), the Glen Rose Formation (Farlow 1987; Pittman 1989) and the Dakota Group (Lockley 1987; Lockley, Yang, et al. 1992). Such an exercise holds the promise of revealing similarities and differences between ichnofaunas from the Barremian-Cenomanian interval in western North America.

Ornithopod Tracks

Large ornithopod tracks and trackway are abundant at the Lakota site (fig. 30.4), and in almost all cases indicate animals that were progressing quadrupedally. The longest trackway segment so far recorded shows four consecutive manus-pes sets. The pes is tridactyl with three oval digital pads and a centrally located subcircular heel pad. The pes axis is rotated inward at about 15°. The manus is oval to subtriangular in shape and has a long axis of about 14 cm (width) and short axis of about 8 cm (length). The manus is situated about 30 cm in front of pes digit III and slightly to the outside (i.e., lateral to the midline). Other trackways show similar configurations and track dimensions (fig. 30.4), but in some cases the manus is located much further from the trackway

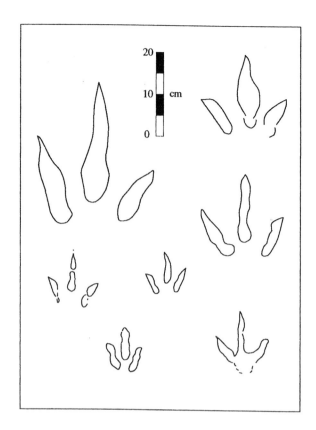

Figure 30.3. Theropod tracks recorded from the Grace Coolidge Creek site, near Hermosa, during the present study (compare with fig. 30.1.)

midline and less anteriorly in relation to the pes. Some trackways show that the manus indented the substrate as it moved forward into the final position where it was implanted, thereby registering an elongate groove or posterior slide mark. Other manus tracks show a slide mark that angles outward or laterally to a much more laterally placed impression. Such slide marks are rare in association with ornithopod manus tracks (Lockley 1987), but have the potential to shed light on ornithopod locomotion (see Lockley and Wright, chap. 29 of this volume).

Several manus impressions are triangular in shape. They consist of anterior medial and posterior lateral indentations marking the poles of the long axis of the impression. Anterolaterally, however, is a third indentation or protrusion in the margin of the manus impression. Opposite, the posteromedial margin of the impression is slightly concave. Such a configuration suggests that the trackmaker was an iguanodontid with manus digits II, III, and IV bound together by integument (Wright 1999). The marginal indentations in anteromedial, anterolateral, and posterolateral positions respectively represent these three digits. Similar tracks are found in the Earliest Cretaceous (Berriasian) of England (Wright 1999; Lockley and Wright, chap. 29 of this volume) and Germany (Lockley and Wright, in prep.). Early Cretaceous tracks from England have recently been assigned provisionally to *Iguandontipus* (Sarjeant et al., 1998) though this ichnogenus was described on the basis of trackways without manus impressions. The tracks from the

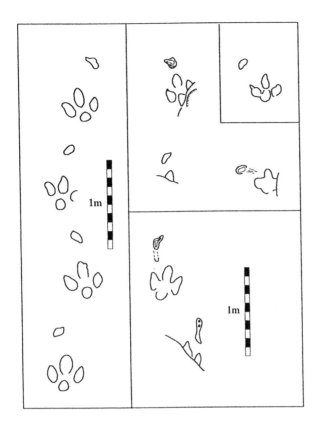

Figure 30.4. Ornithopod tracks
and trackway segments, Grace
Coolidge Creek site. Note that
manus may be placed medially or
laterally and may show slide
marks before registering. Several
manus tracks are subtriangular
and compare with Neocomian
manus tracks from England and
Germany.

Lakota Sandstone, like those from Germany, can be assigned provisionally to *Iguanodontipus* sp. However, as noted by Lockley and Wright
(chap. 29 of this volume), the ichnogenus *Amblydactylus* from the
Aptian of Canada is also valid (Sternberg 1932; Currie and Sarjeant
1979; Currie 1989, 1995) and closer in age to the Lakota material than
it is to the English and German samples. We refer the reader to Lockley and Wright (chap. 29 of this volume) for a fuller review, and in
the meantime advocate caution in the application of ichnotaxonomic
names.

The Lakota ornithopod tracks somewhat resemble *Caririchnium*
from the Cenomanian of Colorado in pes morphology, though the heel
is more rounded (less bilobed). *Caririchnium*, however, also has an oval
manus print with a small inwardly (anteromedially) directed marginal
indentation (Lockley 1987) and no outwardly directed (anterolateral)
protrusion or concave posteromedial margin. Hence, we conclude that
the manus print is significantly different—that is, subtriangular to
crescentic in the Neocomian (Berriasian-Barremian) ichnites, rather
than oval, as in the post-Neocomian (Cenomanian) ichnogenus *Caririchnium* (Lockley and Wright, chap. 29 of this volume). Given the
similarity between the material from South Dakota, England, and
Germany, we conclude that these tracks may be useful in emphasizing
generalized North American–European faunal similarity during the
Neocomian (cf. Lucas 1993). The occurrence of skeletal remains of
Iguanodon lakotaensis from the Lakota Formation (Weishampel and
Bjork 1989), also supports biostratigraphic correlations with Europe.

Paleoecological Observations

Lockley (1991; Lockley et al. 1994) noted that ichnofaunas dominated by ornithopod tracks are associated with siliciclastic facies representing relatively high-latitude (temperate) humid paleoenvironments, such as coal-bearing facies. Theropod tracks in such facies are relatively small and gracile (Matsukawa et al. 1995), as compared with the large, robust theropod tracks associated with sauropod-dominated ichnofacies found mainly in lower-latitude carbonate substrates. It appears that this general pattern is also characteristic of the Lakota assemblage, although, as noted above, work is needed on Cretaceous theropod track assemblages before we can adequately describe the morphological, ichnotaxonomic, and paleoecological changes through time. The relatively small size of many of the theropod tracks from the Lakota is reminiscent of the relationship seen in the Dakota Group of Colorado and adjacent states (Lockley 1987; Lockley, Holbrook, et al. 1992; Lockley and Hunt 1995). Such a relationship prompts one to ask whether the theropods actually preyed on the large ornithopods or on other, smaller prey. It is also interesting to consider whether large vertebrates in humid, well-vegetated paleoenvironments may have been smaller than those found in less forested regions.

To date, all bird tracks so far reported from western North America are also associated with relatively humid siliciclastic facies. It is also known that almost all known bird tracks are those of waterbirds or shorebirds, owing to the ideal circumstances for track preservation found along shorelines (Lockley, Yang, et al. 1992). Given that the Cretaceous radiation of birds appears to have coincided with the radiation of large ornithopods (and not the heyday of Jurassic sauropods), such a relationship is perhaps predictable in terms of biochronology. Nonetheless, it may also have some broad paleoecological significance.

According to Lucas (1993) the Lakota vertebrate fauna is the type for the "Buffalogapian Land Vertebrate age" which is characterized by the dinosaurs *Iguanodon lakotaensis, Camptosaurus depressus, Hypsilophodon wielandi,* and *Sauropelta* sp., along with theropod tracks, and miscellaneous fish and turtle remains. We note here a preponderance of ornithischian (mainly ornithopod dinosaurs). In contrast, it still appears that there are many theropod tracks in the Lakota ichnofauna. We must therefore ask whether theropods are overrepresented in the track assemblages, or whether some of the so-called theropod tracks might be those of ornithopods, such as hypsilophodontids.

Conclusions

The Lakota ichnofauna is long overdue for careful study. Preliminary research indicates an ichnofauna rich in bird, ornithopod, and theropod tracks—the latter perhaps representing several taxa. The bird tracks are provisionally compared with *Aquatilavipes* and, based on the inference of a Barremian age, are evidently the oldest known from North America. The Lakota ichnofauna is possibly as old as any known from the Lower Cretaceous of North America, and so reveals the earliest examples of the trackways of large quadrupedal ornithopods. The diverse theropod track assemblage also warrants further investigation in case other types of tridactyl dinosaurs are represented. It appears also that the Lakota ichnofauna is typical of siliciclastic facies representing temperate humid paleoenvironments in the Cretaceous of North America.

Acknowledgments: We thank David Geary for allowing us access to the Grace Coolidge Creek site. We thank Phil Currie for looking at the bird track specimens and sharing his observations on Cretaceous tracks. We also thank the Black Hills Museum of Natural History and the Geology Department, University of Colorado at Denver, for access to office support, materials, and collections. Casting and curation, undertaken cooperatively between these two institutions, is ongoing as part of a larger, long-term project, so it is premature to provide specimen numbers.

References

Anderson, S. M. 1939. Dinosaur tracks in the Lakota Sandstone of the eastern Black Hills, South Dakota. *Journal of Paleontology* 13: 361–364.

Currie, P. J. 1981. Bird footprints from the Gething Formation (Aptian, Lower Cretaceous) of northeastern British Columbia, Canada. *Journal of Vertebrate Paleontology* 1: 257–264.

Currie, P. J. 1989. Dinosaur footprints of western Canada. In D. D. Gillette and M. G. Lockley (eds.), *Dinosaur Tracks and Traces*, pp. 293–300. Cambridge: Cambridge University Press.

Currie, P. J. 1995. Ornithopod trackways from the Lower Cretaceous of Canada. In W. A. S. Sarjeant (ed.), *Vertebrate Fossils and the Evolution of Scientific Concepts*, pp. 431–443. Reading: Gordon and Breach Publishers.

Currie, P. J., and W. A. S. Sarjeant. 1979. Lower Cretaceous footprints from the Peace River Canyon, B. C., Canada. *Palaeogeography, Palaeoclimatology, Palaeoecology* 28: 103–115.

Farlow, J. O. 1987. *A Guide to the Lower Cretaceous Dinosaur Footprints and Tracksites of the Paluxy River Valley, Somervell County, Texas.* Field trip guide, South-central section, Geological Society of America Annual Meeting.

Fuentes Vidarte, C. 1996. Primeras huellas de Aves en el Weald de Soria (España): Nuevo ichnogenera, *Archaeornithipes* y nueva ichnoespecie *A. meidei. Estudios geologicos* 52: 63–75.

Lee, Y.-N. 1997. Bird and dinosaur footprints in the Woodbine Formation (Cenomanian), Texas. *Cretaceous Research* 18: 849–864.

Lockley, M. G. 1987. Dinosaur footprints from the Dakota Group of eastern Colorado. *Mountain Geologist* 24: 107–122.

Lockley, M. G. 1991. *Tracking Dinosaurs: A New Look at an Ancient World.* Cambridge: Cambridge University Press.

Lockley, M. G., and A. P. Hunt. 1995. *Dinosaur Tracks and Other Fossil Footprints of the Western United States.* New York: Columbia University Press.

Lockley, M. G., J. Holbrook, A. P. Hunt, M. Matsukawa, and C. A. Meyer. 1992. The dinosaur freeway: A preliminary report on the Cretaceous megatracksite, Dakota Group, Rocky Mountain Front Range and Highplains; Colorado, Oklahoma, and New Mexico, pp. 39–54. In R.

Flores (ed.), *Mesozoic of the Western Interior, SEPM Midyear Meeting Fieldtrip Guidebook.*

Lockley, M. G., A. P. Hunt, and C. A. Meyer. 1994. Vertebrate tracks and the ichnofacies concept: Implications for paleoecology and palichnostratigraphy. In S. Donovan (ed.), *The Paleobiology of Trace Fossils,* pp. 241–268. New York: Wiley.

Lockley, M. G., M. Matsukawa, and J. L. Wright. 1999. Is it a bird or is it a . . . ? A new scientific track name suggests that big bird was around 98 million years ago. *Friends of Dinosaur Ridge Annual Report,* pp. 18–20.

Lockley, M. G., S-Y. Yang, M. Matsukawa, R. F. Fleming, F. Lim, and S.-K. Lim. 1992. The track record of Mesozoic birds: Evidence and implications. *Philosophical Transactions of the Royal Society of London* 336: 113–134.

Lucas, S. G. 1993. Vertebrate biochronology of the Jurassic-Cretaceous boundary, North American western interior. *Modern Geology* 18: 371–390.

Matsukawa, M., M. Futakami, M. G. Lockley, C. Peiji, C. Jinhua, C. Zenyao, and U. Bolotsky. 1995. Dinosaur footprints from the Lower Cretaceous of eastern Manchuria, northeast China: Evidence and implications. *Palaios* 10: 3–15.

Mehl, M. G. 1931. Additions to the vertebrate record of the Dakota Sandstone. *American Journal of Science* 21: 441–452.

O'Harra, C. C. 1917. Fossil footprints in the Black Hills. *Pahasapa Quarterly* 6: 20–29.

Pittman, J. 1989. Stratigraphy, lithology depositional environment, and track type of dinosaur track-bearing beds of the Gulf Coastal Plain. In D. D. Gillette and M. G. Lockley (eds.), *Dinosaur Tracks and Traces,* pp. 135–153. Cambridge: Cambridge University Press.

Sarjeant, W. A. S., J. B. Delair, and M. G. Lockley. 1998. The footprints of *Iguanodon:* A history and taxonomic study. *Ichnos* 6: 183–202.

Sohn, I. G. 1979. Nonmarine ostracods in the Lakota Formation (Lower Cretaceous) from South Dakota and Wyoming. *U.S. Geological Survey Professional Paper* 1069: 1–22.

Sternberg, C. M. 1932. Dinosaur tracks from Peace River, British Columbia. *Annual Report, National Museum of Canada* (for 1930): 59–85.

Weishampel, D. B., and P. R. Bjork. 1989. The first indisputable remains of *Iguanodon* (Ornithischia: Ornithopoda) from North America: *Iguanodon lakotaensis.* sp. nov. *Journal of Vertebrate Paleontology* 9: 56–66.

Wright, J. 1999. Ichnological evidence for the use of the forelimb in iguanodontoid locomotion. *Special Papers in Palaeontology* 60: 209–219.

31. New Ichnotaxa of Bird and Mammal Footprints from the Lower Cretaceous (Albian) Gates Formation of Alberta

RICHARD T. McCREA AND
WILLIAM A. S. SARJEANT

Abstract

Recent research on the ichnofauna of the Lower Cretaceous (Albian) Gates Formation near Grande Cache, Alberta, has revealed the presence of numerous bird trackways among the dinosaur track-bearing strata in the Smoky River Coal Mine. This is the second report of bird footprints from the Grande Cache area, but it is the first description of bird footprints from the Gates Formation. Two ichnotaxa of bird footprints are present at the W3 Main site, but only one occurs in abundance. A third type of bird footprint, originally known from a few talus blocks, has now been found *in situ* at the W3 Bird site. One more recently discovered site (W3 Extension) also displays in situ bird footprints. These new discoveries indicate a diverse avifauna in the late Early Cretaceous of Alberta, so far known solely from footprints.

Very small tridactyl mammal footprints with forward-pointing claws were found among bird footprints on a small talus block at the base of the W3 footwall; this is the first record of nonmarsupial mammalian footprints from the Cretaceous. The new avian and mammalian ichnotaxa are described, and the definitions of the avian ichnotaxa *Aquatilavipes, A. swiboldae,* and *Fuscinapeda* are emended.

Institutional Abbreviations: BCPM, British Columbia Provincial Museum, Victoria;; TMP, Tyrrell Museum of Palaeontology, Drumheller, Alberta; UALVP, University of Alberta Laboratory for Vertebrate Paleontology, Edmonton.

Introduction

Purported bird footprints were first discovered in the Upper Cretaceous (Cenomanian) Dunvegan Formation along the Pouce Coupé River, Alberta, by Dr. Charles R. Stelck in 1951 (Currie 1989). No formal descriptions of these prints (UALVP 25271) have yet been published, and we question whether they are truly avian. The first published record of bird footprints in Cretaceous strata of western Canada was from the Gething Formation (Aptian) of the Peace River Canyon in eastern British Columbia. They were small (length 2.0–4.4 cm) and were placed into a new ichnotaxon, *Aquatilavipes swiboldae* (Currie 1981). Ten years later (1991), Darren Tanke (TMP) discovered natural casts of two small tridactyl footprints in situ in a road cut exposure near the Smoky River Coal Mine (Highway 40 site; fig. 31.1). The section of rock containing the footprints was cut out and subsequently lodged at the Royal Tyrrell Museum of Palaeontology (TMP 90.30.1). The presence of a third, faint *A. swiboldae* footprint on this slab was recently discovered by one of us (R. T. McCrea). These footprints were originally thought to be from the Gates Formation (Lockley et al. 1992). However, from the geological maps of the area, no Gates Formation strata are exposed in the vicinity of this tracksite and it was thought by McCrea and Currie (1998) that they were more probably from the Aptian-Albian Cadomin Formation. However, a recent visit to the site revealed that the footprints occur in strata belonging to the Gladstone Formation, which overlies the Cadomin Formation. The Gladstone Formation correlates to the Gething Formation of British Columbia (Langenburg et al. 1987) from which the *Aquatilavipes swiboldae* prints were originally described (Currie 1981). The tridactyl footprints from the Gladstone Formation near Grande Cache, attributed to *Aquatilavipes swiboldae,* were the first confirmed bird footprints to be reported from Alberta.

In the summer of 1998, several expeditions were concentrated on the W3 footwall (W3 Main site; fig. 31.1) within the Smoky River Coal Mine, from which dinosaur footprints had been reported in the early 1990s (McCrea and Currie 1998). On July 15, one of us (R. T. McCrea) observed faint tridactyl footprints on a rippled sandstone layer on the footwall. These footprints are smaller (between 6.4 and 10.1 cm long) than the smallest of the dinosaur footprints occurring on the same footwall; these are between 13.5 and 19.0 cm long. Initially, several individual avian footprints were found. Eventually a trackway consisting of six consecutive footprints was discovered near a large theropod trackway (*Irenesauripus mclearni*). The small tridactyl footprints exhibit wider digital divarications (tables 31.1 and 31.2) than are seen in

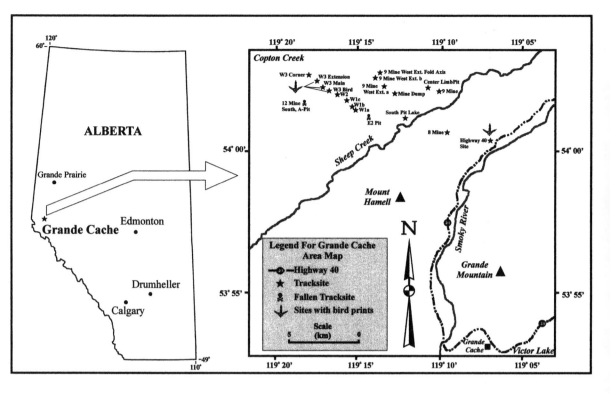

the larger tridactyl (dinosaur) footprints from the same footwall. The gait of the trackmaker was noticeably pigeon-toed and seemed to be accelerating, since stride lengths were observed to increase significantly along the length of the trackway (table 31.1, trackway G5–F6).

The W3 Main site is at an altitude of nearly 1700 meters and is frequently overcast or fogged, which affects one's ability to see all the footprints on the footwall. Also, footwall is oriented so that the sun shines on it for only part of the day. However, the weather and lighting were favorable long enough for us to determine that these new tridactyl footprints were quite abundant, numbering over 750 out of more than 1200 vertebrate footprints mapped within a 500 m² study area of the footwall (McCrea and Sarjeant 1999).

All the small tridactyl footprints were rotated inward toward the midline of the trackway (pigeon-toed). Not all footprints were complete; in several instances, only one or two digital impressions were preserved. The footprints, judging from the length of stride, are those of long-legged birds, perhaps heronlike in form. The prints are wider than they are long; all digits are relatively thick and each bears a short claw. They are here described as a new ichnospecies of *Aquatilavipes,* their inclusion necessitating some revisions to the diagnosis of that ichnogenus. This has also necessitated a revision to the allied ichnogenus *Fuscinapeda.*

Aquatilavipes swiboldae prints were found by Dr. Donald Brinkman (TMP) in two talus blocks at the base of the W3 footwall. One block (TMP 98.89.21) is a natural cast (fig. 31.2a) and the other (TMP 98.89.20) is a natural mold (fig. 31.2b). Since this area has been sub-

Figure 31.1. Sketch map showing the location of the study area and the localities where bird footprints were collected. Modified from Langenberg et al. 1987.

TABLE 31.1

Trackway Location	Print Number	Footprint Length (mm)	Footprint Width (mm)	Digit Length (mm)			Divarification			Pace (mm)	Stride (mm)	Pace Angle
				II	III	IV	II-III	III-IV	TOTAL			
aa6-aa7	1(R)	77	112									
	2	85	115							240		150
	3	80	105							219	445	153
	4	65	110							248	450	157
	5	70	95							230	462	158
	6	65	105	55	65	62	68	54	122	235	455	159
	7	68	95							252	479	151
	8	72	87							240	473	
	x̄	73	100	55	65	62	68	54	122	238	461	155
cc18-dd18	1(R)	70	101	60	70	55	50	70	120			
	2	68	107							220		135
	3	76	95							180	365	150
	4	67	108							235	405	
	x̄	70	103	60	70	55	50	70	120	212	385	143
A9	1(L)	90	110	77	90	68	50	60	110	230		
	2	80	108									
	x̄	85	109	77	90	68	50	60	110	230		
B4-B5	1(R)	64	94	52	64	56	72	63	135			
	2	73	94							261		163
	3	72	77							261	514	178
	4	70	86							254	514	
	x̄	70	88	52	64	56	72	63	135	257	514	171
F3	1(L)	95	107									
	2	85	100							220		140
	3	90	110	71	90	67	55	69	124	210	410	132
	4	87	115							250	420	
	x̄	89	108	71	90	67	55	69	124	227	415	136
G5-F6	1(L)	78	99									
	2	81	100							218		127
	3	78	102							225	396	144
	4	75	104	58	75	55	67	65	132	220	423	149
	5	80	96							225	432	153
	6	82	110							229	442	
	x̄	79	102	58	75	55	67	65	132	223	423	143
G18-F19	1(L)	70	103									
	2	75	103	68	75	65	60	65	125	230		165
	3	72	102							235	460	140
	4	74								190	390	147
	5	74	106							270	430	151
	6	72								225	480	151
	7	76								315	530	
	x̄	73	104	68	75	65	60	65	125	244	458	143
H20-H19	1(L)	63										
	2	75								300		166
	3	70	115	60	70	65	62	59	131	305	600	172
	4	80								260	570	
	x̄	72	115	60	70	65	62	59	131	288	585	169

TABLE 31.2

Trackway Location	Print Number	Footprint Length (mm)	Footprint Width (mm)	Digit Length (mm)			Divarification			Pace (mm)	Stride (mm)	Pace Angle	Footprint Rotation
				II	III	IV	II-III	III-IV	TOTAL				
Trackway	1(R)	90	120	73	90	67	50	69	119				+30
A	2	83	126	79	83	60	52	82	134	208		156	-4
Paratype	3	95	110	55	95	62	63	74	137	273	476	156	+22
Slab	4	101	116	62	101	71	58	68	126	227	489	159	+35
	5	89	116	68	89	67	54	70	124	215	433	163	+14
	6	88	119	65	88	74	61	58	119	215			+23
	7	88	117	68	88	69	58	61	119	221	452		+25
	8												+19
	9	80	123	67	80	69	61	67	128				
	10	89			89	60	55	76	131	234			+41
x̄		89	118	67	89	67	57	69	126	463	463	159	23
Trackway	1(R)	85	107	54	85	60	71	79	150				
B	2												
Paratype	3	83	108	60	83	69	58	55	113		510		+30
Slab	4	82	99	55	82	63	61	53	114	248			+16
	5	85	111	61	85	66	64	55	119	237	476	158	+24
	6												-30
	7	88	95	56	88	64	50	57	107		525		+16
	8	80	110	71	80	67	44	52	96	298			0
	9	88	116	71	80	67	84	53	137	214	510	173	+27
x̄		84	107	61	83	65	62	58	119	249	505	166	14

jected to backfill operations and contains material from other mine sites, it was not certain that these talus blocks originated from the W3 footwall. Recently, a large number of tridactyl prints (natural molds and natural casts) were found in situ on the W3 footwall, away from the main area of study (W3 Bird). These appear referable to *A. swiboldae* and this footwall is the likely source of the talus blocks mentioned above. Avian footprints were also recently found in situ at the W3 Extension site. They are smaller than the other avian footprints from the other W3 footwall sites and have very slender digits, but not enough material has been recovered to describe them.

During the study of the *A. swiboldae* prints on the natural-mold talus block (TMP 98.89.20), one of us (W. A. S.) noticed very small, shallowly impressed mammal footprints at the lower center of the slab. Mammalian footprints were earlier reported from the Gething Formation of the Peace River Canyon, and named *Duquettichnus kooli* by Sarjeant and Thulborn (1986). However, the footprints were markedly larger and so closely comparable to those made by the living Australian brush-tail possum that they are almost certainly marsupial footprints.

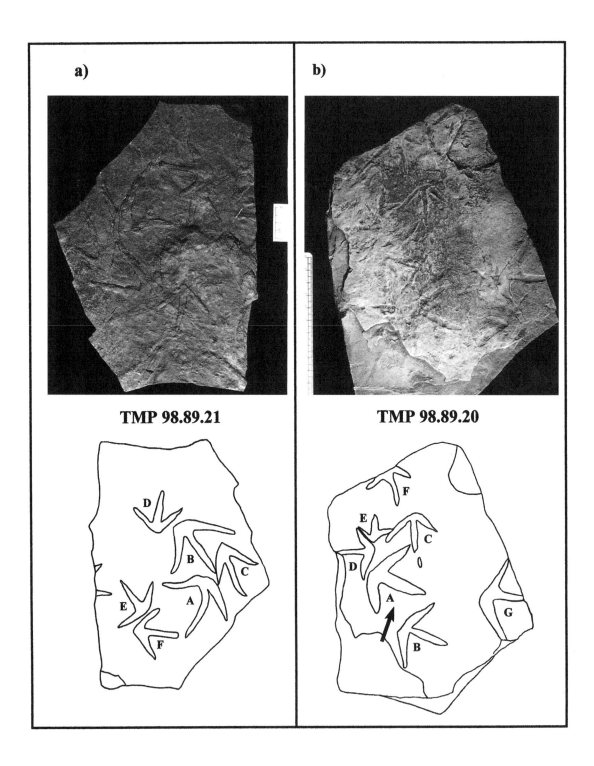

a)

TMP 98.89.21

b)

TMP 98.89.20

The newly discovered imprints are quite different in morphology and are the smallest mammalian footprints yet reported from the Mesozoic.

Terminology

Despite attempts by Casamiquela et al. (1987) and Sarjeant (1989) to standardize the descriptions of ichnofossil taxa, some ambiguities remain, especially between usage by Europeans and North Americans. For that reason, the following clarifications are necessary:

- A *track*way is a series of footprints.
- A *footprint* or *imprint* is a single impression of a foot, isolated or forming part of a trackway. The use of *track* as equivalent to *footprint* is common in North America; however, in Europe (and indeed among game hunters in North America) this term is always used to refer to a series of footprints (equivalent to *trackway*). To avoid confusion the use of the term *track* should be avoided.
- The *total interdigital span* (also known as *total divarication* and *divarication of digits*) is the angle between the axes of the outermost digits (in this case, digits II and IV).
- *The length of the individual digits* are here measured (for the most part) from the tips of the individual digits to the posterior limit of the metatarsal pad. This differs from Currie's (1981) measurements of *Aquatilavipes swiboldae,* which were taken from the tip of the digit to its point of contact with the metatarsal pad. The difficult lighting conditions of the W3 Main site, mentioned above, made it difficult to locate the proximal end of the digits of in situ footprints.
- The *pace angle* is the measurement in degrees to which a footprint is angled outward or inward from the midline of the trackway.

Figures illustrating these methods of footprint and trackway documentation may be found in Leonardi et al. 1987 and Thulborn 1990.

Figure 31.2. (opposite page) (a) Natural cast (TMP 98.89.21), with prints (Aquatilavipes swiboldae Currie, emend. nov.) outlined with a felt marker. (b) Natural mold (TMP 98.89.20), coated with ammonium chloride, exhibiting conspicuous avian footprints (Aquatilavipes swiboldae Currie, emend. nov.) and, at lower center (arrow), tiny mammalian footprints (Tricorynopus? brinkmani ichnosp. nov.). The line drawings are included merely to show the location of faint footprints on the two slabs. Scale along left edge of b in mm.

Systematics

Class Aves
Morphofamily Avipedidae Sarjeant and Langston 1994
Ichnogenus *Aquatilavipes* Currie 1981, emend. nov.

Original Diagnosis: "Made by a bipedal animal with three functional digits. Width greater than length; average divarication of digits II and IV greater than 100°. Digit IV longer than digit II and shorter than digit III. Sharp claw on each digit. No hallux impression" (Currie 1981, 259).

Emended Diagnosis: Footprints of small to large size, showing three digits united proximally, most often in a metatarsal pad ("heel"); webbing and hallux lacking. Digits slim, their maximum width less than 15% of their length; digit III is more than 25% longer than the lateral digits. Total interdigital span greater than 95° and often exceeds 120°. Length of digits II and IV may be similar, but digit IV is frequently somewhat longer. All digits clawed, the claws frequently showing inward flexure in relation to the digit axis. Digital pad impressions may be visible on better-preserved molds or casts —three to four on digit III, two on digits II and IV.

Type Ichnospecies: Aquatilavipes swiboldae Currie 1981. Gething Formation (Early Cretaceous: Aptian), eastern British Columbia.

Remarks: The ichnogeneric diagnosis is here emended to clarify differences from *Fuscinapeda* Sarjeant and Langston 1994. (The diagnosis of that ichnogenus is emended below). An earlier emendation by Lockley et al. (1992, 125), which added to Currie's diagnosis a mention of "faint digital pad impressions," is incorporated, even though their presence or absence depends on the substrate.

As emended here, *Aquatilavipes* differs from *Fuscinapeda* essentially in having more slender digits and from *Aviadactyla* Kordos 1983, in the proximal fusion of the digits and their less sticklike character. It differs from the otherwise very similar *Ludicharadripodiscus* Ellenberger 1980, in the consistent lack of a hallux impression. The digit impressions of *Avipeda* Vialov 1965, emend. Sarjeant and Langston 1994, are shorter and thicker (see also Vialov 1966); those of *Ornithotarnocia* Kordos 1983, show a thicker digit III and a higher degree of asymmetry.

TABLE 31.3

Specimen Number	Print Number	Footprint Length (mm)	Footprint Width (mm)	Digit Length (mm)			Divarification		
				II	III	IV	II-III	III-IV	TOTAL
Natural	A	47	55	34	47	32	46	62	108
Cast	B	44	57	42	44	38	42	48	90
Block	C	42	33	32	42	30	37	34	71
TMP	D	32	37	25	32	27	55	43	98
98.89.21	E	31	43	23	31	24	77	53	130
	F	35	35	26	35	26	47	48	95
Natural	A	53	69	43	53	38	58	68	126
Mould	B	45	63	34	45	38	61	71	132
Block	C	33	45	24	33	33	53	52	105
TMP	D		49	28		28	68	66	134
98.89.20	E	25	31	19	25	20	53	52	105
	F		40						
	G			40			61	62	123
TMP	A	40	55	33	40	27	75	73	138
90.30.1	B	37	57	26	37	35	84	61	145
	C	41	55	29	41	32	64	74	138
TMP 79.23.3 and BCPM 744	76	38	47	22	34	24	48	70	118

Aquatilavipes swiboldae Currie 1981, emend nov.
(fig. 31.2a,b; table 31.3, TMP 79.23.3, BCPM 744)

1981 *Aquatilavipes swiboldae* Currie (pp. 259–261, figs. 1a,c, 2, 3).

1992 *A. swiboldae* Currie emend. Lockley et al. (pp. 115–116, 125, 129, fig. 4).

1994 *A. swiboldae* Currie. Sarjeant and Langston (p. 12).

Original Diagnosis: "Footprints less than 4.5 cm in length, average width 26% greater than length: average divarication of digits II and IV is 113°. Digit III about 50% longer than digit II and 40% longer than digit IV" (Currie 1981, 259).

Emended Diagnosis: A species of *Aquatilavipes* of small size, with slim digits; the thickness of the slimmest digit (III) is less than 8% of its length, the others being somewhat thicker (up to 12.5% of length). All digits terminate in claws, that on digit III being especially acute. The digits were flexible, digits II and III generally curving inward distally, digit IV outward. Digit III is about 50% longer than digit II and 40% longer than digit IV. Total interdigital span varies from 90° to 130°, averaging 113°.

The angle of the footprints to the center of the trackway (footprint rotation) varies, but they tend to be directed inward. The trackway is of moderate breadth.

Description: See Currie 1981 (259–261) for detailed account.

Holotype: Footprint no. 76 (mold and cast). Mold (specimen TMP 79.23.37) lodged in the Royal Tyrrell Museum of Palaeontology, Drumheller, Alberta: cast (specimen BCPM 744) lodged in the British Columbia Provincial Museum, Victoria.

Dimensions: See Currie 1981 (259–260) for details.

Remarks: The diagnosis is here amplified to facilitate comparisons with *A. curriei* ichnosp. nov. All features of the original diagnosis are included.

The dimensions of the specimens from Grande Cache are shown in table 31.3; as noted earlier, these differ from Currie's measurements in that they were taken to the back of the metatarsal pad.

Aquatilavipes curriei McCrea and Sarjeant, ichnosp. nov.
(figs. 31.3–31.10; tables 31.1, 31.2)

Derivation of Name: In honor of Philip J. Currie, in recognition of his contributions to vertebrate paleontology and paleoichnology in western Canada.

Diagnosis: A species of *Aquatilavipes* of moderately large size, the thickness of the digits being around 10% of their length. They terminate in narrow, sharp claws, those of digits II and IV inclined slightly inward toward digit III. Total interdigital span varies between 120° and 135° according to gait and substrate hardness, the angle between digits II and III being consistently larger than between digits III and IV. Digital pads often discernible—three on

Figure 31.3. Aquatilavipes curriei *ichnosp. nov. The holotype cast (TMP 98.89.11), a left pes. The arrow indicates the direction of the illumination.*

digit III, two on digits II and IV. The center of each digit impression may show a groove parallel to the axis of the digit, continuous or discontinuous; this may not be evident in shallower imprints.

The angle of the footprints is always slightly inward toward the center of the trackway; the trackway is quite broad and the pace, though variable, consistently short.

Holotype: Specimen no. TMP 98.89.11; cast of isolated left pes taken from Grid H/G16 (figs. 3–4a). Lodged in the Royal Tyrrell Museum of Palaeontology, Drumheller, Alberta.

Paratype: Specimen no. TMP 98.89.10; cast of trackways (figs. 5–6). Same lodgment.

Type Horizon and Locality: Grande Cache Member of the Gates Formation, early Albian (Lower Cretaceous), Smoky River Coal Mine

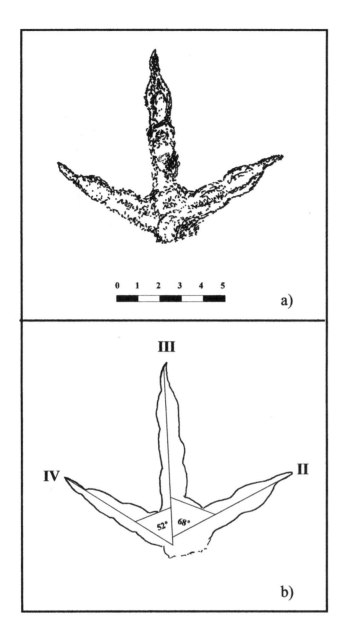

Figure 31.4. Aquatilavipes curriei *ichnosp. nov. (a) The holotype sketch, showing digital pads. Scale in centimeters. (b) Outline of the holotype, showing interdigital angles.*

(Smoky River Coal, Ltd.) about 21 km northwest of Grande Cache, Alberta. Located on the footwall of the W3 Main site below the no. 4 coal seam.

Dimensions: Holotype (by standard measurement): overall length 7.9 mm, overall breadth 9.5 mm; length of digit II, 4.5 mm: III, 6.7 mm: IV, 5.0 mm. Paratype: see table 31.2. Range of dimensions: see tables 31.1, 31.2.

Interdigital Angles: Holotype: see figure 31.4b. Range: see diagnosis.

Remarks: The footprints are present on at least three bedding planes on the W3 footwall (W3 Main tracksite). There is a sandstone layer of medium thickness (19–21 cm, layer C), on which are also found a variety of dinosaur footprints (McCrea and Sarjeant 1999). This

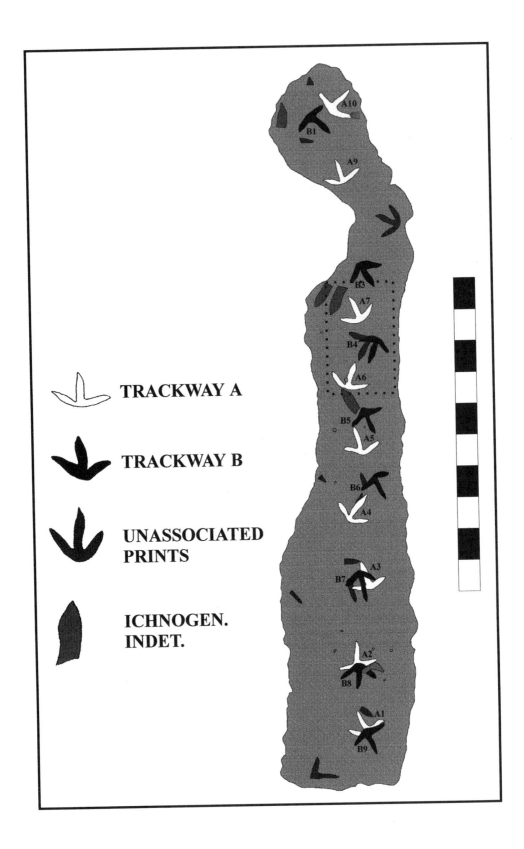

TRACKWAY A

TRACKWAY B

UNASSOCIATED
PRINTS

ICHNOGEN.
INDET.

Figure 31.5. (opposite page) Aquatilavipes curriei ichnosp. nov. Sketch showing the paratype cast (TMP 98.89.10), with two tracks in opposing directions and some isolated, incomplete A. curriei prints as well as some larger avian footprints of uncertain systematic reference. Scale: 1 m. Area represented by dotted line corresponds to the photograph in figure 31.6.

Figure 31.6. Aquatilavipes curriei ichnosp. nov. A part of the paratype cast (fig. 31.5), showing footprints A6 (lower) and A7 (upper) and two other footprints. Scale in centimeters.

layer consists of fine sand (0.15 mm diameter grains) topped with a fine silt layer, indicating the settling of a body of water during lowstand. The overlying layer (layer B) likewise contains *A. curriei* prints and dinosaur footprints, but it is very thin (1–2 cm) and is again composed of fine sandstone, topped with a silt which preserves the bird footprints better than does the underlying sand (McCrea and Sarjeant 1999). Layer B in turn is overlain by layer A, a thin bed (1–2 cm) from which only one bird footprint has been recognized (McCrea and Sarjeant 1999). There are terrestrial plant remains on the track-bearing layers, including carbonized tree trunks, stumps, and cones. The conditions present on the footprint layers were eventually succeeded by a coal swamp, whose deposits make up the 2–3 m thick no. 4 coal seam (Langenburg et al. 1987).

Figure 31.7. Aquatilavipes curriei *ichnosp. nov. Footprint A6 in the paratype cast (fig. 31.5), showing the apparent deformities.*

Ten trackways and over 750 individual footprints of this ichnospecies were studied. The paratype slab (illustrated in figs. 31.5, 31.6) shows that two birds were moving at moderate speed in opposite directions, with a moderately long stride and broad trackways (11.5–14 cm). One print, number A6 on the paratype trackway (figs. 31.7, 31.8), shows craterlike swellings, on the left side of digit III and on the right of the metatarsal pad. These swellings are comparable to the pathological effects produced by bumblefoot in living poultry (Dr. Peter Flood, pers. comm. 1999), but it is perhaps more likely that they result from the activity of infauna in the sediment. Unfortunately, the other prints of this foot in the trackway were not good enough to enable us to distinguish between these alternative hypotheses.

A second trackway (figs. 31.9, 31.10) shows a meandering pattern of imprints, probably indicating a search for food along the edge of a drying-out pool, as evidenced by the numerous invertebrate burrows in the area.

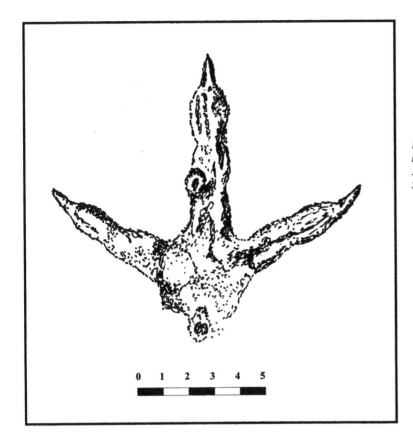

Figure 31.8. Aquatilavipes curriei *ichnosp. nov. Sketch of footprint A6, indicating the "deformities." Scale in centimeters.*

Ichnogenus *Fuscinapeda* Sarjeant and Langston 1994, emend. nov.

1994. *Fuscinapeda* Sarjeant and Langston (pp. 13–14)

Original Diagnosis: "Avian footprints of small to large size, showing three digits, slim or moderately thick (II to IV). Digit III is characteristically more than 25% longer than the lateral digits. Total interdigital span greater than 95° and often exceeds 110°. Digits united proximally, frequently showing a distinct 'heel.' Webbing absent or restricted to the most proximal part of the interdigital angles" (Sarjeant and Langston 1994, 13).

Emended Diagnosis: Tridactyl footprints of small to large size, showing three digits united proximally, most often in a metatarsal pad ("heel"); webbing and hallux lacking. Digits moderately thick to thick, their maximum width exceeding 15% of their length: digit III is more than 25% longer than the lateral digits. Total interdigital span greater than 95° and often exceeds 120°. Length of digits II and III may be similar, but digit IV is frequently somewhat larger. All digits clawed, the claw frequently showing inward flexure in

Figure 31.9. (opposite page)
Aquatilavipes curriei *ichnosp. nov.* Another trackway (grid cc18), photographed obliquely at W3 Main site. (The chalk marks show quarter-meter divisions).

Figure 31.10. Aquatilavipes curriei *ichnosp. nov.* Sketch of the pattern of footprints on the slab illustrated in figure 31.9.

Figure 31.11. (opposite page)
Tricorynopus? brinkmani
ichnosp. nov. Upper left and
right: the holotype impressions,
in two directions of illumination.
Lower left and right: other, less
clearly impressed prints, in two
directions of illumination
(indicated by arrows). Note that
a flaking-off of surface has caused
some prints to be incomplete at
right. Scale in mm.

relation to the digit axis. Digital pad impressions are visible on better-preserved molds or casts—three or four on digit III, two on digits II and IV.

Type Ichnospecies: Fuscinapeda sirin (Vialov 1966) Sarjeant and Langston 1994. Miocene (Helvetian), Ukraine.

Remarks: The diagnosis of this avian ichnogenus is emended to clarify that it differs from *Aquatilavipes* in the greater thickness of the digits.

Avian Footprints, ichnogen. indet.
Figure 31.5 (upper left)

Large, incomplete bird footprints occur on the second of the three bird footprint-bearing layers (layer B). Two impressions of single digits and one imprint exhibiting two unconnected digits may be seen on the *A. curriei* paratype slab, whereas two imprints showing two connected digits were seen (not illustrated) on another area on the W3 footwall, but no complete prints have yet been discovered. The nature and relative orientation of the digits are similar to those of *A. curriei* but represent footprints of a much larger bird, the length and width of the digits indicating that a complete print could be from one and a half to three times the size of *A. curriei* prints (approximately 14–18 cm long). These dimensions approach those of *Magnoavipes lowei,* from the Cenomanian of Texas (Lee 1997) and *Archaeornithipus meijidei,* from the Berriasian of Spain (Fuentes Vidarte 1996). The digits of the large bird prints from Grande Cache are much thicker than the slender digits of *Magnoavipes* and *Archaeornithipus;* however, *Magnoavipes* and *A. curriei* resemble one another in not showing any trace of a hallux impression. Because of the unsatisfactory character of the material discovered so far, we do not consider it proper to formally describe what is likely to prove a new ichnotaxon.

Class Mammalia
Order and Family Indet.
Ichnogenus *Tricorynopus* Sarjeant and Langston 1994
Tricorynopus? brinkmani ichnosp. nov.
(figs. 31.11, 31.12)

Derivation of Name: In tribute to Dr. Donald Brinkman, who discovered the holotype slab.

Diagnosis: Very small digitigrade to semidigitigrade, tridactyl footprints, the imprints of one foot (the presumed pes) being almost twice as large as those of the other foot (the presumed manus). In the presumed manus, the digits radiate symmetrically from the base, with an interdigital span of around 15°; they are moderately thick proximally and become narrower distally. All digits show sharp claws, directed more or less forward. The presumed pes had more flexible and widely spread digits, with an interdigital span of around 60°. In both manus and pes, digit III is longest. In the presumed manus, digits II and IV are of similar length, whereas

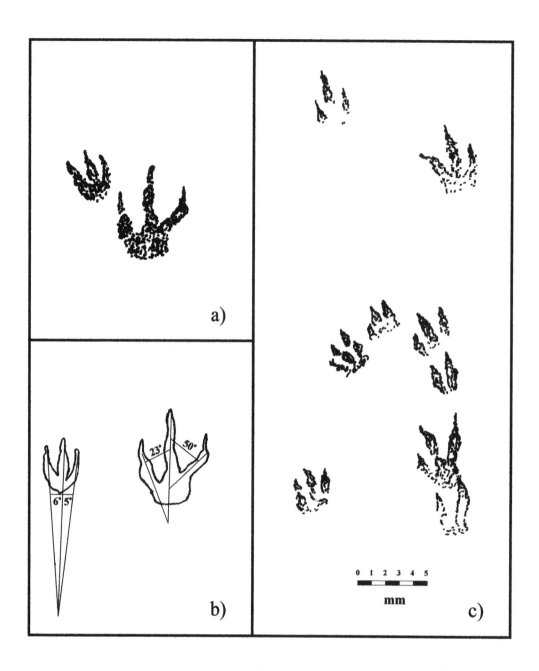

Figure 31.12. (above) Tricorynopus? brinkmani ichnosp. nov. (a) Sketch of the holotype print. Left (and upper): presumed right manus. Right (and lower), presumed right pes. (b) Interdigital angles. Left: presumed manus. Right: presumed pes. Measurements taken along the digit proper: all claws point forward. (c) Sketch of other mammalian prints on the slab (corresponding to fig. 31.12, lower). Note that a flaking-off of surface has caused some prints to be incomplete at right. All scales the same.

digit IV of the presumed pes is longer than digit III and curves outward. Trackway pattern not determined.

Holotype: Imprints at lower center of slab with *Aquatilavipes swiboldae* footprints (TMP 98.89.20), lodged in the Royal Tyrrell Museum of Palaeontology, Drumheller, Alberta (figs. 31.11 upper, 31.12a).

Dimensions: Holotype: presumed manus: length overall 3.5 mm, breadth 3.0 mm. Presumed pes: length overall 7.5 mm, maximum breadth 6.5 mm. Other imprints (figs. 31.11 lower, 31.12c) not capable of measurement.

Type Horizon and Locality: Holotype: Grande Cache Member of the Gates Formation, early Albian (Lower Cretaceous), Smoky River Coal Mine (Smoky River Coal, Ltd.) about 21 km northwest of Grande Cache, Alberta. Discovered in the talus at the base of the W3 footwall.

Remarks: Though a number of these small mammalian footprints are present on the lower central region of the type slab, neither the gait nor any indication of superposition could be discerned. Consequently, the distinction between manus and pes is based wholly on the presumption that the latter is likely to be larger than the former—an assumption recognizably difficult to justify, when so little is known concerning the postcranial morphology of small mammals of the late Mesozoic. A problem was the very light weight of the trackmakers: animals so small—only a few tens of grams—inevitably make very shallow footprints.

These footprints do not altogether accord with the diagnosis of the ichnogenus *Tricorynopus*, in that the presumed manus and presumed pes differ markedly in size. If the discovery of further specimens enables the trackway pattern and the identity of the manus and pes to be determined, it is likely that they will be placed into a new ichnogenus.

The lack of a determinable trackway and the extreme shallowness of the prints make it difficult to make detailed comparisons with any known group of mammals that might have made these footprints. Their size is not very diagnostic, since Lillegraven notes that Mesozoic mammals in general "were in the size range of modern shrews to rats" (1979, 2). It is because of the small size and frailty of their bones that only the teeth—not prone to digestive or erosional decay—are normally preserved.

In attempting a correlation between footprints and potential trackmakers, two methods of comparison are possible. The first is to compare the morphology of the footprint directly with known skeletal material. Since, in the case of Mesozoic mammals, there is an extreme sparsity of postcranial remains, a correlation of this kind cannot presently be done. Another approach involves identifying mammal taxa present in the particular time period during which, and the region where, the footprints were made. However, Clemens et al. caution that "negative evidence has little value for Mesozoic mammals" and that "the absence of a group of mammals at a particular time and place generally cannot be taken as an indication that it did not in fact occur then and there" (1979, 8).

The Cloverly Formation of Montana and Wyoming has yielded a significant amount of Lower Cretaceous mammalian remains, preserved within concretionary nodules (Clemens et al. 1979, 30). Triconodont mammals of the families Amphilestidae and Triconodontidae appear to be the most significant part of this mammalian fauna; however, the amphilestid specimens have a 35 cm body length, excluding the tail (Jenkins and Crompton 1979), and so are probably too big to be the trackmakers; neither is the large triconodont *Gobiconodon* (Jenkins and Schaff 1988). Some of the smaller triconodonts, known only from jaw fragments, such as *Corviconodon* (Cifelli et al. 1998), might be closer to the size of the mammal that produced these footprints. The middle Albian Paluxian Land Mammal Age within the Trinity Group of Texas and Oklahoma (Antlers Formation) contains triconodonts (Triconodontidae), multituberculates, symmetrodonts (Spalacotheriidae), and "Theria of metatherian-eutherian grade" (Aegialodontidae and Pappotheriidae) (Clemens et al. 1979, 30–31). Triconodonts are also known from teeth and lower jaw remains (approximately 21 mm long) from the Lower Cretaceous Arundel Clay of the Patuxent Formation (Cifelli et al. 1999). The Albian-Cenomanian Cedar Mountain Formation of Utah has likewise produced triconodont mammal remains, mainly in the form of teeth and jaw fragments (Cifelli and Madsen 1998). Most of these mammals are known solely from their teeth; few postcranial skeletal remains have been recovered. No Lower Cretaceous mammal remains are known from western Canada (Donald Brinkman, pers. comm. 2000). Consequently, it is beyond our ability to attempt any closer identification of the track-making mammals. It is hoped that continuing work at Grande Cache will lead to the discovery of additional mammal footprints, which may shed more light on the nature of the mammal ichnofauna of the Gates Formation.

Paleoecology of the Tracksites

Lockley and Rainforth (in press) report five bird tracksites in western Canada. With the addition of the Grande Cache bird tracksite, six are now known. In ascending stratigraphic order, these tracks are: *Aquatilavipes swiboldae*, Gething Formation (Aptian), Peace River Canyon, northeastern British Columbia (Currie 1981); *A. swiboldae*, Gladstone Formation (Aptian), near Grande Cache, western Alberta (Lockley et al. 1992); *A. curriei*, *A. swiboldae*, and ichnosp. indet., Gates Formation (Albian), near Grande Cache, western Alberta; *Jindongornipes*-like bird prints, Dunvegan Formation, British Columbia (Lockley and Rainforth in press); ichnosp. indet., St. Mary Formation, (Maastrichtian), southern Alberta (Lockley and Rainforth in press); ichnosp. indet., Horseshoe Canyon Formation (Maastrichtian), eastern Alberta (Lockley and Rainforth in press).

The *A. swiboldae* trackways discovered in the Peace River Canyon were not associated with dinosaur footprints (Currie 1981), though these are present elsewhere in the canyon (Sternberg 1932), nor have dinosaur or bird bones been discovered in adjacent strata. At the W3 Main tracksite, the original discoveries were made in talus blocks; the

A. swiboldae footprints seen at the outcrop were also not associated with dinosaur footprints. In contrast, the *Aquatilavipes curriei* trackways occur in association with an abundance of dinosaur footprints indicating a rich late Early Cretaceous fauna: *Tetrapodosaurus* (ankylosaurs) and *Irenesauripus, Ornithomimipus, Gypsichnites,* and *Irenichnites* (large to small theropods) (McCrea and Sarjeant 1999; McCrea et al. in press). The footprints are preserved in a rippled sandstone surface that contain an abundance of large and small invertebrate traces. Plant remains include not only logs, but also widely spaced tree stumps. Wan (1996) described the diverse Gates flora, which includes ferns, conifers, cycads, and ginkgoes, as well as two species of angiosperms. The paleoenvironment was a coastal plain or deltaic complex (Langenburg et al. 1987).

The *A. curriei* footprints are those of large wading birds, possibly in quest of invertebrates; however, no dabbling marks from the beaks of the birds have yet been recognized. Since there are no mud cracks, it is likely that the footprints were either made in water a few centimeters deep, or that the sediments exposed to the air were so water saturated that they did not dry out completely before burial by later sediments. On the paratype slab (fig. 31.5), the prints from trackway A are much more defined than those of trackway B, even though the trackway B prints overlie those of trackway A. Evidently an interval of time elapsed between the formation of the two trackways during which the substrate became slightly more resistant.

The mammal footprints (TMP 98.89.20) are in part superimposed on avian footprints. However, because they are only seen on a talus block, their temporal relationship to those footprints is unclear.

Bird or Dinosaur Footprints?

Footprint length and width measurements of all bipedal ichnotaxa that occur on the W3 Main footwall within the study area (McCrea and Sarjeant 1999) were compared using footprint length-width ratios. This method of measurement was initially used to distinguish the footprints produced by theropods from those produced by ornithopods, following the procedure of Moratalla et al. (1988). The footprint length-width ratios of the bipedal dinosaur ichnotaxa from W3 Main site are as follows: *Irenesauripus mclearni* (1.20, N = 10), *Ornithomimipus angustus?* (1.09, N = 12), *Irenichnites gracilis* (1.19, N = 11), and *Gypsichnites pascensis* (1.19, N = 27). The averages for all the dinosaur ichnotaxa lie below the 1.25 ratio used by Moratalla et al. (1988) to distinguish between footprints of ornithopods and theropods (theropod > 1.25 > ornithopod). However, the prints of *Aquatilavipes curriei* averaged much lower (0.73, N = 47), which is quite distinct from dinosaurian ichnotaxa ratios (fig. 31.13). Currie's (1981) calculation of average footprint length and width for *A. swiboldae* prints produces a ratio (FW/FL) of 0.80 (N = 44), which is comparable to that of *A. curriei.* He also studied the footprints of some extant paludicolous birds—the killdeer (*Charadrius vociferus*) and the great blue heron (*Ardea herodias*) for comparison with *Aquatilavipes swiboldae* foot-

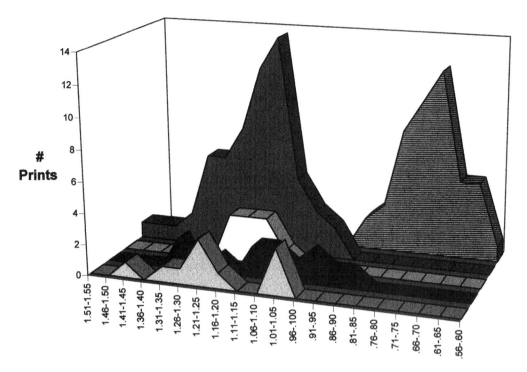

Prints

FL/FW

- ▨ *Irenesauripus mclearni*
- ■ *Ornithomimipus augustus?*
- ☐ *Irenichnites gracilis*
- ■ *Gypsichnites pascensis*
- ▤ *Aquatilavipes curriei*

Figure 31.13. Graph of footprint length/width (FL/FW) ratios of dinosaur and avian ichnotaxa from the W3 Main tracksite, Smoky River Coal Mine, Alberta.

prints. Using the original data tables from Currie's study, we have found that the footprint length/width ratios are: *C. vociferus* (0.88, N = 40), *A. herodias* (0.90, N = 14). These values are slightly higher than those of *Aquatilavipes swiboldae* and *A. curriei*, but still well below the average ratio of the W3 Main dinosaur ichnotaxa. By using footprint length/width ratios, as well as the criteria set out by Lockley et al. (1992), it seems possible to distinguish the footprints made by birds from those made by dinosaurs. Further research needs to be conducted in this area, particularly on the study of modern footprints produced by different taxa of extant birds.

Acknowledgments: We are grateful for the assistance of Smoky River Coal, Ltd., who, realizing the importance of their paleontological resources, continue to allow access to important sites within their mining operation. R. T. M. wishes to acknowledge the efforts of Sandra Jasinoski and Mark Mitchell of the Royal Tyrrell Museum of Paleontology as research assistants during the summer of 1998. Dr. Philip Currie and Dr. Eva Koppelhus were indispensable in collecting footprint data for this research; Dr. Donald Brinkman and Mr. Michael Getty also made important contributions. The difficult task of photographing the shallow mammal prints was admirably carried out by Mr. David Mandeville (Audiovisual Services, University of Saskatchewan). Dr. Peter Flood, Department of Veterinary Anatomy, Western College of Veterinary Medicine, University of Saskatchewan, furnished helpful comments on foot deformities in poultry. Funding and logistical support were generously provided by Smoky River Coal, the Royal Tyrrell

Museum of Paleontology and the Heaton Student Support Grant. This work was completed in the Department of Geological Sciences, University of Saskatchewan.

We also congratulate Dr. Philip Currie on his tireless devotion to vertebrate paleontological research in Canada and abroad, and are happy to acknowledge his many accomplishments. His contributions to this discipline in a scientific research capacity, as well as a very visible public spokesman role, have been most impressive.

References

Casamiquela, R. M., G. R. DeMathieu, H. Haubold, G. Leonardi, and W. A. S. Sarjeant. 1987. Glossary in eight languages and Discussion of the terms and methods. In G. Leonardi (ed.), *Glossary and Manual of Tetrapod Footprint Palaeoichnology*, pp. 21–52. Brasília: Ministerio de Minas Energia, Departamento Nacional de la Produção Mineral.

Cifelli, R. L. and S. K. Madsen. 1998. Triconodont mammals from the medial Cretaceous of Utah. *Journal of Vertebrate Paleontology* 18 (2): 403–411.

Cifelli, R. L., J. R. Wible, and F. A. Jenkins Jr. 1998. Triconodont mammals from the Cloverly Formation (Lower Cretaceous), Montana and Wyoming. *Journal of Vertebrate Paleontology* 18 (1): 237–241.

Cifelli, R. L., T. R. Lipka, C. R. Schaff, and T. B. Rowe. 1999. First Early Cretaceous mammal from the eastern seaboard of the United States. *Journal of Vertebrate Paleontology* 19 (2): 199–203.

Clemens, W. A., J. A. Lillegraven, E. H. Lindsay, and G. G. Simpson. 1979. Where, when, and what: A survey of known Mesozoic mammal distribution. In J. A. Lillegraven, Z. Kielan-Jaworowska, and W. A. Clemens (eds.), *Mesozoic Mammals: The First Two-Thirds of Mammalian History*, pp. 7–58. Berkeley: University of California Press.

Currie, P. J. 1981. Bird footprints from the Gething Formation (Aptian, Lower Cretaceous) of Northeastern British Columbia, Canada. *Journal of Vertebrate Paleontology* 1 (3–4): 257–264.

Currie, P. J. 1989. Dinosaur footprints of western Canada. In D. D. Gillette and M. G. Lockley (eds.), *Dinosaur Tracks and Traces*, pp. 293–300. Cambridge: Cambridge University Press.

Ellenberger, P. 1980. Sur les empreintes de pas de gros mammifères de l'Éocène supérieur de Garrigues-Ste-Eulalie (Gard). *Palaeovertebrata, Mémoire jubilaire R. Lavocat*: 37–78.

Fuentes Vidarte, C. F. 1996. Primeras huellas de Aves en el Weald de Soria (España). Nuevo icnogenero, *Archaeornithipus* y nueva icnoespecie *A. meijidei*. *Estudios geologicos* 52: 63–75.

Jenkins, F. A., Jr., and A. W. Crompton. 1979. Triconodonta. In J. A. Lillegraven, Z. Kielan-Jaworowska, and W. A. Clemens (eds.), *Mesozoic Mammals: The First Two-thirds of Mammalian History*, pp. 74–90. Berkeley: University of California Press.

Jenkins, F. A., Jr., and C. R. Schaff. 1988. The Early Cretaceous mammal *Gobiconodon* (Mammalia, Triconodonta) from the Cloverly Formation in Montana. *Journal of Vertebrate Paleontology* 8: 1–24.

Kordos, L. 1983. Lábnyomok az ipolytarnóci alsó-miocén korú homokkőben [Footprints in Lower Miocene sandstone at Ipolytarnóc, northern Hungary]. *Geologica hungarica, ser. Palaeontologica* 46: 259–415.

Langenburg, C. W., W. Kalkreuth, and C. B. Wrightson. 1987. Deformed Lower Cretaceous coal-bearing strata of the Grande Cache area, Alberta. *Alberta Research Council, Bulletin no. 56*.

Lee, Y.-N. 1997. Bird and dinosaur footprints in the Woodbine Formation (Cenomanian), Texas. *Cretaceous Research* 18: 849–864.

Leonardi, G. 1987. *Glossary and Manual of Tetrapod Footprint Palaeoichnology.* Brasília: Departamento Nacional da la Produçaõ Mineral.

Lillegraven, J. A. 1979. Introduction to J. A. Lillegraven, Z. Kielan-Jaworowska, and W. A. Clemens (eds.), *Mesozoic Mammals: The First Two-Thirds of Mammalian History,* pp. 1–6. Berkeley: University of California Press.

Lockley, M. G., and E. C. Rainforth. In press. The track record of Mesozoic birds and pterosaurs: An ichnological and paleoecological perspective.

Lockley, M. G., S. Y. Yang, M. Matsukawa, F. Fleming, and S. K. Lim. 1992. The track record of Mesozoic birds: Evidence and implications. *Philosophical Transactions of the Royal Society of London,* ser. B, 336: 113–134.

McCrea, R. T., and P. J. Currie. 1998. A preliminary report on dinosaur tracksites in the lower Cretaceous (Albian) Gates Formation near Grande Cache, Alberta. In S. G. Lucas, J. I. Kirkland, and J. W. Estep (eds.), Lower and Middle Cretaceous terrestrial ecosystems. *New Mexico Museum of Natural History and Science Bulletin* 14: 155–162.

McCrea, R. T., and W. A. S. Sarjeant. 1999. A diverse vertebrate ichnofauna from the Lower Cretaceous (Albian) Gates Formation near Grande Cache, Alberta. *Journal of Vertebrate Paleontology, Abstracts* 19 (3): 62A.

McCrea, R. T., M. G. Lockley, and C. A. Meyer. In press. Global distribution of purported ankylosaur track occurrences. In K. Carpenter (ed.), *The Armored Dinosaurs.* Bloomington: Indiana University Press.

Moratalla, J. J., J. L. Sanz, and S. Jiménez. 1988. Multivariate analysis on Lower Cretaceous dinosaur footprints: Discrimination between ornithopods and theropods. *Geobios* 21 (4): 395–408.

Sarjeant, W. A. S. 1989. "Ten paleoichnological commandments": A standardized procedure for the description of fossil vertebrate footprints. In D. D. Gillette and M. G. Lockley (eds.), *Dinosaur Tracks and Traces,* pp. 369–370. Cambridge: Cambridge University Press.

Sarjeant, W. A. S., and W. Langston Jr. 1994. Vertebrate footprints and invertebrate traces from the Chadronian (Late Eocene) of Trans-Pecos Texas. *Texas Memorial Museum Bulletin* 36: 1–86.

Sarjeant, W. A. S., and R. A. Thulborn. 1986. Probable marsupial footprints from the Cretaceous sediments of British Columbia. *Canadian Journal of Earth Sciences* 23: 1223–1227.

Sternberg, C. M. 1932. Dinosaur tracks from Peace River, British Columbia. *National Museum of Canada, Annual Report,* 1930, pp. 59–85.

Thulborn, R. A. 1990. *Dinosaur Tracks.* London: Chapman and Hall.

Vialov, O. S. 1965. *Stratigrafiya neogenovix molass Predcarpatskogo progiba.* Part K. Kiev: Naukova Dumka.

Vialov, O. S. 1966. *Sledy zhiznedeiatel'nosti organizmow i ikh paleontologicheskoe znachenie* [Traces of the activity of organisms and their paleontological meaning], pp. 5–53. Institut Geologii i Geoximi Gway ix Iskopaemuse, Akademya Nauk Ukrainskoi SSR.

Wan Z. 1996. The Lower Cretaceous flora of the Gates Formation from western Canada. Ph.D. thesis, University of Saskatchewan, Saskatoon.

Section VII.
Dinosaurs
and
Human
History

32. Bones of Contention: Charles H. Sternberg's Lost Dinosaurs

DAVID A. E. SPALDING

Abstract

In 1916 fossil collector Charles H. Sternberg and his son Levi left the Geological Survey of Canada and collected for the British Museum (Natural History). Two shipments of dinosaurs were sent across the Atlantic, and the second and better of these, on the SS *Mount Temple*, was sunk by a German raider. Confusion over the nature of the sinking is clarified, and transcriptions of Sternberg's letters show the extent of his personal disappointment and the financial crisis it precipitated in his affairs. The episode sheds light on the Sternberg's work as a freelance fossil collector, his personality, and his relationship with museums. Sternberg's letters also offer specimens from California and give some information about 1917 fieldwork in Texas.

Introduction

Charles H. Sternberg (1850–1943) was one of the leading collectors of North American vertebrate fossils during most of his life. He collected widely in North America, initially with his three sons George Fryer, Charles Mortram, and Levi, and other assistants. Always a freelance collector, he worked for both Cope and Marsh, and sold his fossils to many museums, particularly in North America and Europe. His life has been chronicled by himself in numerous papers and two autobiographical books (Sternberg 1909, 1985), as well as by biographers Riser (1995) and Rogers (1991).

Brief accounts relating to the lost dinosaurs episode appear in Sternberg's 1918 paper and the second edition of his *Hunting Dinosaurs* (1932). Reference to the episode are made in histories of dinosaur study focusing on Canadian work (Russell 1966; Spalding 1999), and on the wider scene (Colbert 1968; Spalding 1993). The incident earns passing reference in many other books on dinosaurs (e.g., de Camp and de Camp 1968; Gross 1985; Psihoyos and Knoebber 1994).

Between 1912 and 1916, Sternberg and (except for brief period when George was employed by Barnum Brown) his sons were working on contract for the Geological Survey of Canada. The survey was at that time responsible for the National Museum of Canada, Ottawa, and supervision was provided by Lawrence Lambe (1863–1919). The Sternbergs were initially in competition with Barnum Brown (1873–1963) of the American Museum of Natural History. This period has become known as the Canadian Dinosaur Rush, from a chapter title in Colbert's 1968 book; it was later referred to as the Great Canadian Dinosaur Rush in Gross 1985.

Brown's Alberta collecting ended in 1915. Next season, the survey decided not to resume fieldwork, though it continued preparation of the material collected. Sternberg was passionate about his fieldwork, and in May 1916 resigned from the survey with his son Levi (1894–1976). This incident broke up the family team, for his older sons, George Fryer Sternberg (1883–1969) and Charles Mortram Sternberg (1885–1981), remained with the survey—the former for a year or two, and the latter for the rest of his working life.

In order to continue collecting in an area he had found very rich in dinosaur material, Sternberg arranged to do contract work for the British Museum (Natural History), to which he had previously sold specimens. Sternberg's client and principal correspondent was Dr. (later Sir) Arthur Smith Woodward (1864–1944), Keeper of the Department of Geology, and senior vertebrate paleontologist at the museum, whose interests ranged from fossil fish to Piltdown Man. Despite the raging of World War I, the museum was able to find funding from the Percy Sladen Memorial Fund, a foundation based at the Linnean Society that encouraged natural history research. Another correspondent was W. D. Lang (1878–1966), invertebrate paleontologist at the museum from 1902 to 1938. Other paleontologists mentioned in the correspondence include William Cutler (1878?–1925), a freelance collector in Alberta and Africa, and Charles Whitney Gilmore (1874–1945), vertebrate paleontologist at the U.S. National Museum. Biographical sources may be found in Sarjeant 1980–96.

Sternberg returned to sites in the Steveville area (now part of Dinosaur Provincial Park, Alberta), where he was working Upper Cretaceous beds then regarded as the Oldman Formation. Charles and Levi found several skeletons, which were sent in two batches to London. The first shipment was successfully transported, but the second was lost when the SS *Mount Temple* was sunk by German action in the mid-Atlantic.

This is not the only occasion on which fossils have been lost while being transported by ship—Sternberg himself had a prior experience with a *Megatherium* that had been sunk and recovered by divers (Sternberg 1985, 271). An earlier collector in Canada, Thomas Chesmer Weston, also lost fossils from the Alberta prairies in 1883 when the *Glenfinlas* sank in Lake Superior (Weston 1899, 152). However, Sternberg's loss on the *Mount Temple* seems to have been the most serious.

Although the material was lost to science, the correspondence associated with the episode provides much information about Sternberg's operations. Sternberg's correspondence relating to the episode is transcribed (with summary and brief quotation of some related documents), followed by discussion.

The Correspondence

A file of papers related to the incident is held by the Natural History Museum in London. Sternberg's letters to Woodward show the story mainly from his perspective. Copies of some other correspondence relating to the incident, involving notably museum staff, the British and Foreign Marine Insurance Company (BFMIC), Ltd. (Liverpool); their agents, Dale and Co. (Montreal), the Canadian Pacific Railway, and the Sladen trustees; as well as a fossil list and a press cutting, are also included in the file. Abbreviations indicated are used in the headings below. Signatures are only given the first time they were used.

Notes on Transcriptions

Sternberg's letters were typed (presumably by himself) on plain paper. He seems to have used a rubber stamp to provide the address, until January 1917, when he has an elaborate letterhead with "Office of Charles H. Sternberg, A.M., 1315 Connecticut Street, Lawrence, Kansas" and a lengthy list of places "My Fossils Have Been Sent To," followed by some information on his experience and publications.

There are numerous spelling errors, irregularities in spacing, and words which run off the page. He has usually corrected the letters with deletions, additions, and underlinings—sometimes with the typewriter and sometimes with a pen. Most punctuation seems to have been added afterward.

Despite these irregularities, the text of the letters is invariably clear, except for occasional handwritten words. In transcription I have aimed at presenting Sternberg's intended letter as mailed while impeding the flow of the text as little as possible by reproducing too many of his mistakes. Thus I have retained Sternberg's paragraphing, capitalization, abbreviations, spelling (without adding *sic*), underlining, and punctuation, both typed and handwritten. Words that have run off the page and are not picked up in the next line are completed in square brackets, th[us]. Handwritten annotations (some initialed by Woodward or Lang) have been made in the margins of some documents. These are transcribed in the appropriate place, and like all handwritten items, are in italics. Addresses are transcribed the first time and thereafter only if they change; dates have been standardized year.month.day in the heading for the letter. Other letters have been transcribed in full or paraphrased (in brackets)—sometimes with brief quotations—according to their importance. My comments and paraphrases are enclosed in curly brackets: {/}.

1916.05.10: BM application to Percy Sladen Memorial Fund
{handwritten draft, perhaps by Smith Woodward}

To employ a highly skilled and experienced collector, Mr Charles H. Sternberg, to obtain remains of Dinosaurian reptiles from the Upper Cretaceous deposits of the Red Deer River region in Alberta, Canada. This is perhaps the richest deposit of nearly complete Dinosaurian skeletons in the world, and has already been explored for several seasons by Mr Sternberg on behalf of the Victoria Museum, Ottawa. He has obtained a most remarkable collection for that museum, but cannot be employed this season on account of the war. He therefore offers his services to the British Museum, but this institution also lacks funds on account of the war. He is personally known to Dr. Woodward and many of his friends, who have proved Mr Sternberg to be a thoroughly honest man. He has several times collected for the British Museum in Kansas and Wyoming, always with excellent results.

The Cretaceous Dinosaurs of Alberta comprise a great variety of the strangest armoured forms related to Triceratops besides other most astonishing developments of the Iguanodont and Megalosaurian groups. A small but valuable collection was made for the British Museum in 1914 by Mr. Wm. E. Cutler (who is now serving with the Canadian troops in France), and this shows the richness of the accumulations and the fine state of preservation of the skeletons.

The American Mus., N.Y. through the aid of private benefactions, has explored the region for several years and obtained remarkable collections for New York; and unless Mr Sternberg can be employed for Britain, all the discoveries this year will be sent to that museum.

Mr Sternberg is willing to work for two months with his complete outfit (including at least 2 skilled assistants) & to send all discoveries to London for the inclusive sum of $2000 (say £400), the first half to be paid to him at the end of his second month (July). If his results were considered satisfactory thus far, he would be willing to continue his work for the other two following months at the same rate ($1000 per month), making a total outlay during the season of $4000 (say £800). Judging from experience with Mr. Cutler's collection in 1914, the freight expenses to London would be about £100.

I propose that the whole collection should be sent to the British Museum, where I feel sure the {word illegible}would agree to clean and prepare all the specimens in return for the gift of the first selection therefrom.

1916.06.04: T. Bailey Saunders (Sladen) to Woodward
{application received} . . . the Trustees are willing to undertake the responsibility for these payments and the necessary freight; if and when the specimens are received at the British Museum and a favourable report made upon them. . . .

1916.09.30: C.H. Sternberg to Woodward
STEVEVILLE, ALBERTA; CANADA
Dr A Smith Woodward,
British Museum of Natural History, LONDON
My dear Sir:-

The second two months ends today, but we have failed to take up the second, rather the third skeleton of a crested dinosaur No. 13: represented at the discovery, by one hind foot (except the phalanges that were nearly all lost) The foot was sticking out of a perpendicular bluff and it has taken unremitting labor during an exceptionally pleasant month, We have not lost a working day. There are still three sections in the quarry we have not wrapped yet If the weather will permit (we are having our first snow storm today), we will get them all wrapped by next tuesday. We have had the most wonderful success <u>three skeletons</u> that can be mounted. But this last one in point of perfection far exceeds the others. The entire trunk with all four limbs and arches in position with the arches column and ribs present preserved in fine sandstone with much of the skin impressio[ns] to be preserved I believe, if care is taken. When we had uncovered the skeleton to the neck, I was sure it was the second best dinosaur discovered here, Brown got the best, But as is so often the case I had the bitter disappointment to find both the neck and skull is missing. We have 12 feet continuous of the tail, Only about three feet of the extreme end missing. Then we have in No. 9. the complete extreme end of the tail. In No 6 we have much of the skull and a complete neck, So by restoring the front part of the head of No 9 and the extreme end of the tail you will have a far better skeleton than any in Ottawa that miss the tail in the best trachodont, and crested duckbill, We have been wonderfully successful No 1 was good, No 9 was better, and No 13 is the best of all. Two skeletons that can be mounted in the last two months. Rather two months and a half, Because we can not possibly get the material out of the breaks and boxed and delivered at the depot before the middle of Oct, I am sure however that I will not lose, as I trust you entirely, knowing that I have done my duty to the very limit of human endurance, and I know you will do yours.

{The section from "We have had the most wonderful success" to "I know you will do yours" is boxed with a marginal comment:} *"This relates to the collection lost in the Mount Temple. ASW"*

I hope the first shipment is enroute, I was ordered to go to Jenner to get my Bill of Lading which I did I think it would be a good thing for you to cable Mr D.C. Coleman Asst Genl Manager And ask him to prepay the premium on $2500 You re paying in London, It is impossible for me to get in rapid touch with him No Wire and mail only twice a week, I instructed him to send me the bill and I would pay but have not heard from him. I enclose a copy of the B.L

{Marginal note beside this paragraph:} *"This lot was duly received by SS. Milwaukee"*

I fear it will be impossible for me to take up the Sand Cr. fossils as the weather is getting so bad we cannot mix plaster, Levi has all the ends of his fingers eaten off by the plaster making the work painful especially in cold weather We have to heat the water for him. But the main reason is, <u>we cannot mix it in freezing weather</u>. And cannot take up a single specimen without plaster.

I hope you will be able to arrange for my comming here next season The big horned and plated dinosaurs are more abundant on Sand Cr. and there are ten times the exposures there, to those here, below the mouth of Berry Cr

In case also you have enough crested dinosaurs and desire a Milodon, Smilodon and great wolf etc from the Tar Pits of California, I believe I could arrange a good exchange for you, They have no Dinosaurs on the Pacific Coast but plenty of sloths and saber toothed tigers.

I would like to build up in the British Museum the third largest collection of Red Deer dinosaurs We can never hope to excell Browns He was here 6 years ago with large party We can however be equal or even superior to Ottawa *if you please* .

I am very anxious to get home as camp life does not agree with me in cold weather But I will not leave until I know this second collection is sure of shipment It will be a much larger one than the first and worth twice as much.

Faithfully yours
Charles H. Sternberg

1916.11.06: C.H. Sternberg to Woodward
Charles H. Sternberg, 1315 Conn. St., Lawrence, Kans.
My dear Sir:-

I am enclosing the Certificate of Insurance No 205006 which I have assigned to you and bill for the same I paid the Premium as pr Dales and Co's receipt. $33 75

Levi wrote me that he had successfuly got all the second shipment—22 boxes—on board the cars on the 21st of Sept last. The conductor took the shipment down to Empress and sent it on its way from there via a passengers train. He got the local bill of lading which he sent to the Division Freight Agt. at Calgary

In return for the same I am to receive a through B/L that I will send you as I did the one for the first shipment. Mr G.D. Robinson Export Agent of the C.P.R. Montreal has just informed me he will arrange to pay the premium on the second lot and will send you bill for the premium and freight. I instructed him to insure at same Rate viz $2500 I am now waiting anxiously every day for returns for the first shipment which I hope you have received

Faithfully yours

{no date} List of Fossil Vertebrates collected by Charles H. Sternberg for Dr. A Smith Woodward for the British Museum of Natural History London from the Belly River Series below Steveville about 2 miles

{Breakdown of contents of 22 boxes}

No. 9 Crested Duckbilled Dinosaur
    Found by Levi Sternberg
    Near head of canyon 2 miles east of Steveville Alberta 100 ft
    below the Prairie.
[list of boxes and contents]
No. 13 Crested Dinosaur (duckbilled) Found by Levi Sternberg Half
    a mile from No 9 and quarter of a mile from No 1 in the
    Steveville badlands
[list of boxes and contents]
{This must be the second shipment sent, as first lot was in 23 boxes; see
letter November 20, 1916}

1916.11.07 C.H. Sternberg to Woodward
My dear Sir:-
    I am pleased to enclose Original Through Bill of Lading for the last
shipment of 22 boxes I have received Notice from Mr Halstead Division Freight Agt. Calgary That the Export Agent Mr Robinson of
Montreal expected to forward this shipment by the SS Mount Temple
Nov 1st So I hope it is about in English waters and you will soon receive
it Mr Robinson wrote me from Montreal he would make arrangements
to have the Marine Insurance paid by them and collected with the
freight charges.
    I am anxiously waiting returns from the first shipment.
                                                        Faithfully Yours

1916.11.20: C.H. Sternberg to Woodward
My dear Sir:-
    I was glad indeed to receive your very welcome letter of the 3rd
telling me of the safe arrival of the first lot of 23 boxes, and sorry I did
not send the B-L sooner. In fact I did not think it necessary at first, but
on my arrival here, I was told that it was customary to send the B L
forward with the material),) I sent the second one as soon as I got it on
the 7th Instant the first I sent some time before.
    I was going to cable you today to ask when I was to receive
returns—Owing to the fact that I was obliged to sell my home in
Ottawa at a great loss, move my family here, and buy my old home at
another expense, I am naturally anxious to get some returns after my
long and strenuous labor for your museum.
    It is certainly a joy to me to know that I have been successful and
that you have received, or will receive so much material All new to your
museum The last lot in my estimation is worth more than double the
first as it contains two nearly complete skeletons of the crested duck-
bills
    My bank at Ottawa was the Bank of Ottawa Elgin St Branch, You
might transfere the account through it, if convenient I had to pay 1 pr
ct discount on Canadian money when I got U.S. money for it
                                                        Faithfully yours

1916.11.24: G.D. Robinson freight agent to Woodward
Canadian Pacific Railway Co.
Export Freight Department,
G.D. Robinson
Export Freight Agent
Montreal

I enclose herewith insurance certificate #27261, $2500.00 covering 22 boxes fossils shipped by Chas. H. Sternberg, from Patricia [a]nd which will clear from Montreal on the SS "Mount Temple" sailing the 25th inst.

The premium on this risk, viz $60.00 has been advanced forward on the bill of lading to be collected in London.

Acknd. 15.12.16
{Wrote to Liverpool office making claim 18.1.17; also to 7.03. Saunders, Sheraton{?}& Co C.P.R. Agent in London.}

1916.12.28: C.H. Sternberg to WoodwardMy dear Sir:-

I have received from Mr T. Bailey Saunders, Percy {Sladen} Memorial Fund a check for $400 this morning. It is certainly welcome though I do not know what it[shipment of fossils] will sell for in New York. My banker does not expect to get more than $1912 for it, and perhaps even less, I had expected to realize $5 00 a pound as I have before the war instead of getting $2000 I will lose $88 Then I have paid more than $30 in interest on money borrowed to carry me while I was waiting for returns, I have not heard a word about the second shipment and what you are going to pay me for that I had no contract with you for more than 4 months or £800 in all. I am extremely anxious to know whether I am to receive more than that, I also sent in my bill for $33 73 Insurance on the first shipment This I have not received. I only feel disappointment that I should lose the exchange [and{?}] cost of collecting on the first shipment, and earnestly hope that in consideration of the much more valuable second two months collection extended to nearly three months I will not lose on that also. It has been 8 months now I have had to borrow money to carry on the work, If I had failed to receive the money it would have ment ruin.

I have not yet lost faith, and hope soon to know what my next check will be.

I would like also to know what the prospect is for the next summers work, I have told you of what we have been forced to leave uncollected in the field I would therefor be much pleased to know whether I can depend on serving you again in the same field next June Rather we could get in the field the middle of May or even the first, hunt the Sand Cr. Beds and take up the material as soon as possible

I am willing to contract for 5 months there at $1000 a month not pounds and will take more help from here so we can take up double likely that we did last year.

Or I will go on the same terms for two or three or four or five months I have heard others mean to go in there next year and it is vital for great succes[s] that I reap the first harvest there. I usually glean the ground as I go.

I know you do not regulate the price of English pounds in the American market nor is it your fault that I lose by the exchange, I cannot help feeling however when I look back on the splendid lot of material I sent you you will do all you can for me with the next payment, I am sending you some photographs Levi took

No 1 <u>Ceratopsia</u> Quarry The bone in front marked around is the so called parietal bone. Notice the horns projecting back of the cross bar of the crest about a foot long

No 2. Quarry 13, Where the 3rd <u>Corythosaurus</u> skeleton came from.

No 3 Quarry No 9: Where the 2nd skeleton came from.

No 4 Quarry 1 showing the splendid Limb bones of No 1 the scattered skeleton[s?]

I am sorry to say, I received a month ago a paper by Barnum Brown, in which he describes for the first time the skull of No 3. (Your specimen is a much finer specimen) He calls it Prosaurolophus I shall anxiously await your next letter As I must now plan for the future. How would you like to send us to the Permian of Texas? It is a long time since I have been there and I can work there as early as Feb or March.

Faithfully yours

Jan 15th. 1917. Difficult to convince Sladen Trust of value of collection. Have Insurance Policy for $2500 for second coll. ASW
Jan 18th, Announced loss and promised payment of insurance directly from Montreal.

1917.01.19: BFMIC to Woodward
{*Mount Temple* please send certificate of insurance—does it include war risk?}
20th Sent certificate B/L
[*on reverse*] *Policy No. 44, Nov 22nd 1916*
Cert No. 37261 "including war risk" shipped on S.S. Mount Temple
Through B/L 1304, Contract no. 133
Calgary 3rd Nov 1916

1917.01.21: Saunders (Sladen) to Smith Woodward
The Percy Sladen Memorial Fund,
c/o The Linnean Society,
Burlington House,
Piccadilly, London, W
Dear Dr Smith Woodward,

I received your letter confirming a rumour, passed on by Mr Bury, that the ship carrying Mr. Sternberg's 2nd collection had been sunk. As you wittily observe, so end the bones of contention. I am sorry if any specimens of great interest are thus lost to the museum.

I have circulated your letter among the trustees and suggested something to them, and as soon as get their reply I will write you again.
Yours sincerely
T. Bailey Saunders

1917.01.22: BFMIC to Woodward
{Accepts that War Risk is covered in the insurance, and asks for second bill of lading and description of shipment}

"According to the reports in the press, The Captain and crew appear to have been landed at St. Vincent"
{It is usual to have the captain's protest or an official statement from the owners as to the loss of a vessel.}
" . . . state to whom you wish the loss paid, when the documents are all in order."

1917.01.24 Cablegram C. H. Sternberg to Woodward
23 JAN 1917

WOODWARD BRITISH MUSEUM LN

SHIPMENT LOST MOUNT TEMPLE COLLECT INSURANCE REMIT ANSWER STERNBERG
24.1.17 Insurance claimed want second bill of lading Woodward

1917.01.22: C. H. Sternberg to Woodward
Office of Charles H. Sternberg, A.M. 1315 Connecticut Street, Lawrence, Kansas.
My Dear Sir:-
On receipt of your wire to the effect that the second shipment had not arrived in London, I at once wrote to the Freight Agent. Mr. John Halstead CPR Calgary with BL. enclosed He at once traced the shipment and wrote to me the 18th of January as follows. "Referring again to your letter of the 10th inst. and my letter of yesterday, I have today received telegram from Export Freight Agent Montreal stating that Steamer Mt Temple from Montreal Nov 26, has not yet arrived at England. I am advised by our Passenger Department here that this vessel has been reported sunk."

This is bitter news for me as well as for you, As I considered the two skeletons in that shipment worth two or three times what the first shipment was, because it contained two skeletons that could be mounted. I will wire today and ask you to collect the insurance $25000 and send it to me at once. The great expense of my expedition makes it necessary for me that I receive immediate returns. I hope if you can influence the Insurance Co to hurry up you will, or that you will get the money from the Sladin Fund and collect from the Insurance Co. Now it occures to me that you cannot afford to lose the magnificent collection that can be made next year on Sand Cr. This is, what I propose You give me a years salary of $500 per month from the first of next May until the following May, and I will not only collect there as long as it is possible to work. but prepare the material during the winter, and when prepared store them at your expense until the waters of the ocean are safe. I cannot bear to think of this awful loss and will devote a year at least to repairing the loss to you, I will take men with me so the man power required will not be missing, No one can be hired there All the able bodied men that can get away from Canada are fighting to put an end to such outrages.

Faithfully yours

Feb 19th Still waiting for B/L

1917.01.23: Draft Woodward? to BFMIC?
{on memo paper headed}:
With Messrs James Powell & Sons
Whitefriars Glass Works
Tudor Street
London E.C.
To Liverpool "Mount Temple"
Return signed subrogation form
Second Bill of Lading not sent
Have written to Consignor, Mr C.H. Sternberg
If B/L returned, London branch of C.P. Ocean Services will endorse it that 22 boxes actually shipped by the Mount Temple.

... his Company accepts the Admiralty statement as to the loss of the steamship, and has reason to believe that the German Admiralty statement is correct that the "Yarrowdale" with the captain and crew of the "Mount Temple" has arrived in Germany.

It is not possible to invoice fossils in their rough state, but I enclose the collectors' enthusiastic though rather rambling letter for your inspection.

When documents are all in order I desire your Montreal office to make payments on my behalf as follows:-.

Mr Charles Sternberg
(address)

U.S.A. Freight Agent, C.P.R. Montreal

2177.20	2500.00
322.80	322.80
2500.00	2177.20

1917.01.24 BFMIC to Woodward
{Enclose 1st Bill of Lading to get endorsed by shipowners that the fossils were "actually shipped on board this vessel." Owners of vessel accept Admiralty's statement, but will their Underwriters?}

1917.01.28: C.H.Sternberg to Woodward
I was delighted to receive after so long a time, your letter of the 15th. I have already written you, what in my judgement would be a good scheme for next years work. I am willing to work as I did last year at a $1000 a month for two or more months. But certainly the material should not be shipped until the sea is reasonably safe. I greatly feared

the first two months collection would not impress the Trustees as much as as the second In my estimation the two skeletons Levi found (and went to the bottom) were worth two or three times what the first collection. As what was lacking in one skeleton was present in the other, Then they <u>were both articulated skeletons</u> and in much better rock than the other material. You must know that when we get material in the fossil beds we have to take it as we find it. and I labored nearly five months for you last year. If all had reached you in safty it would have been of more value than what George Sternberg secured with his large party of men last year. Or I had discovered for the Geological Survey of Canada the year before I labored last year as never before, and it is terrible for me to lose the main fruit of my labor without your Trustees of the Sladin fund making it harder by refusing to continue the work. Because, I am sure we could in all human probability secure a fine collection next year. I know how difficult it is to prepare this material, and it takes years to fully develope it unless as in the case of the two skeletons that went to the bottom where it is easier Further more it has been a financial loss to me any way, as my expenses have been so great. I hope you will arrange it some way so we can recover from the rocks the material lost at sea. I sent the second Bill of Lading to John Halstead Div. Frt. Agt C.P.R Calgary to trace the shipment which he did to Davies Locker I wired him, on receipt of your cable to send it to me at once. It has not yet come I will send it to you the moment I receive it. I do not think I should send the only remaining Bill of Lading I have, There were three sent. Please wire the instant you know, that you cannot give me employment, or you can next season. Now if you can get the money, in case I get satisfactory material, I will go there in May and collect and prepare, giving you the first chance to purchase I can do no more I hope you will succeed in securing at least the choice of the material next year

Faithfully yours

1917.01.30: BFMIC to Woodward
{Asking for indemnity for missing document and form of undertaking as no official statement of loss of steamer. Will then certify amounts for payment.}

{undated} Note in *Science*
{Press cutting attached to letterhead for *Nature,* Macmillan and Co., Ltd., St. Martin's Street, London, W.C.}
"*from Science*" {As Sternberg is the source of the story, the cutting is presumably from *Science* but supplied to the museum by the office of *Nature.*}
Two skeletons of the duck bill dinosaur were lost to science with the sinking recently by a German raider of the ship *Mont Temple,* according to Charles H. Sternberg, of Lawrence, Kans., who found the bones in the red deer country in Alberta, Canada. The prehistoric specimens were thirty-two feet long and were being sent to the British Museum. They filled twenty-two boxes and weighed 20,000 pounds. When the shipments failed to arrive in England, an inquiry was made by Mr.

Sternberg and he received word from the Canadian railroad officials of the fate of the shipment.

1917.02.09: Cablegram C.H. Sternberg to Woodward
WOODWARD BRITISH MUSEUM LN
WHEN WILL YOU REMIT BILL SENT
STERNBERG
Wrote Feb 19th "do not understand this"

1917.02.27: BFMIC to Woodward
{Thanks for Bill of Lading}

1917.03.23: Dale & Co Montreal to BFMIC
{Received claim memorandum on 7th February; complications over payment of insurance on freight costs.}

1917.03.29: BFMIC to Woodward
{Instructed to pay 7th February}
Wrote to Sternberg 30.3.17, ASW

1913 (*1917? W.D.L.*) 03.09: C.H. Sternberg to Woodward
My Dear Sir:-

It was a bitter disappointment to me to receive your letter of Feb 19th to learn that you have not received the second Bill of Lading. I sent you by unregistered mail (I had not received instructions to register it) about the 30th of January. It is hard to me to understand why you did not cable the information contained in this letter when I cabeled "Bill sent" meaning of course the Bill of Lading, I cabled again, asking you when I was to receive my money and you paid not attention to that. You remember that I was not responsible for the shipments after turning them over to the Rwy. Co I was released, and you assumed all risks. It is not right that I should be kept out of my money because of the Insurance Co having not received the second Bill of Lading Neither would it be right for me to be forced to put this bill in the hands of lawyers to collect. You know very well that I did all in my power to do you good service. I should not be forced to suffer on account of the Raider. If you had cabled, that you had not gotten the second Bill. I could have done what I did today. Wire to Mr Halstead to send you another and saved twenty days of time. I am not only paying interest on money that belongs to me but am prevented from going in to the field for lack of it, as I only have a certain amount of credit at my bank. I earnestly hope you will see that I get the money due to me at once, and that you will use the cable and not the mail to assure me.

The Insurance Co have no right to hold up the payment of the Insurance under extisting conditions, You have one original B/L and they know no one could collect, on the second if the first was paid. However, I have wired Mr Halstead to send you another bill of Lading. I am sorry too that the Sladin people are not impressed

If the two magnificent articulated skeletons had not gone to the bottom they would have had $4.000 dollars worth of material, as I sold two specimens not so perfect for $2100.00 and one for $2500, and Further it takes years to prepare Belly River material, I do not think I should in addition to the loss of two specimens be obliged to lose financially. Or be financially ruined, because of the events I am not responsible for. Please wire me if the Sladen Trustees change their minds.

I asked you in the last cable to answer, and you paid no attention to it. I truly hope you will not fail to answer my wires and will wire your self, the instant any thing definite is known, about when I shall receive money, that has been due me over four months Please send be a bill for the Cable fee and I will return it at once

Faithfully yours.

30th cheque was ordered from Liverpool on Feb 7th

1917.03.12: C.H. Sternberg to Woodward
My dear Sir:-
I am glad to receive your favor of February 26th On receipt of the other dated Feb 19th Saying that the Bill of Lading, had not been received I wired Halstead and he wired he had sent a third Bill of Lading last Saturday the 10th I explained in my last why I wanted rapid returns or the knowledge of them, I hope therefore you will cable at my expense, as soon as the Insurance money has been sent, how much and where. I can then borrow money if I need it.

I am sorry the Sladin Memorial Fund is not satisfied I have told you a way that in my estimation a fine collection can be secured of course I can do no more, I did my full duty in the field, and it was not my fault that the most _impressive_ material was discovered the last two months, and that it went to the bottom of the sea. I think that if you see the means of accepting my offer, you better cable me It takes two weeks when not censored and 19 days when it is to hear from you. All my letters but the last have been censored

Hoping for the best
I am faithfully yours

will send my new book {last word illegible}

1917.04.03: C.H. Sternberg to Woodward
My Dear Sir,
Day after day I am waiting for the money I earned, and you promised to pay me, in your letter of June 3rd. 1916 In which you say "As, of course, you do not receive the money in advance, we naturally expect to pay for the fossils at a little higher rate, If your seasons work is worth more than $4000, I shall of course, try to obtain more payment

for you. It is understood that you will deliver everything ready packed to the railroad company, and that we pay insurance and freight and take all risks after the railroad has given you a receipt" All these things, I did faithfully and the last shipment was worth more than all I was to receive, I sent the last Bill of Lading long ago, and yet I do not receive a cent. I spent far more in the field for actual expenses for the nearly five months labor I gave you, I was obliged to mortgage my home, to secure the money above what you have sent me to the limit of my credit and now, I have balanced my account there and find that I have only 163 Dollars with which to carry on my work this summer My son has been in the Permian beds for 6 weeks in Texas I go to join him tomorrow, and hope to return in a month. This expense will reduce my bank, acct. to nothing How am I to go into the field, and continue my life-work unless you pay me what is due to me. It was awful enough to have a German Raider sink the two best specimens of <u>Corythosaurus</u> my party have found in 5 years, As good as the (at least one of them is) the skeleton I sold the Senckenberg Museum for $2500,

It will be still worse to completely ruin me, so I cannot keep at work, I cannot believe it possible, and Now I have told you the financial condition I am in, I am depending on you to see that I get what is mine, I beg of you to wire me when you secure the <u>mone[y]</u>

Further if you cannot send me the money at once, please send me an acknowledgement of the sum you intend to pay, so I may use it as collateral, and can borrow money on it Now the frost is out of the ground I want to make a large collection, and cannot do it without money. As you have always been true to me and I have done all in my power for you, and as I am in no way responsible for the loss of the magnificent material that went to the bottom, I hope you will strain every nerve, Our boys in blue will help you clear the sea of those sea Pirates the scourge and curse of the world, and I hope you will be able at an early date to clear my financial skies that look so dark. I wish you could give me the opportunity to retrieve the loss of last fall. I am sure with five months of constant untiring work I can get you a great collection

Do you want my Permian material, Levi writes he already has 8 skulls and part of two skeletons I cannot describe them yet

Address me at Seymour Texas

Faithfully yours

1917.04.10: BFMIC to Woodward
{copy of letter from Montreal agents}
11th see over

1917.04.11: Draft reply on reverse of above
In reply to your letter of yesterday's date enclosing copy of a letter from Messrs Dale & Co. Ltd Montreal in reference to the insurance of our fossil bones, I have again interviewed the Canadian Pacific Ocean

Services Ltd., and find I misunderstood their letter of 22nd Jan last concerning freight. The fossils sent to us by Mr C.H. Sternberg were valued by him at $2500, (and after dealing with him for many years we have learned to trust his valuations), which {?} I understood from the C.P.R. that he would have to forgo so much of his compensation as would recoup them for freight As that is not so, I shall be glad if Messrs Dale will please pay the whole sum of $2500 to Mr C.H. Sternberg, . . .1917.04.12: BFMIC to Woodward

{acknowledgment}

1917.05.05: C.H. Sternberg to Woodward

My Dear Sir:-

Your favor of March 30th, reached me in Texas, last Tuesday. I at once wrote the Bank of Montreal asking them to learn why the check was not sent, and on reaching home wired the export Agent, who paid the premium on the Policy asking the name of the Insurance Co. and to try and find out why I did not get the check. I received notice from them that Dale & Co had sent me the check the day before, and I received it for the full $2500 today. I cannot tell you what a relief it has been as I feared I would not have the means to collect in Canada the wonderful Dinosaurs of the Red Deer River this season, My son has been a long time at work in the Texas Permian I spent several weeks with him, with remarkable success. Among the genera represented by skulls and parts of skeletons more or less perfect are the following <u>Dimetrodon</u>, <u>Seymouria Baylorensis</u> (of which I secured four skulls with much of the skeletons) <u>Parioticus, Diplocaulus</u> I think there are four skulls with parts of skeleton of <u>D. Magnicornis</u> and other sps. <u>Lysorophus</u> skulls and skeletons <u>Cardicephalus sternbergii</u>. Several skulls and many bones, and several other skulls not identified yet. I was remarkably successful in securing what appears to be a complete skeleton (nearly) in position including head, and hind limbs and feet in situ. The spines are over 3 feet long Some of them and belong to the species <u>D. giganhomogenes</u>. In addition are a number of scattered skeletons of same species, The one I propose to mount this winter will enable me to get the others together. In addition I might say that they are preserved in clay and the silica that has ruined so many Permian vertebrates, slips off easily and makes complete and uninjured bones, As far as I know this <u>main</u> specimen is the most perfect <u>Dimetrodon</u> known, Far ahead of any thing Case describes. My plan is to prepare these specimens I will need the complete skeleton until I have restored others from its study I will then offer it for sale, Mr Gilmour {Gilmore} of the National Museum has been trying for months to have the Director of the National Museum employ us in the field. If he does not succeed, and as an appreciation on my part of our great kindness in securing the $2500, if you desire, I will promise to give you the chance of purchasing the best <u>Dimetrodon</u> skeleton The only one with complete head and feet bones in place and the choice of a series of these rare Permian fossils. In other words I will offer them to you first after they are prepared.

I hope now to go into the Belly River after Dinosaurs, and in case I get a skeleton or some fine skulls, I will inform you. Personally I believe I will do much better financially to run the risk myself. I am so sure of success I shall use every cent of the money from the policy and as much more as I can borrow. I will then offer the prepared material, Ward of Rochester has offered to buy all my material, but I will not sell any fine material to a dealer as long as the Museums of the world stand by my life work and support me. I will be glad to know if you have more hopes of raising the money after the dinosaur skeleton is collected and prepared than under the old plan?

Further: as a small mark of my esteem I want you to retain the book I sent you Hunting Dinosaurs on Red Deer River as a personal gift.

Faithfully yours

1930.06.22: C.H. Sternberg to Keeper of Geology
{offer of Tertiary specimens}

1931.01.21: C.H. Sternberg to British Museum
{offer of Tertiary specimens}
Mr Charles Sternberg is constantly approaching the museum with offers of specimens for purchase. There is no need to take any notice of this appeal. W.D.L{ang}. 8.11.1931

Discussion

Quotations which are not from the correspondence above are sourced.

The Resource

As a result of the work of Brown and Sternberg in the "Canadian Dinosaur Rush" the Upper Cretaceous beds of Alberta (fig. 32.1) were recognized as an exceptionally rich deposit, containing many complete skeletons representing several types of dinosaurs. Many institutions were interested in acquiring material, and Woodward was clearly anxious to obtain dinosaurs from these beds to provide an attractive educational exhibit, to complement the British Museum's other dinosaur material and supplement the much more limited Canadian material already acquired by the museum by the eccentric William Cutler.

Sternberg's Reputation

Sternberg had already collected for the British Museum of Natural History—Riser (1995, 68) refers to correspondence as early as 1903—and naturally he turned to Woodward when his arrangement with the Geological Survey of Canada came to an end. Professionally, Sternberg was highly regarded by Woodward ("highly skilled and experienced" and a "thoroughly honest man"). It seems, however, that museum officials found him difficult to deal with. Sternberg's enthusiastic entrepreneurship probably jarred on the scientific establishment, his willingness to discuss his financial problems was perhaps embarrassing, and no doubt his carelessness with names and the American usage in his letters irritated an England that was still inclined to regard foreigners as odd. Comments in other museum correspondence refers to Sternberg's "rambling letter," while curator Lang later notes that no notice need be taken of his lists of fossils for sale.

More serious were concerns by the museum and foundation about the quality of the first shipment to be received. Sternberg's credibility was on the line, and when the second shipment—worth twice as much as the first, in Sternberg's view—failed to arrive, he had no opportunity to recoup his reputation. The loss of the fossils seems to have been in some ways a relief to the museum, permitting an end to the relationship, for Woodward commented, "so end the bones of contention."

The Economics of Dinosaur Collecting

Dinosaur collecting was (and is) an expensive business. The client institution, one of the largest natural history museums in the world, could only afford to collect overseas during wartime by soliciting funding from a private foundation. The pound was much higher compared to the dollar than it is today—Woodward calculates on five dollars to the pound, so North American products were much more affordable in England than they would be today.

Sternberg was so anxious to collect, he would offer to work for pay that was barely adequate. On the evidence of his son Charles M. Sternberg, he signed up for the Geological Survey of Canada for "about a hundred a month. . . . It was a silly amount." (C. M. Sternberg, interview). His British Museum contract was for $1000 a month for four months, for which he actually worked nearly five. When trying to secure future contracts from Woodward he quotes the same figure (December), later offering as an alternative $500 per month for a year.

Figure 32.1. Corythosaurus *quarry, where the last shipment was excavated.*

Figure 32.2. Exposures along the Red Deer River area, southern Alberta.

The sums named are not Sternberg's personal income (as presumably was the survey figure), but reflect the cost of putting two skilled people (himself and Levi) in the field, supplemented by unskilled labor in the vicinity. Sternberg's outfit (wagon, horses, and equipment) may have been left in Alberta ready for use, but he and Levi were based far away in Ottawa, and the badlands of the province were then more remote from sources of supply than they are now (fig. 32.2).

Based on the sales figures he quotes, Sternberg's costs are not unreasonable. When the museum was unimpressed by the first collection it was perhaps being unfair, because by funding collecting time instead of purchasing already collected specimens it was assuming the risk of an unsuccessful hunt. However, perhaps Sternberg's enthusiasm had led to greater expectations that he was able to realize in the initial shipment.

The Lost Specimens

Sternberg numbered his quarries (and thus the specimens found in them), but confuses the issue by writing about the "second and third" skeletons (presumably in order of discovery). His letters discuss three incomplete specimens of *Corythosaurus,* which could be combined to make a complete display skeleton. No. 6 (with much of the skull and a complete neck) was presumably in the first shipment, because it is not mentioned in the box list. Nos. 9 and 13 made up the lost second shipment.

Sternberg's letters do not describe the skeletons, and most of the box contents are described as "sections" (presumably plastered blocks of associated bones which are perhaps referable to diagrams). Bones specifically referred to are:

No. 9 (Crested duckbilled Dinosaur)—tail, neck, right foot and ischium, sacrum, femur, trunk, ribs, skull. The skull was incomplete (Sternberg 1985, 178)

No. 13, the second "rather the third" crested dinosaur, had "the entire skeleton except a few inches of the tail and THE HEAD" (Sternberg 1985, 180). Its quality "I considered next in perfection to that of Mr Browns" (p. 179)—tail, right tibia and femur, left front foot and a front limb, right ribs and two humerii, left scapula, femora and hind right foot, pelvic arch, ribs. And "there were, in addition large patches of the skin impression" (p. 179).

Sternberg describes nos. 6, 9, and 13 as crested dinosaurs, and in the list and letters as *Corythosaurus.* This genus was named by Barnum Brown and W. D. Matthew (1913), and two of Brown's specimens in the American museum are discussed by Norell et al. (1995). Sternberg received papers from Brown, and so was well aware of this material, and had collected similar specimens himself that Lambe (1914) named as *Stephanosaurus.*

Several species have been attributed to *Corythosaurus,* though now only the type species, *C. casuarius,* is generally accepted (Weishampel and Horner 1990). Without the bones or good photographs of Sternberg's specimen, it is of course impossible to tell if a modern taxonomist would classify it in the same genus. But as it was well known it seems likely that Sternberg was correct, and that it would be the species already known. Weishampel and Horner indicate that in 1990 it was known from "approximately ten articulated skulls and associated postcrania" (1990, 557).

Although the specimens might have yielded further scientific information, the principal loss seems to be to the museum and its many visitors, who were deprived of the opportunity to see this important material.

The Sinking of the Mount Temple

The Canadian Pacific SS *Mount Temple* had been in the public eye during an earlier incident in 1912, when it responded to the SOS of the *Titanic,* but (according to its captain) failed to find the doomed ship at its last reported position, while passengers reported that it was in plain view. Other confusions are associated with its own demise.

Sternberg's second shipment was made on the SS *Mount Temple,* which (based on the letters) was due to leave Montreal on November 25. On January 18 the British Museum wrote to make a claim on the insurance. Somewhere between these dates the *Mount Temple* was sunk in the Atlantic. There were conflicting reports that the captain and crew were landed at St. Vincent (which could be either in the Azores or on the coast of Portugal), or that they had been taken to Germany in another ship, the *Yarrowdale.*

Though Sternberg's *Hunting Dinosaurs* specifies only a "German raider" (1985, 179) and a "German torpedo" (p. 180), some of the literature indicates that the *Mount Temple* was sunk by a U-boat. I have traced this story as far back as the de Camps' *Day of the Dinosaur* (1968). It is a natural assumption that a U-boat sank the *Mount Temple,* based on the prevalence of U-boats in the Atlantic battle and on the reference to torpedoes. The error, unfortunately, has been perpetuated by Riser (1995) and me (Spalding 1993, 1999). However Psihoyoos and Knoebber give additional details about the loss: "On December 6, 1916, the boat was blown up by the German raider *Moewe,* a warship disguised as a common cargo vessel . . . 620 miles (998 km) west of England" (1994, 254). Psihoyos verified the date with the Imperial War Museum (note to Gerhard Maier).

Fuller information comes from a dramatized and unsourced book on the adventures of the *Moewe* (Hoyt 1970). According to this source, the *Moewe* stopped the *Mount Temple,* evacuated the crew, and blew up the ship. The crew were in due course offloaded onto another captured ship, the *Hudson Maru,* and sent to Pernambuco, arriving there on January 16.

Delays in Payment

A letter sent to Sternberg on January 18 notified him that the *Mount Temple* had been reported sunk. Presumably, he received the information not long before January 23, when he cabled the British Museum. By that time an insurance claim had already been made.

Payment of the insurance proved to be a problem. Many questions arose: Was war risk included? Should the cost of freight have been insured? Where were the missing bills of lading? These problems could only be settled slowly. Letters went back and forth between Liverpool and London and across the Atlantic. They were checked by censors (probably automatically because of Sternberg's German name), and were delayed in crossing the Atlantic. The absence of an official protest by the captain (wherever he was) led to delays. Although the museum was speedy in claiming the insurance, it was clearly slow to ask Sternberg for missing documents, and his complaint that they wrote letters instead of cabling is surely justified.

Sternberg did not receive his payment until May 5, nearly four months after news of the loss of the *Mount Temple* reached the parties concerned.

Sternberg's Financial Crisis

Because Sternberg's shipments were insured, the financial impact of the loss should not have been too great. But it is clear from his letters that Sternberg had undertaken a considerable financial risk by moving back to Kansas, being "obliged to sell my home in Ottawa at a great loss, move my family here, and buy my old home at another expense." No longer employed by the Canadian Survey, Sternberg obviously felt the need to return to his familiar base, though it is not clear why this move need be so precipitous as to require losing so much capital on property transactions. By the time payment was received, it is clear that Sternberg's credit was pushed to the limit. If the British Museum (or anyone else) had been willing and able to commit to further collecting in Alberta as Sternberg wished, his expenses would have been less in continuing. No doubt he was also trying to find other clients to support Canadian work, but in the absence of any commitment he had to begin work in Texas, partly because it allowed an earlier start to the field season.

The Impact of the Loss

More serious even than the financial implication was the emotional impact on Sternberg himself. Coming on top of the breakup of the family team, the loss of these specimens was clearly devastating. He had hoped to recoup his credibility with the British Museum after the disappointing first shipment, and wished to undertake future work to "build up in the British Museum the third largest collection of Red Deer dinosaurs." Lack of further interest by the museum was a serious blow after what he must have seen as rejection by the Geological Survey, and the poor relations that had developed with the American Museum after the fractious rivalry of the Canadian Dinosaur Rush.

In his late sixties, Sternberg was still a zealous field man, though "camp life does not agree with me in cold weather." He must have been aware that his future opportunities to pursue his passion were limited, and he perhaps saw this find as the crowning achievement of his career, for his emotional reaction to the loss of "my finest dinosaur" is extreme. All his earlier achievements were forgotten as he laments, "Ten minutes of vandalism destroys all my labor, my hopes, my life almost, because I can never recover from such a blow as this" (Sternberg 1985, 179).

Remarkably, although his sons now found full-time employment in paleontology without him, Sternberg continued to collect and sell specimens into his eightieth year. He was assisted for one more season by Levi Sternberg (who took a leave of absence from the Royal Ontario Museum to do so), and eventually retired to Canada, where he died in Toronto at 93.

Acknowledgments: I am indebted to Phil Currie, who as colleague and friend has helped and encouraged my ongoing interest in dinosaur history in many ways since we were first associated at the Provincial Museum of Alberta a quarter of a century ago. Although my interest in dinosaurs was kindled before I came to Canada (and indeed was one of the factors that brought me across the Atlantic), that interest grew and developed in particular because (partly through Phil's discoveries) I found myself in the middle of a remarkable period of discovery, research, and museum building in which our knowledge of dinosaurs has improved and changed beyond all recognition.

The British Museum (Natural History) has provided access to the Sternberg correspondence and permission to publish. Staff who assisted include the late Dr. Alan Charig, and Sandra Chapman and Sam Collenette. Clive Coy rightly took me to task for my use of the "U-boat" story, and provided related information and useful discussion. Gerhard Maier kindly gave me access to Hoyt's book on the *Moewe*, and other information. Kenneth Carpenter drew my attention to the *Titanic* connection.

References

Archival Sources

Sternberg, C. H., et al. Letters and related documents in Sladen file, British Museum (Natural History).

Sternberg, C. M. Interview, Provincial Archives of Alberta. Transcript in Royal Tyrrell Museum.

Publications

Brown, B., and Matthew, W. D. 1913. *Corythosaurus,* the new duck-billed dinosaur. *American Museum Journal* 15: 427–428.

Colbert, E. H. 1968. *Men and Dinosaurs: The Search in Field and Laboratory.* New York: Dutton.

de Camp, L. S., and C. C. de Camp. 1968. *The Day of the Dinosaur.* New York: Curtis Books, Modern Literary Editions Publishing Co.

Gross, R. 1985. *Dinosaur Country: Unearthing the Badlands' Prehistoric Past.* Saskatoon: Western Producer Prairie Books.

Hoyt, E. P. 1970. *The Elusive Seagull: The Adventures of the WWI German Minelayer, the* Moewe. London: Tandem Publishing.

Lambe, L. M. 1914 On a new genus and species of carnivorous dinosaur from the Belly River formation of Alberta, with a description of *Stephanosaurus marginatus* from the same horizon. *Ottawa Naturalist* 28: 13–20.

Norell, M., E. Gaffney, and L. Dingus. 1995. *Discovering Dinosaurs in the American Museum of Natural History.* New York: Knopf.

Psihoyos, L., and J. Knoebber. 1994. *Hunting Dinosaurs.* New York: Random House.

Riser, M. O. 1995. *The Sternberg Family of Fossil Hunters.* Lewiston, N.Y.: Edwin Mellen Press.

Rogers, K. 1991. *The Sternberg Fossil Hunters: A Dinosaur Dynasty.* Missoula, Mont.: Mountain Press.

Russell, L. S. 1966. Dinosaur Hunting in Western Canada. *Royal Ontario Museum Life Sciences Contribution* 70: 1–37.

Sarjeant, W. A. S., comp. 1980–96. *Geologists and the History of Geology: An International Bibliography from the Origins to 1993.* 10 vols. New York: Arno Press.

Spalding, D. A. E. 1993. *Dinosaur Hunters, One Hundred Fifty Years of Extraordinary Discoveries.* Toronto: Key Porter Books.

Spalding, D. A. E. 1999. *Into the Dinosaurs' Graveyard: Canadian Digs and Discoveries.* Toronto: Doubleday Canada.

Sternberg, C. H. 1909. *The Life of a Fossil Hunter.* New York: Holt. Repr., Bloomington: Indiana University Press, 1990.

Sternberg, C. H. 1918. Five years experience in the fossil beds of Alberta. *Transactions of the Kansas Academy of Science* 28: 205–211.

Sternberg, C. H. 1985 [1917, 1932]. *Hunting Dinosaurs in the Bad Lands of the Red Deer River, Alberta, Canada.* 3d ed. Edmonton: NeWest Press.

Weishampel, D. B., and J. R. Horner. 1990. Hadrosauridae. In D. B. Weishampel, P. Dodson, and H. Osmólska (eds.), *The Dinosauria,* pp. 534–561. Berkeley: University of California Press.

Weston, Thomas Chesmer. 1899. *Reminiscences among the Rocks, in Connection with the Geological Survey of Canada.* Toronto: Warwick Bros. and Rutter.

33. Dinosaurs in Fiction

William A. S. Sarjeant

Abstract

Though the earliest depiction of dinosaurs in fiction dates from 1854, they figured in only two nineteenth-century works. Nor did Conan Doyle's classic *The Lost World* (1912) herald a significant entry into mainstream fiction, though dinosaurs did begin to figure in science fiction and in books and comics for children. Since 1955, however, dinosaurs have been featured ever more prominently and widely. Such writings are important because they have kindled, or expanded, the interest of many persons in paleontology, at the amateur or professional level. Recognizing this, prominent paleontologists—among them Colbert, Halstead, Simpson, Bakker, and Currie—have been increasingly writing books about dinosaurs for juvenile and adult audiences. A particular interest of fictional works is that they not only reflect the changes resulting from our growing knowledge of dinosaur morphology and behavior but, also, in at least a few instances, anticipate later scientific opinion.

Introduction

It was in 1841 that Richard Owen recognized the necessity of bringing together, into one class, the huge creatures whose bones were being discovered in England and France. The splendidly evocative name *Dinosauria,* "fearfully great lizard," was coined by him before his account was published in 1842. Being of such size and having such a name, they were destined to stir the imagination from that time onward.

Yet their entry into literature was slow. Donald Glut (1997, 577) has pointed out that the first mention came in Charles Dickens's *Bleak House* (1852–53), whose opening passage envisions "a Megalosaurus, forty feet long or so, waddling like an elephantine lizard up Holborn Hill" in London. However, dinosaurs gain no further mention in that rather grim novel.

Their next appearance came two years later, in a strange work by John Mill, *The Fossil Spirit: A Boy's Dream of Geology* (fig. 33.1, upper left). To an audience of small boys, a fakir from "Hindostan" recounts the transmigrations that have allowed him to assume many different animal forms at various epochs of geologic time. In the fourth of these, he transformed himself into a huge beast which, as the frontispiece makes evident, accorded with the current concepts of an *Iguanodon*:

> My size was enormous, the bodies of three of the largest elephants were not equal to mine, and my tail, like that of a crocodile, was large and stretched out to the length of twenty feet, whilst my back was more than sixteen feet high. (1854, 45; see fig. 33.8, upper)

The fakir goes on to report his deadly conflict with "the Megalasaurus [*sic*], a carnivorous reptile nearly as large as myself." This is one of the liveliest episodes in what was, alas, a singularly turgid and heavily didactic text.

Surprisingly, having entered literature so early, dinosaurs faded from it thereafter for more than sixty years. Several novels treated with more recently extinct creatures and primitive humans (see Sarjeant 1994), but none with creatures more ancient. Their reentry came splendidly, as a result of a dinner-time challenge by a guest to Arthur Conan Doyle (see Batory and Sarjeant 1994). *The Lost World* (1912; fig. 33.1, lower) recounts an expedition to a physically isolated Guyanan plateau (named Maple White Land after its discoverer), where a Mesozoic fauna still survives in juxtaposition with primitive humans. Yes, the dinosaurs are excellently described, but a large part of the charm of the book is in the interplay between the personalities of expedition members—the two professors of paleontology, hunter Lord John Roxton and narrator-reporter Edward Malone. The two contending professors (the formidable George Edward Challenger—arguably, after Sherlock Holmes, Doyle's finest creation—and the persnickety and ever-dubious Professor Summerlee) have always seemed to me to be echoes of Thomas Henry Huxley and his opponent on Darwinian evolution, Richard Owen: but Dana Batory has, in a work not yet published, made other identifications set upon firmer historic ground.

The Canadian naturalist Charles G. D. Roberts struck a new note when he attempted a fictional portrait of the Mesozoic world in the earliest chapters of *In the Morning of Time* (1919); few subsequent fictional works were to attempt so factual an approach.

In the succeeding years, the flow of works was at first intermittent, then steadier after 1955, until, by the late twentieth century, it has become a flood. A curiosity is a chapter in the first edition of T. H. White's *The Sword in the Stone* (1939) in which "the Wart" (the soon-to-be King Arthur), during one of the educative transformations imposed upon him by Merlin, encounters a snake who tells him of the "war" between *Ceratosaurus nasicornis* and *Atlantosaurus immanis*

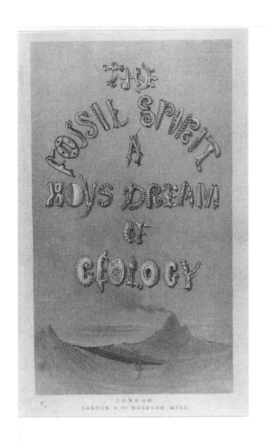

and of how the last *Atlantosaurus* survived to be the dragon slain by St. George. Sadly, through ill-conceived editorial cuts, this chapter disappeared from later editions.

Dinosaurs in an Underworld

The concept of an underworld dates, of course, to classical times and earlier. It first found "scientific" expression in the writings of John Cleves Symmes in 1820 and his belief in an entry to that underworld at the North Pole (see Sarjeant 1994, 319). It gained fictional treatment by Jules Verne (*Voyage au centre de la terre*, 1864) and Edward D. Fawcett (*Swallowed by an Earthquake*, 1894). However, dinosaurs did not enter the subterranean scene until Edgar Rice Burroughs conceived Pellucidar. As I have written earlier, the seven books (1923–1963) in which this imagined underworld was treated "have everything—dinosaurs of course, plus pterodactyls, ape-men, giant ground sloths, brave, handsome males and beauteous, helpless damsels. What more might one desire?" (Sarjeant 1994, 319).

They are also among the earliest works to imagine a world in which dinosaurs have evolved to civilization and dominance—a recurrent theme in later works, and made more feasible when the distinguished vertebrate paleontologist Dale Russell and sculptor Ron Séguin demonstrated, by means of a life-size model, how small theropods such as *Stenonychosaurus* might readily have evolved into "dinosauroids" with all the capacities of humans (Russell and Séguin 1982; see also Mitchell 1998, pl. 1.2).

The Russian geologist Vladimir Obruchev tackled anew the subterranean theme, harking back to Symmes by imagining an Arctic entry to the subterranean world—a world which his scientist heroes name Plutonia (1957; fig. 33.5, upper left). Plutonia contains an even wider mix of inhabitants than do Conan Doyle's Maple White Land or Burroughs's Pellucidar, including not only plesiosaurs, giant tortoises, and pterodactyls, but also Tertiary mammals, mammoths, and primitive humans. The novel ends bleakly. Upon returning to the surface, the scientists become embroiled in the First World War; all their trophies and even their lives are lost.

"The Fires beneath the Earth," a serial in the British *Wizard* comic during the late 1940s, also treated the theme of prehistoric creatures—dinosaurs included—surviving in an underworld. Typically for such productions, the author's name was not stated.

Travelers in Time

John Mill's fakir had dreamed himself back into the ancient past in the form of a variety of long-gone creatures. The great dinosaur hunter Charles Hazelius Sternberg (1850–1943), in the final series of chapters in his otherwise strictly factual *Hunting Dinosaurs in the Badlands of the Red Deer Valley, Alberta* (1917), dreams of traveling back to distant geologic eras in human form. In the chapter "There Were Giants in Those Days" (pp. 128–140), he witnesses the slaying of a *Trachodon* by

Figure 33.1. (opposite page) Upper left: the title page of the earliest novel featuring dinosaurs, John Mill's The Fossil Spirit *(1854). Upper right: the cover of Francis Rolt-Wheeler's* The Monster-Hunters *(1916), with leaping* Dryptosaurus. *Lower: carnosaur footprints on the cover of the illustrated edition of Arthur Conan Doyle's* The Lost World *(1912b).*

a *Tyrannosaurus rex;* when the carnosaur has feasted, it thriftily removes and conserves the victim's skin!

Like the fakir, Sternberg was taken by dreams much further backward in time than the Mesozoic. In contrast, the heroes of three further stories by Edgar Rice Burroughs needed only to explore Caprona, "The Land That Time Forgot" (1924; fig. 33.6, upper left), to witness all stages of evolution from fish to amphibian, reptile, and mammal—and yes, they encounter dinosaurs. However, it is the primitive female humans of that remarkable island that exercise the most powerful attraction! From the protagonists of John Taine's *Before the Dawn* (1934), even less traveling is required, since they witness past times passively by means of a device called a televisor (or electronic analyzer). They identify in particular with a tyrannosaur which they name Belshazzar because of its abilities at feasting—and because of the doom awaiting it.

In C. H. Murray Chapman's *Dragons at Home* (1924), the children who travel back in time from their home fireside (and, later, from London's Natural History Museum) are conducted to the Mesozoic by a pterodactyl; they meet a *Stegosaurus*, an *Iguanodon*, and a "*Ceratops*," among an array of other creatures. The boy Perry, in Francis Rolt-Wheeler's rather-too-didactic *The Monster-Hunters* (1916; fig. 33.1, upper right), dreams of riding an *Anchisaurus*. When in danger of being slain by a voracious *Ceratosaurus*, he is saved by a friendly *Stegosaurus*. Helen, the little girl who is transported from her fireside in Lyell Lodge, in H. C. F. Morant's *Whirlaway* (1937), passes a succession of earlier year-stones, from the Cambrian onward, before going through the Jurassic gate to meet her first dinosaurs. In Lewis Brown's *Yes, Helen, There Were Dinosaurs* (1982), Helen and her uncle Homer Crabtree travel back to the Jurassic less effortfully, in their little Time Car. These children were altogether more fortunate than those in Lady Bray's *Old Time and the Boy; or, Prehistoric Wonderland* (1921), who encountered dinosaurs only in the illustrations to that extraordinarily tedious work.

Time travel by children into the historic and prehistoric past was the theme of a British Broadcasting Corporation radio series, "How Things Began," during the 1940s: I remember with pleasure the episodes in which they encountered dinosaurs, creatures already fascinating to me but much less known to children then than now. This series was, I believe, embodied in an accompanying text, but I have been unable to locate it.

Time saltation by the minds of persons under physical or medical stress is imagined in two stories, respectively by Charles Sheffield and Paul Preiss, in *The Ultimate Dinosaur,* a melange of straight science, scientific exploration, and fiction edited by Preiss and Robert Silverberg (1994).

Lyon Sprague de Camp, a distinguished writer on the history of science and technology as well as of fiction, put a fresh spin on the time travel theme by having a professional hunter conduct wealthy amateurs backward through time to the Mesozoic, to hunt the biggest game of all. This concept, first set forth by de Camp in the short story "A Gun for Dinosaur" (1956), was later developed very effectively into a collec-

Figure 33.2. (opposite page) Upper left: one of the books recounting investigations of reports of living dinosaurs, by Roy P. Mackal (1987). Upper right: cover of the first edition of T. C. Bridges's Men of the Mist *(1923), with leaping carnosaur. Lower left: cover of George G. Simpson's* The Dechronization of Sam Magruder *(1996). Lower right: cover of Robert T. Bakker's* Raptor Red *(1995), with two Utahraptor.*

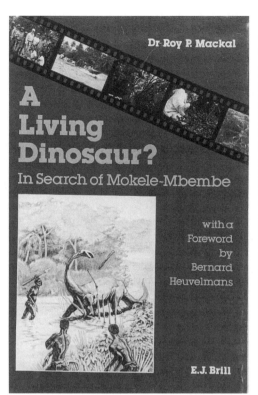

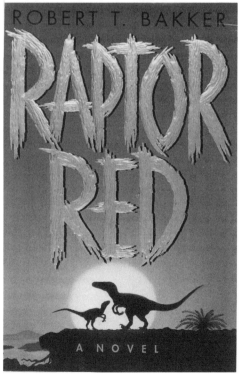

tion of linked tales, *Rivers of Time* (1993; fig. 33.6, lower left), two of which also deal with the Mesozoic. A further story about Rivers's adventures in the Mesozoic, "Crocamander Quest," was included in *The Ultimate Dinosaur* collection (Preiss and Silverberg 1994). The dinosaur-hunting theme is likewise treated in short stories by P. Schuyler Miller and Ray Bradbury, included in the collection *Behold the Mighty Dinosaur* (David Jablonski, ed., 1981) and by Robert Silverberg in a story in *The Ultimate Dinosaur.*

Ray Bradbury's collection *Dinosaur Tales* (1983) includes encounters with dinosaurs via time travel, but also in film and museums. His time travel concept was developed in six books by Stephen Leigh (1992–1995; two coauthored by John J. Miller, 1993, 1995), of which the latest was *Dinosaur Empire* (1995); all envisioned complexities on the "time-crossing roadway." A lesser work is David Bischoff's *Time Machine: Search for Dinosaurs* (1984), in which readers are invited to create their own adventure.

Robert Chilson's *The Shores of Kansas* (1976) has an inventor of time travel facing the perils of commercial greed and sexual predators in his own time and the greater perils of more ferocious predators when he travels into the past.

Another collection of short stories, *Dinosaur Fantastic* (ed. Mike Resnick and Martin H. Greenberg, 1993; fig. 33.6, upper right) contains several short stories in which humans travel into the past, sometimes in their own bodies, sometimes by transmogrification into the bodies of dinosaurs. Michael Bishop's short story "Herding with the Hadrosaurs," in *The Ultimate Dinosaur,* envisions orphaned human children being accepted into a *Corythosaurus* herd. Lee Grimes's *Dinosaur Nexus* (1994; fig. 33.7, upper right) has a team of scientists traveling back to the Late Cretaceous to try to resolve the problem of dinosaur extinction, only to encounter competition from alien scientists with quite different aims.

Surely the most poignant of time travelers to the Mesozoic is Sam Magruder, the account of whose "dechronization" is presented in a posthumously published work (Simpson 1996; fig. 33.2, lower left) by the eminent vertebrate paleontologist George Gaylord Simpson (1902–1984). Having arrived among the dinosaurs, Sam hunts and feeds on them with fair success. However, as time passes, dinosaur-wrought injuries sap his hunting prowess and, unable to return to his own time, he knows he will eventually be slain by a dinosaur. Only the stone slabs, on which Sam has laboriously engraved his story, survive.

Almost as poignant is the fate of the last dinosaur—the long-lived Qfwfq—who, in Italo Calvino's story "The Dinosaurs" (trans. William Weaver, 1968), becomes involved with mammals and even falls in love with one, only to be unkindly rejected. This complex, strange, and ultimately unhappy story is admirably summarized in W. J. T. Mitchell's *The Last Dinosaur Book* (1998, 42–45), in itself an offbeat work that, while roaming from science and history to cartoon, includes several commentaries on dinosaurs in fiction and an array of excellently reproduced illustrations from a variety of sources.

Dinosaurs Surviving Today

In Alan Charig's account of "Disaster Theories of Dinosaur Extinction," he raises the question, "Are dinosaurs really extinct?" critically discussing contemporary claims of the sighting of dinosaurs in remote regions (1995, 310–313; fig. 33.2, upper left). Since such works were not intended as fiction, they are not included here, however dubious their authenticity. In the realm of fiction, Conan Doyle was of course the first to visualize dinosaurs' continuing to live in an isolated environment in our own world, but surprisingly few other writers have utilized that idea. Indeed, I know of only one: T. C. Bridges, whose *Men of the Mist* (1923; fig. 33.2, upper right) hypothesizes the survival of a solitary carnosaur in a fumarole-heated Alaskan valley.

If the dinosaurs did not become extinct, what might have been their role in the resultant, very different world? This is considered by several authors. Dinosaurs contentedly coexisting with man, on a southern hemisphere island continent, are charmingly imagined and superbly illustrated in three books by James Gurney, *Dinotopia: A Land Apart from Time* (1992), *Dinotopia: The World Beneath* (1995), and *Dinotopia: First Flight* (1999), the latter accompanied by a board game. Gurney's vision has been followed up in two novels by Alan Dean Foster, *Dinotopia Lost* (1996) and *The Hand of Dinotopia* (1999).

Robert Mash, in *How to Keep Dinosaurs* (1983), boldly imagines that dinosaurs have survived widely enough to be kept as pets or in zoos, carefully assessing their degree of domesticity or intransigence, the space they will require, and the problems of ensuring they are well fed and happy. Mercedes Lackey and Larry Dixon, in a short story in the *Dinosaur Fantastic* collection (Resnick and Greenberg, eds., 1993), explore the dangers of empathizing with caged dinosaurs. Greg Bear (*Dinosaur Summer*, 1998; fig. 33.4, lower right) intriguingly imagines a post-*Challenger* exploitation of Doyle's Lost World for circus purposes. When the circus craze collapses, his hero Peter Belzoni attempts to take the dinosaurs back to Maple White Land—but finds that is not easy.

In a story I read in the British *Wizard* comic during the late 1940s, the exploitation by an engineering firm of a Trinidad-style lake of pitch caused monsters preserved therein from ever-more-ancient times—dinosaurs included—to successively awaken to new life. Nicholas DiChario, in another story in *Dinosaur Fantastic* (Resnick and Greenberg, eds., 1993), has a sauropod released upon a small American town by an earthquake; unusually, that tale ends without the customary slaying of the beast. Evelyn Lampman's story of *The Shy Stegosaurus of Cricket Creek* (1955; fig. 33.3, upper) is also one that ends happily.

Not so happy is the fate of the dinosaurs biotechnologically regenerated by DNA techniques for show to visitors, on the tropical island imagined by Michael Crichton in *Jurassic Park* (1990; fig. 33.5, lower). Through the film version, that story is well known and has served to greatly increase public interest in dinosaurs in many countries. I enjoyed both book and film; but the sequel, for which Mr. Crichton arrogantly appropriated Conan Doyle's title *The Lost World* (1995), was disappointing and the matching film one of the most hackneyed (and most patently flawed) of all Hollywood epics. A story by Gregory Benford, "Shakers of the Earth," in *The Ultimate Dinosaur* collection ends more happily than these books, with reconstituted sauropods dwelling, free from menacing theropods, in a park in Kansas.

Harry Harrison's Eden trilogy (1984–1988) imagines that the dinosaurs have survived and developed their own special civilization, but that mankind has also evolved and is emerging from a situation of dinosaur dominance to compete successfully with them. The "zoologies" appended to these works, illustrated by Bill Sanderson, give them a particular charm.

Foremost among feats of fictional imagination in this line is Dougal Dixon's *The New Dinosaurs* (1988), a truly remarkable and infinitely painstaking setting forth of the natural history and biogeography of a world that has never been taken over by man or other mammals. In it the dinosaurs, along with a few other surviving Mesozoic animal lineages, have continued to evolve, without ever developing the sort of civilization that Harrison imagined.

Figure 33.3. (opposite page) Dinosaurs for children. Upper: The Shy Stegosaurus of Cricket Creek, as depicted on the cover of Evelyn S. Lampman's book (1955). Lower left: Pataud, le petit dinosaure, the beguiling creation of Darlene Geis (1960). Lower right: a lady's hat in the jaws of a Triceratops, from Janet McNeill's Wait for It (1972).

Dinosaurs for Younger Children

Some of the works about dinosaurs mentioned above were written for children—but for older children, long able to read for (and to) themselves. There is a mounting plethora of dinosaur books for younger children; most strive to be factual, but some tell tales. The ones mentioned below are merely a small sampling.

The earliest I encountered, oddly, was in France—oddly, because French children do not share North American children's fascination with these beasts. Darlene Geis's *Pataud, le petit dinosaure* (1960; fig. 33.3, lower left) charmingly recounts the adventures of a very young sauropod. Gene Darby's *Dinosaur Comes to Town* (1963) features a frightened theropod, who fortunately proves content to eat hamburgers. Dorothy Thompson Landis's *Bronto the Dinosaur* (1967) is a cheerful account of how another "Pataud" slays a dreaded predator: it is billed as "educationally sound" but cannot justly make that claim. (Talking dinosaurs? A baby sauropod felling a mighty theropod?) The dinosaur in Charles Causley's *The Tail of the Trinosaur* (1973) is a somewhat improbable hybrid, whose story is told in lively verse. The belated hatching of dinosaur eggs is crucial to two stories—Oliver Butterworth's *The Enormous Egg* (1956), improbably laid by a hen, and Willis Hall's *The Summer of the Dinosaur* (1977)—while Mordecai Richler's *Jacob Two-Two and the Dinosaur* (1987) tells of a baby dinosaur brought back from Africa by Jacob's parents and mistaken for a lizard until he starts to grow bigger and bigger. Michael Denton's (1981) small Rudi builds *his* dinosaur from egg boxes! The misadventure reported in Janet McNeill's *Wait for It* (1972; fig. 33.3, lower right) happened during a museum visit, when the floral hat of a small girl's aunt became entangled in the jaws of a *Triceratops*.

Marie Halun Bloch's *Footprints in the Swamp* (1985), is unusual in

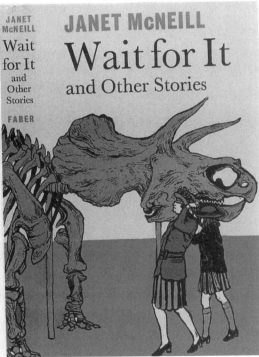

Figure 33.4. (opposite page) More Mesozoic beasts for children. Upper left: a Victorian girl contemplating dinosaurs; cover of Penelope Lively's Fanny and the Monsters *(1979). Upper right: children seeking dinosaur bones; cover of* The Dinosaur Dilemma *(1964). Lower left: the old concept of gliderlike pterodactyls; an illustration in Charles G. D. Roberts's* In the Morning of Time *(1919). Lower right: the rather complex creature on the cover of Greg Bear's* Dinosaur Summer *(1998)—a deinonychosaur-Carnotaurus hybrid?*

that it focuses on the small Mesozoic mammals that were living in the time of the dinosaurs, showing how they were able to survive the environmental changes that finished off the monsters.

Dinosaurs feature in poems for children, such as Jack Prelutsky's *Tyrannosaurus Was a Beast* (1988). They are material for humor in such truly excruciating works as *1001 More Dinosaur Jokes for Kids* by "Alice Saurus" (1994)—though I have to confess that one joke did amuse me!

Closer to reality are Penelope Lively's small Victorian girl, whose imagination is inspired by the skeletons in London's Natural History Museum and the Crystal Palace models of dinosaurs at Sydenham (*Fanny and the Monsters,* 1979; fig. 33.4, upper left), and the Canadian fossil-hunting children in Theresa Heuchert's and Mary C. Wood's *Mystery in the Graveyard of Monsters* (1986). Closest of all are the children in Lois Breitmeyer's and Gladys Leithauser's *The Dinosaur Dilemma* (1964; fig. 33.4, upper right), who join a quest for dinosaur bones in Colorado, only to encounter problems with a high-powered real estate developer, interested not in fossils but only in profit.

Flights into Space and Crime

Dinosaurs, in varying guise, turn up in a variety of works of space fiction. Few such works, though, have any real echoes of the terrestrial condition. In two books Anne McCaffrey visualizes human intervention in the affairs of a *Dinosaur Planet* (1978, 1984a,b; fig. 33.7, upper left) with a fauna puzzlingly like that of earth; the planet Ireta proves indeed to have been "planted" from the ancient Earth by space-traveling scientists anxious to perpetuate that ecosystem before the process of evolution destroys it. Two vertebrate paleontologists, competing for bones on the planet Krishna, are among the manifold imaginings of L. Sprague and Catherine Crook de Camp (*The Bones of Zora,* 1983)—an interstellar echo of the rivalry between Cope and Marsh. However, the most developed science fiction treatment of dinosaurs is to be found in Robert J. Sawyer's three books about the world of Quintaglio (1992–1994; fig. 33.6, lower right), to which dinosaurs have again been transported from earth and on which (unlike Ireta, where the only creatures approaching civilization are pterodactyls) a carnosaur civilization has developed. Indeed, they have progressed to the point at which dinosaur paleontologists are studying stratification and excavating the bones of their ancestors.

Predictably enough, dinosaurs were dragged into the *Star Trek* series; Diane Carey and James I. Kirkland's *Star Trek: First Frontier* (1995) has Capt. Kirk stranded among them, back in the Mesozoic; but in usual space-cowboys-and-Indians fashion, he is brought back safely to his own time.

In contrast, dinosaurs do not figure largely in crime fiction. The murder in Frances and Richard Lockridge's *Dead as a Dinosaur* (1952) does occur in a museum, but not among the dinosaur bones, while the corpse in John Dellinger's *Dinosaur Tracks and Murder* (1995) just happens to be found close to the dinosaur footprints on the Hogback,

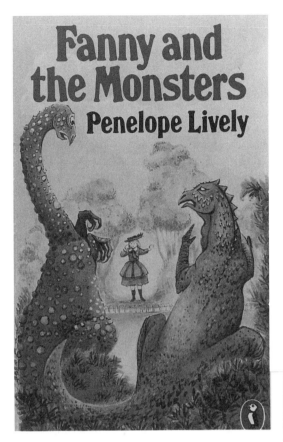

near Denver; there is no other connection. The first relevant crime depicted in fiction is, in fact, an adaptation from film: John Harvey's highly amusing *One of Our Dinosaurs Is Missing* (1976). Garrison Allen's *Dinosaur Cat* (1998) begins, like Dellinger's book, with the finding of a corpse, but this time alongside some sauropod bones; the generic name of these keeps changing during this carelessly written but cheerfully picaresque novel. Sandy Dengler's *The Last Dinosaur* (1994; fig. 33.7, lower right), involves a killing apparently done by a life-size model of *Tyrannosaurus* on a film set in Arizona; but the dinosaur is not guilty!

The best crime fiction novel concerning dinosaurs is the latest, John Paxson's *Bones* (1999; fig. 33.7, lower left). This mystery is set in Montana; the relations between amateur bone hunters and professional paleontologists are well depicted and the crime pivots upon very believable jealousies concerning a crucial scientific discovery.

Approaches to Reality

A fictional work placed in the historic past of dinosaur hunting is Kathryn Lasky's *The Bone Wars* (1988), set in the Judith River region of Montana in the 1870s—a time when the two leading U.S. vertebrate paleontologists were battling for dinosaur bones and government troops were battling the Sioux. In Robert Kroetsch's *Badlands* (1975), an eccentric party of dinosaur hunters travels down the Red Deer River in Alberta, in quest of a discovery that might bring them scientific immortality. In *Pictures from a Trip* (1985), Tim Rumsey's three bone hunters in the American West, one of them blind, share very believable experiences in today's world; the accidental shattering of a *Triceratops* horn (p. 201) must cause a sympathetic shiver in the spine of any paleontologist.

Two Canadian novels are set in museums where dinosaur bones are being curated and displayed—Margaret Atwood's *Life before Man* (1979) and Claudia Casper's *The Reconstruction* (1996)—but, in both cases, the mentions of dinosaurs are only incidental.

There can be no question that fictional portrayals of dinosaurs have aroused interest in many nonscientists; Philip Currie, to whom this volume is dedicated, freely admits that this was the source of his own ever-developing interest in them. Of course, scientists have long been conscious of this. Consequently, the attempt, through fiction, to educate readers on the realities of the Mesozoic, begun by Charles Sternberg and Vladimir Obruchev, has been continued with vigor by other vertebrate paleontologists. Edwin H. Colbert's *The Year of the Dinosaur* (1977; fig. 33.5, upper right), illustrated by his wife, Margaret, tells of twelve months in the life of a brontosaur. Two of four excellent "lives" of extinct animals by Beverly and Jenny Halstead concern, respectively, a brontosaur (1982) and a deinonychosaur (1983). Most recent is Robert T. Bakker's *Raptor Red* (1995; fig. 33.2, lower right), which recounts thrillingly the adventures and misadventures of a migrating female *Utahraptor*.

Most recently, Philip Currie himself has jointly written the first

Figure 33.5. (opposite page) Upper left: cover of the English translation of Vladimir Obruchev's Plutonia *(1957), with aquatic brontosaur. Upper right: cover of Edwin Colbert's* The Year of the Dinosaur *(1977), with a definitely terrestrial brontosaur. Lower: the* Tyrannosaurus *skeleton on the cover of Michael Crichton's* Jurassic Park *(1990), the most commercially successful of all dinosaur novels.*

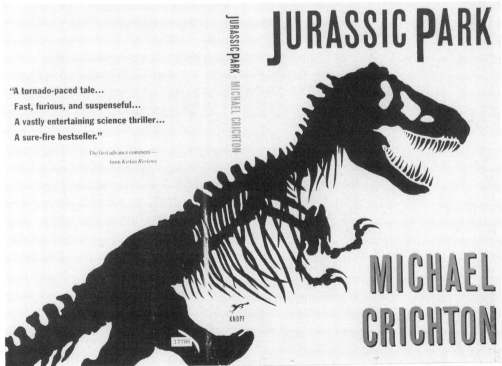

four of a planned series of collaborative works—two being written with Eric Felber and two with Philip's paleobotanist wife, Eva Koppelhus (Felber et al. 1997, 1998; Currie et al. 1998, 1999). The first part of each book realistically depicts the life of a particular dinosaur genus, while the second (and briefer) part is strictly factual. Excellent illustrations by Jan Sovak are a particular treat.

Evolving Concepts

When Mill wrote *The Fossil Spirit* (1854), *Iguanodon* and *Megalosaurus* were conceived of as quadrupeds, having a size truly vast. His text matched the concepts of the period, even to the placement, on the nose of *Iguanodon,* of the spike that later was recognized as one of its thumbs (fig. 33.8, upper). By the time of Conan Doyle's *The Lost World* (1912), it had long been recognized by paleontologists that those dinosaurs were bipeds and, though huge enough, not nearly so gigantic as Richard Owen and his contemporaries had conceived. In contrast, the erroneous idea that dinosaurs were stupid creatures—cold-blooded, slow-moving, or static except when striving to catch prey or to evade capture—lingered long. As Stephen Jay Gould points out in his extended commentary on George Gaylord Simpson's novel (1996), Sam Magruder could only survive, for as long as he did, because his intelligence and agility so amply outmatched those of the dinosaurian predators that menaced him. Indeed, it can be justly claimed that, in endowing dinosaurs with greater intelligence than contemporary scientists would allow, such writers as Edgar Rice Burroughs amply anticipated later scientific deduction.

Most, however, did not. For example, the contemporary concept of *Scelidosaurus* is reflected in Christian O'Connor Morris's illustration to Lady Bray's book (1921; fig. 33.8); it was then considered to be a biped, on the basis of incomplete skeletal material. With fuller knowledge, we now know it to have been a quadruped and almost certainly a primitive ankylosaur.

The belief that sauropods were essentially aquatic creatures also lasted long. Their bulk was considered so large as to require the buoyant support of water and it was even questioned whether they were capable of moving on land at all! A passage in *The Lost World* (1912) may indicate Conan Doyle's belief that they could:

> Once upon a yellow sandbank I saw a creature like a huge swan, with a clumsy body and a high, flexible neck, shuffling about upon the margin. Presently it plunged in, and for some time I could see the arched neck and darting head undulating above the water. Then it dived, and I saw it no more. (1912a, 204; 1912b, 213–214)

However, this creature might have been an elasmosaur, for "Lake Gladys" in Conan Doyle's Maple White Land, though so far from the sea, did contain ichthyosaurs.

Certainly, Obruchev (1957, 219) had no doubt about the mobility of sauropods on land:

The creatures [brontosaurs] then ran off along the shore at a heavy trot, swaying awkwardly on their legs, which were short and feeble in comparison to their massive bodies.

That writer, at least, was ahead of scientific opinion. Only Roland Bird's study of sauropod footprints (1944), and Robert Bakker's subsequent demonstration that sauropod skeletons were emphatically those of habitual land dwellers (1971), caused that old, wrong idea of amphibious sauropods to be jettisoned.

The concept of theropods leaping on their prey dates back to a painting by Charles R. Knight, Dryptosauruses *Fighting* (reproduced in Mitchell 1998, fig. 9.1). The illustration on the cover of Francis Rolt-Wheeler's *The Monster-Hunters* (1916) is essentially a reproduction of that painting. Similar behavior is suggested in T. C. Bridges's work (1923; fig. 33.2, upper right). However, the ability of theropods to leap remains doubtful. Though it remains at least conceivable that coelurosaurs were capable of such behavior, few paleontologists would nowadays envisage an attack of that kind by any larger reptilian predator.

Another wrong idea adopted by Conan Doyle, and by seventy years of subsequent authors, was that pterodactyls were incompetent aeronauts, having flight membranes attached to the flanks and thighs of their scaly bodies and holding the hind limbs outstretched to keep those membranes taut, so that the motion of the wings was minimal and the flight mostly gliderlike. This concept is expressed in *The Lost World*:

> [The creature's] strange shawl suddenly unfurled, spread, and fluttered as a pair of leathery wings. . . . [It was soon] circling slowly round the Queen's Hall with a dry, leathery flapping of its ten-foot wings. . . . (1912a, pp. 299–300; 1912b, p. 310; fig. 33.9, upper)

An illustration in Roberts's *In the Morning of Time* (1919; fig. 33.4, lower left) expresses that mistaken image even more exactly.

Nowadays, we have data to show that the bodies, at least, of pterodactyls had a hairlike cover, that they were relatively efficient flyers, and that they flew with hind limbs drawn up under the body, like birds. However, fiction writers can scarcely be blamed when, even in works so recent and so authoritative as Wellnhofer's *Encyclopedia of Pterosaurs* (1991), the old ideas survived. In contrast, we believe that Conan Doyle was correct when he conceived them to be gregarious creatures:

> The place was a rookery of pterodactyls. There were hundreds of them concentrated within view [including] hideous mothers brooding upon their leathery, yellowish eggs. (1912a, p. 167; 1912b, pp. 175–176)

Obruchev anticipated recent conclusions on the habits of the giant pterodactyl *Quetzalcoatlus* by visualizing pterodactyls as carrion eaters:

> Great activity reigned there. Flying lizards of different sizes hurried to and fro, and had settled on the carcasses of the ceratosaurus and iguanodon. They tore pieces of flesh from the bodies, and devoured them on the spot or carried them off towards the

Figure 33.6. (next page) Dinosaurs in science fiction. Upper left: cover of a later edition (1946) of Edgar Rice Burrough's The Land That Time Forgot, with ape-man, dinosaurs, and all. Upper right: an erudite carnosaur on the cover of the collection Dinosaur Fantastic (1993). Lower left: hunter Reginald Rivers oblivious to the approach of a horned carnosaur; L. Sprague de Camp's Rivers of Time (1993). Lower right: the most erudite carnosaur of all, a Quintaglio paleontologist, from Robert Sawyer's Fossil Hunter (1993)

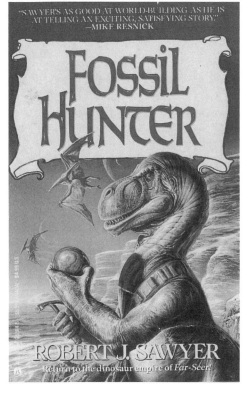

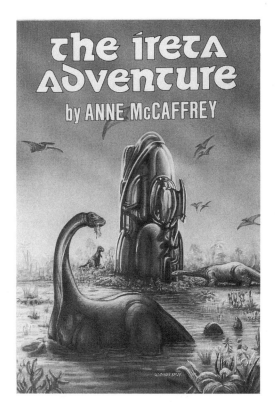

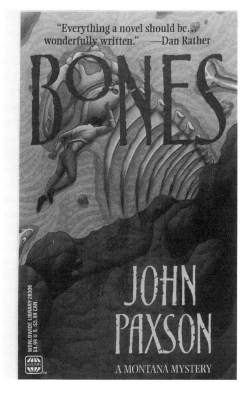

south, to the ravine in the hills, where their nests would probably be located. The screeching, croaking and hissing were ear-splitting. (1957, 167)

Were pterodactyls as intelligent as birds? That is hard to assess, but it seems probable. Anne McCaffery's concept of *Quetzalcoatlus* evolving into a social creature of developed intelligence (1978, 1984a,b) is a not unreasonable extrapolation.

Other hypotheses, innate in the illustrations or explicit in the texts of fictional works, remain hard to assess. C. H. Murray Chapman's *Stegosaurus* that walked bipedally (1924; fig. 33.8, lower) cannot yet be discounted; footprints of stegosaurs are still too rare for us to be certain whether that dinosaur was a habitual biped and occasional quadruped, or vice versa. The function and position of the plates on their backs likewise remain matters for question. Were the plates upright, sloping, or almost horizontal? Was Obruchev right in supposing that those plates were loose?

> The frightened stegosaurus retreated as fast as it could, lurching like an ambling horse, its backbone plates clashing together and making a loud clatter like castanets. (1957, 169)

That idea is unlikely, since the base of those plates is rugose up to 10–15 cm, indicating that they were embedded in the skin to that depth: but it cannot quite be disproved. Likewise, it remains hard to decide whether Morant (1937) was right when visualizing a Triassic dinosaur-like creature with a cover of feathers everywhere but on its head, lower hind legs, and feet. This agreed with contemporary concepts of an ancestor of the birds that flew gliderlike, with feathers on all four wings; indeed, the illustration in Morant (see fig. 33.9, lower) corresponds almost exactly with Zdeněk Burian's illustration of such a creature (Augusta and Burian 1961, 92). That concept was long set aside, but the recent numerous discoveries of variably feathered dinosaurs in China are giving it renewed credibility. Indeed, the latest work by Philip Currie and colleagues (1999) actually portrays the life of one of these, *Sinosauropteryx*.

These recent discoveries are disconcerting in other ways. When Dixon (1988) was imagining the evolutionary developments that might have occurred if there had been no extinctions around the Cretaceous-Tertiary boundary, his visions included "hairy" pterodactyls. However, he could not take into account the imminent realization, from those Chinese fossils, that many small "dinosaurs"—and perhaps even some larger ones, like *Dilophosaurus*—were feathered, without having the least ability to fly. Maybe we will soon be reading tales in which all the carnosaurs—even *Tyrannosaurus!*—had a variable covering of feathers. However, it remains hard to conceive of sauropods with feathers and, as yet, we have no evidence that any ornithischians were feathered. Was the Mesozoic world one of variably feather-covered predators and entirely featherless herbivores? That is an idea which future fiction writers should quickly develop, before further scientific discoveries spoil the fun of such speculations.

Figure 33.7. (previous page) Dinosaurs in science and crime fiction. Upper left: cover of the combined edition (1984) of Anne McCaffrey's two novels set on Ireta, the "dinosaur planet." Upper right: time travel and a contest with reptile-descended aliens in Lee Grimes's Dinosaur Nexus *(1994). Lower left: scientific rivalries lead to murder in John Paxson's* Bones *(1999). Lower right: was the death caused by a gigantic model tyrannosaur? That is a question in Sandy Dengler's* The Last Dinosaur *(1994).*

Figure 33.8. (opposite page) Changing images of dinosaurs. Upper: John Mill's nasal-horned Iguanodon *(1854); middle: Lady Bray's bipedal* Scelidosaurus *(1921); lower: Murray Chapman's bipedal* Stegosaurus *(1924).*

Conclusions

After Charles Dickens's first fictional mention of a dinosaur (1852–53) and John Mill's "Dream" (1854), there followed a remarkably long hiatus before the next—and, arguably, the best—novel on the theme appeared, Conan Doyle's *The Lost World* (1912). Five years later, Charles Sternberg published—rather obscurely—the first account of travel backward in time to the Mesozoic and, two years after that, the first factual portrait of the life of dinosaurs was attempted by Charles G. D. Roberts (1919).

Four fictional treatments in the ninety years following the scientific discovery of the dinosaurs was a slow beginning indeed. Moreover, the pace of publication of novels and short stories treating (or even mentioning) dinosaurs remained slow for a further fifty years. It was only in the 1970s that the trickle truly swelled to a flood. The literary quality of these works is variable but, most often, low; the quality of imagination, in contrast, is high, some novels even approaching epic status. Up to that time, illustrations had been of variable quality but usually not very good. Nowadays they are almost always of high quality. Not only do they embody the latest scientific discoveries, but they add intriguing interpretations of behavior and (in particular) color patterns, which stimulate imaginings of what the Mesozoic world was really like.

Though most often without conscious educational intent—the two series with which Beverly Halstead and Philip Currie have been associated are among the exceptions—such fictional works serve nevertheless to inform the public at large concerning the changing concepts in paleontology. (George Gaylord Simpson's posthumously published novel, by the time it appeared, was already a literary fossil.) Furthermore, they are in general very enjoyable. By stimulating the interest and imaginations of young people, in particular, often they lead to a deeper interest in the fossils to be seen in museums, interpretive centers, and scientific sites. A very respectable roster of paleontologists owed their start to such reading.

It may be long before such works gain recognition in the curricula of English departments at colleges or universities; for reasons incomprehensible to me, the so-called serious English scholars do not rate a creative imagination at all highly if it extends beyond consideration of the human condition. Nevertheless it is, in my own view at least, important because it is educational, enjoyable, and horizon widening, giving to the old bones of the distant past a fresh and vivid life.

Acknowledgments: A shorter version of this chapter was first presented in Calgary, July 1999. Philip Currie was in the audience and, afterward, urged me to publish it. Where could that be done more appropriately than in this tribute to my good friend of many years?

I am indebted to Darren Tanke, Kenneth Carpenter, and an anonymous reviewer for drawing my attention to works unknown to me and for other helpful suggestions; likewise to my friends David Spalding and Tim Tokaryk, for much interchanging of ideas about books, and to my research assistants, Trent Mitchell and Jason Sharp, for aiding in the preparation of this venture into the unreal history of dinosaurs.

Figure 33.9. (opposite page) Changing images of Mesozoic creatures. Upper: a rendition by "JM" of the Albert Hall scene, when Professor Challenger releases his Lost World *pterodactyl (from the* Radio Times, *London, ca. 1946). Lower: Jean Elder's drawing of H. C. F. Morant's Triassic bird ancestor, in* Whirlaway *(1937).*

References

Since the work of artists has been so important in clarifying or developing the ideas of authors, I have named them below, whenever their names are known to me. When works have been published in only a few editions, I have specified those few. In the case of writers such as Dickens, Doyle, and Burroughs, however, the later editions are much too numerous to be listed here.

Allen, G. 1998. *Dinosaur Cat.* New York: Kensington Publishing.

Atwood, M. 1979. *Life before Man.* Toronto: Bantam-Seal.

Augusta, J., and Z. Burian. 1961. *Prehistoric Reptiles and Birds.* Illustrated by Z. Burian. London: Hamlyn.

Bakker, R. T. 1971. Ecology of the Brontosaurs. *Nature* 2: 172–174.

Bakker, R. T. 1995. *Raptor Red.* Illustrated by the author. New York: Bantam Books.

Batory, D. M., and W. A. S. Sarjeant. 1994. "The Terror of Blue John Gap" a geological and literary study. *Journal of the Arthur Conan Doyle Society* 5: 108–125.

Bear, G. 1998. *Dinosaur Summer.* New York: Time Warner.

Bird, R. T. 1944. Did *Brontosaurus* ever walk on land? *Natural History* 53: 63–67. Reprinted in W. A. S. Sarjeant (ed.), *Terrestrial Trace Fossils,* pp. 151–162. Stroudsburg, Pa.: Hutchinson Ross, 1983.

Bischoff, D. 1984. *Time Machine: Search for Dinosaurs.* Illustrated By W. Stout. New York: Bantam Books.

Bloch, M. H. 1985. *Footprints in the Swamp.* Illustrated By R. Shetterly. New York: Atheneum.

Bradbury, R. 1983. *Dinosaur Tales.* New York: Bantam Books.

Bray, Lady F. O. 1921. *Old Time and the Boy; or, Prehistoric Wonderland.* Illustrated by C. O'Connor Morris. London: Allenson.

Breitmeyer, L., and G. Leithauser. 1964. *The Dinosaur Dilemma.* Illustrated by L. Maloy. San Carlos, Calif.: Golden Gate Junior Books.

Bridges, T. C. 1923. *Men of the Mist.* Illustrated by G. H. Evison. London: Harrap. Reprint, London: Collins, ca. 1940.

Brown, L. S. 1982. *Yes, Helen, There Were Dinosaurs: The Story of a Jurassic Time Trip.* Illustrated by the author. Kingston, N.Y.: privately published.

Burroughs, E. R. 1922. *At the Earth's Core.* Illustrated by J. A. St. John. Chicago: McClurg. Reprint, London: Methuen, 1923.

Burroughs, E. R. 1923. *Pellucidar; a Sequel to "At the Earth's Core." Relating the Further Adventures of David Innes in the Land underneath the Earth's Crust.* Illustrated by J. A. St. John. Chicago: McClurg.

Burroughs, E. R. 1924. *The Land That Time Forgot.* London: Methuen. [Comprises "The Land that Time Forgot," orig. publ. in *Blue Book,* New York, August 1918; "The People That Time Forgot," orig. publ. in *Blue Book,* October 1918; and "Out of Time's Abyss," orig. publ. in *Blue Book,* December 1918. These three novelettes were republished in paperback by Ace, New York, ca. 1950.]

Burroughs, E. R. 1929. *Tanar of Pellucidar.* New York: Metropolitan.

Burroughs, E. R. 1930. *Tarzan at the Earth's Core.* New York: Metropolitan.

Burroughs, E. R. 1937. *Back to the Stone Age.* Illustrated by J. C. Burroughs. Tarzana, Calif.: Burroughs.

Burroughs, E. R. 1944. *Land of Terror.* Tarzana, Calif.: Burroughs.

Burroughs, E. R. 1963. *Savage Pellucidar.* New York: Canaveral Press.

Butterworth, O. 1956. *The Enormous Egg*. Illustrated by L. Darling. Boston: Little Brown and Co.

Calvino, I. 1968. The dinosaurs. In *Cosmicomics*. Translated by William Weaver. New York: Harcourt, Brace.

Carey, D., and J. I. Kirkland. 1995. *Star Trek: First Frontier*. New York: Pocket Books.

Casper, C. 1996. *The Reconstruction*. Toronto: Viking. Reprint, New York: St. Martin's Press, 1997.

Causley, C. 1973. *The Tail of the Trinosaur*. Illustrated by J. Gardiner. Leicester, England: Beaver Books.

Charig, A. J. 1995. Disaster Theories of Dinosaur Extinction. In W. A. S. Sarjeant (ed.), *Vertebrate Fossils and the Evolution of Scientific Concepts: Writings in Tribute to Beverly Halstead, by Some of His Many Friends,* pp. 309–328. Reading, England: Gordon and Breach.

Chilson, R. 1976. *The Shores of Kansas*. New York: Popular Library.

Colbert, E. H. 1977. *The Year of the Dinosaur*. Illustrated by M. Colbert. New York: Scribner's.

Crichton, M. 1990. *Jurassic Park*. New York: Knopf.

Crichton, M. 1995. *The Lost World*. New York: Knopf.

Currie, P. J., E. B. Koppelhus, and J. Sovak. 1998. *A Moment in Time with Centrosaurus*. Illustrated by J. Sovak. Calgary: Troödon Productions.

Currie, P. J., E. B. Koppelhus, and J. Sovak. 1999. *A Moment in Time with Sinosauropteryx*. Illustrated by J. Sovak. Calgary: Troödon Productions.

Darby, G. 1963. *Dinosaur Comes to Town*. Illustrated by Art Seiden. Racine, Wis.: Whitman Publishing.

de Camp, L. S. 1956. A gun for dinosaur. *Galaxy Magazine,* March, pp. 6–12. Reprinted in *A Gun for Dinosaur and Other Imaginative Tales*. Garden City, New York: Doubleday, 1963. Reprinted, with other stories, in *Rivers of Time*. New York: Baer/Simon and Schuster, 1993.

de Camp, L. S., and C. C. de Camp. 1983. *The Bones of Zora*. Huntington Woods, Mich.: Phantasia Press.

Dellinger, J. 1995. *Dinosaur Tracks and Murder*. Salt Lake City: Northwest Publishing.

Dengler, S. 1994. *The Last Dinosaur*. Wheaton, Ill.: Victor Books.

Denton, M. 1981. *The Eggbox Brontosaurus*. Illustrated by H. Offen. St. Albans, England: Granada Publishing.

Dickens, C. 1852–53. *Bleak House*. Illustrated by H. K. Browne. London: Bradbury and Evans.

Dixon, D. 1988. *The New Dinosaurs: An Alternative Evolution*. Illustrated by the author. Topsfield, Mass.: Salem House.

Doyle, Sir A. C. 1912a. *The Lost World*. London: Hodder and Stoughton.

Doyle, Sir A. C. 1912b. *The Lost World*. Illus. ed. London: Henry Frowde.

Fawcett, E. D. [1894]. *Swallowed by an Earthquake*. London: Arnold.

Felber, E. P., P. J. Currie, and J. Sovak. 1997. *A Moment in Time with Troödon*. Illustrated by J. Sovak. Calgary: Troödon Productions.

Felber, E. P., P. J. Currie, and J. Sovak. 1998. *A Moment in Time with Albertosaurus*. Illustrated by J. Sovak. Calgary: Troödon Productions.

Foster, A. D. 1996. *Dinotopia Lost*. Atlanta: Turner Publishing.

Foster, A. D. 1999. *The Hand of Dinotopia*. Illustrated by J. Gurney. New York: HarperCollins.

Geis, D. 1960. *Pataud, le petit dinosaure*. Illustrated by Bob Jones. Paris: Gautier Langereau.

Glut, D. 1997. Popular Culture: Literature. In P. J. Currie and K. Padian (eds.), *Encyclopedia of Dinosaurs*. San Diego: Academic Press.

Grimes, L. 1994. *Dinosaur Nexus*. New York: Avon Books.

Gurney, J. 1992. *Dinotopia: A Land Apart from Time*. Illustrated by the author. Atlanta: Turner Publishing.

Gurney, J. 1995. *Dinotopia: The World Beneath*. Illustrated by the author. Atlanta: Turner Publishing.

Gurney, J. 1999. *Dinotopia: First Flight*. Illustrated by the author. New York: HarperCollins. [Includes a board game.]

Hall, W. 1977. *The Summer of the Dinosaur*. Illustrated by J. Griffiths. London: Bodley Head. Reprinted as *Henry Hollings and the Dinosaur*. London: Target Books, 1978.

Halstead, L. B., and J. Halstead. 1982. *A Brontosaur: The Life Story Unearthed*. London: Collins.

Halstead, L. B., and J. Halstead. 1983. *Terrible Claws: The Story of a Carnivorous Dinosaur*. London: Collins.

Harrison, H. 1984. *West of Eden*. Illustrated by B. Sanderson. Toronto: Bantam Books.

Harrison, H. 1986. *Winter in Eden*. Illustrated by B. Sanderson. Toronto: Bantam Books.

Harrison, H. 1988. *Return to Eden*. Illustrated by B. Sanderson. Toronto: Bantam Books.

Harvey, J. 1976. *One of Our Dinosaurs Is Missing*. London: New English Library.

Heuchert, T., and M. C. Wood. 1986. *Mystery in the Graveyard of Monsters*. [Cover title: *In the Graveyard of Monsters*.] Saskatoon: Prairie Lily Cooperative.

Jablonski, D. (ed.). 1981. *Behold the Mighty Dinosaur*. New York: Elsevier/Nelson Books.

Kroetsch, R. 1975. *Badlands*. Toronto: New Press.

Lampman, E. S. 1955. *The Shy Stegosaurus of Cricket Creek*. Illustrated by H. Buel. Garden City, N.Y.: Doubleday.

Landis, D. T. 1967. *Bronto the Dinosaur*. Illustrated by G. Wilde. Chicago: Rand McNally.

Lasky, K. 1988. *The Bone Wars*. New York: Morrow Junior Books.

Leigh, S. 1992. *Dinosaur World*. New York: Avon/Nova.

Leigh, S. 1993. *Dinosaur Planet*. New York: Avon Books.

Leigh, S. 1994. *Dinosaur Warriors*. New York: Avon Books.

Leigh, S. 1995. *Dinosaur Conquest*. New York: Avon Books.

Leigh, S., and J. J. Miller. 1993. *Dinosaur Samurai*. Illustrated by B. Franczak. New York: Avon Books.

Leigh, S., and J. J. Miller. 1995. *Dinosaur Empire*. Illustrated by N. Jainschrigg and C. Skinner. New York: Avon Books.

Lively, P. 1979. *Fanny and the Monsters*. Illustrated by J. Lawrence. London: Heinemann. Reprinted as *Fanny and the Monsters and Other Stories*. Harmondsworth, England: Puffin Books, 1980.

Lockridge, F., and R. Lockridge. 1952. *Dead as a Dinosaur*. New York: Lippincott.

Mackal, R. P. 1987. *A Living Dinosaur? In Search of Mokele-Mbembe*. Foreword by B. Heuvelmans. Leiden: Brill.

Mash, R. 1983. *How to Keep Dinosaurs*. Illustrated by W. Rushton, P. Hood, and D. Wallis. London: Deutsch.

McCaffrey, A. 1978. *Dinosaur Planet*. London: Futura Books.

McCaffrey, A. 1984a. *Dinosaur Planet Survivors*. New York: DelRey Ballantine. [Also published as *The Survivors: Dinosaur Planet II*.]

McCaffrey, A. 1984b. *The Ireta Adventure: Dinosaur Planet and Dinosaur Planet Survivors*. New York: Doubleday.

McNeill, J. 1972. *Wait for It and Other Stories*. London: Faber and Faber.

Mill, J. 1854. *The Fossil Spirit: A Boy's Dream of Geology*. London: Darton.

Mitchell, W. J. T. 1998. *The Last Dinosaur Book: The Life and Times of a Cultural Icon*. Chicago: University of Chicago Press.

Morant, H. C. F. 1937. *Whirlaway*. Illustrated by J. Elder. London: Hutchinson.

Murray Chapman, C. H. [1924]. *Dragons at Home*. Illustrated by the author. London: Wells Gardner and Co.

Obruchev, V. A. 1957. *Plutonia: An Adventure through Prehistory*. Translated by B. Pearce. Illustrated by E. J. Pagram. London: Lawrence and Wishart.

Paxson, J. 1999. *Bones*. Toronto: Worldwide.

Preiss, B., and R. Silverberg (eds.). 1994. *The Ultimate Dinosaur: Past-PresentFuture*. Illustrated by B. Franczak et al. New York: Bantam Books.

Prelutsky, J. 1988. *Tyrannosaurus Was a Beast*. Illustrated by A. Lobel. New York: Mulberry Books.

Resnick, M., and M. H. Greenberg (eds.). 1993. *Dinosaur Fantastic*. New York: Daw Books.

Richler, M. 1987. *Jacob Two-Two and the Dinosaur*. Illustrated by N. Eyolfson. New York: Knopf.

Roberts, C. G. D. 1919. *In the Morning of Time*. Illustrated by F. Gardner. London: Dent.

Rolt-Wheeler, F. W. 1916. *The Monster-Hunters*. Toronto: McClelland, Goodchild, and Stewart; Boston: Lothrop, Lee, and Shepard.

Rumsey, T. 1985. *Pictures from a Trip*. New York: Morrow.

Russell, D., and R. Séguin. 1982. Reconstructions of the small Cretaceous theropod *Stenonychosaurus inequalis* and a hypothetical dinosauroid. Canadian Museum of Nature, *Syllogeus* 37: 1–43.

Sarjeant, W. A. S. 1994. Geology in fiction. In D. F. Branagan and G. H. McNally (eds.), *Useful and Curious Geological Enquiries beyond the World,* pp. 318–337. Springwood, New South Wales: Conference Publications, for the International Commission on the History of Geological Sciences.

Saurus, Alice [pseud.]. 1994. *1001 More Dinosaur Jokes for Kids*. New York: Ballantine Books.

Sawyer, R. J. 1992. *Far-Seer*. New York: Ace Books.

Sawyer, R. J. 1993. *Fossil Hunter*. New York: Ace Books.

Sawyer, R. J. 1994. *Foreigner*. New York: Ace Books.

Simpson, G. G. 1996. *The Dechronization of Sam Magruder*. Edited by J. Simpson Burns, with an afterword by S. J. Gould. New York: St. Martin's Press.

Sternberg, C. H. 1917. *Hunting Dinosaurs in the Bad Lands of the Red Deer River, Alberta, Canada*. San Diego: C. H. Sternberg. Reprint, with introduction by D. A. E. Spalding. Edmonton: NeWest Press, 1985.

Taine, J. 1934. *Before the Dawn*. Baltimore: Williams and Wilkins.

Verne, J. 1864. *Voyage au centre de la terre*. Paris: Hetzel. [Translated into English as *Journey to the Centre of the Earth*. London: Griffith and Farran, 1872.]

Wellnhofer, P. [1991]. *The Illustrated Encyclopedia of Pterosaurs*. London: Salamander Books.

White, T. H. 1939. *The Sword in the Stone*. Decorations by the author; endpapers by R. Lawson. New York: Putnam's Sons.

Publications of
Philip John Currie

Compiled by Clive Coy

Books

Carpenter, K., and P. J. Currie (eds.). 1990. *Dinosaur Systematics: Approaches and Perspectives.* New York: Cambridge University Press.

Currie, P. J. 1991. *The Flying Dinosaurs.* Red Deer, Alta.: Discovery Books, Red Deer College Press.

Currie, P. J. 1994. [*Dinosaur Renaissance.*] Tokyo: Kodansha. (In Japanese.)

Currie, P. J., and Z. V. Spinar. 1994. *The Great Dinosaurs: A Story of the Giants' Evolution.* Stamford, Conn.: Longmeadow Press. London: Sunburst Books.

Currie, P. J., and Z. V. Spinar. 1994. *Velcí dinosauri.* Prague: Aventinum. (Czech edition of *The Great Dinosaurs.*)

Currie, P. J., and Z. V. Spinar. 1994. *Wielkie dinozaury.* Warsaw: Warszawski Dom Wydawniczy. (Polish edition of *The Great Dinosaurs.*)

Currie, P. J. 1995. *Giganten der Lüfte: Das grosse Buch der Flugsaurier.* Würzburg: Arena Verlag. (German edition of *The Flying Dinosaurs.*)

Currie, P. J., and E. B. Koppelhus. 1996. *One Hundred One Questions about Dinosaurs.* New York: Dover.

Currie, P. J., E. Felber, and J. Sovak. 1997. *A Moment in Time with Troodon.* Calgary: Troodon Productions.

Currie, P. J., and K. Padian (eds.). 1997. *Encyclopedia of Dinosaurs.* San Diego: Academic Press.

Currie, P. J., and J. Sovak. 1997. [*The Dinosaur Handbook.*] Tokyo: Yazawa Handbook Series. (In Japanese.)

Currie, P. J., E. Felber, and J. Sovak. 1998. *A Moment in Time with Albertosaurus.* Calgary: Troodon Productions.

Currie, P. J., E. B. Koppelhus, and J. Sovak. 1998. *A Moment in Time with Centrosaurus.* Calgary: Troodon Productions.

Currie, P. J., C. O. Mastin, and J. Sovak. 1998. *The Newest and Coolest Dinosaurs.* Calgary: Grasshopper Books.

Currie, P. J., and Z. V. Spinar. 1998. *Dinosauriers de Heersers van Toen.* Netherlands: R&B Productions. (Dutch edition of *The Great Dinosaurs.*)

Currie, P. J., E. B. Koppelhus, and J. Sovak. 1999. *A Moment in Time with Sinosauropteryx.* Calgary: Troodon Productions.

Scientific Publications

Azuma, Y., and P. J. Currie. 1995. A new giant dromaeosaurid from Japan. *Journal of Vertebrate Paleontology* 15 (supp. to no. 3): 17A. (Abstract.)

Bakker, R. T., M. Williams, and P. J. Currie. 1988. *Nanotyrannus*, a new genus of pygmy tyrannosaur from the Latest Cretaceous of Montana. *Hunteria* 1 (5): 1–30.

Beavan, N. R., P. J. Currie, and A. P. Russell. 1994. Variation in papillar morphology of hadrosaur (Dinosauria: Ornithischia) teeth, possible taxonomic utility. *Journal of Vertebrate Paleontology.* 14 (supp. to no. 3): 16A. (Abstract.)

Burnham, D. A., K. L. Derstler, P. J. Currie, R. T. Bakker, Zhou Z., and J. H. Ostrom. 2000. Remarkable new birdlike dinosaur (Theropoda: Maniraptora) from the Upper Cretaceous of Montana. *University of Kansas Paleontological Contributions* 13: 1–14.

Carpenter, K., and P. J. Currie. 1990. Introduction: On systematics and morphological variation. In K. Carpenter and P. J. Currie (eds.), *Dinosaur Systematics: Approaches and Perspectives,* pp. 1–8. New York: Cambridge University Press.

Carroll, R. L., and P. J. Currie. 1975. Microsaurs as possible apodan ancestors. *Zoological Journal of the Linnean Society.* 57 (3): 229–247.

Carroll, R. L., and P. J. Currie. 1991. The early radiation of diapsid reptiles. In H. P. Schultze and L. Truel (eds.), *Origins of the Higher Groups of Tetrapods: Controversies and Consensus,* pp. 354–424. Ithaca: Cornell University Press.

Claessens, L., S. F. Perry, and P. J. Currie. 1998. Reconstructing theropod lung ventilation. In D. L. Wolberg, K. Gittis, S. Miller, L. Carey, and A. Raynor (eds.), *Dinofest International Symposium, Program and Abstracts,* p. 8. Philadelphia: Academy of Natural Sciences. (Abstract.)

Claessens, L., S. F. Perry, and P. J. Currie. 1998. Using comparative anatomy to reconstruct theropod respiration. *Journal of Vertebrate Paleontology.* 18 (suppl. to no. 3): 34A. (Abstract.)

Clemens, W. A., and P. J. Currie. 1987. Cretaceous and Paleocene terrestrial vertebrate communities. *Geological Association of Canada, Mineralogical Association of Canada Joint Annual Meeting, Program with Abstracts* 12: 33. (Abstract.)

Coria, R. A., and P. J. Currie. 1997. A new theropod from the Rio Limay Formation. *Journal of Vertebrate Paleontology* 17 (suppl. to no. 3): 40A. (Abstract.)

Currie, P. J. 1977. A new haptodontine sphenacodont (Reptilia: Pelycosauria) from the Upper Pennsylvanian of North America. *Journal of Paleontology* 51: 927–942.

Currie, P. J. 1978. The orthometric linear unit. *Journal of Paleontology* 52 (5): 964–971.

Currie, P. J. 1979. Lower Cretaceous dinosaur footprints from the Peace River Canyon, British Columbia, Canada. *Palaeogeography, Palaeoclimatology, Palaeoecology* 28: 103–115.

Currie, P. J. 1980. Dinosaur footprints of the Peace River Canyon, B.C. *Sixteenth Western Inter-University Geol. Conference, Saskatoon:* 27. (Abstract.)

Currie, P. J. 1980. A new younginid (Reptilia: Eosuchia) from the Upper Permian of Madagascar. *Canadian Journal of Earth Sciences* 17 (4): 500–511.

Currie, P. J. 1981. Bird footprints from the Gething Formation (Aptian, Lower Cretaceous) of Northeastern British Columbia, Canada. *Journal of Vertebrate Paleontology*. 1 (3–4): 257–264.

Currie, P. J. 1981. *Hovasaurus boulei,* an aquatic eosuchian from the Upper Permian of Madagascar. *Palaeontographica africana* 24: 99–168.

Currie, P. J. 1981. The osteology and relationships of aquatic eosuchians from the Upper Permian of Africa and Madagascar. Ph.D. thesis, McGill University.

Currie, P. J. 1981. The osteology and relationships of aquatic eosuchians from the Upper Permian of Africa and Madagascar. *Dissertation Abstracts International* 42 (4). (Abstract.)

Currie, P. J. 1981. The vertebrae of *Youngina* (Reptilia: Eosuchia). *Canadian Journal of Earth Sciences* 18 (4): 815–818.

Currie, P. J. 1982. The osteology and relationships of *Tangasaurus mennelli* Haughton (Reptilia, Eosuchia) *Annals of the South African Museum* 86, part 8: 247–265.

Currie, P. J. 1983. Hadrosaur trackways from the Lower Cretaceous of Canada. *Acta palaeontographica polonica* 28 (1–2): 63–73.

Currie, P. J. 1985. Cranial anatomy of *Stenonychosaurus inequalis* (Saurischia, Theropoda) and its bearing on the origin of birds. *Canadian Journal of Earth Sciences* 22: 1643–1658.

Currie, P. J. 1985. Small theropods of Dinosaur Provincial Park, Alberta. *Geological Society of America Abstracts with Program* 17 (4): 215. (Abstract.)

Currie, P. J. 1986. Dinosaur footprints of western Canada. *First International Symposium on Dinosaur Tracks and Traces, Abstracts with Program,* p. 13. Albuquerque: New Mexico Museum of Natural History. (Abstract.)

Currie, P. J. 1987. Bird-like characteristics of the jaws and teeth of troodontid theropods (Dinosauria, Saurischia). *Journal of Vertebrate Paleontology* 7 (1): 72–81.

Currie, P. J. 1987. Discovery of nests of dinosaur eggs with embryos in the Two Medicine Formation of Southern Alberta. *Journal of Vertebrate Paleontology* 7 (suppl. to no. 3): 15A. (Abstract.)

Currie, P. J. 1987. New approaches to studying dinosaurs, Dinosaur Provincial Park. In S. J. Czerkas and E. C. Olson (eds.), *Dinosaurs Past and Present,* pp. 2:100–117. Los Angeles: Los Angeles County Museum of Natural History.

Currie, P. J. 1987. Theropods of the Judith River Formation of Dinosaur Provincial Park. *Fourth Symposium on Mesozoic Terrestrial Ecosystems, Short Papers, Tyrrell Museum of Palaeontology, Occasional Papers* 3: 52–60.

Currie, P. J. 1989. The first records of *Elmisaurus* (Saurischia, Theropoda) from North America. *Canadian Journal of Earth Sciences* 26 (6): 1319–1324.

Currie, P. J. 1989. Dinosaur tracksites of western Canada. In D. D. Gillette and M. G. Lockley (eds.), *Dinosaur Tracks and Traces,* pp. 293–300. New York: Cambridge University Press.

Currie, P. J. 1990. The Elmisauridae. In D. B. Weishampel, P. Dodson, and H. Osmólska (eds.), *The Dinosauria,* pp. 245–248. Berkeley: University of California Press.

Currie, P. J. 1990. The fauna and palaeoecology of the Upper Cretaceous Iren Dabasu Formation of China. In *International Geological Corre-*

lation Program, Project 245: Nonmarine Cretaceous Correlation; Project 262: Tethyan Correlation, p. 8. Bucharest: Institute of Geology and Geophysics. (Abstract.)

Currie, P. J. 1992. Saurischian dinosaurs of the Late Cretaceous of Asia and North America. In N. J. Mateer and P. J. Chen (eds.), *Aspects of Nonmarine Cretaceous Geology,* pp. 237–249. Beijing: China Ocean Press.

Currie, P. J. (ed.). 1993. Results from the Sino-Canadian Dinosaur Project. *Canadian Journal of Earth Sciences.* 30 (10–11): 1997–2272.

Currie, P. J. 1995. New information on the anatomy and relationships of *Dromaeosaurus albertensis* (Dinosauria: Theropoda). *Journal of Vertebrate Paleontology* 15 (3): 576–591.

Currie, P. J. 1995. Ornithopod trackways from the Lower Cretaceous of Canada. In W. A. S. Sarjeant (ed.), *Vertebrate Fossils and the Evolution of Scientific Concepts,* pp. 431–443. Reading: Gordon and Breach.

Currie, P. J. 1995. Phylogeny and systematics of theropods (Dinosauria). *Journal of Vertebrate Paleontology.* 15 (suppl. to no. 3): 25A. (Abstract.)

Currie, P. J. 1995. Wandering dragons: The dinosaurs of Canada and China. In Sun A. and Wang Y. (eds.), *Sixth Symposium on Mesozoic Terrestrial Ecosystems and Biota, Short Papers,* p. 101. Beijing: China Ocean Press. (Abstract.)

Currie, P. J. 1996. Dinosaur eggs, embryos, and babies. In D. L. Wolberg and E. Stump (eds.), *Dinofest International Symposium, Program and Abstracts,* p. 41. Tempe: Arizona State University. (Abstract.)

Currie, P. J. 1996. Out of Africa: Meat-eating dinosaurs that challenge *Tyrannosaurus rex. Science* 272: 971–972.

Currie, P. J. (ed.). 1996. Results from the Sino-Canadian Dinosaur Project, Part 2. *Canadian Journal of Earth Sciences* 33 (4): 511–684.

Currie, P. J. 1997. Chinese dinosaurs. In S. Y. Yang, M. Huh, Y. Lee, and M. G. Lockley (eds.), *International Dinosaur Symposium for the Uhangr Dinosaur Center and Theme Park in Korea, Paleontological Society of Korea, Special Publication.* 2: 93–101.

Currie, P. J. 1998. Feathered dinosaurs. In D. L. Wolberg, K. Gittis, S. Miller, L. Carey, and A. Raynor (eds.), *Dinofest International Symposium, Program and Abstracts,* p. 9. Philadelphia: Academy of Natural Sciences. (Abstract.)

Currie, P. J. 1999. Skeletal anatomy of the feathered dinosaurs of China. In *New Perspectives on the Origin and Early Evolution of Birds,* p. 9. New Haven: Yale Peabody Museum of Natural History and the Department of Geology and Geophysics, Yale University. (Abstract.)

Currie, P. J., and R. L. Carroll. 1984. Ontogenetic changes in the eosuchian reptile *Thadeosaurus. Journal of Vertebrate Paleontology* 4 (1): 68–84.

Currie, P. J., and P. Dodson. 1984. Mass death of a herd of ceratopsian dinosaurs. In W. E. Reif and F. Westphal (eds.), *Third Symposium on Terrestrial Ecosystems, Short Papers,* p. 61–66. Tübingen: Attempto Verlag.

Currie, P. J., and P. Dodson. 1990. The Neoceratopsia. In D. B. Weishampel, P. Dodson, and H. Osmólska (eds.), *The Dinosauria,* p. 593–618. Berkeley: University of California Press.

Currie, P. J., and D. A. Eberth. 1993. Palaeontology, sedimentology, and palaeoecology of the Iren Dabasu Formation (Upper Cretaceous), Inner Mongolia, People's Republic of China. *Cretaceous Research* 14: 127–144.

Currie, P. J., S. J. Godfrey, and L. Nessov. 1993. New caenagnathid (Dinosauria: Theropoda) specimens from the Upper Cretaceous of North America and Asia. *Canadian Journal of Earth Sciences* 30: 2255–2272.

Currie, P. J., and J. R. Horner. 1998. Lambeosaurine hadrosaur embryos (Reptilia: Ornithischia). *Journal of Vertebrate Paleontology* 8 (suppl. to no. 3): 13A. (Abstract.)

Currie, P. J., and A. R. Jacobsen. 1995. An azhdarchid pterosaur eaten by a velociraptorine theropod. *Canadian Journal of Earth Sciences* 32: 922–925.

Currie, P. J., and T. Jerzykiewicz. 1990. The dinosaur fauna of the Djadokhta Formation of Northern China. In V. A. Krassilov (ed.), *International Geological Correlation Program, Project 245: Nonmarine Cretaceous Correlations,* p. 14. Vladivostok: U.S.S.R. Academy of Sciences, Far Eastern Branch, Institute of Biology and Pedology. (Abstract.)

Currie, P. J., E. B. Koppelhus, and A. F. Muhammad. 1995. Stomach contents of a hadrosaur from the Dinosaur Park Formation (Campanian, Upper Cretaceous) of Alberta, Canada. In Sun A. and Wang Y. (eds.), *Sixth Symposium on Mesozoic Terrestrial Ecosystems and Biota, Short Papers,* p. 111–114. Beijing: China Ocean Press.

Currie, P. J., and E. H. Koster (eds.). 1987. *Fourth Symposium on Mesozoic Terrestrial Ecosystems. Short Papers. Tyrrell Museum of Palaeontology. Occasional Papers,* no. 3.

Currie, P. J., G. C. Nadon, and M. G. Lockley. 1991. Dinosaur footprints with skin impressions from the Cretaceous of Alberta and Colorado. *Canadian Journal of Earth Sciences* 28: 102–115.

Currie, P. J., M. A. Norell, Ji Q., and Ji. S.-A. 1998. The anatomy of two feathered theropods from Liaoning, China. *Journal of Vertebrate Paleontology* 18 (suppl. to no. 3): 36A. (Abstract.)

Currie, P. J., and K. Padian. 1983. A new pterosaur record from the Judith River (Oldman) Formation of Alberta. *Journal of Paleontology* 57: 599–600.

Currie, P. J., and J. H. Peng. 1993. A juvenile specimen of *Saurornithoides mongoliensis* from the Upper Cretaceous of Northern China. *Canadian Journal of Earth Sciences* 30: 2224–2230.

Currie, P. J., K. Rigby Jr., and R. E. Sloan. 1990. Theropod teeth from the Judith River Formation of Southern Alberta, Canada. In K. Carpenter and P. J. Currie (eds.), *Dinosaur Systematics: Approaches and Perspectives,* pp. 107–125. New York: Cambridge University Press.

Currie, P. J., and D. A. Russell. 1982. A giant pterosaur (Reptilia: Archosauria) from the Judith River (Oldman) Formation of Alberta. *Canadian Journal of Earth Sciences* 19 (4): 894–897.

Currie, P. J., and D. A. Russell. 1985. Egg-stealing dinosaurs from the Cretaceous of Alberta. *Proceedings of the Pacific Division, American Association for the Advancement of Science* 4 (1): 25. (Abstract.)

Currie, P. J., and D. A. Russell. 1988. Osteology and relationships of *Chirostenotes pergracilis* (Saurischia, Theropoda) from the Judith River (Oldman) Formation of Alberta, Canada. *Canadian Journal of Earth Sciences* 25: 972–986.

Currie, P. J., and W. A. S. Sarjeant. 1977. Dinosaur tracks from Cretaceous sediments of Peace River Canyon Near Hudson Hope, British Columbia. *American Association of Petroleum Geologists Bulletin.* 61 (5): 778. (Abstract.)

Currie, P. J., and W. A. S. Sarjeant. 1979. The osteology of haptodontine sphenacodonts (Reptilia: Pelycosauria). *Palaeontographica A.* 163: 130–168.

Currie, P. J., P. Vickers, and T. H. Rich. 1996. Possible oviraptorosaur (Theropoda, Dinosauria) specimens from the Early Cretaceous Otway Group of Dinosaur Cove, Australia. *Alcheringa* 20: 73–79.

Currie, P. J., and Zhao X. J. 1991. Two new theropods from the Jurassic of Xinjiang, People's Republic of China. *Journal of Vertebrate Paleontology* 11 (suppl. to no. 3): 24A. (Abstract.)

Currie, P. J., and Zhao X. J. 1993. A new large theropod (Dinosauria, Theropoda) from the Jurassic of Xinjiang, People's Republic of China. *Canadian Journal of Earth Sciences* 30: 2027–2036.

Currie, P. J., and Zhao X. J. 1993. A new troodontid (Dinosauria, Theropoda) braincase from the Judith River Formation (Campanian) of Alberta. *Canadian Journal of Earth Sciences* 30: 2231–2247.

Dodson, P., and P. J. Currie. 1988. The smallest ceratopsid skull—Judith River Formation of Alberta. *Canadian Journal of Earth Sciences* 25 (6): 926–930.

Dong Z. M., and P. J. Currie. 1993. Protoceratopsian embryos from Inner Mongolia, China. *Canadian Journal of Earth Sciences* 30: 2248–2254.

Dong Z. M., and P. J. Currie. 1995. On the discovery of an oviraptorid skeleton on a nest of eggs. *Journal of Vertebrate Paleontology* 15 (suppl. to no. 3): 26A. (Abstract.)

Dong Z. M., and P. J. Currie. 1996. On the discovery of an oviraptorid skeleton on a nest of eggs at Bayan Mandahu, Inner Mongolia, People's Republic of China. *Canadian Journal of Earth Sciences* 33 (4): 631–636.

Farlow, J. O., D. L. Brinkman, W. L. Abler, and P. J. Currie. 1991. Size, shape, and serration density of theropod dinosaur lateral teeth. *Modern Geology* 16: 161–198.

Fiorillo, A. R., and P. J. Currie. 1994. Theropod teeth from the Judith River Formation (Upper Cretaceous) of south-central Montana. *Journal of Vertebrate Paleontology.* 14: 74–80.

Forster, J. S., P. J. Currie, J. A. Davies, R. Siegele, S. G. Wallace, and D. Zelenitsky. 1996. Elastic recoil detection (ERD) with extremely heavy ions. *Nuclear Instruments and Methods in Physics Research B* 113: 308–311.

Godfrey, S. J., and P. J. Currie. 1994. A xiphisternal from the Dinosaur Park Formation (Campanian, Upper Cretaceous) of Alberta, Canada. *Canadian Journal of Earth Sciences* 31: 1661–1663.

Horner, J. R., and P. J. Currie. 1994. Embryonic and neonatal morphology and ontogeny of a new species of *Hypacrosaurus* (Ornithischia, Lambeosauridae) from Montana and Alberta. In K. Carpenter, K. Hirsch, and J. Horner (eds.), *Dinosaur Eggs and Babies,* pp. 312–336. Cambridge: Cambridge University Press.

Jerzykiewicz, T., P. J. Currie, D. A. Eberth, P. A. Johnston, E. H. Koster, and J. J. Zheng. 1993. Djadokhta Formation correlative strata in Chinese Inner Mongolia: An overview of the stratigraphy, sedimentary geology, and paleontology and comparisons with the type locality in the pre-Altai Gobi. *Canadian Journal of Earth Sciences* 30: 2180–2195.

Jerzykiewicz, T., P. J. Currie, P. A. Johnston, E. H. Koster, and R. Gradzinski. 1989. Upper Cretaceous dinosaur-bearing eolianites in the Mongolian Basin. *Twenty-Eighth International Geological Congress, Washington, D.C.* 2:122–123. (Abstract.)

Koppelhus, E. B., P. J. Currie, and A. F. Muhammad. 1995. Can a palynological analysis be used to determine if stomach contents of a hadrosaur from the Dinosaur Park Formation (Campanian: Upper Cretaceous) of Alberta, Canada, is the dinosaur's last meal or not? *American Association of Stratigraphic Palynologists, Twenty-Eighth Annual Meeting, Program and Abstracts* A-11. (Abstract.)

Koster, E. H., and P. J. Currie. 1987. Upper Cretaceous coastal plain sediments at Dinosaur Provincial Park, Southeast Alberta. *Geological Society of America, Rocky Mountain Section, Decade of North American Geology Centennial Field Guide* 2: 9–14.

Makovicky, P., and P. J. Currie. 1996. Discovery of a furcula in tyrannosaurid theropods. *Journal of Vertebrate Paleontology* 16 (suppl. to no. 3): 50A. (Abstract.)

Makovicky, P. J., and P. J. Currie. 1997. Discovery of a furcula in tyrannosaurid theropods, and its functional and phylogenetic implications. *First European Workshop on Vertebrate Palaeontology, Copenhagen. Extended Abstracts and Short Papers. Geological Society of Denmark. On Line Series no. 1:* (http://www.purl.dk//net/9710-0100).

Makovicky, P., and P. J. Currie. 1998. The presence of a furcula in tyrannosaurid theropods, and its phylogenetic and functional implications. *Journal of Vertebrate Paleontology* 18: 143–149.

McCrea, R. T., and P. J. Currie. 1998. A preliminary report on dinosaur tracksites in the Lower Cretaceous (Albian) Gates Formation Near Grande Cache, Alberta. *New Mexico Museum of Natural History and Science Bulletin* 14: 155–162.

Myhrvold, N. P., and P. J. Currie. 1997. Supersonic sauropods? Tail dynamics in the diplodocids. *Paleobiology* 23: 393–409.

Myhrvold, N. P., and P. J. Currie. 1998. Supersonic sauropods? Tail dynamics in the diplodocids. In D. L. Wolberg, K. Gittis, S. Miller, L. Carey, and A. Raynor (eds.), *Dinofest International Symposium, Program and Abstracts,* p. 41. Philadelphia: Academy of Natural Sciences. (Abstract.)

Qiang J., P. J. Currie, M. A. Norell, and Ji S.-A. 1998. Two feathered dinosaurs from northeastern China. *Nature* 393: 753–761.

Ruben, J. A., W. J. Hillenius, N. R. Geist, A. Leitch, T. D. Jones, P. J. Currie, J. R. Horner, and G. Espe III. 1996. The metabolic status of some Late Cretaceous dinosaurs. *Science* 273: 1204–1207.

Ryan, M. J., and P. J. Currie. 1996. First report of Protoceratopsidae (Neoceratopsia) from the Late Campanian Judith River Group, Alberta, Canada. *Journal of Vertebrate Paleontology* 16 (suppl. to no. 3): 61A. (Abstract.)

Ryan, M. J., and P. J. Currie. 1998. First report of protoceratopsians (Neoceratopsia) from the Late Cretaceous Judith River Group, Alberta, Canada. *Canadian Journal of Earth Sciences* 35: 820–826.

Ryan, M. J., P. J. Currie, J. D. Gardiner, and J. M. Lavigne. 1997. Baby hadrosaurid material associated with an unusually high abundance of *Troodon* teeth from the Horseshoe Canyon Formation (Early Maastrichtian), Alberta, Canada. *Journal of Vertebrate Paleontology* 17: 72A. (Abstract.)

Tanke, D. H., and P. J. Currie. 1995. Intraspecific fighting behavior inferred from toothmark trauma on skulls and teeth of large carnosaurs (Dinosauria). *Journal of Vertebrate Paleontology* 15 (suppl. to no. 3): 55A. (Abstract.)

Tanke, D. H., P. J. Currie, and P. L. Larson. 1992. Once bitten, twice shy: Predator toothmarks on oreodont (Mammalia: Merycoidodontidea) skulls, Middle and Upper Oligocene Brule Formation of South Dakota and Nebraska. *Journal of Vertebrate Paleontology* 12 (suppl. to no. 3): 54A. (Abstract.)

Tazaki, K., M. Aratani, S. Noda, P. J. Currie, and W. S. Fyfe. 1994. Microstructure and chemical composition of duckbilled dinosaur eggshell. *Science Reports of Kanazawa University* 34: 17–37.

Varricchio, D., and P. J. Currie. 1991. New theropod finds from the Two Medicine Formation (Campanian) of Montana. *Journal of Vertebrate Paleontology* 12 (suppl. to no. 3): 59A.

Vickaryous, M. K., A. P. Russell, P. J. Currie, K. Carpenter, and J. I. Kirkland. 1998. The cranial sculpturing of ankylosaurs (Dinosauria: Ornithischia): Reappraisal of developmental hypotheses. *Journal of Vertebrate Paleontology* 8 (suppl. to no. 3): 83A. (Abstract.)

Wilson, M. C., and P. J. Currie. 1985. *Stenonychosaurus inequalis* (Saurischia: Theropoda) from the Judith River (Oldman) Formation of Alberta: New findings on metatarsal structure. *Canadian Journal of Earth Sciences* 22: 1813–1817.

Wu X., D. B. Brinkman, A. P. Russell, Dong Z., P. J. Currie, Hou L., and Cui G. 1993. Oldest known amphisbaenian from the Upper Cretaceous of Chinese Inner Mongolia. *Nature* 366: 57–59.

Zelenitsky, D., L. V. Hills, and P. J. Currie. 1996. Parataxonomic classification of ornithoid eggshell fragments from the Oldman Formation (Judith River Group, Upper Cretaceous), Southern Alberta. *Canadian Journal of Earth Sciences* 33: 1655–1667.

Selected Nontechnical Publications

Braman, D. R., P. J. Currie, L. Hills, R. Revel, D. Russell, A. Sweet, and M. Wilson. 1984. *Plains Region, Campanian to Paleocene*. Sixth International Palynology Conference, Calgary. Field Trip no. 1.

Currie, P. J. (ed. and publ.). 1965–1972. *ERBivore*. (Popular magazine devoted to the works of Edgar Rice Burroughs.)

Currie, P. J. 1980. Mesozoic vertebrate life in Alberta and British Columbia. *Mesozoic Vertebrate Life* 1: 27–40.

Currie, P. J. 1981. Hunting dinosaurs in Alberta's huge bonebed. *Canadian Geographic* 101 (4): 34–39.

Currie, P. J. 1981. The Provincial Museum of Alberta: dinosaurs in the public eye. *Geoscience Canada* 8 (1): 33–35.

Currie, P. J. 1982. Geological Association of Canada. Paleontology Division Fieldtrip: Dinosaur Provincial Park.

Currie, P. J. 1984. I dinosauri del Canada. In *Sulle orme dei dinosauri*. Venice: Erizzo.

Currie, P. J. 1984. Fossils and the law. *Fossils Quarterly* (Canadian issue) 3 (2): 3–9.

Currie, P. J. 1985. Dinosaur Provincial Park. *New Canadian Encyclopedia*. Edmonton: Hurtig.

Currie, P. J. 1985. Dinosaurs. *New Canadian Encyclopedia*. Edmonton: Hurtig.

Currie, P. J. 1986. Dinosaur fauna. In B. G. Naylor (ed.), *Dinosaur Systematics Symposium, Field Trip Guidebook to Dinosaur Provincial Park*, pp. 17–23. Tyrrell Museum of Palaeontology, Drumheller, Alta.

Currie, P. J. 1988. Dinosaur hunters. In *The Valley of the Dinosaurs—Its Families and Coal Mines*, pp. i–xi. East Coulee Community Association, East Coulee, Alta.

Currie, P. J. 1988. The discovery of dinosaur eggs at Devil's Coulee.

Alberta: Studies in the Arts and Sciences (University of Alberta Press, Edmonton) 1 (1): 3–10.

Currie, P. J. 1989. Dragons and dinosaurs, the dinosaur project discovers ancient ties between east and west. *Earth Science* 42 (2): 10–13.

Currie, P. J. 1989. Long distance dinosaurs. *Natural History* 6 (89): 60–65.

Currie, P. J. 1989. Research at the Tyrrell Museum of Palaeontology. *Cab and Crystal* 2 (4): 10–11.

Currie, P. J. 1989. Theropod dinosaurs of the Cretaceous. In *The Age of Dinosaurs. Paleontological Society, Short Courses in Paleontology* 2: 113–120.

Currie, P. J. 1990. Dinosaur hunters. *Dinogramme* 4 (2): 1–2.

Currie, P. J. 1990. Dinosaurs. *Colliers Encyclopedia, International Year Book 1989*, pp. 62–71. New York: Macmillan Educational Co.

Currie, P. J. 1990. Dinosaurs. *Funk and Wagnall's Year Book*. New York: Funk and Wagnall.

Currie, P. J. 1990. Foreword to *The Last Great Dinosaurs* by M. Reid, pp. vii–viii. Red Deer, Alta.: Discovery Books.

Currie, P. J. 1990. Review of *Digging into the Past* by Edwin Colbert. *Copeia* 19 (1): 255–256.

Currie, P. J. 1991. The Sino-Canadian dinosaur expeditions, 1986–1990. *Geotimes* 36 (4): 18–21.

Currie, P. J. 1992. China-Canada-Alberta-Ex Terra. In von D. Hauff (ed.), *Alberta's Parks, Our Legacy*, pp. 179–182. Edmonton: Alberta Parks Foundation.

Currie, P. J. 1992. Dinosaur. In *1993 Yearbook of Science and Technology*. New York: McGraw-Hill.

Currie, P. J. 1992. Foreword to *Dinosaurian Faunas of China* by Dong Z. Berlin: Springer Verlag.

Currie, P. J. 1992. Migrating dinosaurs. In B. Preiss and R. Silverberg (eds.), *The Ultimate Dinosaur*, pp. 183–195. New York: Bantam Books.

Currie, P. J. 1993. [Black Beauty.] *Dinosaur Frontlines* 4: 22–36. (In Japanese.)

Currie, P. J. 1993. Dinosaur. In *1994 Yearbook of Science and Technology*, pp. 121–123. New York: McGraw-Hill.

Currie, P. J. 1993. Dinosaurs and the development of the Royal Tyrrell Museum of Palaeontology, Drumheller, Canada. *Deciphering the Natural World and Role of Collections and Museums*, pp. 43–45. Copenhagen: Geologisk Museum.

Currie, P. J. 1993. Dinosaurs from Dinosaur Provincial Park. *Dinosaur Provincial Park Times* (Alberta Recreation and Parks) Spring issue: 3.

Currie, P. J. 1993. 1992 Fieldwork. *Dinosaur Provincial Park Times* (Alberta Recreation and Parks) Spring issue: 3.

Currie, P. J. 1993. On mahars, gryfs, and the paleontology of ERB. *Burroughs Bulletin*, n.s., 16: 21–24.

Currie, P. J. 1993. [The search for dinosaur fossils in China. The dinosaurs of Canada and China. Warm-Blooded Dinosaurs.] *Newton—Graphic Science Magazine*. 13 (8): 56–57 (In Japanese.)

Currie, P. J. 1993. *Troodon*, the Cretaceous intellect with too many names. *Dinonews* 6: 6–8.

Currie, P. J. 1994. Communication in dinosaurs. In *Voices from Dinosaurs*. Toshiba-EMI, Japan. (Liner notes to compact audio disk.)

Currie, P. J. 1994. [Dinosaur Renaissance.] *HON* (Tokyo: Kodansha) 8: 33–35. (In Japanese.)

Currie, P. J. 1994. Dinosaurs of Pellucidar. *Burroughs Bulletin*, n.s., 17: 5–9.

Currie, P. J. 1994. Fieldwork at Dinosaur Provincial Park [1993]. *Dinosaur Provincial Park Times* (Alberta Recreation and Parks) Spring issue: 3.

Currie, P. J. 1994. [Herding behavior and its implications for migration in dinosaurs.] *Dinosaur Frontline* 7: 74–85. (In Japanese.)

Currie, P. J. 1994. Hunting ancient dragons in China and Canada. In G. D. Rosenberg and D. L. Wolberg (eds.), *Dino Fest: Proceedings of a Conference for the General Public.* Paleontology Society Special Publication 7: 387–396.

Currie, P. J. 1994. [The life and behavior of *Tyrannosaurus rex.*] *Newton—Graphic Science Magazine* 14 (7): 58–59. (In Japanese.)

Currie, P. J. 1994. Smoked cod au gratin; poached eggs. In *The Great Canadian Literary Cookbook*, pp. 35–37. Sechelt, B.C.: Festival of the Written Arts.

Currie, P. J. 1995. Fieldwork at Dinosaur Provincial Park (1994). *Dinosaur Provincial Park Times* (Alberta Recreation and Parks) Spring issue: 3.

Currie, P. J. 1995. [The origin and evolution of the Theropoda.] In *The T. rex World Exposition, Guide Book* (exhibition catalogue), pp. 64–77. Tokyo: Gakken. (In Japanese.)

Currie, P. J. 1995. Preface to *Dinosaurs of the Tetori Group in Japan*, p. 5. Fukui Prefectural Museum. (In Japanese.)

Currie, P. J. 1995. [The Relationship of dinosaurs and birds.] In *Dinosaurs of the Tetori Group in Japan*, pp. 30–33. Fukui Prefectural Museum. (In Japanese.)

Currie, P. J. 1995. Review of "Lies of the rich and shameless" by Giles Quartet. *Drumheller Mail* December 27, sec. 2, p. 5.

Currie, P. J. 1996. Dinosaurs in *The Land That Time Forgot. Burroughs Bulletin*, n.s., 25: 12–16.

Currie, P. J. 1996. Fantastic flying fossils. *Calgary Herald*, March 2, p. B4.

Currie, P. J. 1996. [Feathered dinosaurs and the origin of birds.] *Newton—Graphic Science Magazine* 17 (2): 114–119. (In Japanese.)

Currie, P. J. 1996. 1995 Fieldwork at Dinosaur Provincial Park. *Dinosaur Provincial Park Times* (Alberta Recreation and Parks) Spring issue: 3.

Currie, P. J. 1996. The great dinosaur egg hunt. *National Geographic Magazine* 189 (5): 96–111.

Currie, P. J. 1997. Braincase anatomy. In P. J. Currie and K. Padian (eds.), *The Encyclopedia of Dinosaurs*, pp. 81–85. San Diego: Academic Press.

Currie, P. J. 1997. Dromaeosauridae. In P. J. Currie and K. Padian (eds.), *The Encyclopedia of Dinosaurs*, pp. 194–195. San Diego: Academic Press.

Currie, P. J. 1997. Elmisauridae. In P. J. Currie and K. Padian (eds.), *The Encyclopedia of Dinosaurs*, pp. 209–210. San Diego: Academic Press.

Currie, P. J. 1997. Erenhot Dinosaur Museum. In P. J. Currie and K. Padian (eds.), *The Encyclopedia of Dinosaurs*, pp. 210–211. San Diego: Academic Press.

Currie, P. J. 1997. Feathered dinosaurs. In P. J. Currie and K. Padian (eds.), *The Encyclopedia of Dinosaurs*, p. 241. San Diego: Academic Press.

Currie, P. J. 1997. Gastroliths. In P. J. Currie and K. Padian (eds.), *The Encyclopedia of Dinosaurs*, p. 270. San Diego: Academic Press.

Currie, P. J. 1997. Graduate studies. In P. J. Currie and K. Padian (eds.), *The Encyclopedia of Dinosaurs*, pp. 280–281. San Diego: Academic Press..

Currie, P. J. 1997. Paleontological Museum, Ulaan Baatar. In P. J. Currie and K. Padian (eds.), *The Encyclopedia of Dinosaurs,* pp. 524–525. San Diego: Academic Press.

Currie, P. J. 1997. Preface to *Tyrannosaurus rex: A Highly Important and Virtually Complete Fossil Skeleton.* Sotheby's Auction Catalogue, sale 7045.

Currie, P. J. 1997. Raptors. In P. J. Currie and K. Padian (eds.), *The Encyclopedia of Dinosaurs,* p. 626. San Diego: Academic Press.

Currie, P. J. 1997. Sino-Canadian dinosaur project. In P. J. Currie and K. Padian (eds.), *The Encyclopedia of Dinosaurs,* p. 661. San Diego: Academic Press.

Currie, P. J. 1997. Sino-Soviet expeditions. In P. J. Currie and K. Padian (eds.), *The Encyclopedia of Dinosaurs,* pp. 661–662. San Diego: Academic Press.

Currie, P. J. 1997. Theropoda. In P. J. Currie and K. Padian (eds.), *The Encyclopedia of Dinosaurs,* pp. 731–737. San Diego: Academic Press.

Currie, P. J. 1997. Theropods. In J. Farlow and M. Brett-Surman (eds.), *The Complete Dinosaur,* pp. 216–233. Bloomington: Indiana University Press.

Currie, P. J. 1998. *Caudipteryx* revealed. *National Geographic Magazine* 194 (1): 86–89.

Currie, P. J. 1999. Foreword to *Into the Dinosaurs' Graveyard: Canadian Digs and Discoveries* by D. Spalding. pp. xiii–xiv. Toronto: Doubleday Canada.

Currie, P. J. 2000. Foreword to *Dinosaur Imagery: The Science of Lost Worlds and Jurassic Art—The Lazendorf Collection,* p. ix–xi. San Diego: Academic Press.

Currie, P. J., Dong Z., and D. A. Russell. 1988. The Dinosaur Project: An international cooperative program on dinosaurs. *Vertebrata PalAsiatica.* 26 (3): 235–240.

Currie, P. J., Dong Z., and D. A. Russell. 1989. The 1988 field program of the Dinosaur Project, an international cooperative program on dinosaurs. *Vertebrata PalAsiatica* 27 (3): 293–295.

Currie, P. J., and A. Garneau. 1984. Alberta's new fossil museum opens in 1985. *Fossils Quarterly* (Canadian issue) 3 (2): 11–18.

Currie, P. J., L. Hoffman, and B. Reeves. 1980. *Alberta's Prehistoric Past, Teacher's Guide.* Alberta Heritage Learning Resources Project. Edmonton: Alberta Education.

Currie, P. J., and S. Sampson. 1996. On the trail of Cretaceous Dinosaurs. In R. Ludvigsen (ed.), *Life in Stone: A Natural History of British Columbia's Fossils,* pp. 143–155. Vancouver: University of British Columbia Press.

Currie, P. J., and J. Sovak. 1995. *Jurassic Dinosaurs.* Mineola, N.Y.: Dover Publications. (Trading cards.)

Currie, P. J., and J. Sovak. 1996. *Cretaceous Dinosaurs.* Mineola, N.Y.: Dover Publications. (Trading cards.)

Koster, E. H., and P. J. Currie. 1986. Sedimentological background. In B. G. Naylor (ed.), *Dinosaur Systematics Symposium, Field Trip Guidebook to Dinosaur Provincial Park,* pp. 6–16. Drumheller, Alta.: Tyrrell Museum of Palaeontology.

Koster, E., P. J. Currie, D. Eberth, D. Brinkman, P. Johnston, and D. Braman. 1987. *Sedimentology and Palaeontology of the Upper Cretaceous Judith River / Bearpaw Formations at Dinosaur Provincial Park, Alberta.* Fieldtrip Guidebook for the Geological Association of Canada.

Index

Illustrations are indicated by italicized page numbers.

pathology in, 350; in *The Sword in the Stone*, 505–507
Cerebrum, 26–27; endocranial allometry of theropod, 26–28
Cerradicas, Spain, quadrupedal ornithopod trackways from, 431, *432, 437–438, 438–439*
Cerro del Pueblo Formation, *313–314, 315–316*
Champsosaurus, juvenile, 207
Chaoyangsaurus youngi: character states of, 261; phylogeny of, 253, *254, 255*
Chapman, C. H. Murray, 508, 522, *522–523*
Charadrius vociferus, 475
Charig, Alan, 511
Chasmosaurinae: from Alberta, 280, 291, 293, 294; in Judithian time, 315; *Triceratops* in, 322–323
Chasmosaurus: from Alberta, 291; forelimbs of, 93; mass death assemblages of, 268
Chasmosaurus belli from Alberta, 280, 291
Chasmosaurus canadensis from Alberta, 291
Chasmosaurus kaiseni from Alberta, 291
Chasmosaurus russelli from Alberta, 280, 291
Chelonia. *See* Turtles
Chen Pei-ji, 118
Chevrons. *See* Tail; Vertebrae
Chile, quadrupedal ornithopod trackways from, 439
Chilson, Robert, 510
China: *Alectrosaurus olseni* from, 68; ankylosaur from, 237–241; *Bienosaurus* gen. nov. from, 237–241; dinosaur collecting in, xiv; feathered dinosaurs from, xiv, 118–125; ornithopods from, 184; pathological theropod tooth from, 347; quadrupedal ornithopod trackways from, 429, *433, 434, 438–439*; *Shashanosaurus huoyanshanensis* from, 71–72; tyrannosaurids from, 64, 66; *Tyrannosaurus bataar* from, 70
Chirostenotes: from Alberta, 291–292, 293, 297n5; *Elmisaurus elegans* versus, 53; PCA of feet of, *416, 420–421*; stress fractures of, 333
Chirostenotes elegans: from Alberta, 280, 291–292; *Caenagnathus sternbergi* versus, 48; provenance of, 43–46
Chirostenotes pergracilis: from Alberta, 280, 281, 291, 293–294, 297n5; *Elmisaurus elegans* versus, 48–49, 52, 53, 54, *54*; hierarchical cluster analysis of feet of, *417, 418*; pedal phalanges of, 53, *54*; provenance of, 43–46; sexual dimorphism in, 54–55

Choteau, Montana: dinosaurs from, 305; Two Medicine Formation at, 299, *299*, 300
Chubutisaurus: caudal vertebrae of, 161; *Venenosaurus* gen. nov. versus, 148, 155, 158
Chubutisaurus insignis in Titanosauriformes, 154
Chure, Dan, 10
Cladistic analysis of Tyrannosauridae, 64, 65–66, 66–72
Claggett Formation, *Maiasaura* from, 303
Claggett Sea, transgressions of, 301, 307
Clarendonian faunas, 323
Clarendonian-Holocene turnover, 323–324
Clavicles: of *Caudipteryx, pl. 2C, pl. 2D*, 124, 130; of *Protarchaeopteryx, pl. 1D, pl. 1E*, 120–121, 130; of *Segisaurus*, 122; of *Sinosauropteryx*, 130; of tetanurans, 130; of theropods, 120
Cleveland-Lloyd Quarry: pathological *Allosaurus* from, 340–341; pathological eggshells from, *384, 388–389*; tracks above, 404–405
Cloverly Formation: mammals from, 474; ornithopods from, 183, 184; oviraptorosaurs from, 44–45; Poison Strip Sandstone and, 142; Poison Strip Sandstone Member versus, 187
Cluster analysis of theropod feet, 412–413, *415–421*
CMNH 9380, forelimbs of, 92, 93–94
Cnemial crest of *Quilmesaurus* gen. nov., 3, 6, 7
Cnidaria from TMH quarry, 222, 223
Coelophysis: hierarchical cluster analysis of feet of, *417*; PCA of feet of, *416, 420–421*; stress fractures of, 333
Coelophysis bauri, hierarchical cluster analysis of feet of, *417, 418*
Coeluridae, stress fractures of, 333
Coelurosauria: brains of, 19, 29, 30; feathered dinosaurs in, 118–119; pathological tracks of, 348; phylogeny of, 131–132; shoulder girdle of, 129–130; Tyrannosauridae as, 65
Coelurus, 17; PCA of feet of, *416, 420–421*; Tyrannosauridae versus, 65
Coelurus fragilis, hierarchical cluster analysis of feet of, *417, 418*
Colbert, Edwin Harris, 516, *516–517*
Collagenous fibers, theropod feather impressions as, 120–121
College of Eastern Utah, 142
Colorado: fossil eggshells from, 385, *386–387*; *Ignotornis* from, 446; juvenile ornithopod from, 197–205; *Laelaps trihedrodon* from,

Edmontonia longiceps from Alberta, 281, 293, 297

Edmontonia rugosidens: from Alberta, 280, 290; from Two Medicine Formation, 300, 303, 305

Edmontonian time: dinosaur immigration to North America during, 321; North American dinosaurs during, 311, *315–316, 315–317,* 320, 324

Edmontosaurus: from Alberta, 281, 292, 295; in Edmontonian time, 317; as endemic, 322–323; forelimbs of, 93; in Lancian time, 313, 317; Talkeetna Mountains hadrosaur versus, 224, 225, 227

Edmontosaurus regalis from Alberta, 280, 292

Egg Mountain, Montana, fossil eggshell from, *388–389*

Eggs: abnormal condor, 380–381, *381,* 382, 385; feathers and incubation of, 117, 128, 132; retention in oviduct of, 381–382; from Two Medicine Formation, 299–300

Eggshells: abnormal thickness of, 380–384; diagenetically altered, 387, *388–389;* from Dinosaur Provincial Park, 206, 208, 209, 212; external abnormalities of, 379, *380;* multilayered abnormal, 382–384, *383, 384, 386–387,* 386–387, 389; pathological, 378–389; from poultry industry, 379; recognizing abnormal, 384–387; stacked, 386–387, *386–387*

Einiosaurus, 54, *270;* agonistic encounters among, *270;* in Judithian time, 315; mass death assemblages of, 268

Einiosaurus procurvicornis from Two Medicine Formation, 300, 303, 305

El Picacho Formation, *317–318*

El Rhaz Formation, ornithopods from, 184

Elasmosauridae, gastroliths from, 167, 168

Elder, Jean, *524–525*

Elkhorn Volcanics, Two Medicine Formation and, 302, 307

Elmisauridae: from Alberta, 297n5; feet of, 42, 55; pedal phalanges of, 53, *54, 55;* taxonomy of, 43–46, 48–49

Elmisaurus: from Alberta, 297n5; stress fractures of, 333; taxonomy of, 48–49

Elmisaurus elegans: from Alberta, 292, 297n5; metatarsal of, 42, *49,* 49–50, *54;* pedal phalanges of, 50–54, *51;* provenance of, 43–46, 54–55; sexual dimorphism in, 54–55; taxonomy of, 48–49

Elmisaurus rarus: Elmisaurus elegans versus, 48–49, 52, 53, *54;* feet of,

43; pedal phalanges of, 53, *54;* taxonomy of, 49

Emausaurus, Bienosaurus gen. nov. versus, 241

Embryonic dinosaurs, 211. *See also* Juvenile dinosaurs

Enantiornithes, phylogeny of, *132*

Encyclopedia of Pterosaurs (Wellnhofer), 519

Endemism of North American herbivorous dinosaurs, 311–312, 322–323

Endocasts. *See* Brain endocasts

Endocranium: allometry of theropod, 25–30; of *Carcharodontosaurus saharicus,* 19–31; defined, 19–20

Endolymphatic duct of *Carcharodontosaurus saharicus,* 24

Endothermy, feathers and, 127

England: ornithopods from, 184; quadrupedal ornithopod trackways from, 429–431, *430,* 437–438, *438–439,* 448–449; *Scelidosaurus* from, 238

Enormous Egg, The (Butterworth), 512

Environments of neoceratopsians, 268–269

Eoceratops from Alberta, 291

Eoceratops canadensis from Alberta, 291

Eolambia caroljonesa, 184; discovery of, 185

Eoraptor, PCA of feet of, *416,* 420–421

Eoraptor lunensis, hierarchical cluster analysis of feet of, *417, 418, 419*

Eosinophilic granuloma, 373

Epanterias, 16

Epanterias amplexus, 16, 17

Erlikosauridae from Alberta, 291

Erlikosaurus: from Alberta, 280, 291; mandible of, 39

Etiologies: of juxtacortical lesions, 364, 365, 371–376; of theropod pathologies, 352–354

Eubostrychoceras: Talkeetna Mountains hadrosaur taphonomy and, 228; from TMH quarry, 222, *223*

Eubostrychoceras japonicum, age of Talkeetna Mountains hadrosaur and, 222

Eubrontes, pathology in, 348

Euhelopus, Venenosaurus gen. nov. versus, 154

Eumaniraptora, feathers of, 119

Eumeces, endocranium of, 28

Euoplocephalus: from Alberta, 290, 292; AMNH 5245 as, 245; forelimbs of, 93; from Two Medicine Formation, 305

Euoplocephalus tutus from Alberta, 280, 290, 292

Euornithopoda, Planicoxa gen. nov. as, 193

Europe: abnormal eggshell from Late Cretaceous of, 381; hadrosaurs from, 220; land connection to North America from, 185; ornithopod trackways from, 428; ornithopods from, 183, 184

Evanston Formation, *317–318*

Ewing's sarcoma, 373

Exostosis, *369*

Extant vertebrates, socioecology of, 264–267

Extraspherulitic growth units in abnormal eggshell, *380, 385–386, 386–387*

Facial nerve: of *Carcharodontosaurus,* 23; of *Montanoceratops,* 251

Fanny and the Monsters (Lively), 514, *514–515*

Farlow, James O., 408

Fascitis, periostitis and, 364

Fawcett, Edward D., 507

Feathered dinosaurs, *pl. 5, pl. 18,* xiv, *522, 524–525;* discovery of, 118; origin of flight and, 117–133

Feathers: of *Caudipteryx, pl. 2E, pl. 2F, pl. 3A, pl. 3B, 123–124, 124–125;* of *Caudipteryx zoui, pl. 18;* discovery of dinosaurs with, 118; evolution of, 117, 126–127; exaptations of, 127–129; flight and, 127, 128–129, 129–131; histology of fossil, 126; phylogeny of, 131–133, *132;* of *Protarchaeopteryx, pl. 1F, pl. 1G, pl. 2A, pl. 2B, 123–124, 123–124, 125;* of *Sinosauropteryx, pl. 1B, 123–124;* of *Sinosauropteryx prima, pl. 17;* of therizinosauroids, 118, 119

Feeding behavior: of hadrosaurs, 212–213; of neoceratopsians, 263, 268–269; tooth marks and, 58; tooth surface scratches and, 84–86; *Tyrannosaurus rex* forelimbs and, 90, 113

Feet: of *Acrocanthosaurus, 411;* of *Alioramus remotus,* 69; avulsion injuries in, 334; of Elmisauridae, 42, 43, 46; of *Elmisaurus elegans,* 48–54, *49, 51, 54;* of juvenile hadrosaurs, 216; of juvenile ornithopod, 197; morphometric analysis of, 410–413; of oviraptorosaurs, 52–53; pathological *Gorgosaurus libratus,* 344; pathological ornithomimid, 343; pathological oviraptorid, 343; pathological *Poekilopleuron bucklandii,* 340; pathological theropod, 346, 349, 351, 352, *356;* of *Planicoxa* gen. nov., 193; shapes of theropod, 415–421; stress fractures in, 331, 332–334; of Talkeetna Mountains hadrosaur, *229, 230, 230–231;* trackmaking and, 408–425. *See also* Tracks; Trackways

Felber, Eric, 518

Females. *See* Sexual dimorphism

Femur: of *Creosaurus trigonodon, 16–17;* of *Epanterias amplexus,* 16–17; of juvenile hadrosaurs, *210,* 211, 216, 217; of juvenile ornithopod, 197, 199, *201, 203;* of Kirtland Shale aublysodontine, 68; of *Laelaps trihedrodon, 15–16;* of *Othnielia rex,* 204; pathological theropod, *356;* pathologies of theropod, 337, 338; of *Planicoxa* gen. nov., 183, *190, 191, 192–193;* of *Protarchaeopteryx,* 121; of *Quilmesaurus* gen. nov., 3, *4–6, 5, 6, 7;* of *Sinosauropteryx, 123–124;* of Talkeetna Mountains hadrosaur, 229

Fenestra pseudorotunda of *Carcharodontosaurus,* 25

Fibula: of juvenile hadrosaurs, 216; of juvenile ornithopod, 197, 202; pathological *Allosaurus fragilis,* 340; pathological theropod, 351, *356;* of Talkeetna Mountains hadrosaur, *229, 230*

Fiction, dinosaurs in, 504–525

Field Museum of Natural History. *See* FMNH PR 2081

Fighting (Knight painting), *506–507,* 519

Fish from TMH quarry, 222, 223, 232, 233

Flight: feathers and, 127, 128–129, 129–131; origin of avian, 117, 126–131, 131–133

Flight stroke, evolution of, 129–130

Floodplains as optimal track environments, 397

Fluctuating asymmetry in theropods, 357–358

Fluvial systems as optimal track environment, 397

FMNH PR 2081, 113; avulsion injuries of, 334; forelimbs of, 91–92, 93–94, *95, 96, 97, 99, 100, 100,* 101, *102–104, 103, 104–105,* 106–107, 109, 113; healed injuries in, 345

Footprint defined, 459

Footprints. *See* Tracks

Footprints in the Swamp (Bloch), 512–514

Foramen magnum: of *Carcharodontosaurus,* 25; of *Montanoceratops cerorhynchus,* 245–247

Foraminifera: age of Talkeetna Mountains hadrosaur and, 220–222; from TMH quarry, 223

Force-based system (FBS) in forelimb biomechanics, 107

Ford, Tracy L., 331

Forebrain, endocranial allometry of, 26–27

Forelimbs: of *Alectrosaurus olseni,* 68;

229; of *Tyrannosaurus rex*, 91–
92, 101, *102*; of *Venenosaurus*
gen. nov., 144, 148, *148*
Range of motion (ROM) of *Tyranno-
saurus rex* forelimb, 111–113,
112
Rank indicators: among extant
vertebrates, 266; among
neoceratopsians, 270
Rapid City Regional Airport, theropod
tracks at, 444, *444, 445*
Raptor Red (Bakker), *508–509, 516*
Rebbachisaurus, gastroliths from, 167,
168
Recessus scalae tympani of *Carcharodontosaurus saharicus*, 24–25
Reconstruction, The (Casper), 516
Rectrices: of *Caudipteryx*, 124–125; of
Protarchaeopteryx, 123–124
Red Deer River Valley, basal neoceratopsians from, 244
Reflectance value (RV) ranges of
gastroliths, 176–177
Regumiel de la Sierra, Spain, quadru-
pedal ornithopod trackways
from, 431–432, *432*, 437–438,
438–439
Remiges: of *Archaeopteryx*, 129; in
birds, 129; of *Caudipteryx*, 125,
129
Reptiles: brains of nonavian, 19, 30;
endocrania of nonavian, 26–28
Resistive force (RF) in forelimb
biomechanics, 107–111, *108*
Resistive force arm (RFA) in forelimb
biomechanics, 107–111, *108*
Resnick, Mike, 510, 511, *519–520*
Resource exploitation by neoceratopsians, 268–269
Retarded growth among neoceratopsians, 270
Ribs: of *Apatosaurus excelsus, 155*; of
juvenile ornithopod, 197, *201,
202*; pathological *Acrocanthosaurus atokensis*, 342;
pathological dromaeosaurid,
342–343; pathological *Gorgosaurus libratus*, 344; pathological
Megalosaurus bucklandii, 340;
pathological *Sinraptor dongi*,
341–342; pathological theropod,
349, *355*; pathologies of
theropod, 338; of Talkeetna
Mountains hadrosaur, *229*, 230;
of *Venenosaurus* gen. nov., 153–
154, *154*
Ricardoestesia from Alberta, 280, 288,
289, 292, 294, 295, 297
Ricardoestesia gilmorei from Alberta,
280, 281, 288, 292, 295, 297n3
Ricardoestesia sp. nov. from Alberta,
281, 288, 292, 294, 295
Richler, Mordecai, 512
Ringbone Formation, *313–314*
Rio Negro Province, Argentina,
Quilmesaurus gen. nov. from, 3

Rivers of Time (de Camp), 510, *519–
520*
Roberts, Charles G. D., 505, *514–515*,
519, 525
Rocky Mountains: dinosaur provinci-
ality and, 319–320; Two
Medicine Formation and, 298–
299, *299*
Rolt-Wheeler, Francis, *506–507*, 508,
519
Rothschild, Bruce, 331, 338, 364
Royal Ontario Museum, 502
Royal Society of Canada, Phil Currie
in, xv
Royal Tyrrell Museum of Paleontol-
ogy, 224, 282, 454; juvenile
hadrosaur material at, 206–213,
215–218; Phil Currie at, xiv;
tracks at, 455–474. *See also*
TMP88.121.39; TMP98.14.1
Ruby Ranch Member, stratigraphy of,
140, *140–141*, 142, *186, 187*
Rumsey, Tim, 516
Runzel marks, tracks and, 401
Russell, Anthony P., 279
Russell, Dale A., xiv
Ryan, Michael J., 279

Sacrum. *See* Vertebrae
St. Mary River Formation, *315–316*;
basal neoceratopsians from, 244,
252; bird tracks from, 474;
dinosaurs from, 281, 295, 297;
Montanoceratops cerorhynchus
from, 243; quadrupedal
ornithopod trackways from, *435,
436, 438–439*; tracks from, *400,
402*
Saltasaurus: caudal vertebrae of, 161;
Venenosaurus gen. nov. versus,
143, 145, 147, 148, 150, 152,
157–158
Sampson, Scott D., 263
San Carlos Formation, *313–314*
Sanders, Frank, 166
Sanderson, Bill, 512
Santonian, North American titano-
saurs during, 321
Sao Khua Formation, *Siamotyrannus
isanensis* from, 71
Sarjeant, William A. S., 453
Saskatchewan, *Tyrannosaurus rex*
from, 71
Saunders, T. Bailey: letter to A. S.
Woodward by, 489; letter to A. S.
Woodward from, 484
Saurian Hill locality, 11, 14
Saurischia. *See* Prosauropoda;
Sauropoda; Theropoda
Sauroidichnites abnormis, pathology
in, 348
Saurolophus: from Alberta, 292; as
immigrant, 321
Saurolophus osborni from Alberta,
280, 292
Sauropelta: from Cloverly Formation,

dinosaurs of, 481–502; reputation of, 497–498
Sternberg, Charles Mortram, 481, 482, 498
Sternberg, George Fryer, 481, 482
Sternberg, Levi, 481, 485, 499, 502
Sternum: of *Archaeopteryx,* 130; of *Caudipteryx,* 130; of *Cedarosaurus weiskopfae,* 169; of *Protarchaeopteryx,* 120–121, 122, 130; of tetanurans, 130
Steveville, Alberta, Charles H. Sternberg at, 482
Stokesosaurus, 17
Stress fractures: features of, 331–332; of theropods, 331–335
Struthio, beak of, 38
Struthiomimus, pl. 9: from Alberta, 291, 293, 297; angular of, *40;* lower jaw of, 34, *40;* PCA of feet of, *416;* stress fractures of, 333
Struthiomimus altus: from Alberta, 280, 281, 291, 293; angular of, 39, *40;* hierarchical cluster analysis of feet of, *417;* lower jaw of, 38, *40*
Struthiomimus brevetertius: from Alberta, 293; lower jaw of, 34
Struthiomimus currelli, fluctuating asymmetry in, 357
Struthiomimus ingens from Alberta, 293
Struthiomimus samueli: from Alberta, 291; lower jaw of, 34
"*Struthiomimus sedens,*" hierarchical cluster analysis of feet of, *417, 418, 419*
"*Stygivenator,*" taxonomy of, 68
Styracosaurus: from Alberta, 282, 290; mass death assemblages of, 268
Styracosaurus albertensis from Alberta, 280, 290
Styracosaurus ovatus from Two Medicine Formation, 300, 303, 305
Subadult dinosaurs. *See* Juvenile dinosaurs
Subashi Formation, *Shanshanosaurus huoyanshanensis* from, 65–66, 71–72
Summer of the Dinosaur, The (Hall), 512
Supradentary of *Gallimimus bullatus,* 38
Supraoccipital of *Montanoceratops cerorhynchus,* 245–247
Surangular: of *Gallimimus bullatus,* 37, *37;* of juvenile hadrosaurs, 215
Surface periosteal reaction, 366
Surfacer software, brain endocast volume via, 28
Swallowed by an Earthquake (Fawcett), 507
Sweetgrass arch, *299*

Sword in the Stone, The (White), 505–507
Symmes, John Cleves, 507
Symmetrodonta, 474
Synapomorphies: of Ceratopsia, 262; of Tyrannosauridae, 64, 66, 76–83; of *Tyrannosaurus,* 70–71
Syncerus, sexual maturation in, 265
Syntarsus: forelimbs of, 93; fused metatarsals of, 339; sexual dimorphism in, 97
Syntarsus rhodesiensis: fluctuating asymmetry in, 357; fractures in, 358; mandible of, 39; pathology in, 340, 350

Tail: articulations in sauropod, 159–162; of *Caudipteryx,* 124–125; of juvenile hadrosaurs, 218; pathological theropod, 351; of *Planicoxa* gen. nov., 191; of Talkeetna Mountains hadrosaur, 225, *229;* of *Venenosaurus* gen. nov., 139–140, 143–147, *145, 146*
Tail of the Trinosaur, The (Causley), 512
Talkeetna Mountains, Alaska, *220–221*
Talkeetna Mountains hadrosaur (TMH), 219–234, *224, 226, 227, 229*
Tanke, Darren H., xvii, 206, 331, 454
Taphonomy: of Talkeetna Mountains hadrosaur, 219–234; of TMH quarry, 228–234; of Tony's Bone Bed, 186–187; of tracks, 395–405
Tarbosaurus: brains of, 20; stress fractures of, 333
Tarbosaurus bataar, taxonomy of, 70–71
Tarbosaurus efremovi, taxonomy of, 70
Tarsals: pathological theropod, 351, *356. See also* Astragalus
Tatisaurus: Bienosaurus gen. nov. versus, 239, 240, 241; from Lufeng Basin, 237
Tawasaurus: Bienosaurus gen. nov. versus, 239; from Lufeng Basin, 237
Taxonomy: of Aublysodontinae, 64, 67–68; of Caenagnathidae, 43–46; of Elmisauridae, 43–46, 48–49; of Oviraptorosauria, 43–46; theropod pathologies and, 348–359; of Tyrannosauridae, 64–73; of Tyrannosaurinae, 64. *See also* Ichnotaxonomy
Teeth: of Aublysodontinae, 64, 66–67, 67–68; of *Bienosaurus* gen. nov., 238–239, *239, 240;* of *Ceratosaurus,* 17; of juvenile hadrosaurs, 215; of *Laelaps trihedrodon,* 11–13, *12–13,* 14–15,

Indexer: George Olshevsky (Phil Currie's dino-pal for more than 22 years)

DARREN H. TANKE
is a Technician in the Dinosaur Research Program at the Royal Tyrrell
Museum of Palaeontology in Alberta. He has worked with Philip since
1979. This is his first book project.

KENNETH CARPENTER
is an authority on dinosaurs and Mesozoic marine reptiles and is af-
filiated with the Denver Museum of Natural History. He is author of
Eggs, Nests, and Baby Dinosaurs (Indiana University Press) and has edited important collections of papers dealing with dinosaurs, including
Dinosaur Systematics: Approaches and Perspectives (with Philip J.
Currie) and *The Armored Dinosaurs* (forthcoming from Indiana University Press).